The Humongous Book of Statistics Problems

Translated for People who don't Speak Math!!

by W. Michael Kelley and Robert A. Donnelly Jr., Ph.D.

ALPHA

A member of Penguin Group (USA) Inc.

ALPHA BOOKS

Published by the Penguin Group

Penguin Group (USA) Inc., 375 Hudson Street, New York, New York 10014, USA

Penguin Group (Canada), 90 Eglinton Avenue East, Suite 700, Toronto, Ontario M4P 2Y3, Canada (a division of Pearson Penguin Canada Inc.)

Penguin Books Ltd., 80 Strand, London WC2R 0RL, England

Penguin Ireland, 25 St. Stephen's Green, Dublin 2, Ireland (a division of Penguin Books Ltd.)

Penguin Group (Australia), 250 Camberwell Road, Camberwell, Victoria 3124, Australia (a division of Pearson Australia Group Pty. Ltd.)

Penguin Books India Pvt. Ltd., 11 Community Centre, Panchsheel Park, New Delhi—110 017, India

Penguin Group (NZ), 67 Apollo Drive, Rosedale, North Shore, Auckland 1311, New Zealand (a division of Pearson New Zealand Ltd.)

Penguin Books (South Africa) (Pty.) Ltd., 24 Sturdee Avenue, Rosebank, Johannesburg 2196, South Africa

Penguin Books Ltd., Registered Offices: 80 Strand, London WC2R 0RL, England

International Standard Book Number: 978-1-59257-865-8
Library of Congress Catalog Card Number: 2006926601

11   10   09   8   7   6   5   4   3   2   1

Interpretation of the printing code: The rightmost number of the first series of numbers is the year of the book's printing; the rightmost number of the second series of numbers is the number of the book's printing. For example, a printing code of 09-1 shows that the first printing occurred in 2009.

Printed in the United States of America

Note: This publication contains the opinions and ideas of its author. It is intended to provide helpful and informative material on the subject matter covered. It is sold with the understanding that the author and publisher are not engaged in rendering professional services in the book. If the reader requires personal assistance or advice, a competent professional should be consulted.

The author and publisher specifically disclaim any responsibility for any liability, loss, or risk, personal or otherwise, which is incurred as a consequence, directly or indirectly, of the use and application of any of the contents of this book.

Most Alpha books are available at special quantity discounts for bulk purchases for sales promotions, premiums, fund-raising, or educational use. Special books, or book excerpts, can also be created to fit specific needs.

For details, write: Special Markets, Alpha Books, 375 Hudson Street, New York, NY 10014.

Contents

Introduction

Introduction

Are you in a statistics class? Yes? Then you need this book. Here's why:

- Fact #1: <u>The best way to learn statistics is by working out statistics problems.</u> There's no denying it. If you could figure this class out just by reading the textbook or taking good notes in class, everybody would pass with flying colors. Unfortunately, the harsh truth is that you have to buckle down and work problems out until your fingers are numb.

- Fact #2: <u>Most textbooks only tell you what the answers to their practice problems are but not how to do them!</u> Sure your textbook may have 175 problems for every topic, but most of them only give you the answers. That means if you don't get the answer right you're totally screwed! Knowing you're wrong is no help at all if you don't know WHY you're wrong. Statistics textbooks sit on a huge throne, like the Great and Terrible Oz and say, "Nope, try again," and you do. Over and over, usually getting the problem wrong. What a delightful way to learn! (Let's not even get into why they only tell you the answers to the odd problems. Does that mean the book's actual author didn't even feel like working out the even ones?)

- Fact #3: <u>Even when math books try to show you the steps for a problem, they do a lousy job.</u> Math people love to skip steps. You'll be following along fine with an explanation and then all of a sudden bam, you're lost. You'll think to yourself, "How did they do that?" or "Where the heck did that 42 come from? It wasn't there in the last step!" Why do almost all of these books assume that in order to work out a problem on page 200, you'd better know pages 1 through 199 like the back of your hand? You don't want to spend the rest of your life on homework! You just want to know why you keep getting a negative number when you're trying to calculate the probability of drawing a full house from a deck of cards, which you will learn in Chapter 5.

All of our notes are off to the side like this and point to the parts of the book we're trying to explain.

- Fact #4: <u>Reading lists of facts is fun for a while, but then it gets old.</u> Let's cut to the chase. Just about every single kind of statistics problem you could possibly run into is in here—after all, this book is humongous! If a thousand problems aren't enough, then you've got some kind of crazy stats hunger, my friend, and I'd seek professional help. This practice book was good at first, but to make it great, we went through and worked out all the problems and took notes in the margins when we thought something was confusing or needed a little more explanation. We also drew little skulls next to the hardest problems, so you'd know not to freak out if they were too challenging. After all, if you're working on a problem and you're totally stumped, isn't it better to know that the problem is supposed to be hard? It's reassuring, at least for us.

We think you'll be pleasantly surprised by how detailed the answer explanations are, and we hope you'll find our little notes helpful along the way. Call us crazy, but we think that people who want to learn statistics and are willing to spend the time drilling their way through practice problems should actually be able to figure the problems out and learn as they go, but that's just our 2¢.

One final word of warning. A lot of statistics classes use calculators that work all of the formulas out for you. In this book, we're showing you how the formulas work, but there's a lot of arithmetic involved, so we round the decimals off. This means the calculator answer may differ a teeny bit from our answers, but this way you actually get to see the steps and understand what's going on.

Good luck and make sure to come visit our websites at www.stat-guide.com and www.calculus-help.com. If you feel so inclined, drop us an e-mail and give us your 2¢. (Not literally, though—real pennies clog up the Internet pipes.)

—Mike Kelley and Bob Donnelly

Acknowledgments

Special thanks to the technical reviewer, Kitty Vogel, an expert who double-checked the accuracy of what you'll learn here. Kitty has taught A.P. Statistics since its inception, and is more passionate about stats than anyone I've ever met. She is an extremely talented educator, and it's almost a waste of her impressive skill set to merely proofread this book, but I am appreciative nonetheless.

Trademarks

All terms mentioned in this book that are known to be or are suspected of being trademarks or service marks have been appropriately capitalized. Alpha Books and Penguin Group (USA) Inc. cannot attest to the accuracy of this information. Use of a term in this book should not be regarded as affecting the validity of any trademark or service mark.

Dedication

Bob: This book would not have been possible without the loving support of my wife and best friend, Debbie. Your encouragement and belief in me were a constant source of inspiration. Thank you for your unending patience with me during this project. I love you always.

Mike: For Lisa, Nick, Erin, and Sara, the four reasons anything in my life is worth doing.

Chapter 1
DISPLAYING DESCRIPTIVE STATISTICS

Summarizing data in tables, charts, and graphs

The main focus of descriptive statistics is to summarize and present data. This chapter demonstrates a variety of techniques available to display descriptive statistics. Presenting data graphically allows the user to extract information more efficiently.

There are many different tools available for displaying descriptive statistics. Frequency distributions are a simple way to summarize raw data in tables and make the information more useful. Histograms convert these tables into charts and provide a picture of the data. Bar and pie charts offer a variety of ways to display categorical data. Finally, line and scatter charts allow you to view relationships between two variables in a graphical format.

Frequency Distributions

Showing your data in a table

> *Note: Problems 1.1–1.3 refer to the data set below, the daily demand for hammers at a hardware store over the last 20 days.*
>
> | Daily Demand | | | | |
> |---|---|---|---|---|
> | 2 | 1 | 0 | 2 | 1 |
> | 3 | 0 | 2 | 4 | 0 |
> | 3 | 2 | 3 | 4 | 2 |
> | 2 | 2 | 4 | 3 | 0 |
>
> **1.1** Develop a frequency distribution summarizing this data.

A frequency distribution is a two-column table. In the left column, list each value in the data set from least to greatest. Count the number of times each value appears and record those totals in the right column.

Daily Demand	Frequency
0	4
1	2
2	7
3	4
4	3
Total	**20**

> *Note: Problems 1.1–1.3 refer to the data set in Problem 1.1, the daily demand for hammers at a hardware store over the last 20 days.*
>
> **1.2** Develop a relative frequency distribution for the data.

Divide the frequency of each daily demand by the total number of data values (20).

Daily Demand	Frequency	Relative Frequency
0	4	$4 \div 20 = 0.20$
1	2	$2 \div 20 = 0.10$
2	7	$7 \div 20 = 0.35$
3	4	$4 \div 20 = 0.20$
4	3	$3 \div 20 = 0.15$
Total	**20**	**1.00**

The sum of the relative frequencies should always equal 1.00.

Note: Problems 1.1–1.3 refer to the data set in Problem 1.1, the daily demand for hammers at a hardware store over the last 20 days.

1.3 Develop a cumulative relative frequency distribution for the data.

Daily Demand	Relative Frequency	Cumulative Relative Frequency
0	4 ÷ 20 = 0.20	0.20
1	2 ÷ 20 = 0.10	0.20 + 0.10 = 0.30
2	7 ÷ 20 = 0.35	0.30 + 0.35 = 0.65
3	4 ÷ 20 = 0.20	0.65 + 0.20 = 0.85
4	3 ÷ 20 = 0.15	0.85 + 0.15 = 1.00
Total	**1.00**	

> The cumulative relative frequency for a particular row is the relative frequency for that row plus the cumulative relative frequency for the previous row. The last cumulative relative frequency should always be 1.00.

Note: Problems 1.4–1.6 refer to the data set below, the number of calls per day made from a cell phone for the past 30 days.

Cell Phone Calls per Day					
4	5	1	0	7	8
3	6	8	3	0	9
2	12	14	5	5	10
7	2	11	9	4	3
1	5	7	3	5	6

1.4 Develop a frequency distribution summarizing the data.

Because this data has many possible outcomes, you should group the number of calls per day into groups, which are known as *classes*. One option is the $2^k \geq n$ rule to determine the number of classes, where k equals the number of classes and n equals the number of data points. Given $n = 30$, the best value for k is 5.

Calculate the width W of each class.

> Because $2^4 = 16 < 30$, $k = 4$ is not large enough. Instead, use $k = 5$, which satisfies the $2^k \geq n$ rule because $2^5 = 32 \geq 30$.

$$W = \frac{\text{largest value} - \text{smallest value}}{\text{number of classes}} = \frac{14 - 0}{5} = 2.8 \approx 3$$

Set the size of each class to 3 and list the classes in the left column of the frequency distribution. Count the number of values contained in each group and list those values in the right column.

Each class includes three values. This first class contains values 0, 1, and 2.

Calls per Day	Frequency
0–2	6
3–5	11
6–8	7
9–11	4
12–14	2
Total	**30**

Note: Problems 1.4–1.6 refer to the data set in Problem 1.4, the number of calls per day made from a cell phone for the past 30 days.

1.5 Develop a relative frequency distribution for the data.

Divide the frequency of each class by the total number of data values (30).

Calls per Day	Frequency	Relative Frequency
0–2	6	$6 \div 30 = 0.200$
3–5	11	$11 \div 30 = 0.367$
6–8	7	$7 \div 30 = 0.233$
9–11	4	$4 \div 30 = 0.133$
12–14	2	$2 \div 30 = 0.067$
Total	**30**	**1.00**

Because of rounding, in some cases the total relative frequency may not add up to exactly 1.00.

Note: Problems 1.4–1.6 refer to the data set in Problem 1.4, the number of calls per day made from a cell phone for the past 30 days.

1.6 Develop a cumulative relative frequency distribution for the data.

The cumulative relative frequency for a particular row is the relative frequency (calculated in Problem 1.5) for that row plus the cumulative relative frequency for the previous row.

Calls per Day	Relative Frequency	Cumulative Relative Frequency
0–2	$6 \div 30 = 0.200$	0.200
3–5	$11 \div 30 = 0.367$	$0.200 + 0.367 = 0.567$
6–8	$7 \div 30 = 0.233$	$0.567 + 0.233 = 0.800$
9–11	$4 \div 30 = 0.133$	$0.800 + 0.133 = 0.933$
12–14	$2 \div 30 = 0.067$	$0.933 + 0.067 = 1.000$
Total	**1.00**	

Histograms

Frequency distributions in a chart

1.7 Develop a histogram for the data set below, a grade distribution for a statistics class.

Grade	Number of Students
A	9
B	12
C	6
D	2
F	1
Total	**30**

The height of each bar in the histogram reflects the frequency of each grade.

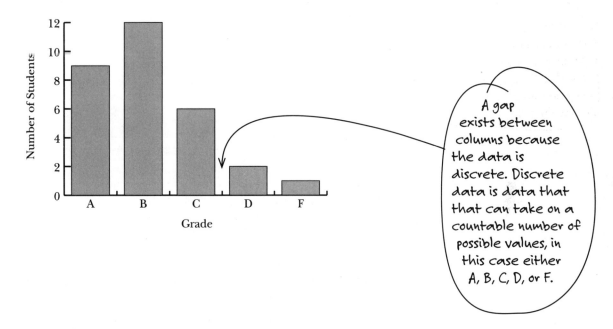

A gap exists between columns because the data is discrete. Discrete data is data that that can take on a countable number of possible values, in this case either A, B, C, D, or F.

1.8 Develop a histogram for the frequency distribution below, the commuting distance for 50 employees of a particular company.

Commuting Miles	Frequency
0–under 4	3
4–under 8	10
8–under 12	6
12–under 16	16
16–under 20	6
20–under 25	9
Total	**50**

The height of each bar in the histogram reflects the frequency for each group of commuting distances.

No gap exists between columns because the data is continuous. Continuous data can assume any value in an interval. A person can drive any distance btween 0 and 24 miles, so you can't leave any gaps.

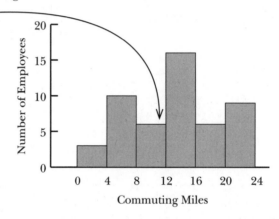

1.9 Develop a histogram for the data set below, the mileage of a specific car with a full tank of gas.

Miles per Tank									
302	315	265	296	289	301	308	280	285	318
267	300	309	312	299	316	301	286	281	311
272	295	305	283	309	313	278	284	296	291
310	302	282	287	307	305	314	318	308	280

First, develop a frequency distribution for the data. Using the $2^k \geq n$ rule, set $k = 6$ because $2^6 = 64 \geq 40$. Calculate the width W of each class.

$$W = \frac{\text{largest value} - \text{smallest value}}{\text{number of classes}} = \frac{318 - 265}{6} = 8.833 \approx 10$$

Set the size of each class equal to 10 and count the number of values contained in each class.

Miles per Tank	Frequency
260–under 270	2
270–under 280	2
280–under 290	10
290–under 300	5
300–under 310	12
310–under 320	9
Total	**40**

The height of each bar in the histogram reflects the frequency for each group of miles per tank of gas.

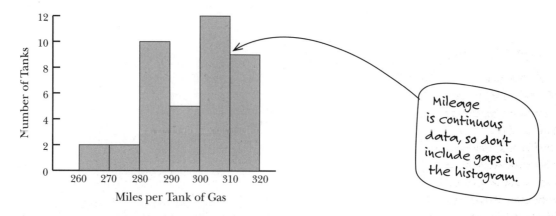

Mileage is continuous data, so don't include gaps in the histogram.

1.10 Develop a histogram for the data set below, the number of home runs hit by 40 Major League Baseball players during the 2008 season.

Home Runs									
48	40	38	37	37	37	37	37	36	36
35	34	34	34	33	33	33	33	33	33
33	32	32	32	32	32	31	31	29	29
29	29	28	28	27	27	27	27	27	26

Develop a frequency distribution for the data. Apply the $2^k \geq n$ rule and set $k = 6$ because $2^6 = 64 \geq 40$. Calculate the width W of each class.

$$W = \frac{\text{largest value} - \text{smallest value}}{\text{number of classes}} = \frac{48 - 26}{6} = 3.667 \approx 4$$

Set the size of each class equal to 4 and count the number of values contained in each class.

Home Runs	Frequency
25–28	8
29–32	11
33–36	13
37–40	7
41–44	0
45–48	1
Total	**40**

The height of each bar in the histogram reflects the frequency for each group of home runs.

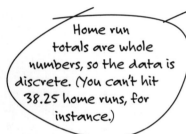

Home run totals are whole numbers, so the data is discrete. (You can't hit 38.25 home runs, for instance.)

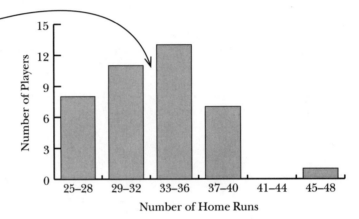

Bar Charts

Setting the bar for visual data.

1.11 Construct a column bar chart for the data bellow, an individual's credit card balance at the end of the last 8 months.

Month	Balance ($)
1	375
2	514
3	834
4	603
5	882
6	468
7	775
8	585

A column bar chart uses vertical bars to represent categorical data. The height of each bar corresponds to the value of each category.

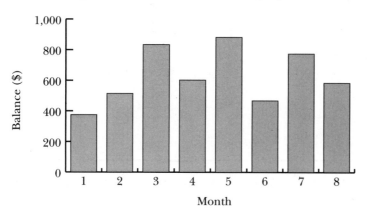

> Categorical data is data that is organized in discrete groups, such as the months in this problem.

1.12 Construct a column bar chart for the data below, a company's monthly sales totals.

Month	Sales ($)
1	10,734
2	8,726
3	14,387
4	11,213
5	9,008
6	8,430

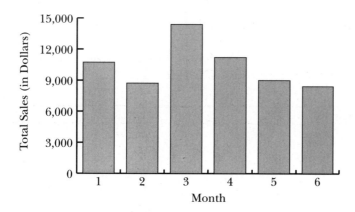

1.13 Construct a horizontal bar chart for the data set below, weekly donations collected at a local church.

Week	Donations ($)
1	2,070
2	2,247
3	1,850
4	2,771
5	1,955
6	2,412
7	1,782

This problem asks you to make a horizontal bar chart, which looks like the charts in problems 1.11 and 1.12 turned on their sides.

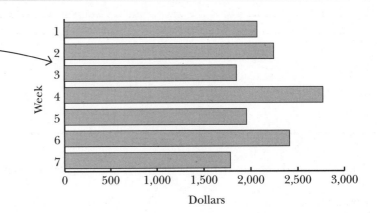

1.14 Construct a column bar chart for the data set below, the number of wins for each team in the National League East Division in the 2008 Major League Baseball season.

Team	Wins
Phillies	92
Mets	89
Marlins	84
Braves	72
Nationals	59

A column bar chart uses vertical bars to represent categorical data.

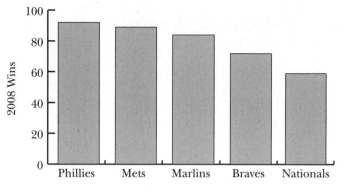

Note: Problems 1.15–1.16 refer to the data set below, weekly sales data in units for two stores.

Week	Store 1	Store 2
1	502	438
2	428	509
3	683	562
4	419	575

> The sales data for each week compares the sales for each store.

1.15 Construct a grouped column bar chart for the data, grouping by week.

Because there are two data values for each time period (a value for Store 1 and a value for Store 2), you should use a grouped column bar chart.

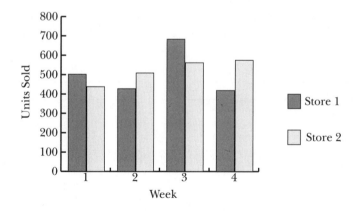

Note: Problems 1.15–1.16 refer to the data set in Problem 1.15, weekly sales data in units for two stores.

1.16 Construct a stacked column bar chart for the data, grouping by store.

Each column represents the total units sold each week between the two stores.

When you stack the sales data of the stores on top of each other, the total height of the stack represents the total sales for both stores.

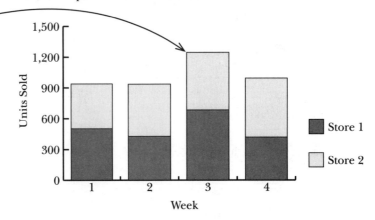

Note: Problems 1.17–1.20 refer to the data set below, the investment portfolio for three different investors in thousands of dollars.

	Investor 1	Investor 2	Investor 3
Savings	7.2	15.0	12.9
Bonds	3.8	9.6	7.4
Stocks	11.7	8.0	6.8

1.17 Construct a grouped horizontal bar chart, grouping by investor.

Three horizontal bars are arranged side-by-side for each investor, indicating the amount of each investment type.

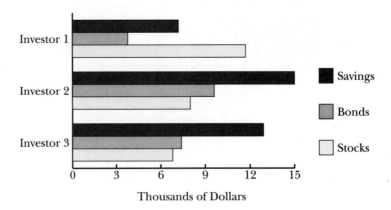

Note: Problems 1.17–1.20 refer to the data set in Problem 1.17, showing the investment portfolio for three different investors in thousands of dollars.

1.18 Construct a grouped horizontal bar chart, grouping by investment type.

Arrange three horizontal bars representing the investors, side-by-side, for each investment type.

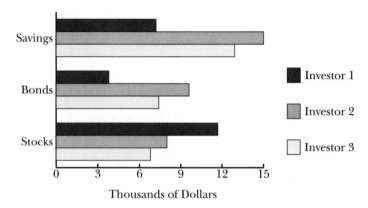

Thousands of Dollars

Note. Problems 1.17–1.20 refer to the data set in Problem 1.17, the investment portfolio for three different investors in thousands of dollars.

1.19 Construct a stacked horizontal bar chart, grouping by investor.

Each investor is represented by three horizontally stacked bars that indicate that investor's total investments by type.

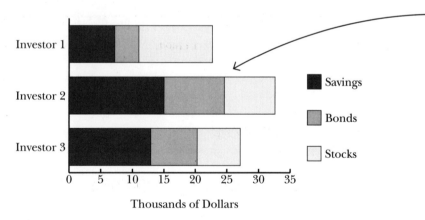

Thousands of Dollars

Each bar in this stacked chart represents the total investments of a single investor. The total length of each bar represents the total amount invested by each person.

Note: Problems 1.17–1.20 refer to the data set in Problem 1.17, the investment portfolio for three different investors in thousands of dollars.

1.20 Construct a stacked horizontal bar chart, grouping by investment type.

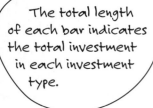

The total length of each bar indicates the total investment in each investment type.

Represent each investment type using three horizontally stacked bars.

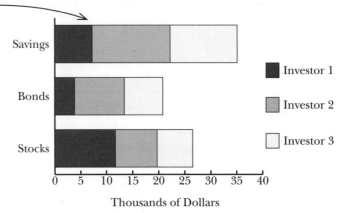

Thousands of Dollars

Pie Charts
Showing your categorical data in a circle

1.21 Construct a pie chart for the data set below, a grade distribution for a college class.

Grade	Number of Students
A	9
B	12
C	7
D	2
Total	**30**

Convert the frequency distribution to a relative frequency distribution, as explained in Problems 1.2 and 1.5.

Grade	Number of Students	Relative Frequency
A	9	9 ÷ 30 = 0.30
B	12	12 ÷ 30 = 0.40
C	7	7 ÷ 30 = 0.23
D	2	2 ÷ 30 = 0.07
Total	**30**	**1.00**

Multiply each relative frequency distribution by 360 to calculate the corresponding central angle for each category in the pie chart. A central angle has a vertex at the center of the circle and sides that intersect the circle, defining the boundaries of each category in a pie chart.

This is the number of degrees in a circle.

Grade	Relative Frequency	Central Angle
A	0.30	$0.30 \times 360 = 108°$
B	0.40	$0.40 \times 360 = 144°$
C	0.23	$0.23 \times 360 = 83°$
D	0.07	$0.07 \times 360 = 25°$
Total	**1.00**	**360°**

The central angle determines the size of each pie segment.

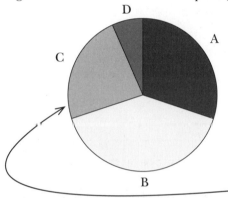

The central angle for this category is 83°.

1.22 Construct a pie chart for the data in the table below, the number of total wins recorded by the Green Bay Packers football team in five recent seasons.

Year	Number of Wins
2003	10
2004	10
2005	4
2006	8
2007	13

Convert the frequency distribution to a relative frequency distribution.

Year	Number of Wins	Relative Frequency
2003	10	$10 \div 45 = 0.22$
2004	10	$10 \div 45 = 0.22$
2005	4	$4 \div 45 = 0.09$
2006	8	$8 \div 45 = 0.18$
2007	13	$13 \div 45 = 0.29$
Total	**45**	**1.00**

Multiply each relative frequency distribution by 360 to calculate the central angle of each category in the pie chart.

Year	Relative Frequency	Central Angle
2003	0.22	$0.22 \times 360 = 80°$
2004	0.22	$0.22 \times 360 = 80°$
2005	0.09	$0.09 \times 360 = 32°$
2006	0.18	$0.18 \times 360 = 64°$
2007	0.29	$0.29 \times 360 = 104°$
Total	**1.00**	**360°**

The central angle determines the size of each pie segment.

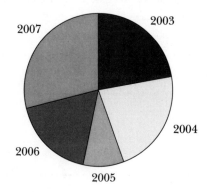

1.23 Construct a pie chart for the data in the table below, an individual investor's portfolio.

Investment	Dollars
Savings	9,000
Bonds	12,800
CDs	21,700
Stocks	34,500

The total investment is $78,000. Divide the figure for each category by this number to determine the percentage of the total investment each category represents.

Investment	Dollars	Percentage
Savings	9,000	9,000 ÷ 78,000 = 0.12
Bonds	12,800	12,800 ÷ 78,000 = 0.16
CDs	21,700	21,700 ÷ 78,000 = 0.28
Stocks	34,500	34,500 ÷ 78,000 = 0.44
Total	**78,000**	**1.00**

Multiply each percentage by 360 to calculate the central angle for each category in the pie chart.

Investment	Percentage	Central Angle
Savings	0.12	0.12 × 360 = 43°
Bonds	0.16	0.16 × 360 = 58°
CDs	0.28	0.28 × 360 = 101°
Stocks	0.44	0.44 × 360 = 158°
Total	**1.00**	**360°**

Use the central angles calculated above to draw appropriately sized sectors of the pie chart. If you have difficulty visualizing angles, use a protractor.

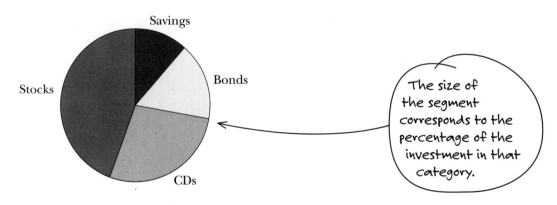

The size of the segment corresponds to the percentage of the investment in that category.

1.24 Construct a pie chart for the frequency distribution below, the daily high temperature (in degrees Fahrenheit) in a particular city over the last 40 days.

Daily High Temperature	Frequency
40–under 45	6
45–under 50	12
50–under 55	17
55–under 60	5
Total	**40**

Determine the relative frequency distribution for each temperature range.

Daily High Temperature	Frequency	Relative Frequency
40–under 45	6	$6 \div 40 = 0.150$
45–under 50	12	$12 \div 40 = 0.300$
50–under 55	17	$17 \div 40 = 0.425$
55–under 60	5	$5 \div 40 = 0.125$
Total	**40**	**1.000**

Calculate the central angle for each category in the pie chart.

Daily High Temperature	Relative Frequency	Central Angle
40–under 45	0.150	$0.150. \times 360 = 54°$
45–under 50	0.300	$0.300 \times 360 = 108°$
50–under 55	0.425	$0.425 \times 360 = 153°$
55–under 60	0.125	$0.125 \times 360 = 45°$
Total	**1.000**	**360°**

Use the central angles to construct appropriately sized sectors of the pie chart.

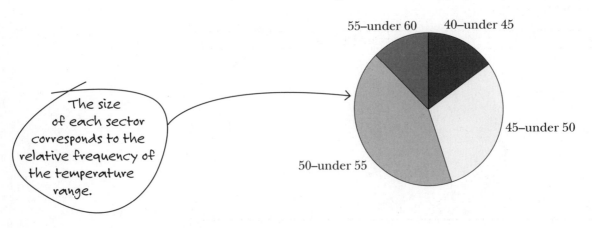

The size of each sector corresponds to the relative frequency of the temperature range.

Line Charts

Data over time in a chart

1.25 Construct a line chart for the data in the table below, the number of wins recorded by the Philadelphia Phillies for seven seasons.

Year	Number of Wins
2002	80
2003	86
2004	86
2005	88
2006	85
2007	89
2008	92

Place the time variable (year) on the *x*-axis and place the variable of interest (wins) on the *y*-axis.

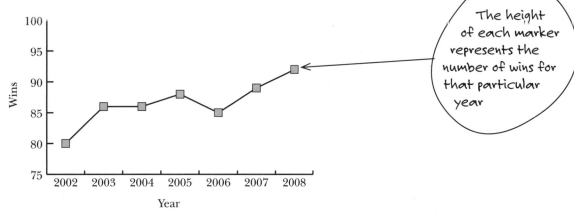

The height of each marker represents the number of wins for that particular year

1.26 Construct a line chart for the data in the table below, the percent change in annual profit for a company by year.

Year	Percent Change
2001	3.8%
2002	−2.1%
2003	−3.6%
2004	3.0%
2005	4.0%
2006	0.6%
2007	2.4%

Place the time variable (year) on the *x*-axis and place the variable of interest (percent change) on the *y*-axis.

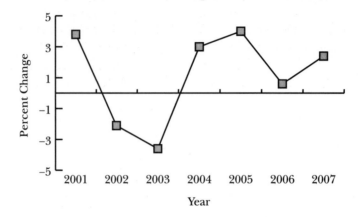

1.27 Construct a line chart for the data in the table below, the population of Delaware by decade during the 1800s.

Year	Population
1800	64,273
1810	72,674
1820	72,749
1830	76,748
1840	78,085
1850	91,532
1860	112,532

Place the time variable (year) on the *x*-axis and place the variable of interest (population) on the *y*-axis.

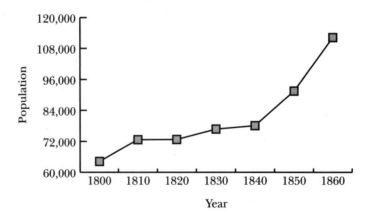

Scatter Charts

Illustrate relationships between two variables

1.28 Construct a scatter chart for the data in the table below, the number of hours eight students studied for an exam and the scores they earned on the exam.

Study Hours	Exam Score
5	84
7	92
4.5	82
7	80
8	90
6.5	78
5.5	74
4	75

> The dependent variable is the variable that changes as a result of changes in the independent variable. Studying longer should give you a higher grade. The reverse doesn't make sense—getting a higher test score doesn't result in you studying longer for that test.

Place the independent variable (study hours) on the *x*-axis and the dependent variable (exam score) on the *y*-axis.

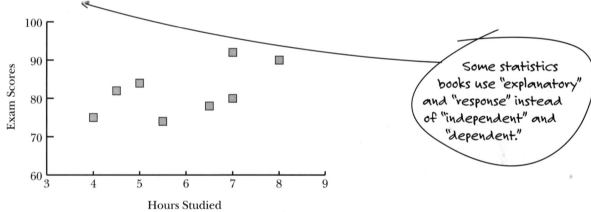

> Some statistics books use "explanatory" and "response" instead of "independent" and "dependent."

1.29 Construct a scatter chart for the data below, the mileage and selling price of eight used cars.

Mileage	Selling Price
21,800	$16,000
34,000	$11,500
41,700	$13,400
53,500	$14,800
65,800	$10,500
72,100	$12,300
76,500	$8,200
84,700	$9,500

Place the independent variable (mileage) on the *x*-axis and the dependent variable (selling price) on the *y*-axis.

An increase in the independent variable (mileage) appears to cause the dependent variable (selling price) to decrease, as anyone who has recently purchased a used car would expect.

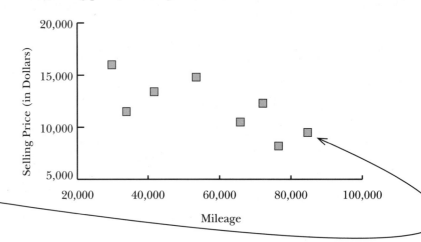

1.30 Construct a scatter chart for the data in the table below, eight graduate students' grade point averages (GPA) and entrance exam scores for M.B.A. programs (GMAT).

GPA	GMAT
3.7	660
3.8	580
3.2	450
4.0	710
3.5	550
3.1	600
3.3	510
3.6	750

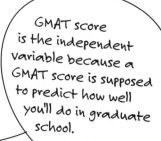

GMAT score is the independent variable because a GMAT score is supposed to predict how well you'll do in graduate school.

Place the independent variable (GMAT) on the *x*-axis and the dependent variable (GPA) on the *y*-axis.

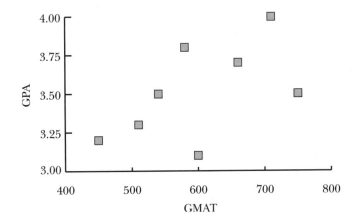

Chapter 2
CALCULATING DESCRIPTIVE STATISTICS: MEASURES OF CENTRAL TENDENCY

Finding the center of the data

One of the most common roles served by descriptive statistics is determining the central tendency of data. This chapter investigates the primary means by which the "center" of a data set can be described, including the mean, median, mode, and percentiles.

Data may be reported in either raw or summarized form. To understand the central tendency of raw data, the mean, median, mode, and weighted mean can be very helpful. The position of raw data can be described using percentiles. Summarized data, in the form of frequency distributions, can be described using a weighted mean technique.

Mean

The average

2.1 The table below lists the number of students enrolled in the five different statistics courses offered by a college. Calculate the mean number of students per class.

Number of Students				
18	22	25	26	15

A population represents all of the available data: in this case, all of the statistics classes that are offered. A sample is a subset of a population.

The average (or arithmetic mean) μ of a data set is the sum of the data values $\sum x$ divided by the population size N.

$$\mu = \frac{\sum x}{N} = \frac{18+22+25+26+15}{5} = \frac{106}{5} = 21.2$$

2.2 The table below lists the number of students enrolled in the five different statistics courses offered by a college. Calculate the mean number of students per class.

Number of Students				
18	22	25	26	75

The mean μ of a data set is the sum of the data divided by the population size.

$$\mu = \frac{\sum x}{N} = \frac{18+22+25+26+75}{5} = \frac{166}{5} = 33.2$$

Notice that a mean can be heavily influenced by an extreme value in the data. The only difference between the enrollment numbers in Problems 2.1 and 2.2 is the size of the last class. That class increased from 15 to 75 students and caused the mean class size to grow from 21.2 to 33.2.

2.3 The table below lists the time, in minutes, it takes seven random customers to check out at a local grocery store. Calculate the mean time it takes a customer to check out.

Number of Minutes						
6	4	2	5	4	1	5

This data set is considered a sample because only a portion of all the customers is included in the data set. (The store had better serve more than seven customers total!)

Divide the sum of the data $\sum x$ by the sample size n.

$$\bar{x} = \frac{\sum x}{n} = \frac{6+4+2+5+4+1+5}{7} = \frac{27}{7} = 3.9 \text{ minutes}$$

2.4 A consumer group tested the gas mileage of two car models in five trials. Which model averages more miles per gallon?

	Trial 1	Trial 2	Trial 3	Trial 4	Trial 5
Model A	22	24	19	25	20
Model B	21	25	27	23	21

Calculate the sample means for each model separately.

$$\text{Model A: } \bar{x} = \frac{\sum x}{n} = \frac{22+24+19+25+20}{5} = \frac{110}{5} = 22.0 \text{ miles per gallon}$$

$$\text{Model B: } \bar{x} = \frac{\sum x}{n} = \frac{21+25+27+23+21}{5} = \frac{117}{5} = 23.4 \text{ miles per gallon}$$

Model B averages more miles per gallon than Model A.

2.5 The table below reports the number of minutes eight randomly selected airline flights were either early (negative values) or late (positive values) arriving at their destinations. Calculate the sample mean.

Number of Minutes Early or Late							
12	–10	32	–4	0	16	5	18

Combine the positive and negative values and divide by the sample size.

$$\bar{x} = \frac{\sum x}{n} = \frac{12+(-10)+32+(-4)+0+16+5+18}{8} = \frac{69}{8} = 8.6 \text{ minutes}$$

The average flight is 8.6 minutes late.

Adding 12 and –10 is the same as subtracting 10 from 12: 12 + (–10) = 12 – 10.

2.6 The following table lists the daily percent increase (or decrease) of a stock price over a five-day period. Calculate the mean daily change in the stock price.

Percent Increase or Decrease				
0.018	–0.026	–0.057	–0.039	0.016

If the average was –8.6, it would mean that the average flight was 8.6 minutes early.

If the sum $\sum x$ is less than zero, the average change is a decrease, whereas a positive sum indicates an average increase.

$$\bar{x} = \frac{\sum x}{n} = \frac{0.018+(-0.026)+(-0.057)+(-0.039)+0.016}{5} = \frac{-0.088}{5} = -0.0176$$

The stock price decreased an average of 1.76 percent per day.

2.7 The table below lists the points scored by three basketball players over six games. Identify the player with the highest average points per game.

Player	Points Scored per Game					
Kevin	17	16	8	21	25	18
Paul	20	18	7	36	34	15
Ray	18	9	17	19	17	12

Calculate each player's scoring average separately.

$$\text{Kevin: } \bar{x} = \frac{\sum x}{n} = \frac{17+16+8+21+25+18}{6} = \frac{105}{6} = 17.5 \text{ points per game}$$

$$\text{Paul: } \bar{x} = \frac{\sum x}{n} = \frac{20+18+7+36+34+15}{6} = \frac{130}{6} = 21.7 \text{ points per game}$$

$$\text{Ray: } \bar{x} = \frac{\sum x}{n} = \frac{18+9+17+19+17+12}{6} = \frac{92}{6} = 15.3 \text{ points per game}$$

Paul averaged the most points per game, 21.7.

Note: Problems 2.8–2.9 refer to the data set below, the daily demand for tires at a particular store over a seven-day period.

Day	Demand
January 4	18
January 5	23
January 6	14
January 7	15
January 8	25
January 9	34
January 10	29

2.8 The store develops a demand forecast that is based on the average demand from the previous three days. Calculate a demand forecast for January 11.

Calculate the average demand for January 8–10.

$$\bar{x} = \frac{\sum x}{n} = \frac{25+34+29}{3} = \frac{88}{3} = 29.3$$

The demand forecast for January 11 is 29.3 tires.

Note: Problems 2.8–2.9 refer to the data set in Problem 2.8, daily demand for tires at a particular store over a seven-day period.

2.9 Assume that the store calculates a demand forecast every day based on the average demand from the previous three days. Based on the given data, on which days was the demand forecast the most and the least accurate?

To forecast demand, you need data for the three previous days. January 7 is the first day for which there is enough data. To calculate the forecast for January 4, for instance, you'd need sales information for January 1-3.

You can only determine the accuracy of the demand forecast from January 7 to January 10. To calculate each forecast, calculate the mean sales for the preceding three days.

$$\text{January 7 demand forecast:} \quad \frac{18+23+14}{3} = \frac{55}{3} = 18.3$$

$$\text{January 8 demand forecast:} \quad \frac{23+14+15}{3} = \frac{52}{3} = 17.3$$

$$\text{January 9 demand forecast:} \quad \frac{14+15+25}{3} = \frac{54}{3} = 18$$

$$\text{January 10 demand forecast:} \quad \frac{15+25+34}{3} = \frac{74}{3} = 24.6$$

To determine how accurate each forecast is, calculate the absolute value of the difference between the forecasted demand and the actual demand for each day. For instance, the projected demand for January 7 is 18.3 (as calculated above), but the actual demand was only 15 tires. Thus, the prediction was inaccurate by a margin of $|18.3-15| = 3.3$ tires.

The problem doesn't ask whether the prediction was too high or too low—just how far off it was from the actual demand. That means you're looking for the "biggest" number, whether it's positive or negative, so take the absolute value of the difference.

$$\text{January 7 prediction error:} \quad |18.3-15| = 3.3$$
$$\text{January 8 prediction error:} \quad |17.3-25| = 7.7$$
$$\text{January 9 prediction error:} \quad |18-34| = 16$$
$$\text{January 10 prediction error:} \quad |24.6-29| = 4.4$$

The demand forecast was most accurate on January 7 and least accurate on January 9.

2.10 The average exam score for 10 students in a statistics class was 85. One student dropped the class, changing the average exam score of the remaining students to 90. What was the exam score of the student who dropped the class?

Calculate the sum $\sum x$ of the exam scores for the original 10 students.

$$\sum x = \bar{x}n$$
$$= (85)(10)$$
$$= 850$$

To get this, cross-multiply the familiar mean equation $\bar{x} = \dfrac{\sum x}{n}$ to solve it for $\sum x$.

The same equation can be used to calculate the sum of the exam scores for the remaining nine students.

$$\sum x = \bar{x}n$$
$$= (90)(9)$$
$$= 810$$

The difference of the sums is 850 – 810 = 40, so the student who dropped the course had an exam score of 40.

2.11   The table below lists the ages of the players on a basketball team.

Players' Ages									
30	27	36	24	26	24	28	25	29	31

After the oldest player was released and replaced by a new player, the average age of the team was lowered by one year. Determine the age of the new player.

Calculate the sum $\sum x$ of the original players' ages and the mean age $\bar{x}$.

$$\sum x = 30 + 27 + 36 + 24 + 26 + 24 + 28 + 25 + 29 + 31 = 280$$

$$\bar{x} = \frac{\sum x}{n} = \frac{280}{10} = 28$$

Once the team releases its oldest player, the average age decreases by one: 28 – 1 = 27. Calculate the new sum of the players' ages.

$$\sum x = \bar{x}n = (27)(10) = 270$$

The total of the team's ages is reduced by 280 – 270 = 10 years, so the new player is 10 years younger than the 36-year-old player who was released. The replacement player is 36 – 10 = 26 years old.

Median
Right smack in the middle

2.12   The table below lists exam scores for nine students. Identify the median test score.

Exam Scores								
85	96	75	82	93	88	80	90	71

Write the exam scores in order, from least to greatest.

Sorted Exam Scores								
71	75	80	82	85	88	90	93	96

The median is the "middle number" when you list the data in order from least to greatest. P = 50 here because 50 percent of the data is less than the middle value.

Identify the index point i for the median, using the formula $i = \frac{P}{100}(n)$, where $P = 50$ and n is the number of data points.

$$i = \frac{P}{100}(n) = \left(\frac{50}{100}\right)(9) = \left(\frac{1}{2}\right)(9) = 4.5$$

The index point identifies the position of the median. In this case, the index point is not an integer, so round it up to the nearest integer, 5. Thus, the median is the value in the fifth position of the sorted data set: 85.

The median is 85, because it's right in the middle of the data—just like a highway median is right in the middle of the road. There are four values less than 85 and four values greater than 85.

2.13 The table below lists exam scores for nine students. Determine the median test score.

Exam Scores								
85	96	75	82	93	88	80	90	21

Rewrite the data in order, from least to greatest.

Sorted Exam Scores								
21	75	80	82	85	88	90	93	96

Identify the index point i by substituting $P = 50$ and $n = 9$ into the formula $i = \frac{P}{100}(n)$.

$$i = \left(\frac{50}{100}\right)(9) = \left(\frac{1}{2}\right)(9) = 4.5$$

The median is 85, the fifth sorted data point.

The data values in this problem are the same as the data values in Problem 2.12, except for this number. This number is 71 in Problem 2.12.

2.14 The table below lists a city's daily high temperature for an 11-day period. Identify the median high temperature.

Daily High Temperature										
62	68	73	71	82	77	74	68	65	68	69

Write the temperatures in order, from least to greatest.

Sorted Daily High Temperature										
62	65	68	68	68	69	71	73	74	77	82

Identify the index point.

$$i = \frac{P}{100}(n) = \left(\frac{50}{100}\right)(11) = \left(\frac{1}{2}\right)(11) = 5.5$$

The index point is not an integer, so round it to the nearest greatest integer, 6. The median of the data is 69, the sixth data point.

The median, unlike the mean, is not influenced by extreme values. Changing the lowest exam score from 71 to 21 did not affect the median exam score.

When there are an odd number of data points, the median will always be the middle number. In this example $n = 11$, so there are five data points below the median and five data points above it.

Midrange
Halfway between the endpoints

2.15 A company asked customers to rate a service on a scale of 1 to 100 and recorded the 24 most recent ratings in the table below. Identify the midrange of the ratings.

Customer Ratings							
92	80	86	76	81	65	71	96
90	62	69	75	70	83	94	85
89	85	78	73	88	95	75	80

The highest customer rating is 96; the lowest is 62. To calculate the midrange (MR), determine the mean of these extreme data values.

The midrange is a quick way to estimate the center of the data without adding all the values (for the mean) or sorting them from low to high (for the median).

$$MR = \frac{\text{highest value} + \text{lowest value}}{2} = \frac{96 + 62}{2} = \frac{158}{2} = 79$$

2.16 The table below represents the amount of time, in seconds, it took to complete 15 ATM transactions. Identify the midrange of the ATM transaction time.

Seconds per ATM Transaction				
152	97	126	115	135
105	86	140	139	111
123	147	90	101	133

Identify the highest and lowest transaction times (152 and 86, respectively) and calculate the mean of those values.

One drawback of the midrange is that it is heavily influenced by extreme data values. For example, changing 152 to 552 changes the midrange from 119 seconds to $\frac{552 + 86}{2} = 319$ seconds.

$$MR = \frac{\text{highest value} + \text{lowest value}}{2} = \frac{152 + 86}{2} = \frac{238}{2} = 119 \text{ seconds}$$

2.17 The table below indicates the number of minutes eight randomly selected airline flights were either early (negative values) or late (positive values) arriving at their destinations. Calculate the midrange of the data set.

Number of Minutes Early or Late							
12	–10	32	–4	0	16	5	18

Calculate the mean of the highest (32) and lowest (–10) values.

$$MR = \frac{\text{highest value} + \text{lowest value}}{2} = \frac{32 + (-10)}{2} = \frac{22}{2} = 11 \text{ minutes late}$$

2.18 The table below lists the average daily high temperature, in degrees Celsius, for the first six months of the year in Barrow, Alaska. Calculate the midrange for daily high temperature during this time period.

Average High Temperature					
−22	−23	−22	−14	−4	−2

$$MR = \frac{\text{highest value} + \text{lowest value}}{2} = \frac{(-2)+(-23)}{2} = \frac{-25}{2} = -12.5°\text{C}$$

Mode

The most frequent value

2.19 The table below lists daily demand for hammers at a hardware store over the last 20 days. Identify the mode.

Daily Demand for Hammers									
2	1	0	2	1	3	0	2	4	0
3	2	3	4	2	2	2	4	3	0

Organize the demand for hammers in a frequency distribution.

See Problem 1.1 for a review of frequency distributions.

Daily Demand	Frequency
0	4
1	2
2	7
3	4
4	3
Total	20

The mode of a data set is the value that appears most frequently. In this example, a daily demand of 2 hammers is more common than a daily demand of 0, 1, 3, or 4 hammers. Therefore, the mode is 2 hammers per day.

The mode is the daily demand that appears most frequently (2), not the frequency of that demand (7).

2.20 A grocery store records the time, in minutes, required to check out nine randomly selected customers. Identify the mode of this data set.

Minutes per Customer								
3	4	2	5	4	1	5	5	4

Construct a frequency distribution for the data set.

Minutes	Frequency
1	1
2	1
3	1
4	3
5	3
Total	9

Bimodal means there are two modes. The numbers 4 and 5 occur more than any other numbers, and they occur exactly the same number of times.

This data set is bimodal; the modes are 4 and 5 minutes.

2.21 The table below lists the ages of five employees in a particular company. Identify the mode of this data set.

Employee Age				
27	46	37	29	58

Organize the ages of the employees in a frequency distribution.

Employee Age	Frequency
27	1
29	1
37	1
46	1
58	1
Total	5

No data value appears more frequently than any other.

There is no mode for this data.

2.22 The table below lists the number of minutes the last 10 customers to call a company waited on hold for technical support over the phone. Describe the central tendency of the data in two different, and appropriate, ways.

Minutes on Hold									
17	8	10	20	16	5	18	12	15	6

Because there do not appear to be any extreme values in this data set, the mean is an appropriate choice to measure central tendency.

An extreme value, also known as an outlier, exists when one data point is significantly higher or lower than the rest of the data set. Outliers will be covered in more detail in Chapter 3.

$$\bar{x} = \frac{\sum x}{n} = \frac{17+8+10+20+16+5+18+12+15+6}{10} = \frac{127}{10} = 12.7 \text{ minutes}$$

You can also use the median to describe the central tendency of the data. Rewrite the data in order, from least to greatest.

Sorted Minutes on Hold									
5	6	8	10	**12**	**15**	16	17	18	20

The sample size of the data set is 10, so the median is the average of the fifth and sixth sorted data values.

$$\text{median} = \frac{12+15}{2} = \frac{27}{2} = 13.5 \text{ minutes}$$

When the sample size is even, the median is the average of the two middle numbers in the sample. There are four numbers before the fifth and sixth positions and four numbers after.

The mode is not a good choice to describe the central tendency of the data, because there is no mode in this data set.

2.23 The table below lists the exam scores for a class of eight students. Choose the most appropriate measure of central tendency: the mean, median, or mode.

Exam Scores							
87	96	91	85	93	15	80	90

Because an extreme value appears to exist in the exam scores (15), the median is likely the best choice to measure central tendency for this data set. Rewrite the exam scores in order, from lowest to highest.

Fifteen is really different from the rest of the data values—they're all 80 or higher.

Sorted Exam Scores							
15	80	85	87	90	91	93	96

Identify the index point i for the median, using the formula $i = \frac{P}{100}(n)$, where $P = 50$ and n is the number of data points.

$$i = \frac{P}{100}(n) = \left(\frac{50}{100}\right)(8) = \left(\frac{1}{2}\right)(8) = 4$$

Because i is an integer, the median is the midpoint of the values in position $i = 4$ and position $i + 1 = 5$.

$$\frac{87 + 90}{2} = 88.5$$

A good measure of central tendency for these exam scores is the median: 88.5.

The mean exam score is 79.6, which is lower than all but the lowest exam score, so it's hardly the "center" of the data. There is no mode for this data set.

Percentile

On a scale from 1 to 100

2.24 The energy consumption per person (in millions of BTUs) for various states in 2005 is shown in the table below. Identify the 30th percentile.

State	BTUs per Person (in millions)
Alabama	467
Alaska	1,194
Arizona	249
Arkansas	410
California	232
Colorado	305
Connecticut	258
Delaware	372
Florida	257

List the BTU values in order, from least to greatest.

Sorted BTUs per Person
232 249 257 258 305 372 410 467 1,194

Calculate the index point for the 30th percentile by substituting $P = 30$ and $n = 9$ into the formula $i = \frac{P}{100}(n)$.

$$i = \frac{P}{100}(n) = \left(\frac{30}{100}\right)(9) = 2.7$$

This is the same formula you used to calculate the median in Problems 2.12–2.14.

Because i is not an integer, you should round it to the next highest integer, 3. The 30th percentile is the data value in the third position of the sorted data: 257.

2.25 The table below lists the populations of the 11 largest U.S. cities in 2007. Identify the 65th percentile for the data.

City	Population (in millions)
New York, NY	8.27
Los Angeles, CA	3.83
Chicago, IL	2.84
Houston, TX	2.21
Phoenix, AZ	1.55
Philadelphia, PA	1.45
San Antonio, TX	1.33
San Diego, CA	1.27
Dallas, TX	1.24
San Jose, CA	0.94
Detroit, MI	0.92

Notice that this table is sorted from highest population to lowest population (not lowest to highest). Once you calculate the index point, you'll need to count up from the bottom of the chart.

Calculate the index point for the 65th percentile.

$$i = \frac{P}{100}(n) = \left(\frac{65}{100}\right)(11) = 7.15$$

Round the index point up to 8 (the smallest integer greater than 7.15). The 65th percentile is the eighth highest value on the chart: 2.21 million.

2.26 The table below lists the top-grossing films in the United States (as of November 2008), adjusted for inflation. Identify the 40th percentile.

Film	2008 Dollars (in millions)
Gone with the Wind (1939)	1,431
Star Wars (1977)	1,261
The Sound of Music (1965)	1,008
E.T.: The Extra Terrestrial (1982)	1,004
The Ten Commandments (1956)	927
Titanic (1997)	909
Jaws (1975)	907
Doctor Zhivago (1965)	879
The Exorcist (1973)	782
Snow White (1937)	772

Calculate the index point for the 40th percentile.

$$i = \frac{P}{100}(n) = \left(\frac{40}{100}\right)(10) = 4$$

Because i is an integer, the 40th percentile is the average of the values in position i and position $i + 1$. The value at the fourth position from the bottom of the table is 907 and the value at the fifth position is 909. The midpoint of these two values is 908 million dollars, which is the 40th percentile.

2.27 The table below lists the total points scored by each of the 32 NFL teams during the 2007 season. Identify the 15th percentile.

Total Points							
219	226	252	259	263	267	267	268
275	283	301	320	334	334	334	336
346	365	373	379	379	380	393	393
402	404	411	412	435	455	456	589

Calculate the index point i for the fifteenth percentile.

$$i = \frac{P}{100}(n) = \left(\frac{15}{100}\right)(32) = 4.8$$

Round the value of i up to the nearest integer, 5. The fifteenth percentile is the fifth sorted data point: 263.

2.28 The following table lists the systolic blood pressure of 15 adults. Calculate the percentile rank of the value 126.

Systolic Blood Pressure							
136	122	130	114	121	110	113	103
126	121	120	130	115	125	116	

Begin by listing the blood pressures in order, from lowest to highest.

Sorted Blood Pressures							
103	110	113	114	115	116	120	121
121	122	125	126	130	130	136	

The data value 126 is the twelfth position, so there are 11 values that are less than 126.

There are eleven blood pressures listed that are less than 126. Substitute 11 and the total number of data values (15) into the percentile formula that follows.

$$\text{percentile} = \frac{(\text{number of values less than } x) + 0.5}{\text{total number of values}}(100)$$

$$= \frac{11 + 0.5}{15}(100)$$

$$= (0.767)(100)$$

$$= 76.7\%$$

Some books don't add 0.5 in this step.

The systolic blood pressure reading of 126 is at the 76.7th percentile rank.

2.29 The table below represents the tips left by 20 customers for the waitstaff at a local restaurant. Find the percentile rank of a $10.00 tip.

Tips					
$18.50	$10.25	$12.35	$7.75	$14.00	$9.50
$11.45	$16.00	$15.50	$12.75	$10.50	$12.40
$13.75	$10.75	$8.25	$17.85	$16.50	$17.60
$14.10	$10.00				

List the tips in order, from lowest to highest.

Sorted Tips					
$7.75	$8.25	$9.50	$10.00	$10.25	$10.50
$10.75	$11.45	$12.35	$12.40	$12.75	$13.75
$14.00	$14.10	$15.50	$16.00	$16.50	$17.60
$17.85	$18.50				

Apply the percentile formula introduced in Problem 2.28.

$$\text{percentile} = \frac{(\text{number of values below } x) + 0.5}{\text{total number of values}}(100)$$

$$= \frac{3 + 0.5}{20}(100)$$

$$= (0.175)(100)$$

$$= 17.5\%$$

The $10.00 tip is at the 17.5th percentile rank.

2.30 The table below lists the total runs scored by each of the 16 National League teams in Major League Baseball during the 2008 season. Calculate the percentile rank for the number of runs scored by the Pittsburgh Pirates.

Team	Runs	Team	Runs
Cubs	855	Pirates	735
Mets	799	Diamondbacks	720
Phillies	799	Astros	712
Cardinals	779	Reds	704
Marlins	770	Dodgers	700
Braves	753	Nationals	641
Brewers	750	Giants	640
Rockies	747	Padres	637

Apply the percentile formula, given $x = 735$.

The table is sorted from most runs to least, so count the number of teams below the Pirates to figure out how many teams scored fewer points than they did.

$$\text{percentile} = \frac{(\text{number of values below } x) + 0.5}{\text{total number of values}}(100)$$
$$= \frac{7 + 0.5}{16}(100)$$
$$= 0.46875(100)$$
$$= 46.9\%$$

The Pirates ended the 2008 season in the 46.9th percentile for runs scored.

2.31 The table below lists the SAT scores of 27 college students. Calculate the percentile rank for an SAT score of 947.

SAT Scores

387	454	490	564	629	665	668	673	683
696	807	894	919	941	947	974	1038	1074
1165	1177	1188	1245	1346	1419	1471	1535	1559

List the SAT scores in order, from lowest to highest.

Fourteen students in this sample population scored lower than 947. Substitute this value into the percentile formula.

$$\text{percentile} = \frac{(\text{number of values below } 947) + 0.5}{\text{total number of values}}(100)$$
$$= \frac{14 + 0.5}{27}(100)$$
$$= 0.537(100)$$
$$= 53.7\%$$

The SAT score of 947 is in the 53.7th percentile for this sample.

2.32 The following tables record the times of swimmers in the 50-meter freestyle and 100-meter backstroke events at a recent meet. A particular swimmer swam the freestyle event in 31 seconds and the backstroke in 75 seconds. In which event did the swimmer perform better relative to the competition?

50-Meter Freestyle Times							
27	28	31	32	32	33	34	34

100-Meter Backstroke Times									
69	70	72	75	76	78	78	78	79	79

To measure relative performance between the two events, calculate the percentile rank for each time individually.

$$\text{freestyle percentile} = \frac{(\text{number of values below } 31) + 0.5}{\text{total number of values}}(100)$$

$$= \frac{5 + 0.5}{8}(100)$$

$$= 0.6875(100)$$

$$= 68.75\%$$

Because lower swim times are better than higher times, the number of values "below" 31 seconds is the number of swimmers who took longer to finish the race. A good swim time should give you a high percentile rank, not a low one.

Notice that there are more participants in the backstroke (10) than in the freestyle (8).

$$\text{backstroke percentile} = \frac{(\text{number of values below } 75) + 0.5}{\text{total number of values}}(100)$$

$$= \frac{6 + 0.5}{10}(100)$$

$$= 0.65(100)$$

$$= 65\%$$

A higher percentile rank indicates that the swimmer defeated a proportionately larger number of competitors. Therefore, the swimmer performed better in the freestyle event.

2.33 The following tables list the exam scores from an English class and a chemistry class. A particular student scored an 88 on the English test and a 76 on the chemistry test. Determine which exam score is higher relative to her classmates.

English Exam Scores												
75	76	78	78	80	80	80	80	81	81	84	85	85
85	86	86	88	90	90	91	92	94	95	96	98	

Chemistry Exam Scores												
54	56	59	60	60	63	64	65	68	68	69	70	71
72	74	75	75	76	78	80						

Calculate the student's percentile ranks for the exams separately, noting that population sizes for the classes vary.

$$\text{English percentile} = \frac{(\text{number of values below } 88) + 0.5}{\text{total number of values}}(100)$$

$$= \frac{16 + 0.5}{25}(100)$$

$$= (0.66)(100)$$

$$= 66\%$$

$$\text{chemistry percentile} = \frac{(\text{number of values below } 76) + 0.5}{\text{total number of values}}(100)$$

$$= \frac{17 + 0.5}{20}(100)$$

$$= 0.875(100)$$

$$= 87.5\%$$

The student performed better on the chemistry exam relative to her classmates, even though her chemistry exam score was lower than her English exam score.

Weighted Mean

Averaging using different weights

2.34 A statistics professor administered an exam to three different classes. The 8:00 class had 20 students and an average exam score of 82.4. The 9:00 class had 24 students and an average of 78.9. The 29 students in the 10:00 class averaged 87.3. Calculate the overall average for all three classes combined.

The weighted mean is the mean of each average score multiplied by the number of students in the class. In the formula below, w_i represents the weight of each group (in this case the size of each class) and x_i represents the average score of the corresponding class.

The i's represent the different classes, so $w_1 = 20$ and $x_1 = 82.4$ are w and x values for class #1, $w_2 = 24$ and $x_2 = 78.9$ represent class #2, and $w_3 = 29$ and $x_3 = 87.3$ represent class #3.

$$\bar{x} = \frac{\sum w_i x_i}{\sum w_i}$$

$$= \frac{(20)(82.4) + (24)(78.9) + (29)(87.3)}{20 + 24 + 29}$$

$$= \frac{6,073.3}{73}$$

$$= 83.2$$

The overall average is 83.2.

2.35 Assume a student's final statistics grade for the semester is based on a combination of a final exam score, a project score, and a homework score, each weighted according to the table below. Calculate the final grade of a student with an exam score of 94, a project score of 89, and a homework score of 83.

Type	Weight (percent)
Exam	50
Project	35
Homework	15

Multiply each of the students' scores by the corresponding weights and then average those products.

$$\bar{x} = \frac{\sum w_i x_i}{\sum w_i}$$

$$= \frac{(50)(94) + (35)(89) + (15)(83)}{50 + 35 + 15}$$

$$= \frac{9,060}{100}$$

$$= 90.6$$

If you got an answer of 88.7, you forgot to weight the categories. This student did well on the exam, which is worth more than the other categories. That's why the final grade is higher than two of the three individual category scores.

The student's final grade is 90.6.

2.36 A company has four locations at which customers were surveyed for their satisfaction ratings. The table below lists the average customer rating for each of the four locations and the number of customers that responded to the survey at that location. Calculate the total average customer rating for the company.

Location	Average Rating	Number of Customers
1	7.8	117
2	8.5	86
3	6.6	68
4	7.4	90

The average rating must be a weighted mean because each location collected a different number of customer surveys. The weights are the number of customer surveys turned in for each location.

Apply the weighted mean formula, averaging the products of the ratings and the number of survey respondents for each location.

$$\bar{x} = \frac{\sum w_i x_i}{\sum w_i}$$
$$= \frac{(117)(7.8) + (86)(8.5) + (68)(6.6) + (90)(7.4)}{117 + 86 + 68 + 90}$$
$$= \frac{2,758.4}{361}$$
$$= 7.6$$

The average customer rating for all four locations is 7.6.

2.37 The table below lists the grades and credit hours earned one semester by a college student. Assuming A, B, and C grades correspond to 4, 3, and 2 grade points respectively, calculate the student's grade point average for the semester.

Course	Credit Hours	Final Grade
Math	3	A
English	3	C
Chemistry	4	A
Business	4	B

The number of credit hours per course vary, so you have to use a weighted mean. An A in a 3-credit-hour course is worth less than an A in a 4-credit-hour course.

Multiply the grade point equivalent of each letter grade (4, 3, or 2) by the number of credit hours for that course. The student's grade point average is the mean of those three products.

$$\bar{x} = \frac{\sum w_i x_i}{\sum w_i}$$

$$= \frac{(3)(4.0) + (3)(2.0) + (4)(4.0) + (4)(3.0)}{3 + 3 + 4 + 4}$$

$$= \frac{46}{14}$$

$$= 3.29$$

The student's grade point average is 3.29.

Mean of a Frequency Distribution

Averaging discrete data

2.38 The table below records the results of a survey that asked respondents how many cats lived in their households. Calculate the average number of cats per household.

Number of Cats	Number of Households
0	58
1	22
2	15
3	8
4	2

The number of cats in each household varies, so you should apply the weighted mean formula. Unlike the weighted mean problems in the preceding section, the problems in this section are weighted according to the frequencies of each category. Here, survey responses of zero through four cats are multiplied by the frequency with which each number was reported, and the products are averaged.

$$\bar{x} = \frac{\sum w_i x_i}{\sum w_i}$$

$$= \frac{(58)(0) + (22)(1) + (15)(2) + (8)(3) + (2)(4)}{58 + 22 + 15 + 8 + 2}$$

$$= \frac{84}{105}$$

$$= 0.80$$

This is a mean, so it doesn't need to be a whole number.

The average number of cats per household is 0.80.

2.39 A six-year-old company surveyed its employees to determine how long each person had worked there; the results are reported in the following table. Calculate the average length of time the employees have worked for the company.

Years of Service	Number of Employees
1	5
2	7
3	10
4	8
5	12
6	3

Apply the weighted mean formula, weighting each year of service value by the corresponding frequency.

The denominator of the weighted mean formula is the sum of the weights. In this problem, the weights are how often (by employee) each count (years worked) occurs.

$$\bar{x} = \frac{\sum w_i x_i}{\sum w_i}$$

$$= \frac{(5)(1)+(7)(2)+(10)(3)+(8)(4)+(12)(5)+(3)(6)}{5+7+10+8+12+3}$$

$$= \frac{159}{45}$$

$$= 3.53$$

The mean length of employment is approximately 3.53 years.

2.40 An airline recorded the number of no-shows (people who fail to arrive at the gate on time to board the plane) for its last 120 flights. The frequencies are listed in the following table. Calculate the average number of no-shows per flight.

Number of No-Shows	Number of Flights
0	37
1	31
2	20
3	16
4	12
5	4

Calculate the mean number of no-shows, weighting each value (zero through five) by the corresponding frequency.

$$\bar{x} = \frac{\sum w_i x_i}{\sum w_i}$$

$$= \frac{(37)(0)+(31)(1)+(20)(2)+(16)(3)+(12)(4)+(4)(5)}{37+31+20+16+12+4}$$

$$= \frac{187}{120}$$

$$= 1.56 \text{ no-shows per flight}$$

Mean of a Grouped Frequency Distribution

Calculating the mean of grouped data

2.41 The table below lists the frequencies of the grouped scores for the 2008 Masters Golf Tournament. Use a weighted average to approximate the mean golf score shot during the tournament.

Final Score	Frequency
280–283	2
284–287	8
288–291	14
292–295	14
296–299	5
300–303	2

Identify the midpoint of each range of scores.

Final Score	Midpoint	Frequency
280–283	281.5	2
284–287	285.5	8
288–291	289.5	14
292–295	293.5	14
296–299	297.5	5
300–303	301.5	2

Calculate the mean of the endpoints:
$$\frac{280+283}{2} = \frac{563}{2} = 281.5$$

Calculate the weighted mean by multiplying the midpoints calculated above by the corresponding frequencies and then averaging those products.

$$\bar{x} = \frac{\sum w_i x_i}{\sum w_i}$$

$$= \frac{(2)(281.5)+(8)(285.5)+(14)(289.5)+(14)(293.5)+(5)(297.5)+(2)(301.5)}{2+8+14+14+5+2}$$

$$= \frac{13,099.5}{45}$$

$$= 291.1$$

The approximate mean score for the 2008 Masters was approximately 291.1.

2.42 The following table divides the employees at a company into categories according to their ages. Approximate the mean employee age.

Age Range	Number of Employees
20–24	8
25–29	37
30–34	25
35–39	48
40–44	27
45–49	10

Identify the midpoint of each age range.

Age Range	Midpoint	Number of Employees
20–24	22	8
25–29	27	37
30–34	32	25
35–39	37	48
40–44	42	27
45–49	47	10

Multiply the midpoints of each range by the corresponding frequencies and then calculate the mean of the products.

$$\bar{x} = \frac{\sum w_i x_i}{\sum w_i}$$

$$= \frac{(8)(22)+(37)(27)+(25)(32)+(48)(37)+(27)(42)+(10)(47)}{8+37+25+48+27+10}$$

$$= \frac{5,355}{155}$$

$$= 34.5 \text{ years}$$

The approximate mean employee age is approximately 34.5 years.

2.43 The NFL has played a 16-game schedule every year from 1978 to 2007, excluding the strike-shortened 1982 season. The table below summarizes the number of games won by the Green Bay Packers during this time period. Approximate the mean number of games the Packers won per year.

Wins per Season	Frequency
3–4	4
5–6	4
7–8	7
9–10	7
11–12	4
13–14	3

Identify the midpoint of each range in the table.

Wins per Season	Midpoint	Frequency
3–4	3.5	4
5–6	5.5	4
7–8	7.5	7
9–10	9.5	7
11–12	11.5	4
13–14	13.5	3

Calculate the weighted mean.

$$\bar{x} = \frac{\sum w_i x_i}{\sum w_i}$$

$$= \frac{(4)(3.5) + (4)(5.5) + (7)(7.5) + (7)(9.5) + (4)(11.5) + (3)(13.5)}{4 + 4 + 7 + 7 + 4 + 3}$$

$$= \frac{241.5}{29}$$

$$= 8.33 \text{ wins per season}$$

2.44 The following table summarizes the grade point averages (GPAs) of graduate students in a statistics class. Approximate the mean GPA of the class.

GPA	Frequency
3.0–under 3.2	5
3.2–under 3.4	9
3.4–under 3.6	6
3.6–under 3.8	16
3.8–4.0	2

Calculate the midpoint of each range of GPAs.

GPA	Midpoint	Frequency
3.0–under 3.2	3.1	5
3.2–under 3.4	3.3	9
3.4–under 3.6	3.5	6
3.6–under 3.8	3.7	16
3.8–4.0	3.9	2

To calculate the mean GPA, apply the weighted mean formula.

$$\bar{x} = \frac{\sum w_i x_i}{\sum w_i}$$

$$= \frac{(5)(3.1)+(9)(3.3)+(6)(3.5)+(16)(3.7)+(2)(3.9)}{5+9+6+16+2}$$

$$= \frac{133.2}{38}$$

$$= 3.51$$

The approximate mean GPA of the graduate students is approximately 3.51.

Chapter 3
CALCULATING DESCRIPTIVE STATISTICS: MEASURES OF VARIATION

Determining the dispersion of the data

Chapter 2 investigated the central tendency, a descriptive statistic used to characterize a collection of data based upon a value (the mean, median, or mode) that was representative of the data set as a whole. In this chapter, you will explore a different category of descriptive statistic, called the *variance*, that describes how closely spaced—or spread out—the data values are.

When you summarize data using a single number, you lose important information about that data. There's no law that says you can only use one kind of descriptive statistic to characterize a data set, and in this chapter you'll learn a few more (including range, standard deviation, and variance).

Range

How wide is your data?

3.1 The following table lists the balance due on a credit card over a five-month period. Calculate the range of the data set.

Credit Card Balance				
$485	$610	$1,075	$737	$519

> The range is always a positive number.

To calculate the range of a data set, subtract its lowest value from its highest.

$$\text{range} = \$1,075 - \$485 = \$590$$

3.2 The following table lists the balance due on a credit card over a five-month period. Calculate the range of the data set.

Credit Card Balance				
$485	$610	$7,075	$737	$519

These data values match the data values in Problem 3.1, with one exception. The third value has been increased from $1,075 (in Problem 3.1) to $7,075.

$$\text{range} = \$7,075 - \$485 = \$6,590$$

> The range can be heavily influenced by one extreme value. In this case, the range skyrocketed from $590 (in Problem 3.1) to $6,590, even though only one data value was changed.

3.3 The following table lists the number of minutes eight randomly selected airline flights were either early (negative values) or late (positive values) arriving at their destinations. Calculate the range of this sample.

Number of Minutes Early or Late							
12	–10	32	–4	0	16	5	18

The highest value in this data set is 32 minutes and the lowest value is –10 minutes.

$$\text{range} = 32 - (-10) = 32 + 10 = 42$$

3.4 The following table lists daily high temperatures (in degrees Celsius) for Pevek, Russia. Determine the range of these temperatures.

Daily High Temperature									
–15	–6	–2	–8	–18	–21	–24	–25	–24	–25

The highest temperature is –2°C and the lowest temperature is –25°C.

$$\text{range} = -2 - (-25) = -2 + 25 = 23°C$$

3.5 The following table lists the total points scored per season by the Chicago Bears and the Dallas Cowboys over a five-year span. Compare the ranges of the data sets to determine which team scored more consistently.

Year	Dallas	Chicago
2007	455	334
2006	425	427
2005	325	260
2004	293	231
2003	289	283

Dallas: range = 455 − 289 = 166 points
Chicago: range = 427 − 231 = 196 points

The Dallas Cowboys scored more consistently season-to-season because their range is smaller than the range for the Chicago Bears.

> A team is consistent when it scores about the same number of points every year. The Bears have a wider range of points, meaning their annual score totals are a bit more unpredictable.

3.6 The following table lists the weight loss (negative values) or weight gain (positive values) in pounds for nine individuals who participated in a weight loss program. Calculate the range of final weight loss results.

Daily High Temperature								
−3	−6	−3	−5	0	−7	4	−1	−4

range = 4 − (−7) = 4 + 7 = 11 pounds

3.7 The following table lists recent golf scores for three friends. Identify the most and least consistent golfers of the group, according to the ranges of their scores.

Golfer	Golf Scores							
Sam	102	98	105	105	100	103	100	99
Debbie	79	85	86	80	96	91	87	
Jeff	86	94	81	90	95	82	88	

> Sam played more often than Debbie and Jeff, but that's okay. You don't need the same number of data points when you're comparing ranges. (And it looks like Sam needs the practice anyway.)

Sam :   range = 105 − 98 = 7
Debbie: range = 96 − 79 = 17
Jeff :   range = 95 − 81 = 14

Sam is the most consistent golfer because he has the smallest range of the three friends. Debbie is the least consistent golfer because her range is the widest.

> Sam's the worst golfer of the three, but that has nothing to do with his consistency.

Interquartile Range
Finding the middle 50 percent of the data

The index method uses the formula $i = \frac{P}{100}(n)$, where P is the percentage and n is the sample size. See Problems 2.12–2.14 for more information.

Note: Problems 3.8–3.12 refer to the data set below, the number of patient visits per week at a chiropractor's office over a ten-week period.

Number of Patients per Week									
75	86	87	90	94	102	105	109	110	120

3.8 Calculate the first quartile of the data using the index method.

The first quartile, known as Q_1, is the twenty-fifth percentile of the data set.

$$i = \frac{P}{100}(n) = \left(\frac{25}{100}\right)(10) = \left(\frac{1}{4}\right)(10) = \frac{10}{4} = \frac{5}{2} = 2.5$$

If you need to review percentiles, flip back to Problems 2.24–2.33.

Because i is not an integer, the next integer greater than i corresponds to the position of the first quartile. Thus, Q_1 is in the third position for this data set: $Q_1 = 87$ patient visits.

Make sure the data is sorted from least to greatest, so that the third highest value is actually in the third position.

Note: Problems 3.8–3.12 refer to the data set in Problem 3.8, the number of patient visits per week at a chiropractor's office over a ten-week period.

3.9 Calculate the second quartile of the data.

The second quartile, known as Q_2, is the median of the data.

$$i = \frac{P}{100}(n) = \left(\frac{50}{100}\right)(10) = \left(\frac{1}{2}\right)(10) = 5$$

The median is the middle of a sorted data set—the fiftieth percentile—so set $P = 50$ in the index point formula.

Because i is an integer, the median is the average of the values in position $i = 5$ and position $i + 1 = 6$.

$$Q_2 = \frac{94 + 102}{2} = \frac{196}{2} = 98 \text{ patient visits}$$

Note: Problems 3.8–3.12 refer to the data set in Problem 3.8, the number of patient visits per week at a chiropractor's office over a ten-week period.

3.10 Calculate the third quartile of the data.

The third quartile, known as Q_3, is the seventy-fifth percentile of the data set.

$$i = \frac{P}{100}(n) = \left(\frac{75}{100}\right)(10) = \left(\frac{3}{4}\right)(10) = \frac{30}{4} = \frac{15}{2} = 7.5$$

You get $i = 7.5$, a decimal value, so round it up to the next integer: $i = 8$.

The third quartile is the data value in the eighth position: $Q_3 = 109$ patient visits.

Note: Problems 3.8–3.12 refer to the data set in Problem 3.8, the number of patient visits per week at a chiropractor's office over a ten-week period.

3.11 Calculate the interquartile range (IQR) of the data.

The interquartile range represents the middle 50 percent of the data, and is equal to the difference between the third and first quartiles. Recall that $Q_1 = 87$ and $Q_3 = 109$ (according to Problems 3.8 and 3.10, respectively).

$$IQR = Q_3 - Q_1 = 109 - 87 = 22 \text{ patient visits}$$

Note: Problems 3.8–3.12 refer to the data set in Problem 3.8, the number of patient visits per week at a chiropractor's office over a ten-week period.

3.12 Calculate the interquartile range using the median method and compare it to the IQR computed in Problem 3.11.

The median method allows you to identify the quartiles of the data given only the median, and it does not require you to use the index point formula $i = \dfrac{P}{100}(n)$. According to Problem 3.9, the median of the data is 98.

To find the first quartile Q_1, list the sorted data less than the median value of 98.

75   86   87   90   94

The first quartile is the median of this data subset. There are an even number of values (5), so the median of the data (and hence the first quartile of the complete data set) is the middle number: $Q_1 = 87$.

Similarly, the third quartile is the middle number when the data values greater than the median are listed.

102   105   109   110   120

Thus, $Q_3 = 109$. The interquartile range is the difference between the third and first quartiles.

$$IQR = Q_3 - Q_1 = 109 - 87 = 22 \text{ patient visits}$$

The index and median methods result in the same IQR value, 22.

> The IQR is still defined the same way in the median method: $IQR = Q_3 - Q_1$. However, you'll calculate the first and third quartiles slightly differently than you did in Problems 3.8 and 3.10.

> For this data set, the index method and the median method give you the same IQR. However, it doesn't always happen that way (as you'll see in Problems 3.13–3.14).

Note: Problems 3.13–3.14 refer to the data set below, the number of pages per book from a random sample of nine paperback novels.

Number of Pages per Paperback Novel								
322	340	351	365	402	460	498	525	567

3.13 Calculate the interquartile range using the index method.

Calculate the index point of Q_1.

$$i = \frac{P}{100}(n) = \left(\frac{25}{100}\right)(9) = \left(\frac{1}{4}\right)(9) = \frac{9}{4} = 2.25$$

> This book sorted the data from least to greatest for you again. How considerate!

You can't have an i that's a decimal, because i refers to a specific position. Make sure you round to the next greatest integer, not the closest integer.

Round i up to the nearest integer: $i = 3$. Hence, $Q_1 = 351$. Now calculate the index point of Q_3.

$$i = \frac{P}{100}(n) = \left(\frac{75}{100}\right)(9) = \left(\frac{3}{4}\right)(9) = \frac{27}{4} = 6.75$$

The third quartile is in the seventh position of the sorted data set: $Q_3 = 498$. Calculate the interquartile range.

$$IQR = Q_3 - Q_1 = 498 - 351 = 147 \text{ pages}$$

According to the index method, the IQR of the data is 147 pages.

Note: Problems 3.13–3.14 refer to the data set in Problem 3.13, the number of pages per book from a random sample of nine paperback novels.

3.14 Calculate the interquartile range using the median method, and compare it to the IQR computed in Problem 3.13.

There is an odd number of data points (9), so the median is the middle value, in the fifth position: $Q_2 = 402$ pages. List the sorted data values that are less than the median.

322   **340**   **351**   365

The first quartile is the median of these four values. Because the number of data points is even, the median is the average of the two middle numbers.

$$Q_1 = \frac{340 + 351}{2} = \frac{691}{2} = 345.5$$

Now list the sorted data values that are greater than the median.

460   **498**   **525**   567

Again, the median is the average of the two middle numbers.

$$Q_3 = \frac{498 + 525}{2} = \frac{1,023}{2} = 511.5$$

Calculate the interquartile range.

$$IQR = Q_3 - Q_1 = 511.5 - 345.5 = 166 \text{ pages}$$

The IQR is 147 pages according to the index method and 166 pages according to the median method. Be aware that different textbooks use different methods to calculate quartiles.

And calculating quartiles differently results in different results for the IQR, which is based on Q_1 and Q_3.

3.15 The following table lists the average monthly crude oil price per barrel for the first seven months of 2007 and 2008. Use the median method of calculating the interquartile range to determine the year in which prices were more consistent.

2007	2008
$46.53	$84.70
$51.36	$86.64
$52.64	$96.87
$55.43	$104.31
$58.08	$117.40
$59.25	$126.16
$65.96	$126.33

Determine the median for 2007. There are seven data points for each year, so the median is the value in the fourth position: $Q_2 = \$55.43$. The first quartile for 2007 is the median of the prices less than the median price of $55.43: $Q_1 = \$51.36$. The third quartile for 2007 is the median of the prices greater than the median price of $55.43: $Q_3 = \$59.25$. Calculate the IQR of the 2007 oil prices.

$$2007 \text{ IQR} = Q_3 - Q_1 = \$59.25 - \$51.36 = \$7.89$$

The median of the 2008 oil prices is $104.31. Identify the first and third quartiles of the 2008 data: $Q_1 = \$86.64$ and $Q_3 = 126.16$. Calculate the corresponding IQR.

$$2008 \text{ IQR} = Q_3 - Q_1 = \$126.16 - \$86.64 = \$39.52$$

The oil prices in 2007 were more consistent than the oil prices in 2008.

> A smaller inter-quartile range means more consistency.

3.16 A cereal producer uses a filling process designed to add 16 ounces of cereal to each box. In order to meet quality control standards, the interquartile range must be less than 0.40 ounces, centered around the target weight of 16.00 ounces.

The following table lists the weights of 24 cereal boxes. Use the index method of calculating the IQR to determine whether this sample meets the quality control standard.

Sorted Weight of Cereal per Box					
15.70	15.70	15.72	15.73	15.75	15.76
15.78	15.84	15.90	15.95	15.98	16.02
16.05	16.06	16.10	16.15	16.15	16.22
16.30	16.32	16.32	16.35	16.36	16.36

Identify the index point of Q_1, the first quartile.

$$i = \frac{P}{100}(n) = \left(\frac{25}{100}\right)(24) = \left(\frac{1}{4}\right)(24) = \frac{24}{4} = 6$$

The first quartile is the average of the values in positions six and seven of the sorted data set.

$$Q_1 = \frac{15.76 + 15.78}{2} = \frac{31.54}{2} = 15.77$$

Identify the index point of Q_3, the third quartile.

$$i = \frac{P}{100}(n) = \left(\frac{75}{100}\right)(24) = \left(\frac{3}{4}\right)(24) = \frac{72}{4} = 18$$

The third quartile is the average of the values in positions 18 and 19 of the sorted data set.

$$Q_3 = \frac{16.22 + 16.30}{2} = \frac{32.52}{2} = 16.26$$

Calculate the interquartile range.

$$IQR = Q_3 - Q_1 = 16.26 - 15.77 = 0.49$$

This sample does not meet the quality standard, because the interquartile range is greater than the standard of 0.40 ounces.

Outliers
Separating the good data from the bad

An outlier is an extremely high or extremely low data value, as compared to the rest of the data. Outliers can make some descriptive statistics (like the mean and range) very misleading.

3.17 The following table lists the number of days that 15 houses in a particular area were on the market waiting to be sold. Use the index method of calculating the IQR to determine whether the data set contains any outliers.

Sorted Days on the Market per House							
9	10	21	36	37	40	46	50
53	59	61	64	75	94	115	

Determine the position of Q_1 in the sorted data set.

$$i = \frac{P}{100}(n) = \left(\frac{25}{100}\right)(15) = \left(\frac{1}{4}\right)(15) = \frac{15}{4} = 3.75$$

Remember: when the index point is not an integer, round up to the next integer.

The first quartile is the fourth data value: $Q_1 = 36$. Determine the position of Q_3 using the index equation.

$$i = \frac{P}{100}(n) = \left(\frac{75}{100}\right)(15) = \left(\frac{3}{4}\right)(15) = \frac{45}{4} = 11.25$$

The third quartile is in position 12: $Q_3 = 64$. Calculate the interquartile range.

$$IQR = Q_3 - Q_1 = 64 - 36 = 28 \text{ days}$$

Calculate the lower limit for outliers using the formula $Q_1 - 1.5(IQR)$.

$$\text{lower limit} = 36 - 1.5(28) = 36 - 42 = -6$$

Zero is the smallest number greater than −6 that makes sense. If a house is not on the market for an entire day, it's technically been for sale for zero days.

A house cannot be on the market for a negative number of days, so the lower limit for outliers is zero days. Calculate the upper limit for outliers using a similar formula: $Q_3 + 1.5(IQR)$.

$$\text{upper limit} = 64 + 1.5(28) = 64 + 42 = 106$$

Any data value less than the lower limit or greater than the upper limit is considered an outlier. There are no data points less than the lower limit. However, the value 115 is greater than the upper limit, and therefore is considered an outlier.

3.18 The following table lists the *Monday Night Football* TV ratings from Nielsen Media Research for 13 games during the 2007 season. Use the median method of calculating the IQR to determine whether the data set contains any outliers.

Sorted *MNF* Nielsen Ratings for 2007 Season						
8.5	8.5	9.0	9.6	9.9	10.8	11.1
11.6	11.8	12.5	13.0	13.1	14.0	

The 13 data values are listed in order, from least to greatest. The median is in position seven: $Q_2 = 11.1$. The first quartile is the median of the values less than 11.1.

$$Q_1 = \frac{9.0 + 9.6}{2} = \frac{18.6}{2} = 9.3$$

The third quartile is the median of the values greater than 11.1.

$$Q_3 = \frac{12.5 + 13.0}{2} = \frac{25.5}{2} = 12.75$$

Calculate the interquartile range.

$$IQR = Q_3 - Q_1 = 12.75 - 9.3 = 3.45$$

Calculate the lower and upper limits for outliers.

$$\text{lower limit} = Q_1 - 1.5(\text{IQR}) = 9.3 - 1.5(3.45) = 4.125$$

$$\text{upper limit} = Q_3 + 1.5(\text{IQR}) = 12.75 + 1.5(3.45) = 17.925$$

There are no outliers in the ratings because none of the ratings are less than 4.125 or greater than 17.925.

3.19 The following table lists the number of minutes 18 randomly selected airline flights were either early (negative values) or late (positive values) arriving at their destinations. Use the index method of calculating the IQR to determine whether the data set contains outliers.

Sorted Number of Minutes Early or Late					
−42	−25	−17	−10	−4	−4
0	6	8	12	17	18
18	20	33	52	61	64

Calculate the index point for the first quartile.

$$i = \frac{P}{100}(n) = \left(\frac{25}{100}\right)(18) = \left(\frac{1}{4}\right)(18) = \frac{18}{4} = \frac{9}{2} = 4.5$$

The first quartile is in the fifth position: $Q_1 = -4$. Calculate the index point for the third quartile.

$$i = \frac{P}{100}(n) = \left(\frac{75}{100}\right)(18) = \left(\frac{3}{4}\right)(18) = \frac{54}{4} = \frac{27}{2} = 13.5$$

The third quartile is in position 14: $Q_3 = 20$. Calculate the interquartile range.

$$\text{IQR} = Q_3 - Q_1 = 20 - (-4) = 20 + 4 = 24 \text{ minutes}$$

Calculate the lower and upper limits for outliers.

$$\text{lower limit} = Q_1 - 1.5(\text{IQR}) = (-4) - 1.5(24) = (-4) - 36 = -40$$

$$\text{upper limit} = Q_3 + 1.5(\text{IQR}) = 20 + 1.5(24) = 20 + 36 = 56$$

You can have multiple outliers, including values at both the beginning and the end of the sorted data.

The value −42 is an outlier because it is less than the lower limit of −40 minutes. Similarly, the values 61 and 64 are outliers because they exceed the upper limit of 56 minutes.

3.20 The following table lists the amount an individual is under (negative values) or over (positive values) budget each month for a year. Use the index method of calculating the IQR to identify outliers, then eliminate them to calculate the mean of the data.

Sorted Over/Under Monthly Budget					
–$825	–$675	–$212	–$136	–$86	$24
$157	$180	$237	$247	$519	$882

Determine the index position of the first quartile.

$$i = \frac{P}{100}(n) = \left(\frac{25}{100}\right)(12) = \left(\frac{1}{4}\right)(12) = \frac{12}{4} = 3$$

The first quartile is the average of the third and fourth values of the sorted data.

$$Q_1 = \frac{(-212)+(-136)}{2} = \frac{-348}{2} = -174$$

Calculate the index point for the third quartile.

$$i = \frac{P}{100}(n) = \left(\frac{75}{100}\right)(12) = \left(\frac{3}{4}\right)(12) = \frac{36}{4} = 9$$

The third quartile is the average of the ninth and tenth values.

$$Q_3 = \frac{237+247}{2} = \frac{484}{2} = 242$$

Calculate the interquartile range.

$$\text{IQR} = Q_3 - Q_1 = 242 - (-174) = 416$$

Calculate the lower and upper limits for outliers.

$$\text{lower limit} = Q_1 - 1.5(\text{IQR}) = -174 - 1.5(416) = -\$798$$
$$\text{upper limit} = Q_3 + 1.5(\text{IQR}) = 242 + 1.5(416) = \$866$$

There are two outliers that might be excluded from the data before the mean is calculated. The value –$825 is less than the lower limit of –$798, and $882 is greater than the upper limit of $866. Calculate the mean of the data excluding the outliers.

The decision to include or exclude outliers is often a judgment call. It's okay to exclude them if they are not representative of the data set. For instance, perhaps an unexpected home repair caused the individual to go $882 over budget one month.

$$\bar{x} = \frac{\sum x}{n}$$

$$= \frac{(-675)+(-212)+(-136)+(-86)+24+157+180+237+247+519}{10}$$

$$= \frac{255}{10}$$

$$= \$25.50$$

The mean (excluding the outliers) is \$25.50 per month over budget.

Visualizing Distributions

Box-and-whisker plots and distribution diagrams

The second quartile is more commonly called the median.

3.21   Describe the five-number summary for a box-and-whisker plot.

The five-number summary for a box-and-whisker plot consists of five data points: the smallest data value; the first, second, and third quartiles; and the largest data value.

3.22   The following table lists the number of children who attend an after-school program over an 11-day period. Construct a box-and-whisker plot for the data using the index method.

Sorted Number of Children per Day										
20	22	29	37	49	56	64	70	70	87	92

The median of the data is the sixth number of the eleven sorted data values: $Q_2 = 56$. Determine the positions of the first and third quartiles.

$$i = \frac{P}{100}(n) = \left(\frac{25}{100}\right)(11) = \left(\frac{1}{4}\right)(11) = \frac{11}{4} = 2.75$$

$$i = \frac{P}{100}(n) = \left(\frac{75}{100}\right)(11) = \left(\frac{3}{4}\right)(11) = \frac{33}{4} = 8.25$$

The first and third quartiles are in positions three and nine of the sorted data, respectively: $Q_1 = 29$ and $Q_3 = 70$. Calculate the interquartile range.

$$\text{IQR} = Q_3 - Q_1 = 70 - 29 = 41$$

Calculate the lower and upper limits for outliers.

$$\text{lower limit} = Q_1 - 1.5(\text{IQR}) = 29 - 1.5(41) = -32.5$$

$$\text{upper limit} = Q_3 + 1.5(\text{IQR}) = 70 + 1.5(41) = 131.5$$

There are no outliers in this data set. Thus, the five-number summary is 20, 29, 56, 70, and 92. The *box* of a box-and-whisker plot is a rectangle bounded on the left by $Q_1 = 29$ and on the right by $Q_3 = 70$. Divide the rectangle at the median $Q_2 = 56$. The *whiskers* of a box-and-whisker plot are horizontal lines that extend from the rectangle to the extreme values 20 and 92, as illustrated below.

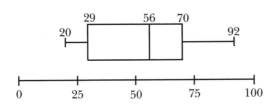

If one (or more) of the numbers at the ends of the data set were big or small enough to classify as outliers, some texrtbooks exclude them as the upper or lower bounds in the five-number summary.

3.23 Describe and sketch a right-skewed distribution, identifying the relative positions of the mean and median.

In a right-skewed distribution, most of the data is concentrated on the left side of the distribution. Therefore, the right tail of the distribution is longer than the left tail and the mean is greater than the median.

Which is the exact opposite of what you'd expect.

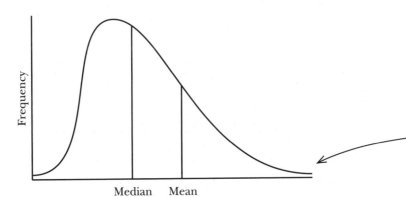

If you were skiing down the long tail of a right-skewed distribution, you would be moving toward the right.

3.24 Describe and sketch a symmetrical bell-shaped distribution, identifying the relative positions of the mean and median.

In a symmetrical bell-shaped distribution, the data values are evenly distributed on both sides of the center. Most of the data values are relatively close to the mean and median, which are approximately equal and near the center of the distribution.

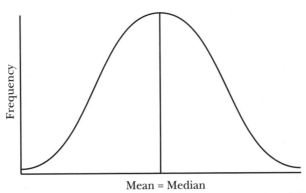

3.25 Describe and sketch a left-skewed distribution, identifying the relative positions of the mean and median.

In a left-skewed distribution, most of the data is concentrated on the right side of the distribution. The left tail of the distribution is longer than the right tail, and the median is greater than the mean.

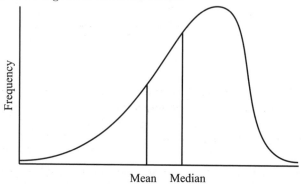

Note: Problems 3.26–3.27 refer to the data set below, the number of days it takes an author to write each chapter of a 15-chapter book.

Sorted Number of Days per Chapter							
9	13	13	13	14	15	15	15
25	25	25	26	36	36	49	

3.26 Construct a box-and-whisker plot for the data using the index method.

The median of the data is the eighth of the fifteen sorted data values: $Q_2 = 15$. Determine the positions of the first and third quartiles.

$$i = \frac{P}{100}(n) = \left(\frac{25}{100}\right)(15) = \left(\frac{1}{4}\right)(15) = \frac{15}{4} = 3.75$$

$$i = \frac{P}{100}(n) = \left(\frac{75}{100}\right)(15) = \left(\frac{3}{4}\right)(15) = \frac{45}{4} = 11.25$$

The first and third quartiles are in positions four and twelve of the data, respectively: $Q_1 = 13$ and $Q_3 = 26$. Calculate the interquartile range.

$$\text{IQR} = Q_3 - Q_1 = 26 - 13 = 13$$

Calculate the lower and upper limits for outliers.

$$\text{lower limit} = Q_1 - 1.5(\text{IQR}) = 13 - 1.5(13) = -6.5$$

$$\text{upper limit} = Q_3 + 1.5(\text{IQR}) = 26 + 1.5(13) = 45.5$$

Note that you should consider zero the lower limit for outliers, because –6.5 is not a valid length of time. The data contains one outlier: 49. Hence, the five-number summary is 9, 13, 15, 26, and 49.

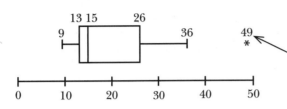

> If your textbook excludes outliers, then the five-number summary would be 9, 13, 15, 26, and 36.

> Use an asterisk to indicate an outlier in a box-and-whisker plot.

Note: Problems 3.26–3.27 refer to the data set in Problem 3.26, the number of days it takes an author to write each chapter of a 15-chapter book.

3.27 Describe the shape of the distribution.

According to Problem 3.26, the median of the data set is $Q_2 = 15$. Calculate the mean of the data, excluding the outlier 49 identified in Problem 3.26. As an aside, note that there may be circumstances in which the outlier is a legitimate data value that should be included in the mean calculation. Whether to include an outlier is a judgment call best made by a person familiar with the circumstances under which the data is analyzed.

$$\bar{x} = \frac{\sum x}{n}$$
$$= \frac{9+13+13+13+14+15+15+15+25+25+25+26+36+36}{14}$$
$$= \frac{280}{14}$$
$$= 20$$

> See Problem 3.23.

Because the mean (20) is greater than the median (15), the distribution is most likely right-skewed.

Note: Problems 3.28–3.29 refer to the following data set, the distances a car drives (in miles) on a full tank of gas after 15 fill-ups at a gas station.

Sorted Distance per Tank of Gas							
215	229	236	239	240	244	247	255
262	264	271	279	280	282	285	

3.28 Construct a box-and-whisker plot for the data, using a five-number summary.

The median of the data is the number in the eighth position: $Q_2 = 255$. Calculate the index points for the first and third quartiles.

$$i = \frac{P}{100}(n) = \left(\frac{25}{100}\right)(15) = \left(\frac{1}{4}\right)(15) = \frac{15}{4} = 3.75$$

$$i = \frac{P}{100}(n) = \left(\frac{75}{100}\right)(15) = \left(\frac{3}{4}\right)(15) = \frac{45}{4} = 11.25$$

The first and third quartiles are in positions four and twelve, respectively: $Q_1 = 239$ and $Q_3 = 279$.

The five-number summary is 215, 239, 255, 279, and 285.

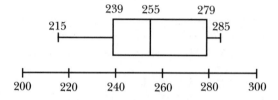

Note: Problems 3.28–3.29 refer to the data set in Problem 3.28, the distances a car drives (in miles) on a full tank of gas after 15 fill-ups at a gas station.

3.29 Describe the shape of the distribution.

According to Problem 3.28, the median of the data is $Q_2 = 255$. Calculate the mean.

$$\bar{x} = \frac{\sum x}{n}$$

$$= \frac{215 + 229 + 236 + 239 + 240 + 244 + 247 + 255 + 262 + 264 + 271 + 279 + 280 + 282 + 285}{15}$$

$$= \frac{3,828}{15}$$

$$= 255.2$$

Because the median (255) is approximately equal to the mean (255.2), the distribution is most likely symmetrical.

Stem-and-Leaf Plot

The flower power of data

3.30 Describe the structure of a stem-and-leaf diagram.

A stem-and-leaf diagram displays the distribution of a data set by separating each value into a stem and a leaf. The stem is the first digit (or digits) of the number and the leaf is the last digit. Leaves with a common stem are grouped together in ascending order. For example, the numbers 28, 34, 42, 47, and 49 would be displayed in the following manner.

```
2 | 8
3 | 4
4 | 2 7 9
```

The left column is the stem and the right column contains the leaves. The last row has three leaves, so the data has three values with a stem of 4—three values in the forties.

Some textbooks require you to include a key for stem-and-leaf plots so that you can correctly interpret them. The key for this plot would be "2|8 = 28."

3.31 The following table lists the total annual snowfall (in inches) for 30 cities. Construct a stem-and-leaf plot for the data.

Sorted Inches of Snowfall									
11	12	14	17	20	20	22	25	25	26
26	28	30	32	32	34	35	35	38	39
39	41	41	43	45	46	48	49	50	56

The stem of each data value is its tens digit, and the leaf is the ones digit.

For example, the tens digit of 25 is 2 and the ones digit is 5.

```
1 | 1 2 4 7
2 | 0 0 2 5 5 6 6 8
3 | 0 2 2 4 5 5 8 9 9
4 | 1 1 3 5 6 8 9
5 | 0 6
```

3.32 Identify the data values represented by the following stem-and-leaf plot, which lists the number of new cars sold at a local dealership over the last 20 months. Assume that 2|9 = 29.

```
0 | 8 9 9
1 |
2 | 0 1 3 3 4 4 5 7 8 9
3 | 0 2 3 3 5
4 | 1 3
```

Even though there aren't any data points with a stem of one, it should still be included as a stem in the stem-and-leaf plot.

The stem of each data value represents its tens digit, and the leaf of each data value represents its ones digit.

Sorted Number of Cars Sold per Month									
8	9	9	20	21	23	23	24	24	25
27	28	29	30	32	33	33	35	41	43

The data is sorted from highest to lowest (because more strikeouts is better than fewer when you're a pitcher), but the stem-and-leaf diagram is still organized from lowest to highest number of strikeouts.

3.33 The following table lists the number of strikeouts posted by the top 30 strikeout pitchers in the 2008 Major League Baseball season. Construct a stem-and-leaf plot for the data and identify the mode.

Sorted Number of Strikeouts

265	251	231	214	206	206	206	206	201	200
196	187	186	186	184	183	183	183	181	180
175	173	172	172	170	166	166	165	163	163

A leaf consists of a single digit, so the stems for these data values consist of the hundreds *and* tens digits. Thus, 16|3 = 163.

```
16 | 3 3 5 6 6
17 | 0 2 2 3 5
18 | 0 1 3 3 3 4 6 6 7
19 | 6
20 | 0 1 6 6 6 6
21 | 4
22 |
23 | 1
24 |
25 | 1
26 | 5
```

To find the mode, look for the most equal digits in a row. There are three 3's in the third row, but there are more 6's in the fifth row—four of them to be exact. Therefore, the data set contains four 206's.

Notice that stem 20 contains four equal leaves. Thus, the mode of the data is 206.

Note: Problems 3.34–3.35 refer to the following data set, the top 30 NFL quarterback ratings during the 2007 season.

Sorted Quarterback Ratings

117.2	104.1	102.2	98.0	97.4	95.7
94.6	91.4	89.9	89.8	89.4	88.1
87.2	86.7	86.1	84.8	82.5	82.4
80.9	77.6	77.2	76.8	75.6	75.2
73.9	71.1	71.0	70.8	70.4	70.3

3.34 Construct a stem-and-leaf plot for the data, such that each stem represents 10 possible unique data values.

36 is the biggest data value that's not an outlier.

The leaf is usually the rightmost digit in the data value, which in this case would be the number in the tenths place. However, this results in stems ranging from 70 to 117, including many stems that would not correspond to leaves. Thus it is better to round the data to the nearest integer values.

Sorted Quarterback Ratings Rounded					
117	104	102	98	97	96
95	91	90	90	89	88
87	87	86	85	83	82
81	78	77	77	76	75
74	71	71	71	70	70

The hundreds and the tens values of each rounded data point make up the stem; the ones value is the leaf. Note that each stem represents 10 possible unique data values. For instance, the stem 7 could contain data values 70, 71, 72, 73, 74, 75, 76, 77, 78, and 79.

```
 7  | 0 0 1 1 1 4 5 6 7 7 8
 8  | 1 2 3 5 6 7 7 8 9
 9  | 0 0 1 5 6 7 8
10  | 2 4
11  | 7
```

Note: Problems 3.34–3.35 refer to the data set in Problem 3.34, the top 30 NFL quarterback ratings during the 2007 season.

3.35 Construct a stem-and-leaf plot for the data such that each stem represents five possible unique data values. ←

Each of the following stems contains a parenthetical number, either (0) or (5). This number represents the smallest possible leaf value for that stem. For instance, stem 7(0) could contain data values 70, 71, 72, 73, and 74; stem 7(5) could contain data values 75, 76, 77, 78, and 79.

```
 7(0) | 0 0 1 1 1 4
 7(5) | 5 6 7 7 8
 8(0) | 1 2 3
 8(5) | 5 6 7 7 8 9
 9(0) | 0 0 1
 9(5) | 5 6 7 8
10(0) | 2 4
10(5) |
11(0) |
11(5) | 7
```

Splitting the stems in half is a good idea when some stems have a lot of leaves and others don't have as many. You're trying to see how spread out the data is, and sometimes you need to spread out the stems to do that.

A back-to-back stem-and-leaf diagram is handy when you're comparing two similar data sets. Place a common stem down the center. One distribution has leaves on the left, and the other has leaves on the right.

3.36 The following two tables list the numbers of home runs hit by the leaders in this category in the National League and the American League for the 2008 Major League Baseball season. Construct a back-to-back stem-and-leaf diagram comparing the two leagues. What conclusions can you draw based on this diagram?

Sorted National League Home Run Leaders

48	40	38	37	37	37	36	34	33	33
33	33	32	32	29	29	29	28	28	27
27	27	26	26	25	25	25	25	25	25

Sorted American League Home Run Leaders

37	36	35	34	34	33	33	32	32	32
31	29	27	27	25	25	24	23	23	23
23	23	23	22	22	22	21	21	21	21

More than half of the home run leaders in each league hit between 21 and 29 home runs. Thus, a stem of 2 would have a disproportionately large number of leaves. Split the stems as instructed in Problem 3.35, so that each represents five possible values.

The numbers on this side are written in reverse order, so that the leaves grow "outward" from the stem on both sides.

National League		American League
	2(0)	1 1 1 1 2 2 2 3 3 3 3 3 3 4
9 9 9 8 8 7 7 7 6 6 5 5 5 5 5 5	2(5)	5 5 7 7 9
4 3 3 3 3 2 2	3(0)	1 2 2 2 3 3 4 4
8 7 7 7 6	3(5)	5 6 7
0	4(0)	
8	4(5)	

The majority of the National League's batters hit between 25 and 29 home runs. Most of the American League leaders hit between 20 and 24 home runs, and none of them hit 40 or more.

Variance and Standard Deviation of a Population

The most common ways to measure dispersion

> Note: Problems 3.37–3.40 refer to the data set below, the number of hurricanes that struck the continental United States each decade during the twentieth century.

Decade	Number of Hurricanes
1901–1910	18
1911–1920	21
1921–1930	13
1931–1940	19
1941–1950	24
1951–1960	17
1961–1970	14
1971–1980	12
1981–1990	15
1991–2000	14

3.37 Calculate the variance of the data using the standard method.

This data set is considered a population because all of the hurricanes for each decade of the twentieth century are included—not just a sample. The standard method of computing population variance σ^2 is the following equation, in which x represents each data value, μ represents the population mean, and N represents the number of data values.

$$\sigma^2 = \frac{\sum(x-\mu)^2}{N}$$

> Subtract the mean from every single data value, and then square each difference. Add up those squares and divide the sum by N.

Calculate the mean.

$$\mu = \frac{\sum x}{N}$$
$$= \frac{18+21+13+19+24+17+14+12+15+14}{10}$$
$$= \frac{167}{10}$$
$$= 16.7$$

Subtract the mean from each data value and calculate the squares of the differences $(x-\mu)^2$, as demonstrated in the following table. Then, calculate the sum of squares.

x	$x - \mu$	$(x - \mu)^2$
18	$18 - 16.7 = 1.3$	1.69
21	$21 - 16.7 = 4.3$	18.49
13	$13 - 16.7 = -3.7$	13.69
19	$19 - 16.7 = 2.3$	5.29
24	$24 - 16.7 = 7.3$	53.29
17	$17 - 16.7 = 0.3$	0.09
14	$14 - 16.7 = -2.7$	7.29
12	$12 - 16.7 = -4.7$	22.09
15	$15 - 16.7 = -1.7$	2.89
14	$14 - 16.7 = -2.7$	7.29
Total		**132.1**

All the numbers in this column are positive, because they've all been squared.

The variance σ^2 is the sum of the squares calculated above (132.1) divided by the population size.

$$\sigma^2 = \frac{\sum (x - \mu)^2}{N} = \frac{132.1}{10} = 13.21$$

The general rule for rounding is to use one more decimal place in your calculations than the raw data contained. However, judgment also comes into play when making these decisions. When calculations are performed manually, rounding needs to occur and the final result may be affected by the number of decimal places used.

And the more data points you have, the less work the shortcut method is.

Note: Problems 3.37–3.40 refer to the data set in Problem 3.37, the number of hurricanes that struck the continental United States each decade during the twentieth century.

3.38 Calculate the variance of the data using the shortcut method.

The shortcut version of the variance formula provides the same result as the standard method but requires fewer computations.

$$\sigma^2 = \frac{\sum x^2 - \frac{\left(\sum x\right)^2}{N}}{N}$$

To calculate $\sum x^2$ (the sum of the squares), square each data value and then add up the squares. To calculate $\left(\sum x\right)^2$ (the square of the sum), add up all the data values first and then square that sum.

The following table contains the square of each data value.

x	x^2
18	324
21	441
13	169
19	361
24	576
17	289
14	196
12	144
15	225
14	196
$\sum x = 167$	$\sum x^2 = 2{,}921$

Substitute $\sum x = 167$ and $\sum x^2 = 2{,}921$ into the variance shortcut formula.

$$\sigma^2 = \frac{\sum x^2 - \dfrac{\left(\sum x\right)^2}{N}}{N}$$

$$= \frac{2{,}921 - \dfrac{(167)^2}{10}}{10}$$

$$= \frac{2{,}921 - \dfrac{(27{,}889)}{10}}{10}$$

$$= \frac{2{,}921 - 2{,}788.9}{10}$$

$$= \frac{132.1}{10}$$

$$= 13.21$$

The standard method and the shortcut method for calculating the population variance provide the same result for this data set: $\sigma^2 = 13.21$ hurricanes.

Note: Problems 3.37–3.40 refer to the data set in Problem 3.37, the number of hurricanes that struck the continental United States each decade during the twentieth century.

3.39 Calculate the standard deviation of the data.

The standard deviation σ is the square root of the population variance: $\sigma = \sqrt{\sigma^2}$. According to Problem 3.37, that variance was $\sigma^2 = 13.21$.

$$\sigma = \sqrt{\sigma^2} = \sqrt{13.21} = 3.63 \text{ hurricanes}$$

> *The CV is useful when you're comparing two data sets that aren't exactly alike, especially if the different data sets aren't measured using the same units.*

> *The coefficient of variation equals the standard deviation divided by the mean times 100.*

> *You have to use a different formula to calculate the variance of a sample. See Problem 3.45.*

Note: Problems 3.37–3.40 refer to the data set in Problem 3.37, the number of hurricanes that struck the continental United States each decade during the twentieth century.

3.40 Calculate the coefficient of variation (CV) for the data.

The coefficient of variation measures the percentage of variation in the data relative to the mean of the data. Use the formula below to calculate the CV for a population.

$$CV = \frac{\sigma}{\mu}(100\%)$$

Recall that $\sigma = 3.63$ (according to Problem 3.39) and $\mu = 16.7$ (according to Problem 3.37).

$$CV = \frac{3.63}{16.7}(100\%) = 21.7\%$$

Note: Problems 3.41–3.44 refer to the data set below, the number of students enrolled in all five of a college's statistics classes.

Number of Students				
18	22	25	26	15

3.41 Calculate the variance of the data using the standard method.

This data set is considered a population because it represents all of the statistics classes at the college.

$$\sigma^2 = \frac{\sum(x-\mu)^2}{N}$$

Calculate the mean.

$$\mu = \frac{\sum x}{N} = \frac{18+22+25+26+15}{5} = \frac{106}{5} = 21.2$$

Subtract the mean from each data value, square the difference, and calculate the sum of the squares.

x	$x - \mu$	$(x-\mu)^2$
18	$18 - 21.2 = -3.2$	10.24
22	$22 - 21.2 = 0.8$	0.64
25	$25 - 21.2 = 3.8$	14.44
26	$26 - 21.2 = 4.8$	23.04
15	$15 - 21.2 = -6.2$	38.44
Total		**86.8**

Calculate the variance.

$$\sigma^2 = \frac{\sum(x - \mu)^2}{N} = \frac{86.8}{5} = 17.36$$

Note: Problems 3.41–3.44 refer to the data set in Problem 3.41, the number of students enrolled in all five of a college's statistics classes.

3.42 Calculate the variance of the data using the shortcut method.

To apply the shortcut method, you must first compute the sum of the squares of the data values and the square of the sum of the data values.

x	x^2
18	324
22	484
25	625
26	676
15	225
$\sum x = 106$	$\sum x^2 = 2{,}334$

$$\sigma^2 = \frac{\sum x^2 - \frac{\left(\sum x\right)^2}{N}}{N}$$

$$= \frac{2{,}334 - \frac{(106)^2}{5}}{5}$$

$$= \frac{2{,}334 - \frac{(11{,}236)}{5}}{5}$$

$$= \frac{2{,}334 - 2{,}247.2}{5}$$

$$= \frac{86.8}{5}$$

$$= 17.36$$

Both the standard and shortcut methods produce the same value for the variance: $\sigma^2 = 17.36$.

Note: Problems 3.41–3.44 refer to the data set in Problem 3.41, the number of students enrolled in all five of a college's statistics classes.

3.43 Calculate the standard deviation of the data.

The population standard deviation σ is the square root of the population variance: $\sigma = \sqrt{\sigma^2}$. According to Problem 3.42, the variance of the data is 17.36.

$$\sigma = \sqrt{\sigma^2} = \sqrt{17.36} = 4.17 \text{ students}$$

Note: Problems 3.41–3.44 refer to the data set in Problem 3.41, the number of students enrolled in all five of a college's statistics classes.

3.44 Calculate the coefficient of variation (CV) for the data.

The coefficient of variation is equal to 100 times the quotient of the standard deviation and the mean. Recall that $\sigma = 4.17$ (according to Problem 3.43) and $\mu = 21.2$ (according to Problem 3.41).

$$CV = \frac{\sigma}{\mu}(100\%) = \frac{4.17}{21.2}(100\%) = 19.7\%$$

Note: Problems 3.45–3.48 refer to the following data set, the number of students absent from a school each day last week.

15	8	6	22	5

3.45 Calculate the variance of the sample using the standard method.

Be careful! The variance formula for a sample is slightly different from the variance formula for a population. Double-check your denominators.

This data set is considered a sample because only five school days from the entire school year are included. Calculate the mean of the sample.

$$\bar{x} = \frac{\sum x}{n}$$
$$= \frac{15 + 8 + 6 + 22 + 5}{5}$$
$$= \frac{56}{5}$$
$$= 11.2$$

Subtract the mean from each value and square each difference.

x	$x - \bar{x}$	$\left(x - \bar{x}\right)^2$
15	$15 - 11.2 = 3.8$	14.44
8	$8 - 11.2 = -3.2$	10.24

x	$x - \bar{x}$	$\left(x - \bar{x}\right)^2$
6	$6 - 11.2 = -5.2$	27.04
22	$22 - 11.2 = 10.8$	116.64
5	$5 - 11.2 = -6.2$	38.44
Total		**206.8**

Apply the sample variance formula.

$$s^2 = \frac{\sum \left(x - \bar{x}\right)^2}{n-1} = \frac{206.8}{5-1} = \frac{206.8}{4} = 51.7$$

> When calculating the variance of a sample, you divide by $n - 1$, which is one less than the number of data points. That's different than the variance of a population, where you divide by N, the actual number of data points.

Note: Problems 3.45–3.48 refer to the data set in Problem 3.45, the number of students who were absent from school each day last week.

3.46 Use the shortcut method to verify the variance calculated in Problem 3.45.

Calculate the sum of the data values and the sum of the squares of the data values.

x	x^2
15	225
8	64
6	36
22	484
5	25
$\sum x = 56$	$\sum x^2 = 834$

Apply the shortcut formula to calculate the variance of the sample.

$$s^2 = \frac{\sum x^2 - \dfrac{\left(\sum x\right)^2}{n}}{n-1}$$

$$= \frac{834 - \dfrac{(56)^2}{5}}{5-1}$$

$$= \frac{834 - \dfrac{(3,136)}{5}}{4}$$

$$= \frac{834 - 627.2}{4}$$

$$= \frac{206.8}{4}$$

$$= 51.7$$

> Square $\sum x$ first, then divide by n, then subtract what you get from $\sum x^2$.

Note: Problems 3.45–3.48 refer to the data set in Problem 3.45, the number of students who were absent from school each day last week.

3.47 Calculate the standard deviation of the sample.

The standard deviation s is the square root of the variance s^2. According to Problems 3.45 and 3.46, $s^2 = 51.7$.

$$s = \sqrt{s^2} = \sqrt{51.7} = 7.19 \text{ students}$$

The variance and standard deviation are always positive.

Note: Problems 3.45–3.48 refer to the data set in Problem 3.45, the number of students who were absent from school each day last week.

3.48 Calculate the coefficient of variation (CV) for the number of absent students.

Divide the standard deviation of the sample ($s = 7.19$, according to Problem 3.47) by the sample mean ($\bar{x} = 11.2$, according to Problem 3.45).

$$\text{CV} = \frac{s}{(\bar{x})}(100\%) = \frac{7.19}{11.2}(100\%) = 64.2\%$$

3.49 A certain cell phone plan includes a fixed number of calling minutes per customer. The following table lists the number of minutes a particular customer was over (positive values) or under (negative values) that quota during the first seven months of his cell phone agreement. Calculate the sample standard deviation of the data.

−15.5	25	0	−10.5	−17.5	10	−23

Though both the standard and shortcut methods will produce the same variance, the shortcut method requires fewer computations. Calculate the sum of the data values, as well as the sum of the squares of the data values.

	x	x^2
	−15.5	240.25
	25	625
	0	0
	−10.5	110.25
	−17.5	306.25
	10	100
	−23	529
Total	−31.5	1,910.75

Apply the shortcut formula for the variance of a sample.

$$s = \sqrt{\frac{\sum x^2 - \frac{\left(\sum x\right)^2}{n}}{n-1}}$$

$$= \sqrt{\frac{1{,}910.75 - \frac{(-31.5)^2}{7}}{7-1}}$$

$$= \sqrt{\frac{1{,}910.75 - \frac{(992.25)}{7}}{6}}$$

$$= \sqrt{\frac{1{,}910.75 - 141.75}{6}}$$

$$= \sqrt{\frac{1{,}769}{6}}$$

$$= \sqrt{294.83}$$

$$= 17.17$$

> You're asked to find the standard deviation (s), which is the square root of the variance (s^2). That's why the variance formula has a square root symbol over it.

Note: Problems 3.50–3.51 refer to the data set below, the number of home runs hit by New York Yankees players Derek Jeter and Alex Rodriguez for eight consecutive Major League Baseball seasons.

Year	Jeter	Rodriguez
2001	21	52
2002	18	57
2003	10	47
2004	23	36
2005	19	48
2006	14	35
2007	12	54
2008	11	35

3.50 Calculate the standard deviations for home runs hit by each player.

Calculate the sums and the sums of the squares of the data values for each player independently.

	Jeter		Rodriguez	
	x	x^2	x	x^2
	21	441	52	2,704
	18	324	57	3,249
	10	100	47	2,209
	23	529	36	1,296
	19	361	48	2,304
	14	196	35	1,225
	12	144	54	2,916
	11	121	35	1,225
Total	**128**	**2,216**	**364**	**17,128**

Apply the shortcut method to calculate the standard deviations of Jeter's and Rodriguez's annual home run totals: s_J and s_R, respectively.

$$s_J = \sqrt{\frac{2{,}216 - \frac{(128)^2}{8}}{8-1}} \qquad s_R = \sqrt{\frac{17{,}128 - \frac{(364)^2}{8}}{8-1}}$$

$$= \sqrt{\frac{2{,}216 - \frac{(16{,}384)}{8}}{7}} \qquad = \sqrt{\frac{17{,}128 - \frac{(132{,}496)}{8}}{7}}$$

$$= \sqrt{\frac{2{,}216 - 2{,}048}{7}} \qquad = \sqrt{\frac{17{,}128 - 16{,}562}{7}}$$

$$= \sqrt{\frac{168}{7}} \qquad = \sqrt{\frac{566}{7}}$$

$$= \sqrt{24} \qquad = \sqrt{80.86}$$

$$= 4.90 \qquad = 8.99$$

> The more consistent a player, the less his home run number varies year to year. The lower the CV, the lower the variance, and thus the greater the consistency.

> **Note: Problems 3.50-3.51 refer to the data set in Problem 3.50, the number of home runs hit each season by Derek Jeter and Alex Rodriguez between 2001–2008.**

3.51 Use the standard deviations calculated in Problem 3.50 to determine which player was a more consistent home run hitter. Justify your answer.

Calculate the average number of home runs hit by Jeter $\left(\overline{x}_J\right)$ and Rodriguez $\left(\overline{x}_R\right)$.

$$\overline{x}_J = \frac{\sum x}{n} \qquad \overline{x}_R = \frac{\sum x}{n}$$

$$= \frac{21+18+10+23+19+14+12+11}{8} \qquad = \frac{52+57+47+36+48+35+54+35}{8}$$

$$= \frac{128}{8} \qquad = \frac{364}{8}$$

$$= 16 \qquad = 45.5$$

Calculate the coefficients of variation for the home run totals of Jeter (CV$_J$) and Rodriguez (CV$_R$).

$$CV_J = \frac{s}{\left(\overline{x}\right)}(100\%) \qquad CV_R = \frac{s}{\left(\overline{x}\right)}(100\%)$$

$$= \frac{4.90}{16}(100\%) \qquad = \frac{8.99}{45.5}(100\%)$$

$$= 30.6\% \qquad = 19.8\%$$

Alex Rodriguez's home run record has a higher standard deviation but is more consistent, because it has a lower coefficient of variation (19.8%) than Derek Jeter's record (30.6%) over this time period. Standard deviation is affected by the relative size of the numbers. Because Rodriguez averages nearly three times as many home runs as Jeter, the resulting larger standard deviation is unsurprising. ←

Use the coefficient of variation to compare standard deviations, because it takes differences in the means into account.

Variance and Standard Deviation for Grouped Data

Calculating dispersions for frequency distributions

Note: Problems 3.52–3.53 refer to the data set below, the frequencies of the grouped scores for the 2008 Masters Golf Tournament.

Final Score	Frequency
280–283	2
284–287	8
288–291	14
292–295	14
296–299	5
300–303	2

3.52 Calculate the sample variance for the golf scores shot during the tournament, using the shortcut method.

The given grouped data table does not provide the actual scores shot by the golfers at the Masters—only ranges of scores and the number of scores in each range. It is possible to calculate the variance of the data, though more accuracy would be guaranteed if the actual scores (rather than the ranges) were provided.

Calculate the midpoint x_m of each range.

Here's the formula for the variance of a grouped data sample:

$$s^2 = \frac{\sum(f \cdot x_m^2) - \frac{\left[\sum(f \cdot x_m)\right]^2}{n}}{n-1}$$

The midpoints are x_m and f stands for frequency—the number of golfers that belong to each range.

Final Score	Midpoint x_m
280–283	281.5
284–287	285.5
288–291	289.5
292–295	293.5
296–299	297.5
300–303	301.5

To calculate the midpoint of the range 280-283, add the endpoints and divide by 2:

$$(280 + 283) \div 2$$
$$= 563 \div 2$$
$$= 281.5$$

In the table below, column A lists the midpoints calculated above and column B lists the frequencies f of the ranges with the corresponding midpoints. Column C contains the products of columns A and B. Column D contains the squares of the values in column A, and column E is the product of column D and the frequency f from column B. The sums of columns B, C, and E appear at the bottoms of the columns.

	A	B	C	D	E
	x_m	f	$f \cdot x_m$	$x_m^{\,2}$	$f \cdot x_m^{\,2}$
	281.5	2	563	79,242.25	158,484.5
	285.5	8	2,284	81,510.25	652,082
	289.5	14	4,053	83,810.25	1,173,343.5
	293.5	14	4,109	86,142.25	1,205,991.5
	297.5	5	1,487.5	88,506.25	442,531.25
	301.5	2	603	90,902.25	181,804.5
Total		**45**	**13,099.5**		**3,814,237.25**

Substitute $\sum (f \cdot x_m) = 13,099.5$, $\sum (f \cdot x_m^{\,2}) = 3,814,237.25$, and $n = 45$ into the variance formula for grouped data.

$$s^2 = \frac{\sum (f \cdot x_m^{\,2}) - \dfrac{\left[\sum (f \cdot x_m) \right]^2}{n}}{n-1}$$

$$s^2 = \frac{3,814,237.25 - \dfrac{(13,099.5)^2}{45}}{45-1}$$

$$= \frac{3,814,237.25 - \dfrac{171,596,900.3}{45}}{45-1}$$

$$= \frac{3,814,237.25 - 3,813,264.45}{44}$$

$$= \frac{972.8}{44}$$

$$= 22.11$$

The variance of the 2008 Masters scores is $s^2 = 22.11$.

Note: Problems 3.52–3.53 refer to the data set in Problem 3.52, the frequencies of the grouped scores for the 2008 Masters Golf Tournament.

3.53 Calculate the sample standard deviation for the golf scores shot during the tournament.

The sample standard deviation s for grouped data is the square root of the variance s^2. According to Problem 3.52, $s^2 = 22.11$.

$$s = \sqrt{s^2} = \sqrt{22.11} = 4.70$$

Note: Problems 3.54–3.55 refer to the data set below, the number of employees of a particular organization in different age ranges.

Age Range	Number of Employees
20–24	8
25–29	37
30–34	25
35–39	48
40–44	27
45–49	10

3.54 Calculate the sample standard deviation for this grouped data, using the shortcut method.

Identify the midpoint x_m of each age range.

Age Range	Midpoint x_m
20–24	22
25–29	27
30–34	32
35–39	37
40–44	42
45–49	47

In order to apply the shortcut method for the variance of a sample, you need to calculate the products of the midpoints and their corresponding frequencies (column C in the table that follows), the products of the squares of the midpoints and their corresponding frequencies (column E), and the sums of each.

If you have the option to use either the standard or the shortcut method to calculate a standard deviation, use the shortcut method.

	A	B	C	D	E
	x_m	f	$f \cdot x_m$	x_m^2	$f \cdot x_m^2$
	22	8	176	484	3,872
	27	37	999	729	26,973
	32	25	800	1,024	25,600
	37	48	1,776	1,369	65,712
	42	27	1,134	1,764	47,628
	47	10	470	2,209	22,090
Total		**155**	**5,355**		**191,875**

Substitute the sums at the bottoms of columns B, C, and E for n, $\sum \left(f \cdot x_m \right)$, and $\sum \left(f \cdot x_m^2 \right)$, respectively, into the formula for the standard deviation of a grouped sample.

$$s = \sqrt{\dfrac{\sum \left(f \cdot x_m^2 \right) - \dfrac{\left[\sum \left(f \cdot x_m \right) \right]^2}{n}}{n-1}}$$

$$= \sqrt{\dfrac{191,875 - \dfrac{\left(5,355 \right)^2}{155}}{155-1}}$$

$$= \sqrt{\dfrac{191,875 - \dfrac{28,676,025}{155}}{155-1}}$$

$$= \sqrt{\dfrac{191,875 - 185,006.613}{154}}$$

$$= \sqrt{\dfrac{6868.387}{154}}$$

$$= \sqrt{44.600}$$

$$= 6.68$$

The units for variance are the square of the units of the data. In this case, the units would be years^2, which doesn't make a lot of sense. That's why units are usually omitted when you're dealing with variance.

Note: Problems 3.54–3.55 refer to the data set in Problem 3.54, the number of employees in each age group in a particular organization.

3.55 Use the standard deviation calculated in Problem 3.54 to calculate the sample variance for employee age in the organization.

The sample variance is the square of the sample standard deviation. According to Problem 3.54, $s = 6.68$.

$$s^2 = \left(6.68 \right)^2 = 44.62$$

Chebyshev's Theorem

Putting the standard deviation to work

3.56 Define Chebyshev's Theorem.

Regardless of how the data are distributed, at least $\left(1-\dfrac{1}{k^2}\right)(100\%)$ of the values will fall within k standard deviations of the mean, where k is a number greater than one. To illustrate the theorem, substitute $k=2$ into the expression and simplify.

$$\left(1-\frac{1}{2^2}\right)(100\%)=\left(1-\frac{1}{4}\right)(100\%)=(0.75)(100\%)=75\%$$

Therefore, 75% of the values lie within $k=2$ standard deviations of the mean.

Chebyshev (also known as Tchebysheff) was a Russian mathematician who lived from 1821 to 1894

Chebyshev's Theorem applies to all distributions, whether they are symmetrical, left-skewed, or right-skewed.

3.57 Using Chebyshev's Theorem, determine the minimum percentage of observations from a distribution that would be expected to fall within 3, 3.5, and 4 standard deviations of the mean.

k does not have to be an integer

Substitute $k=3$ into Chebyshev's Theorem.

$$\left(1-\frac{1}{k^2}\right)(100\%)=\left(1-\frac{1}{3^2}\right)(100\%)=\left(1-\frac{1}{9}\right)(100\%)=88.9\%$$

At least 88.9% of the observations from a distribution will lie within 3 standard deviations of the mean. Repeat the process, substituting $k=3.5$ and $k=4$ into Chebyshev's Theorem.

$$\left(1-\frac{1}{k^2}\right)(100\%)=\left(1-\frac{1}{3.5^2}\right)(100\%)=\left(1-\frac{1}{12.25}\right)(100\%)=91.8\%$$

$$\left(1-\frac{1}{k^2}\right)(100\%)=\left(1-\frac{1}{4^2}\right)(100\%)=\left(1-\frac{1}{16}\right)(100\%)=93.8\%$$

At least 91.8% of the observations will lie within 3.5 standard deviations of the mean, and at least 93.8% of the observations will lie within 4 standard deviations of the mean.

Note: Problems 3.58–3.60 refer to a distribution of home sales prices with a mean of $300,000 and a standard deviation of $50,000.

3.58 Determine the price range in which at least 75% of the houses sold.

As demonstrated in Problem 3.56, at least 75% of the observations for a distribution will fall within $k = 2$ standard deviations of the mean. Add two standard deviations to the mean to identify the upper bound of the price range $(\mu + k\sigma)$ and subtract two standard deviations from the mean to identify the lower bound of the price range $(\mu - k\sigma)$.

$$
\begin{aligned}
\mu + k\sigma &= \$300,000 + (2)(\$50,000) & \mu - k\sigma &= \$300,000 - (2)(\$50,000) \\
&= \$300,000 + \$100,000 & &= \$300,000 - \$100,000 \\
&= \$400,000 & &= \$200,000
\end{aligned}
$$

The prices of at least 75% of the houses are between $200,000 and $400,000.

Note: Problems 3.58–3.60 refer to a distribution of home sales prices with a mean of $300,000 and a standard deviation of $50,000.

3.59 Determine the minimum percentage of the houses that should sell for prices between $150,000 and $450,000.

The interval in question must be symmetrical around the mean—the mean needs to be in the middle of the range.

According to Problem 3.58, the upper boundary of Chebyshev's Theorem is equal to $\mu + k\sigma$. Set this expression equal to upper boundary given by the problem, substitute the mean and standard deviation of the home sales prices into the equation, and solve for k.

$$
\begin{aligned}
\mu + k\sigma &= 450,000 \\
300,000 + k(50,000) &= 450,000 \\
50,000k &= 450,000 - 300,000 \\
50,000k &= 150,000 \\
k &= \frac{150,000}{50,000} \\
k &= 3
\end{aligned}
$$

According to Problem 3.57, at least 88.9% of the observations from a distribution will lie within three standard deviations of the mean. Therefore, the minimum percentage of the houses that should sell for prices between $150,000 and $450,000 is 88.9%.

Note: Problems 3.58–3.60 refer to a distribution of home sales prices with a mean of $300,000 and a standard deviation of $50,000.

3.60 Determine the minimum percentage of the houses that should sell for prices between $170,000 and $430,000.

Use the procedure outlined in Problem 3.59 to calculate k: substitute the mean and standard deviation of the home sales prices into the equation for the upper boundary and solve for k.

$$\mu + k\sigma = 430,000$$
$$300,000 + k(50,000) = 430,000$$
$$50,000k = 430,000 - 300,000$$
$$50,000k = 130,000$$
$$k = \frac{130,000}{50,000}$$
$$k = 2.6$$

Apply Chebyshev's Theorem.

$$\left(1 - \frac{1}{k^2}\right)(100\%) = \left(1 - \frac{1}{(2.6)^2}\right)(100\%)$$
$$= \left(1 - \frac{1}{(6.76)}\right)(100\%)$$
$$= (1 - 0.148)(100\%)$$
$$= 85.2\%$$

At least 85.2% of the selling prices should fall within the range of $170,000 to $430,000.

Note: Problems 3.61–3.62 refer to the following table, the number of home runs hit by the leaders in this category for the National League during the 2001 Major League Baseball season. The mean of the data is 37.9 and the standard deviation is 11.

Sorted National League Home Run Leaders									
73	64	57	49	49	45	41	39	38	38
37	37	37	36	36	34	34	34	34	34
33	31	31	30	30	29	27	27	27	25

3.61 Verify that Chebyshev's Theorem holds true for two standard deviations around the mean.

Calculate the lower and upper boundaries of the range.

$$\mu + k\sigma = 37.9 + (2)(11) \qquad \mu - k\sigma = 37.9 - (2)(11)$$
$$= 37.9 + 22 \qquad\qquad = 37.9 - 22$$
$$= 59.9 \qquad\qquad\qquad = 15.9$$

For you base-ball historians, those players are Barry Bonds and Sammy Sosa.

All but the 2 most proficient home run hitters, of the 30 in the data table, are included in this interval. Calculate this percentage.

$$\frac{28}{30} = 93.3\%$$

According to Problem 3.56

Chebyshev's Theorem states that at least 75% of the players' records will fall within two standard deviations of the mean. Therefore, Chebyshev's Theorem holds true in this example.

Note: Problems 3.61–3.62 refer to the table in Problem 3.61. The mean of the data is 37.9 and the standard deviation is 11.

3.62 Verify that Chebyshev's Theorem holds true for three standard deviations around the mean.

Calculate the lower and upper boundaries of the range.

$$\mu + k\sigma = 37.9 + 3(11) \qquad \mu - k\sigma = 37.9 - (3)(11)$$
$$= 37.9 + 33 \qquad\qquad = 37.9 - 33$$
$$= 70.9 \qquad\qquad\qquad = 4.9$$

The number 88.9% comes from Problem 3.57.

Of the 30 players whose home run totals are listed, all but the top player belong to the interval bounded below by 4.9 and above by 70.9. Calculate the percentage of players within three standard deviations of the mean.

$$\frac{29}{30} = 96.7\%$$

Chebyshev's Theorem holds true for this data with $k = 3$, because at least 88.9% of the players' records are within three standard deviations of the mean.

Chapter 4
INTRODUCTION TO PROBABILITY

What are the chances?

This chapter explores the foundational concepts of probability, the measurement of uncertainty reached through statistical analysis. Probability and statistics are inexorably tied together mathematically, as many of the theorems in subsequent chapters are based at least in part in probability.

This chapter starts with the basics of probability, defining sample space, events, and outcomes. You'll progress through important concepts like the addition rules (and the multiplication rule) for probability, as well as conditional probability and Bayes' Theorem. If you've ever wondered what the odds were of pulling a certain card from a standard deck, this is the chapter for you.

Types of Probability

Starting with the basics

4.1 Define each of the following probability terms, using the example of rolling a pair of standard six-sided dice and adding the numbers that result: *experiment, outcome, sample space,* and *event.*

An experiment is the process of measuring or observing an activity for the purpose of collecting data. Rolling a pair of dice would be considered an experiment. An outcome is a particular result of an experiment. For example, if you were to roll a pair of threes, then the outcome would be $3 + 3 = 6$.

A sample space consists of all the possible outcomes of the experiment. In the example of two standard dice, the smallest possible outcome would be rolling a pair of ones ($1 + 1 = 2$); the largest outcome would be a pair of sixes ($6 + 6 = 12$). Thus, the sample space for the experiment would be $\{2, 3, 4, 5, 6, 7, 8, 9, 10, 11,$ and $12\}$.

> It's possible to get every number between 2 and 12 when rolling a pair of dice. Notice that the sample space is written inside a pair of braces.

An event is a subset of the sample space that is of particular interest to the experiment. For instance, one event could be rolling a total of two, three, four, or five with a pair of dice. Usually, your task in a probability problem is to determine the likelihood that a particular event will occur with respect to the sample space (for example, identifying how often you will roll a total of two, three, four, or five given two standard dice).

4.2 Define *classical probability* and provide an example.

Classical probability is computed by dividing the number of ways a particular event may occur by the total number of outcomes the experiment may produce.

> $P(A)$ = the probability that Event A will occur.

$$P(A) = \frac{\text{number of possible outcomes in which event } A \text{ occurs}}{\text{total number of outcomes in the sample space}}$$

Classical probability requires an understanding of the underlying process so that the number of outcomes associated with an event can be counted. For instance, a standard deck contains 52 cards. Of those 52 cards, 13 are diamonds. To determine the probability of drawing a diamond from a shuffled deck of cards, divide the number of diamond cards by the total number of cards in the deck.

$$P(A) = \frac{13}{52} = \frac{1}{4}$$

4.3 Define *empirical probability* and provide an example.

Empirical probability relies on relative frequency distributions to determine the probability of events. It is often used when there is little understanding of the underlying process, so data is gathered about the events of interest instead. For example, consider the following grade distribution for a statistics class.

Grade	Relative Frequency
A	0.15
B	0.40
C	0.25
D	0.15
F	0.05
Total	**1.00**

The probability that a randomly selected student received a B grade is 40 percent.

> There may always be 13 diamonds in a deck of cards, but it's unlikely that the same 40 percent of the class will always get a B on a statistics exam, so instead you predict the probability based on the data you collect.

4.4 Define *subjective probability* and provide an example.

Subjective probability is used when classical and empirical probabilities are not available. Under these circumstances, you rely on experience and intuition to estimate probabilities. Subjective probability would be used to answer the question, "What is the probability that the New York Jets will make the NFL playoffs next year?" The response may be based in part on data from past seasons, but because information about the upcoming season is not known, the assessment will be subjective.

4.5 Of the numbers below, which could be valid measures of probability?

(a) 0.16

(b) −0.7

(c) 0

(d) 54%

(e) 1.06

(f) 118%

(g) $\dfrac{2}{3}$

(h) 1

(i) $-\dfrac{1}{6}$

> Inclusive means "includes the boundaries," so zero and 1 are valid probability values (as are 0% and 100%).

Probabilities can be represented numerically as real numbers between zero and one, inclusive. Therefore, (a), (c), (g), and (h) are valid representations of a probability. Probabilities are neither negative, so (b) and (i) are invalid, nor greater than one, so (e) is invalid.

Probabilities can also be expressed as percentages between 0% and 100%, inclusive. Therefore, (d) is valid. However, (f) is invalid because $118 > 100$.

> That means giving 110% in the big game is not a valid measure of probability, even though most postgame locker-room interviews would have you believe otherwise.

4.6 Classify each of the following as an example of classical, empirical, or subjective probability.

(a) The probability that the baseball player Ryan Howard will get a hit during his next at bat.

(b) The probability of drawing an ace from a deck of cards.

(c) The probability that a friend of yours will shoot lower than 100 during her next round of golf.

(d) The probability of winning the next state lottery drawing.

(e) The probability that the price of gasoline will exceed a certain price per gallon in six months.

(a) Empirical probability. Howard's batting average for this season provides historical data upon which you can base your conclusion. If his batting average is .251, then there is a 25.1% chance that his next at bat will result in a hit.

(b) Classical probability. Standard decks of cards are constructed in a uniform, predictable way. Therefore, you can be certain about the outcomes and the sample space.

> If she has kept careful records of her scores, you could argue that this is empirical probability instead, like part (a).

(c) Subjective probability. Unless your friend has kept careful and extensive records of her past golf scores, assessing this probability will be subjective.

(d) Classical probability. The probability of winning the next state lottery drawing can be calculated by dividing the chances your numbers will be drawn by the number of possible lottery ticket outcomes.

> See Problem 4.4.

(e) Subjective probability. Much like when you predict the future success of a football franchise, historic data here does not necessarily reflect the trends and patterns of future data. Too many immeasurable and dynamic factors affect the price of gas to predict what it will be in a week, let alone in six months.

4.7 Given the probability that a randomly selected student in a class is a female is 56%, determine the probability that the selected student is a male.

The sum of probabilities for all possible events must equal one. In this experiment, there are only two possible outcomes, choosing a male student or choosing a female student. (Note that percentages are converted to decimals before substituting the probabilities into the equation below.)

> The complement of event A is "when anything else happens except for A." The complement of picking a female student would be picking a male student.

$$P(\text{male}) + P(\text{female}) = 1.0$$
$$P(\text{male}) = 1.0 - P(\text{female})$$
$$P(\text{male}) = 1.0 - 0.56$$
$$P(\text{male}) = 0.44$$

This is known as the complement rule in probability. The probability of the complement of an event is one minus the probability of the event.

4.8 A customer survey asked respondents to indicate their highest level of education. The only three choices in the survey are high school, college, and other. If 31% indicated high school and 49% indicated college, determine the percentage of respondents who chose the "other" category.

The sum of the probabilities of all three possible outcomes must equal one.

$$P(\text{college}) + P(\text{high school}) + P(\text{other}) = 1.0$$
$$P(\text{other}) = 1.0 - P(\text{college}) - P(\text{high school})$$
$$P(\text{other}) = 1.0 - 0.49 - 0.31$$
$$P(\text{other}) = 0.20$$

The percentage of respondents who indicated "other" as their education category is 20%.

4.9 Define *mutually exclusive events*. Provide an example of two events that are mutually exclusive and two events that are not.

Two events, A and B, are considered mutually exclusive if the occurrence of one event prevents the occurrence of the other. Consider the two events below when rolling a pair of dice.

$$A = \text{the total is 11}$$
$$B = \text{a pair is rolled}$$

These two events cannot occur at the same time and are therefore mutually exclusive. Consider the two events below that are *not* mutually exclusive.

$$C = \text{the total is 12}$$
$$D = \text{a pair is rolled}$$

There's no pair of numbers that adds up to 11. The numbers would have to be 5.5 and 5.5.

Rolling a pair of sixes results in a total of 12. Because both events can occur at the same time, they are not mutually exclusive.

4.10 Define *independent events*. Provide an example of two events that are independent and two events that are not.

Two events, A and B, are considered independent if the occurrence of A has no effect on the probability of B occurring. Consider the events below, given an experiment in which you flip a coin and roll one six-sided die.

$$A = \text{coin is heads}$$
$$B = \text{a two is rolled}$$

Because the probability of rolling a two has no effect on the outcome of a coin flip, events A and B are independent. The two events below, however, are not independent.

C = all college students arrive to statistics class on time

D = the roads leading to the college are icy and dangerous

Commuting students may not be able to arrive on time for class if the roads are treacherous. Because event C can be influenced by event D, the events are not independent.

> *If the class were given online (and traveling to class was therefore not an issue), then the events would be independent.*

4.11 A card is chosen randomly from a standard deck, recorded, and then replaced. A second card is then drawn and recorded. Consider the events below.

A = first card is an ace of spades

B = second card is an ace of spades

Are these events independent? Are they mutually exclusive?

Because the card is replaced after the first drawing, both events can occur. Therefore, the events are *not* mutually exclusive. The probability of drawing the ace of spades (or any single card in the deck, for that matter) is the same each time a card is drawn. Thus, the events are independent.

4.12 A card is chosen randomly from a standard deck, recorded, and *not* replaced. A second card is then drawn and recorded. Consider the events below.

A = first card is an ace of spades

B = second card is an ace of spades

Are these events independent? Are they mutually exclusive?

Events A and B are mutually exclusive. If the first card you draw is the ace of spades and the card is not returned to the deck, then event B cannot occur. The events are not independent. If event A occurs, then event B cannot occur.

However, if event A does *not* occur, then the probability of event B is $\frac{1}{51}$.

> *In other words, if the first card you draw is something other than the ace of spades.*

Note: Problems 4.13–4.16 refer to the data set below, the results of a survey asking families how many cats they own.

Number of Cats	Relative Frequency of Households
0	0.30
1	0.36
2	0.25
3	0.07
4	0.02
Total	**1.00**

The sum of the probabilities for all possible outcomes must equal one.

4.13 Determine the probability that a randomly selected household in the survey had fewer than two cats.

Households with fewer than two cats have either one or zero cats. Add the probabilities of both outcomes.

$$P(\text{less than two cats}) = P(\text{zero cats}) + P(\text{one cat})$$
$$= 0.30 + 0.36$$
$$= 0.66$$

There is a 66% chance that a randomly selected household from the survey had less than two cats.

Note: Problems 4.13–4.16 refer to the data set in Problem 4.13, the results of a survey asking families how many cats they own.

4.14 Determine the probability that a household in the survey had two or fewer cats.

Households with two or fewer cats have two, one, or zero cats. Add the probabilities of all three outcomes, based on the survey data.

$$P(\text{two or fewer cats}) = P(\text{zero cats}) + P(\text{one cat}) + P(\text{two cats})$$
$$= 0.30 + 0.36 + 0.25$$
$$= 0.91$$

The complement of having two or fewer cats is having three or more cats. A total of 9% of respondents had three or more cats, so 100% – 9% = 91% of the respondents had two or fewer cats.

There is a 91% chance that a randomly selected household from the survey had two or fewer cats.

Note: Problems 4.13–4.16 refer to the data set in Problem 4.13, the results of a survey asking families how many cats they own.

4.15 Determine the probability that a randomly selected household from the survey had more than one cat.

Add the probabilities of a house having two, three, or four cats.

$$P(\text{more than one cat}) = P(\text{two cats}) + P(\text{three cats}) + P(\text{four cats})$$
$$= 0.25 + 0.07 + 0.02$$
$$= 0.34$$

There is a 34% chance that a randomly selected household from the survey had more than one cat.

Note: Problems 4.13–4.16 refer to the data set in Problem 4.13, the results of a survey asking families how many cats they own.

4.16 Determine the probability that a randomly selected household from the survey had one or more cats.

According to Problem 4.15, the probability of the household owning more than one cat was 0.34. To compute the probability of owning *one or more* cats, add the probability of owning one cat.

$$P(\text{one or more cats}) = P(\text{more than one cat}) + P(\text{one cat})$$
$$= 0.34 + 0.36$$
$$= 0.70$$

The complement of owning one or more cats is owning zero cats (0.30). That means there's a 1 − 0.30 = 0.70 probability of owning one or more cats.

There is a 70% chance that a randomly selected household from the survey had one or more cats.

Note: Problems 4.17–4.19 refer to the data set below, the relative frequency of executive salaries at a particular organization.

Event	Salary Range	Relative Frequency
A	Under $60,000	0.09
B	$60,000–under $70,000	0.21
C	$70,000–under $80,000	0.28
D	$80,000–under $90,000	0.15
E	$90,000–under $100,000	0.23
F	$100,000 or more	0.04
Total		**1.00**

4.17 Determine the probability that a randomly selected executive has a salary greater than or equal to $70,000 but less than $100,000.

Salaries between $70,000 and $100,000 comprise events C, D, and E.

$$P(C) + P(D) + P(E) = 0.28 + 0.15 + 0.23 = 0.66$$

There is a 66% probability that a randomly selected executive will have a salary greater than or equal to $70,000 but less than $100,000.

Note: Problems 4.17–4.19 refer to the data set in Problem 4.17, the relative frequency of executive salaries at a particular organization.

4.18 Determine the probability that a randomly selected executive has a salary that is either less than $60,000 or greater than or equal to $90,000.

Events A, E, and F describe salaries less than $60,000 or greater than or equal to $90,000.

$$P(A) + P(E) + P(F) = 0.09 + 0.23 + 0.04 = 0.36$$

There is a 36% probability that a randomly selected executive will have a salary that is either less than $60,000 or greater than or equal to $90,000.

Note: Problems 4.17–4.19 refer to the data set in Problem 4.17, the relative frequency of executive salaries at a particular organization.

4.19 Are events A through F mutually exclusive?

Yes, events A through F are mutually exclusive. Every positive real number belongs to exactly one of the categories, so no matter what salary an executive may have, it will correspond to exactly one event. ←

> The categories don't overlap, because the upper boundaries of events A, B, C, D, and E are "under" the lower boundary of the next event.

Note: Problems 4.20–4.22 refer to the following data, the relative frequency for the daily demand for computers at a local electronics store.

Daily Demand	Relative Frequency
0	0.16
1	0.12
2	0.24
3	0.14
4	0.17
5	0.06
6	0.09
7	0.02
Total	**1.00**

> You're calculating empirical probability based on historical data.

4.20 Predict the probability that tomorrow's demand will be at least four computers. ←

A demand for "at least four computers" means, in this instance, selling four, five, six, or seven computers.

$$P(\text{at least } 4) = P(4) + P(5) + P(6) + P(7)$$
$$= 0.17 + 0.06 + 0.09 + 0.02$$
$$= 0.34$$

There is a 34% chance that at least four computers will be sold tomorrow.

Note: Problems 4.20–4.22 refer to the data in Problem 4.20, the relative frequency for the daily demand for computers at a local electronics store.

4.21 Determine the probability that tomorrow's demand will be no more than two computers.

The phrase "no more than two" equates to selling zero, one, or two computers.

$$P(\text{no more than } 2) = P(0) + P(1) + P(2)$$
$$= 0.16 + 0.12 + 0.24$$
$$= 0.52$$

The probability of selling no more than two computers tomorrow is 52%.

Note: Problems 4.20–4.22 refer to the data in Problem 4.20, the relative frequency for the daily demand for computers at a local electronics store.

4.22 Are the eight events in this problem mutually exclusive?

Yes, these events are mutually exclusive. For any particular day, only one level of demand can occur—only one number can represent each day's computer sales. Because the events do not overlap, each demand can belong to only one range.

Addition Rules for Probability
Combining probabilities using "or"

4.23 A recent survey found that 62% of the households surveyed had Internet access, 68% had cable TV, and 43% had both. Determine the probability that a randomly selected household in the survey had either Internet access or cable.

The addition rule for probability determines the probability that either event A or event B will occur.

$$P(A \text{ or } B) = P(A) + P(B) - P(A \text{ or } B)$$

Consider events A and B, defined below.

> $P(A \text{ and } B)$ is the probability that events A and B occur at the same time. You have to subtract it to avoid counting the same households multiple times.

$$A = \text{household has Internet}$$
$$B = \text{household has cable}$$

Apply the addition rule for probability.

$$P(A \text{ or } B) = P(A) + P(B) - P(A \text{ and } B)$$
$$= 0.62 + 0.68 - 0.43$$
$$= 0.87 \leftarrow$$

The probability that a randomly selected household had either Internet access or cable is 87%. ←

> You didn't have to subtract P(A or B) in Problems 4.13–4.22 because the events in those problems were mutually exclusive. When the events can overlap, you need to subtract that overlap.

Note: Problems 4.24–4.25 refer to a local university, at which 62% of the students are undergraduates, 55% of the students are male, and 48% of the undergraduate students are male.

4.24 Determine the probability that a randomly selected student is either male or an undergraduate.

> The household may have both Internet and cable— that's okay. There's only a 13% chance that it will have neither.

Consider events A and B, as defined below.

$$A = \text{student is an undergraduate}$$
$$B = \text{student is a male}$$

Apply the addition rule for probability.

$$P(A \text{ or } B) = P(A) + P(B) - P(A \text{ and } B)$$
$$= 0.62 + 0.55 - 0.48$$
$$= 0.69$$

The probability that a randomly selected student is either male or an undergraduate is 69%.

Note: Problems 4.24–4.25 refer to a local university, at which 62% of the students are undergraduates, 55% of the students are male, and 48% of the undergraduate students are male.

4.25 Illustrate the probabilities using a Venn diagram.

The left circle represents the undergraduate students and the right circle represents male students in the diagram below. The intersection of the two circles, the shaded region of the diagram, represents the male undergraduate students.

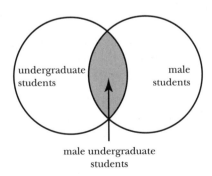

male undergraduate students

Note: Problems 4.26–4.29 refer to the data set below, the number of cars of various types at a local dealership.

	Sedan	SUV
New	24	15
Used	9	12

4.26 Determine the probability that a randomly selected car is new.

Calculate the total number of cars at the dealership.

$$24 + 15 + 9 + 12 = 60$$

Of the 60 cars, $24 + 15 = 39$ are new. Divide the number of new cars by the total number of cars to calculate the probability of randomly selecting a new car.

$$P(\text{new}) = \frac{39}{60} = 0.65$$

There is a 65% chance of randomly selecting a new car.

> Some textbooks refer to this as a simple or a marginal probability.

Note: Problems 4.26–4.29 refer to the data set in Problem 4.26, the number of cars of various types at a local dealership.

4.27 Determine the probability that a randomly selected car is a sedan.

According to Problem 4.26, there are 60 cars at the dealership, of which $24 + 9 = 33$ are sedans. Divide the number of sedans by the number of available vehicles.

$$P(\text{sedan}) = \frac{33}{60} = 0.55$$

There is a 55% chance of randomly selecting a sedan.

Note: Problems 4.26–4.29 refer to the data set in Problem 4.26, the number of cars of various types at a local dealership.

4.28 Determine the probability that a randomly selected car is a new sedan.

Of the 60 cars at the dealership, 24 are new sedans. Divide the 24 possible outcomes by the sample space of 60 cars.

$$P(\text{new and sedan}) = \frac{24}{60} = 0.40$$

There is a 40% chance of randomly selecting a new sedan.

Note: Problems 4.26–4.29 refer to the data set in Problem 4.26, the number of cars of various types at a local dealership.

4.29 Determine the probability that a randomly selected car is either used or an SUV. ⟵

Apply the addition rule for probability. ⟵

$$P(\text{used or SUV}) = P(\text{used}) + P(\text{SUV}) - P(\text{used SUV})$$

Of the 60 cars at the dealership, 21 are used, 27 are SUVs, and 12 are used SUVs. Calculate the probability of randomly selecting a car from each category.

$$P(\text{used}) = \frac{21}{60} = 0.35$$

$$P(\text{SUV}) = \frac{27}{60} = 0.45$$

$$P(\text{used SUV}) = \frac{12}{60} = 0.20$$

Substitute these values into the addition rule equation.

$$P(\text{used or SUV}) = 0.35 + 0.45 - 0.20$$
$$= 0.60$$

The probability that the selected car is either used or an SUV is 60%.

> Some textbooks use "union notation" to represent the probability of either event A or event B occurring. In this problem, you would write P(new ∪ sedan). SUV

> It only makes sense to apply this rule when the events are not mutually exclusive. Otherwise, P(A and B) would be zero, and subtracting zero wouldn't be very useful.

Note: Problems 4.30–4.37 refer to the following table, which lists the number of medals won by countries during the 2008 Beijing Summer Olympics.

	Gold	Silver	Bronze	Total
China	51	21	28	100
United States	36	38	36	110
Russia	23	21	28	72
Other	192	223	261	676
Total	**302**	**303**	**353**	**958**

4.30 Determine the probability that a randomly selected medal was won by Russia.

Russian won 72 of the 958 total medals.

$$P(\text{Russia}) = \frac{72}{958} = 0.075$$

There is a 7.5% probability that a randomly selected medal was won by Russia.

Note: Problems 4.30–4.37 refer to the data set in Problem 4.30, the number of medals won by countries during the 2008 Beijing Summer Olympics.

4.31 Determine the probability that a randomly selected medal was silver.

The silver medals accounted for 303 of the 958 total medals.

$$P(\text{silver}) = \frac{303}{958} = 0.316$$

There is a 31.6% probability that a randomly selected medal was silver.

Note: Problems 4.30–4.37 refer to the data set in Problem 4.30, the number of medals won by countries during the 2008 Beijing Summer Olympics.

4.32 Determine the probability that a randomly selected medal was a silver medal awarded to Russia.

> Don't use the addition rule for probability here. You're not being asked how many medals were either silver or awarded to Russia.

Of the 958 medals awarded, 21 silver medals were earned by Russia.

$$P(\text{Russia and silver}) = \frac{21}{958} = 0.022$$

> The probability of two events both occurring is called the joint probability, or the intersection of two events. It's written $P(A \cap B)$.

There is a 2.2% chance that a randomly selected medal was a silver medal awarded to Russia.

Note: Problems 4.30–4.37 refer to the data set in Problem 4.30, the number of medals won by countries during the 2008 Beijing Summer Olympics.

4.33 Use the probabilities calculated in Problems 4.30–4.32 to determine the probability that a randomly selected medal was either a silver medal or awarded to Russia.

According to Problems 4.30–4.32, $P(\text{Russia}) = 0.075$, $P(\text{silver}) = 0.316$, and $P(\text{Russia and silver}) = 0.022$. Apply the addition rule for probability.

$$P(\text{Russia or silver}) = P(\text{Russia}) + P(\text{silver}) - P(\text{Russia and silver})$$
$$= 0.075 + 0.316 - 0.022$$
$$= 0.369$$

There is a 36.9% chance that a randomly selected medal was either awarded to Russia or was silver.

Note: Problems 4.30–4.37 refer to the data set in Problem 4.30, the number of medals won by countries during the 2008 Beijing Summer Olympics.

4.34 Are the events "silver medal" and "Russia" mutually exclusive?

No, they are not mutually exclusive. It is possible for both events to occur simultaneously. A medal can be both silver and awarded to Russia. If events A and B are mutually exclusive, then $P(A \text{ and } B) = 0$. However, according to Problem 4.32, $P(\text{Russia and silver}) = 0.022$.

Note: Problems 4.30–4.37 refer to the data set in Problem 4.30, the number of medals won by countries during the 2008 Beijing Summer Olympics.

4.35 Determine the probability that a randomly selected medal was either a gold medal or awarded to the United States.

In order to calculate $P(\text{United States or gold})$, you must first calculate $P(\text{United States})$, $P(\text{gold})$, and $P(\text{United States and gold})$.

$$P(\text{United States}) = \frac{110}{958} = 0.115$$

$$P(\text{gold}) = \frac{302}{958} = 0.315$$

$$P(\text{United States and gold}) = \frac{36}{958} = 0.038$$

Apply the addition rule for probability.

$$P(\text{United States or gold}) = P(\text{United States}) + P(\text{gold}) - P(\text{United States and gold})$$
$$= 0.115 + 0.315 - 0.038$$
$$= 0.392$$

The probability that a randomly selected medal was awarded to the United States or was gold is 39.2%.

Note: Problems 4.30–4.37 refer to the data set in Problem 4.30, the number of medals won by countries during the 2008 Beijing Summer Olympics.

4.36 Are the events "gold medal" and "silver medal" mutually exclusive?

Yes, they are mutually exclusive. A medal cannot be both gold and silver. The probability of randomly selecting such a medal would be zero:

$$P(\text{gold and silver}) = 0.$$

In this experiment, you choose a single medal at random. A single medal could be gold, could be silver, or could be neither (don't forget bronze). However, a single medal can't be gold and silver.

Note: Problems 4.30–4.37 refer to the data set in Problem 4.30, the number of medals won by countries during the 2008 Beijing Summer Olympics.

4.37 Determine the probability that a randomly selected medal is either gold or silver.

Calculate the probabilities that a randomly selected medal is gold or is silver.

$$P(\text{gold}) = \frac{302}{958} = 0.315$$

$$P(\text{silver}) = \frac{303}{958} = 0.316$$

Recall that $P(A \text{ or } B) = P(A) + P(B)$ when A and B are mutually exclusive.

$$P(\text{gold or silver}) = P(\text{gold}) + P(\text{silver})$$
$$= 0.315 + 0.316$$
$$= 0.631$$

> You can still apply the addition rule for probability. You'll just be subtracting P(gold and silver) = 0.

There is a 63.1% chance that a randomly selected medal is gold or silver.

Note: In Problems 4.38–4.41, a single card is drawn from a standard 52-card deck.

4.38 Determine the probability that the card drawn is an ace, a two, or a three.

There are four cards of every rank in the deck. Therefore, the probability of selecting a specific rank from the deck is $\frac{4}{52} = \frac{1}{13}$. Drawing an ace, a two, and a three are mutually exclusive events.

$$P(2 \text{ or } 3 \text{ or } 4) = P(2) + P(3) + P(4)$$
$$= \frac{1}{13} + \frac{1}{13} + \frac{1}{13}$$
$$= \frac{3}{13}$$
$$= 0.231$$

> That means you don't have to subtract P(A and B) when you apply the addition rule for probability.

There is a 23.1% chance that the single card will be an ace, a two, or a three.

Note: In Problems 4.38–4.41, a single card is drawn from a standard 52-card deck.

4.39 Determine the probability that the card drawn is a diamond, spade, or club.

The complement of drawing a diamond, spade, or club is drawing the only remaining suit from the deck, a heart. Calculate the probability of drawing a heart.

$$P(\text{heart}) = \frac{13}{52} = \frac{1}{4} = 0.25$$

> A standard deck contains four suits (hearts, clubs, spades, and diamonds). There are 13 cards of each suit in the deck.

Recall that the probability of an event is equal to one minus the probability of its complement.

$$P(\text{diamond, spade, or club}) = 1 - P(\text{heart})$$
$$= 1 - 0.25$$
$$= 0.75$$

There is a 75% chance that the single card will be a diamond, spade, or club.

Note: In Problems 4.38–4.41, a single card is drawn from a standard 52-card deck.

4.40 Determine the probability that the single card drawn is a four, a five, or a spade.

Calculate the probabilities of selecting a four, a five, or a spade.

$$P(4) = \frac{4}{52} \qquad P(5) = \frac{4}{52} \qquad P(\text{spade}) = \frac{13}{52}$$

These three events are not mutually exclusive—it is possible to draw a four or a five that is also a spade. Calculate the probabilities of selecting the four or five of spades.

$$P(4 \text{ of spades}) = \frac{1}{52} \qquad P(5 \text{ of spades}) = \frac{1}{52}$$

Apply the addition rule for probability, accounting for the cards that are outcomes of both events.

$$P(4 \text{ or } 5 \text{ or spade}) = P(4) + P(5) + P(\text{spade}) - P(4 \text{ and spade}) - P(5 \text{ and spade})$$
$$= \frac{4}{52} + \frac{4}{52} + \frac{13}{52} - \frac{1}{52} - \frac{1}{52}$$
$$= \frac{19}{52}$$
$$= 0.365$$

There is a 36.5% chance that the card selected will be a four, a five, or a spade.

> A standard deck contains one of each card.

> There are 19 cards that would qualify (S, H, D, and C are the suits): 2S, 3S, 4S, 5S, 6S, 7S, 8S, 9S, 10S, JS, QS, KS, AS, 4H, 4D, 4C, 5H, 5D, 5C.

Note: In Problems 4.38–4.41, a single card is drawn from a standard 52-card deck.

4.41 Determine the probability that the single card drawn is a seven, an eight, a diamond, or a heart.

Four cards in the deck are sevens, four cards are eights, 13 cards are diamonds, and 13 cards are hearts. In the equation below, let D represent diamonds and H represent hearts.

$$P(7 \text{ or } 8 \text{ or } D \text{ or } H) = P(7) + P(8) + P(D) + P(H) - P(7 \text{ and } D) -$$
$$P(7 \text{ and } H) - P(8 \text{ and } D) - P(8 \text{ and } H)$$
$$= \frac{4}{52} + \frac{4}{52} + \frac{13}{52} + \frac{13}{52} - \frac{1}{52} - \frac{1}{52} - \frac{1}{52} - \frac{1}{52}$$
$$= \frac{30}{52}$$
$$= 0.577$$

There is a 57.7% chance that the card selected will be a seven, an eight, a diamond, or a heart.

Conditional Probability
Probabilities that depend on other events

Note: Problems 4.42–4.47 refer to the following data, the total number of wins recorded by two friends playing tennis against each other, based on the length of time they warmed up before the match.

Warm-up Time	Deb Wins	Bob Wins	Total
Short	4	6	10
Long	16	24	40
Total	**20**	**30**	**50**

4.42 Determine the probability that Deb wins the next match if she only has a short time to warm up.

What's the probability of A, assuming B happened?

Conditional probability describes how likely some event A is to occur if you assume that some event B has already happened: $P(A \mid B)$. The formula below is used to calculate conditional probability.

Divide by the probability of the event that is given.

$$P(A \mid B) = \frac{P(A \text{ and } B)}{P(B)} = \frac{P(A \cap B)}{P(B)}$$

In this problem, you assume that the warm-up time was short, so given a short warm-up time, you are asked to calculate the probability of Deb winning the match.

$$P(\text{Deb}|\text{Short}) = \frac{P(\text{Deb and short})}{P(\text{short})}$$

Of the 50 games played, 10 were preceded by a short warm-up period. Assuming a short warm-up, Deb won only four games.

$$P(\text{Deb and short}) = \frac{4}{50} = 0.08$$

$$P(\text{short}) = \frac{10}{50} = 0.20$$

Substitute these values into the above conditional probability formula.

$$P(\text{Deb}|\text{Short}) = \frac{0.08}{0.20} = 0.40$$

The probability that Deb will win the next match given a short warm-up period is 40%.

Note: Problems 4.42–4.47 refer to the data set in Problem 4.42, the total number of wins recorded by two friends playing tennis against each other, based on the length of time they warmed up before the match.

4.43 Assuming Deb won the last match, determine the likelihood that the warm-up period before the match was short.

Apply the formula for conditional probability using the following events: $A =$ the warm-up was short; $B =$ Deb won the match.

$$P(A|B) = \frac{P(A \text{ and } B)}{P(B)}$$

$$P(\text{short}|\text{Deb}) = \frac{P(\text{short and Deb})}{P(\text{Deb})}$$

You're dividing the probability that a random match had a short warm-up and was won by Deb by the probability that a random match was won by Deb.

Of the 50 matches played, 20 had a short warm-up. Only four of the matches with short warm-ups were won by Deb. Calculate the corresponding probabilities.

$$P(\text{short}) = \frac{20}{50} = 0.40$$

$$P(\text{short and Deb}) = \frac{4}{50} = 0.08$$

Substitute these values into the conditional probability formula above.

$$P(\text{short} \mid \text{Deb}) = \frac{P(\text{short and Deb})}{P(\text{Deb})}$$
$$= \frac{0.08}{0.40}$$
$$= 0.20$$

In general,
$P(A|B) \neq P(B|A)$.

Assuming Deb won the last match, there is a 20% chance that the preceding warm-up period was short. Compare the answers to Problems 4.42 and 4.43 to verify that $P(\text{Deb} \mid \text{short}) \neq P(\text{short} \mid \text{Deb})$.

Note: Problems 4.42–4.47 refer to the data set in Problem 4.42, the total number of wins recorded by two friends playing tennis against each other, based on the length of time they warmed up before the match.

4.44 Given that the warm-up time was short, determine the probability that Bob wins the next match.

According to the historical data, there is a $\frac{10}{50} = 0.20$ probability that a short warm-up will precede a match; there is a $\frac{6}{50} = 0.12$ probability that the match will be short and Bob will win. Calculate $P(\text{Bob} \mid \text{short})$.

If you're calculating $P(A|B)$, A is the event you're trying to calculate the probability of, and B is the event you're assuming is true. The order is very important.

$$P(\text{Bob} \mid \text{short}) = \frac{P(\text{Bob and short})}{P(\text{short})}$$
$$= \frac{0.12}{0.20}$$
$$= 0.60$$

The probability that Bob will win the next match given the warm-up is short is 60%.

Note: Problems 4.42–4.47 refer to the data set in Problem 4.42, the total number of wins recorded by two friends playing tennis against each other, based on the length of time they warmed up before the match.

4.45 Assuming the warm-up time is long, determine the probability that Bob wins the next match.

There is a $\frac{40}{50} = 0.80$ probability that a random match will be preceded by a long warm-up; there is a $\frac{24}{50} = 0.48$ probability that a random match will be long and Bob will win. Calculate $P(\text{Bob} \mid \text{long})$.

$$P(\text{Bob} \mid \text{long}) = \frac{P(\text{Bob and long})}{P(\text{long})}$$
$$= \frac{0.48}{0.80}$$
$$= 0.60$$

The probability that Bob will win the next match, assuming the warm-up is long, is 60%.

Note: Problems 4.42–4.47 refer to the data set in Problem 4.42, the total number of wins recorded by two friends playing tennis against each other, based on the length of time they warmed up before the match.

4.46 Deb claims she has a better chance of winning the match if the warm-up is long. Is there any validity to her claim?

If Deb's claim is true, then the statement below must be true.

$$P(\text{Deb} \mid \text{long}) > P(\text{Deb})$$

The probability of Deb winning when the warm-up is long is greater than the probability of Deb winning in general.

Calculate the probability of Deb winning regardless of the warm-up time.

$$P(\text{Deb}) = \frac{20}{50} = 0.40$$

Calculate the conditional probability of Deb winning given the warm-up is long.

$$P(\text{Deb} \mid \text{long}) = \frac{P(\text{Deb and long})}{P(\text{long})}$$
$$= \frac{16 / 50}{40 / 50}$$
$$= \frac{0.32}{0.80}$$
$$= 0.40$$

This makes sense if you look at Problem 4.45. When the warm-up was long, Bob won 60% of the time. Bob and Deb are playing against each other, which means Deb has to win 40% of those matches.

Because $P(\text{Deb} \mid \text{long}) = P(\text{Deb})$, Deb's claim is invalid.

Note: Problems 4.42–4.47 refer to the data set in Problem 4.42, the total number of wins recorded by two friends playing tennis against each other, based on the length of time they warmed up before the match.

4.47 Are the events "Deb" and "long" independent?

Events "Deb" and "long" are independent of each other because the probability of Deb winning is not affected by the long warm-up (according to Problem 4.46).

Events *A* and *B* are independent of each other if $P(A \mid B) = P(A)$ and $P(B \mid A) = P(B)$ are true statements.

$P(\text{Deb} \mid \text{long})$ has to be greater than $P(\text{Deb})$, not equal to it. She wins the same percentage of matches, regardless of the length of the warm-up period.

Note: Problems 4.48–4.50 refer to the data set below, the number of customers who have overdue accounts, according to credit card type and the number of days d the account is overdue.

Days Overdue	Card Type			
	Standard	**Gold**	**Platinum**	**Total**
$d < 30$	154	117	56	**327**
$31 \leq d \leq 60$	87	101	10	198
$61 \leq d \leq 90$	33	49	12	**94**
$d > 90$	10	15	17	42
Total	**284**	**282**	**95**	**661**

4.48 What is the probability that a randomly selected account 31–60 days overdue is a gold card?

> Conditional probability P(A|B), with events A = "card is gold" and B = "card is 31-60 days overdue."

Calculate the probability of selecting a gold account at random if you choose from accounts that are between 31 and 60 days overdue.

$$P(\text{gold} \mid 31 \leq d \leq 60) = \frac{P(\text{gold and } 31 \leq d \leq 60)}{P(31 \leq d \leq 60)}$$

$$= \frac{101 / 661}{198 / 661}$$

$$= \frac{0.153}{0.300}$$

$$= 0.510$$

Assuming the account is 31–60 days overdue, there is a 51.0% chance that the card is gold.

Note: Problems 4.48–4.50 refer to the data in Problem 4.48, the number of customers who have overdue accounts, according to credit card type and the number of days d the account is overdue.

4.49 What is the probability that a randomly selected gold card account is 61–90 days overdue?

Calculate the probability of selecting an account between 61 and 90 days overdue if you choose randomly from gold card accounts.

$$P(61 \leq d \leq 90 \mid \text{gold}) = \frac{P(61 \leq d \leq 90 \text{ and gold})}{P(\text{gold})}$$

$$= \frac{49 / 661}{282 / 661}$$

$$= \frac{0.0741}{0.4266}$$

$$= 0.174$$

Note: Problems 4.48–4.50 refer to the data in Problem 4.48, the number of customers who have overdue accounts, according to credit card type and the number of days d the account is overdue.

4.50 Determine whether the events "$31 \leq d \leq 60$" and "platinum" are independent.

> In other words, $P(A) = P(A|B)$ and $P(B) = P(B|A)$. The probability of A occurring shouldn't change if you assume B, and vice versa.

If the events are independent, then the probability of each occurring should be equal to the conditional probability of each, assuming the other. Calculate the probability of selecting an account that is 31–60 days overdue assuming the account is platinum.

$$P(31 \leq d \leq 60 \,|\, \text{platinum}) = \frac{P(31 \leq d \leq 60 \text{ and platinum})}{P(\text{platinum})}$$
$$= \frac{10/661}{95/661}$$
$$= \frac{0.0151}{0.1437}$$
$$= 0.105$$

Calculate the probability that a randomly selected account is between 31 and 60 days overdue.

$$P(31 \leq d \leq 60) = \frac{198}{661} = 0.300$$

> You could also compare P(platinum |31 ≤ d ≤ 60) to P(platinum) to show the events are not independent.

Because $P(31 \leq d \leq 60 \,|\, \text{platinum}) \neq P(31 \leq d \leq 60)$, the events are not independent.

4.51 At a local restaurant, 20% of the customers order take-out. If 7% of all customers order take-out and choose a hamburger, determine the probability that a customer who orders take-out will order a hamburger.

The probabilities below are given in the problem.

$$P(\text{take-out}) = 0.20$$
$$P(\text{take-out and hamburger}) = 0.07$$

You are asked to compute the conditional probability $P(\text{hamburger} \,|\, \text{take-out})$.

> You know the customer has ordered take-out. You don't know if she ordered a hamburger or not. This problem is making me hungry.

$$P(\text{hamburger} \,|\, \text{take-out}) = \frac{P(\text{hamburger and take-out})}{P(\text{take-out})}$$
$$= \frac{0.07}{0.20}$$
$$= 0.35$$

There is a 35% chance that a customer who orders take-out will order a hamburger.

4.52 Thirty-four percent of customers who purchased from an e-commerce site had orders exceeding $100. Given 22% of the customers have orders exceeding $100 and also use the site's sponsored credit card for payment, determine the probability that a customer whose order exceeds $100 will use the sponsored credit card for the payment.

The probabilities below can be gleaned from the problem.

$$P(\text{order} > \$100) = 0.34$$
$$P(\text{order} > \$100 \text{ and credit card used}) = 0.22$$

You are asked to compute the conditional probability $P(\text{credit card used} \mid \text{order} > \$100)$.

$$P(\text{credit card used} \mid \text{order} > \$100) = \frac{P(\text{credit card used and order} > \$100)}{P(\text{order} > \$100)}$$
$$= \frac{0.22}{0.34}$$
$$= 0.647$$

There is a 64.7% chance that a customer who places an order greater than $100 will use the sponsored credit card for payment.

Note: Problems 4.53–4.54 refer to an electronics store. According to the store's historical records, 65% of its digital camera customers are male, 18% of its digital camera customers purchase the extended warranty, and 10% of its digital camera customers are female and purchase the extended warranty.

4.53 Determine the probability that a male digital camera customer will purchase the extended warranty.

The probabilities below are provided by the problem.

$$P(\text{male}) = 0.65$$
$$P(\text{warranty}) = 0.18$$
$$P(\text{female and warranty}) = 0.10$$

This works a lot like the tennis victory table in Problem 4.42.

You are asked to calculate the conditional probability $P(\text{warranty} \mid \text{male})$. However, you are not given the value $P(\text{warranty and male})$. In order to identify this value, set up a table that contains the given information.

Gender	Warranty	No Warranty	Total
Male			**0.65**
Female	0.10		
Total	**0.18**		

The complement of a male customer is a female customer. Therefore, $P(\text{female}) = 1 - P(\text{male}) = 1 - 0.65 = 0.35$. Similarly, the complement of a customer who purchases a warranty is a customer who does not.

$$P(\text{no warranty}) = 1 - P(\text{warranty})$$
$$= 1 - 0.18$$
$$= 0.82$$

Insert these probabilities into the table.

Gender	Warranty	No Warranty	Total
Male			**0.65**
Female	0.10		**0.35**
Total	**0.18**	**0.82**	**1.00**

Complete the table, noting that each row and column must have the indicated totals.

Gender	Warranty	No Warranty	Total
Male	0.08	0.57	0.65
Female	0.10	0.25	0.35
Total	**0.18**	**0.82**	**1.00**

0.18 − 0.10 = 0.08
0.65 − 0.08 = 0.57
0.82 − 0.57 = 0.25

Now that you can determine the value of $P(\text{warranty and male})$, calculate $P(\text{warranty} \mid \text{male})$.

$$P(\text{warranty} \mid \text{male}) = \frac{P(\text{warranty and male})}{P(\text{male})}$$
$$= \frac{0.08}{0.65}$$
$$= 0.123$$

There is a 12.3% chance that a male customer will purchase the extended warranty.

Note: Problems 4.53–4.54 refer to an electronics store. According to the store's historical records, 65% of its digital camera customers are male, 18% of its digital camera customers purchase the extended warranty, and 10% of its digital camera customers are female and purchase the extended warranty.

4.54 Are female customers more or less likely to purchase the extended warranty? Justify your answer.

Use the completed chart in Problem 4.53 to calculate $P(\text{warranty} \mid \text{female})$.

$$P(\text{warranty} \mid \text{female}) = \frac{P(\text{warranty and female})}{P(\text{female})}$$
$$= \frac{0.10}{0.35}$$
$$= 0.286$$

A female customer will purchase the extended warranty approximately 28.6% of the time, which is considerably higher than the 12.3% chance that a male will purchase the warranty (as computed in Problem 4.53).

This means that the events "warranty" and "male" are not independent. (Neither are the events "warranty" and "female.")

Note: Problems 4.55–4.57 refer to the following data, collected by a major airline that tracked the on-time status of 500 flights originating in Los Angeles and New York.

- 50 flights were early
- 275 flights were on time
- 285 flights originated in Los Angeles
- 150 flights originated in Los Angeles and were on time
- 60 flights originated in New York and were late

4.55 Calculate the probability that a late-arriving flight originated in New York.

Construct a table that organizes the given information, listing the number of flights originating from each airport that arrived early, on time, and late.

Status	NY	LA	Total
Early			50
On time		150	275
Late	60		
Total		285	500

Complete the table by ensuring that the sum of each row is the number in the rightmost column and the sum of each column is the corresponding number in the last row.

Status	NY	LA	Total
Early	30	20	50
On time	125	150	275
Late	60	115	175
Total	215	285	500

The problem indicates that the flight is late; calculate the conditional probability that the flight originated from New York.

$$P(\text{New York} \mid \text{late}) = \frac{P(\text{New York and late})}{P(\text{late})}$$
$$= \frac{60/500}{175/500}$$
$$= \frac{0.12}{0.35}$$
$$= 0.343$$

A randomly selected late flight has a 34.3% chance of having originated in New York.

Note: Problems 4.55–4.57 refer to the data in Problem 4.55, collected by a major airline that tracked the on-time status of 500 flights originating in Los Angeles and New York.

4.56 Calculate the probability that a flight originating in Los Angeles arrived at its destination on time.

Consider the table below, completed in Problem 4.55.

Status	NY	LA	Total
Early	30	20	50
On time	125	150	275
Late	60	115	175
Total	215	285	500

Calculate the conditional probability of the flight arriving on time, assuming it originated in Los Angeles.

$$P(\text{on time} \mid \text{Los Angeles}) = \frac{P(\text{on time and Los Angeles})}{P(\text{Los Angeles})}$$
$$= \frac{150/500}{285/500}$$
$$= \frac{0.3}{0.57}$$
$$= 0.526$$

A flight originating in Los Angeles has a 52.6% chance of arriving on time.

Note: Problems 4.55–4.57 refer to the data in Problem 4.55, collected by a major airline that tracked the on-time status of 500 flights originating in Los Angeles and New York.

4.57 Are the events "Early" and "LA" independent?

Consider the table below, completed in Problem 4.55.

Status	NY	LA	Total
Early	30	20	**50**
On time	125	150	**275**
Late	60	115	**175**
Total	**215**	**285**	**500**

The events "early" and "LA" are independent if $P(\text{early} \mid \text{Los Angeles}) = P(\text{early})$ and $P(\text{Los Angeles} \mid \text{early}) = P(\text{Los Angeles})$. According to the calculations that follow, the first of those statements is not true.

> The statement $P(\text{Los Angeles} \mid \text{early}) = P(\text{Los Angeles})$ isn't true either.

$$P(\text{early} \mid \text{Los Angeles}) = \frac{P(\text{early and Los Angeles})}{P(\text{Los Angeles})}$$
$$= \frac{20/500}{285/500}$$
$$= \frac{0.04}{0.57}$$
$$= 0.070$$

$$P(\text{early}) = \frac{50}{500}$$
$$= 0.10$$

Because these probabilities are not equal, the events are not independent.

The Multiplication Rule for Probability
Two or more events occurring at the same time

4.58 A card is drawn from a standard deck and not replaced. A second card is then drawn. What is the probability that both cards are clubs?

Define the events below.

$$A = \text{first card is a club}$$
$$B = \text{second card is a club}$$

> If the first card was a club, there's a $12/51 = 23.5\%$ chance of drawing a club as the second card. If the first card was not a club, there's a $13/51 = 25.5\%$ chance of drawing a club as the second card.

A and B are not independent, because the first card is not replaced. Apply the multiplication rule for events that are not independent: $P(A \text{ and } B) = P(A)P(B \mid A)$.

> You could write P(A and B) as $P(A \cap B)$.

$$P(A \text{ and } B) = P(A)P(B|A)$$
$$= \left(\frac{13}{52}\right)\left(\frac{12}{51}\right)$$
$$= \frac{156}{2,652}$$
$$= 0.059$$

The probability of drawing two clubs from the deck (without replacing the first card) is 5.9%.

4.59 A card is drawn from a standard deck and replaced. A second card is then drawn. What is the probability that both cards are hearts?

Define the events.

$$A = \text{first card is a heart}$$
$$B = \text{second card is a heart}$$

The events are independent because the first card is replaced. Apply the multiplication rule for independent events.

$$P(A \text{ and } B) = P(A)P(B) \leftarrow$$
$$= \left(\frac{13}{52}\right)\left(\frac{13}{52}\right)$$
$$= \frac{169}{2,704}$$
$$= 0.0625$$

There's a 13/52 chance of drawing a heart as the first card. If you put the first card back, the deck returns to 52 cards, and there's a 13/52 chance again to draw another heart.

The probability of drawing two hearts from the deck with replacement is 6.25%.

When two events are independent, calculate the probability of both occurring by multiplying the probabilities of each occurring separately.

4.60 Voter records for a large county indicate that 46% of registered voters are Republicans. If three voters are selected randomly, determine the probability that all three are Republican.

Define the events.

$$A = \text{first voter is a Republican}$$
$$B = \text{second voter is a Republican}$$
$$C = \text{third voter is a Republican}$$

Pulling one voter out of a big crowd won't drop the overall percentage of Republicans very much, if at all.

It is acceptable to assume that the events are independent. Apply the correct multiplication rule.

$$P(A \text{ and } B \text{ and } C) = P(A)P(B)P(C)$$
$$= (0.46)(0.46)(0.46)$$
$$= 0.097$$

The probability of selecting three Republicans at random is 9.7%.

The one without conditional probability in it.

Note: Problems 4.61–4.63 refer to the semester grades of 20 students in an M.B.A. class: seven students earned an A, ten students earned a B, and three students earned a C.

4.61 If three students are selected (without replacement), determine the probability that all three students earned an A.

One of the 20 students is selected and not replaced.

Because of the small population size, the events are not independent. Each time you select a student, the population size decreases by 5 percent, which affects the probability of selecting subsequent A students. Thus, you cannot calculate the probability of selecting three students using the multiplication rule for independent events.

The answer is not $\left(\dfrac{7}{20}\right)\left(\dfrac{7}{20}\right)\left(\dfrac{7}{20}\right) = 4.3\%$

The probability that the first student selected is an A student is $\dfrac{7}{20}$. That leaves $20 - 1 = 19$ students in the class and $7 - 1 = 6$ students who earned an A. Thus, the probability of selecting a second A student is $\dfrac{6}{19}$. Similarly, there is a $\dfrac{5}{18}$ probability of selecting a third A student. To determine the probability of selecting three A students, multiply each of the probabilities.

$$\left(\frac{7}{20}\right)\left(\frac{6}{19}\right)\left(\frac{5}{18}\right) = \frac{210}{6,840} = 0.031$$

Note: Problems 4.61–4.63 refer to the semester grades of 20 students in an M.B.A. class: seven students earned an A, ten students earned a B, and three students earned a C.

4.62 If three students are selected (without replacement), determine the probability that none of the students earned an A.

Seven students earned an A, so $20 - 7 = 13$ students didn't.

The probability that the first student selected did not earn an A is $\dfrac{13}{20}$. The probability of selecting a second B or C student is $\dfrac{12}{19}$, and the probability of selecting a third is $\dfrac{11}{18}$. Multiply the three probabilities to determine how likely it is that you will randomly select three students who did not earn an A.

$$\left(\frac{13}{20}\right)\left(\frac{12}{19}\right)\left(\frac{11}{18}\right) = \frac{1,716}{6,840} = 0.251$$

Note: Problems 4.61–4.63 refer to the semester grades of 20 students in an M.B.A. class: seven students earned an A, ten students earned a B, and three students earned a C.

4.63 If three students are selected (without replacement), determine the probability that at least one of the students earned an A. ←

You could select one, two, or three students with an A to satisfy this problem.

Consider the complement of the event described here. The complement of selecting at least one A student is selecting zero A students. According to Problem 4.62, the probability that three randomly selected students have not earned an A is 0.251.

Thus, according to the complement rule, the probability of selecting at least one A student is $1 - 0.251 = 0.749$.

Note: Problems 4.64–4.66 refer to a statistic reporting that 68 percent of adult males in China smoke.

4.64 Calculate the probability that five randomly selected adult males from China are smokers.

Because of the large population from which you are drawing, selecting the five individuals can be considered independent events. Thus, each time a male adult is chosen, there is a 0.68 probability that he smokes.

$$P(\text{choosing five smokers}) = (0.68)(0.68)(0.68)(0.68)(0.68)$$
$$= (0.68)^5$$
$$= 0.145$$

Note: Problems 4.64–4.66 refer to a statistic reporting that 68 percent of adult males in China smoke.

4.65 Calculate the probability that five randomly selected adult males from China are nonsmokers.

As in Problem 4.64, you can assume that selecting each individual is an independent event. If the probability of selecting a smoker is 0.68, then the complement (the probability of selecting a nonsmoker) is $1 - 0.68 = 0.32$. Calculate the probability that all five randomly selected males are nonsmokers.

$$P(\text{choosing five nonsmokers}) = (0.32)(0.32)(0.32)(0.32)(0.32)$$
$$= (0.32)^5$$
$$= 0.0033$$

Note: Problems 4.64–4.66 refer to a statistic reporting that 68 percent of adult males in China smoke.

4.66 If five adult males from China are randomly selected, determine the probability that at least one of the five is a smoker.

The complement of randomly selecting at least one smoker is randomly selecting zero smokers. According to Problem 4.65, the probability of selecting five nonsmokers at random is 0.0033. Apply the complement rule to calculate the probability of selecting at least one smoker.

$$P(\text{at least one smoker}) = 1 - 0.0033 = 0.9967$$

> Thomas Bayes (1701–1761) was a mathematician and a published Presbyterian minister who used mathematics to study religion. Holy theorem!

Bayes' Theorem

Another way to calculate conditional probabilities

Note: Problems 4.67–4.68 refer to the data set below, the number of cars of various types at a local dealership.

	Sedan	SUV	Total
New	24	15	**39**
Used	9	12	**21**
Total	**33**	**27**	**60**

4.67 Use Bayes' Theorem to calculate the probability that a randomly selected car is new, given that it is a sedan. Verify the result by computing the conditional probability directly.

> There's also a much longer version of Bayes' Theorem:
> $$P(A|B) = \frac{P(A)\,P(B|A)}{P(A)P(B|A) + P(A')P(B|A')}$$
> See Problem 4.69.

Bayes' Theorem provides an alternative method of calculating conditional probability, according to the formula below.

$$P(A|B) = \frac{P(A)P(B|A)}{P(B)}$$

Apply Bayes' Theorem using the events A = new and B = sedan.

$$P(\text{new}|\text{sedan}) = \frac{P(\text{new})\,P(\text{sedan}|\text{new})}{P(\text{sedan})}$$

> These are just your basic, nonconditional probabilities. Some books call them prior probabilities in the context of Bayes' Theorem.

Calculate the marginal probabilities P(new) and P(sedan), as well as the conditional probability P(sedan | new).

> Some books call this a revised posterior probability.

$$P(\text{new}) = \frac{39}{60} \qquad P(\text{sedan}) = \frac{33}{60} \qquad P(\text{sedan}|\text{new}) = \frac{24}{39}$$
$$= 0.65 \qquad\qquad = 0.55 \qquad\qquad\qquad = 0.6154$$

Substitute these values into Bayes' Theorem.

$$P(\text{new}\,|\,\text{sedan}) = \frac{P(\text{new})P(\text{sedan}\,|\,\text{new})}{P(\text{sedan})}$$

$$= \frac{(0.65)(0.6154)}{0.55}$$

$$= 0.727$$

To verify the result, notice that there are a total of 33 sedans at the dealership, of which 24 are new. Thus, the probability of randomly selecting a new car from the collection of sedans is $P(\text{new}\,|\,\text{sedan}) = \dfrac{24}{33} = 0.727$.

Note: Problems 4.67–4.68 refer to the data set in Problem 4.67, the number of cars of various types at a local dealership.

4.68 Use Bayes' Theorem to calculate the probability that a randomly selected new car is a sedan.

Apply Bayes' Theorem using the events A = sedan and B = new.

$$P(A\,|\,B) = \frac{P(A)P(B\,|\,A)}{P(B)}$$

$$P(\text{sedan}\,|\,\text{new}) = \frac{P(\text{sedan})P(\text{new}\,|\,\text{sedan})}{P(\text{new})}$$

$$P(\text{sedan}\,|\,\text{new}) = \frac{(33\,/\,60)(24\,/\,33)}{39\,/\,60}$$

$$P(\text{sedan}\,|\,\text{new}) = \frac{(0.55)(0.7273)}{0.65}$$

$$P(\text{sedan}\,|\,\text{new}) = 0.615$$

4.69 A college graduate believes he has a 60% chance of getting a particular job. Historically, 75% of the candidates who got a similar job had two interviews; 45% of the unsuccessful candidates had two interviews. Apply Bayes' Theorem to calculate the probability that this candidate will be hired, assuming he had two interviews.

Define the events below.

$$H = \text{the candidate was hired}$$
$$SI = \text{the candidate had two interviews}$$

You are asked to calculate the probability of the candidate getting the job, assuming he has two interviews: $P(H\,|\,SI)$. You cannot apply the short version of Bayes' Theorem, because you do not know the value of $P(SI)$. Instead, you must apply the long version of the formula.

$$P(H|SI) = \frac{P(H)P(SI|H)}{P(H)P(SI|H) + P(H')P(SI|H')}$$

Notice that the formula contains H', the complement of H. If event H = "the candidate is hired," then H' = "the candidate is not hired." Recall that the candidate believes $P(H) = 0.60$. According to the complement rule, $P(H') = 1 - 0.60 = 0.40$.

$$P(H|SI) = \frac{P(H)P(SI|H)}{P(H)P(SI|H) + P(H')P(SI|H')}$$

$$= \frac{(0.60)(0.75)}{(0.60)(0.75) + (0.40)(0.45)}$$

$$= \frac{0.45}{0.45 + 0.18}$$

$$= \frac{0.45}{0.63}$$

$$= 0.714$$

The probability of being hired, given a second interview, is 71.4%.

Chapter 5
COUNTING PRINCIPLES AND PROBABILITY DISTRIBUTIONS

Odds you can count on

This chapter will build on the fundamental concepts of probability explored in Chapter 4, with an ultimate goal of analyzing probability distributions, a collection of discrete probabilities for an event. As a means to that end, the chapter introduces three additional probability concepts: the Fundamental Counting Principle, permutations, and combinations.

If you've worked through Chapter 4 (and if you haven't, you should do that before starting this chapter), you can find the probability of choosing certain cards from a standard deck (a diamond, a queen, or a red card, for example). In this chapter, you'll deal with slightly more complicated experiments that require you to understand permutations and combinations. The chapter ends with probability distributions, which are essentially collections of probabilities.

Fundamental Counting Principle

How probable is it that two separate events occur?

5.1 Define the Fundamental Counting Principle and provide an example.

According to the Fundamental Counting Principle (FCP), if event A can occur m possible ways and event B can occur n possible ways, there are mn different ways both events can occur. For example, if an ice-cream store offers nine different flavors and three different sizes, there are $9(3) = 27$ possible combinations of flavors and sizes.

5.2 The menu of a particular restaurant lists three appetizers, eight entrées, four desserts, and three drinks. Assuming a meal consists of one appetizer, one entrée, one dessert, and one drink, how many different meals can be ordered?

Multiply the ways each component of the meal can be ordered to calculate the number of possible meal combinations.

$$\text{Possible meals} = (\text{appetizer choices})(\text{entrée choices})(\text{dessert choices})(\text{drink choices})$$
$$= (3)(8)(4)(3)$$
$$= 288$$

> The Fundamental Counting Principle can be used with more than just two events (like the m and n example in Problem 5.1). In this problem, there are four events you end up multiplying together.

5.3 A particular state license plate contains three letters (A–Z) followed by four digits (1–9). To avoid the possibility of mistaking one for the other, the number zero and the letter O are not used. How many unique license plates can be created?

A license plate contains seven characters. There are 25 choices for the first three characters and nine choices for the last four characters. In the diagram below, each character of the license plate is accompanied by the possible ways that character can be chosen.

Letter	Letter	Letter	Digit	Digit	Digit	Digit
25	25	25	9	9	9	9

According to the Fundamental Counting Principle, the number of possible license plates is equal to the product of the possible ways each character can be chosen.

$$\text{number of possible license plates} = (25)(25)(25)(9)(9)(9)(9)$$
$$= (25)^3 (9)^4$$
$$= 102,515,625$$

> There are 26 letters of the alphabet and 10 digits (0–9), but one letter and one number are not eligible to appear.

5.4 If a specific area code has eight three-digit exchanges, how many seven-digit phone numbers are available in that area code?

Each phone number contains one exchange and four digits. There are eight possible exchanges and ten choices (0–9) for each digit.

Exchange	Digit 1	Digit 2	Digit 3	Digit 4
8	10	10	10	10

There are (8)(10)(10)(10)(10) = 80,000 possible phone numbers.

> A typical phone number is AAA-EEE-XXXX, where AAA is the three-digit area code, EEE is the three-digit exchange, and XXXX are the numbers assigned to those exchanges.

5.5 The starting five players of a basketball team are announced one by one at the beginning of a game. Calculate the total number of different ways the order of players can be announced.

In Problems 5.3 and 5.4, it was acceptable to repeat a choice. For instance, a letter can repeat in a license plate and a phone number can contain two of the same digit. However, in this problem, repetition is not allowed.

> You don't want to announce the same player twice—you want to announce each of the five players' names once.

There are five different positions in which the starters' names are announced (the first player introduced would be assigned to position one). Once a player has been introduced, that player cannot be assigned another position in that particular sequence. This is known as selection without replacement.

Therefore, there are five players to choose from for the first position, four players for the second position, three for the third position, two for the fourth position, and only one for the final position.

First player announced	Second player announced	Third player announced	Fourth player announced	Fifth player announced
5	4	3	2	1

There are 5! = (5)(4)(3)(2)(1) = 120 different ways to announce the five starting players' names.

> This is 5! (read "five factorial"). To calculate a factorial, multiply the number by every integer less than that number, all the way down to one. For example, 7! = (7)(6)(5)(4)(3)(2)(1) = 5,040.

5.6 Calculate the total number of ways eight people can be seated at a table that has eight seats.

If an event fills n positions with n different choices without replacement, then the total number of ways the event can be completed is $n!$. In this problem, eight people are placed into eight seats without repetition, so there are 8! possible seating arrangements.

$$8! = (8)(7)(6)(5)(4)(3)(2)(1) = 40,320$$

> A person can be in only one seat at a time.

5.7 A multiple-choice test consists of ten questions, each with four choices. Calculate the probability that a student who randomly guesses the answer to each question will get all of the questions correct.

Let's say the four choices for each question are A, B, C, and D. If you pick A for question one, that doesn't mean you're not allowed to pick A again. (Otherwise, you'd run out of choices by the fifth question.)

There are four ways to choose the answer for each of the ten questions. Note that the factorial method described in Problems 5.5 and 5.6 is not used, because this is an example of selection with replacement. Apply the Fundamental Counting Principle to calculate the total number of ways the student could complete the test.

$$(4)(4)(4)(4)(4)(4)(4)(4)(4)(4) = 4^{10} = 1,048,576$$

There is only one correct sequence of answers; divide that one correct sequence by the number of possible sequences.

$$\frac{1}{1,048,576} = 0.00000095$$

There is a 0.000095% chance that the student will randomly choose the correct answer for all ten questions.

A tree diagram for a potential family tree

5.8 A couple wishes to have three children and wants to determine how the children could potentially be born in terms of gender and birth order. Calculate the number of possible ways the children could be born and illustrate your answer using a tree diagram.

The couple wishes to have three children. There are two possible genders for each child in the birth order. Apply the Fundamental Counting Principle.

First child	Second child	Third child
2	2	2

There are $(2)(2)(2) = 2^3 = 8$ ways in which the children could be born. In the diagram below, each branch represents one possible outcome. Beginning at the left, the path divides at the birth of each child, branching upward for a boy and downward for a girl.

That may sound sexist, but don't read anything into it. This is just a silly chart, not a proclamation that having three girls would be the worst possible outcome (because it's at the bottom of the list).

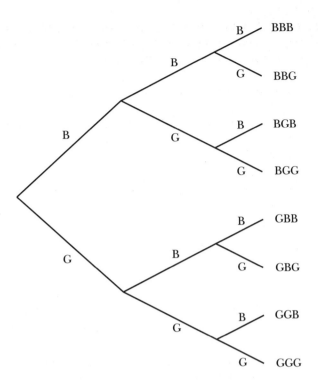

There are eight possible ways the children could be born (BBB, BBG, BGB, BGG, GBB, GBG, GGB, and GGG)

Permutations

How many ways can you arrange a collection of things?

5.9 Define a permutation and provide an example.

Combinatorics defines a permutation as a sequence of objects in which is order is a defining factor. For instance, if you are given the set {A, B, C} and are asked to identify unique permutations, choosing two elements at a time, *AB* and *BA* are considered unique permutations. Although both contain the same two elements, the order in which the elements appear distinguishes them.

You are commonly asked to calculate the number of permutations that exist for a set containing *n* elements if you choose *r* of them at a time. In the example above, you choose $r = 2$ of the $n = 3$ letters. The number of possible permutations is defined as $_nP_r$ or $P(n,r)$ and is calculated using the formula below.

$$_nP_r = \frac{n!}{(n-r)!}$$

To calculate the total number of ways *A*, *B*, and *C* can be arranged in order, two at a time, evaluate $_3P_2$.

> A theoretical mathematics discipline that (for the purposes of probability and statistics) studies the ways a set of objects can be grouped

$$_3P_2 = \frac{3!}{(3-2)!} = \frac{3!}{1!} = \frac{3 \cdot 2 \cdot 1}{1} = \frac{6}{1} = 6$$

The six possible permutations are *AB*, *AC*, *BA*, *BC*, *CA*, and *CB*.

5.10 If a salesperson is responsible for nine stores, how many different ways can she schedule visits with five stores this week?

You are asked to calculate the number of ways you can permute nine objects, choosing five of them at a time. Calculate $_9P_5$.

$$_nP_r = \frac{n!}{(n-r)!}$$
$$_9P_5 = \frac{9!}{(9-5)!}$$
$$_9P_5 = \frac{9!}{4!}$$

> The order is important, as the order she selects the stores dictates the order in which she'll visit them.

To reduce the fraction, expand the factorial in the numerator in order to eliminate the denominator.

$$_9P_5 = \frac{9 \cdot 8 \cdot 7 \cdot 6 \cdot 5 \cdot 4!}{4!}$$
$$= 9 \cdot 8 \cdot 7 \cdot 6 \cdot 5$$
$$= 15,120$$

> Write 9! as (9)(8)(7)(6)(5)(4!). It still has the same value—after all, you could write 4! as (4)(3)(2)(1)—but this way you can cancel out 4! in the numerator and denominator.

There are 15,120 different ways in which the salesperson can schedule visits to five of the nine stores this week.

5.11 Calculate the number of ways eight swimmers can place first, second, or third in a race.

A result in which swimmers *A*, *B*, and *C* take first, second, and third place, respectively, is considered different from a result in which the same swimmers finish in a different order. Hence, a permutation should be calculated.

$$_8P_3 = \frac{8!}{(8-3)!} = \frac{8!}{5!} = \frac{8 \cdot 7 \cdot 6 \cdot 5!}{5!} = 8 \cdot 7 \cdot 6 = 336$$

> Wondering why you should care that the order of the swimmers matters? If the order didn't matter, you'd use the combination $_8C_3$ instead. See Problems 5.14–5.27.

There are 336 different ways in which eight swimmers can finish first, second, or third.

5.12 A combination lock has a total of 40 numbers on its face and will unlock given the proper three-number sequence. How many unique combinations are possible if no numbers are repeated?

Even though this problem uses the term "combination," it refers to the number sequence used to gain access to the lock, not a mathematical combination. The order of the three numbers in the unlocking sequence are important, so calculate $_{40}P_3$

$$_{40}P_3 = \frac{40!}{(40-3)!} = \frac{40!}{37!} = \frac{40 \cdot 39 \cdot 38 \cdot 37!}{37!} = 40 \cdot 39 \cdot 38 = 59,280$$

There are 59,280 unique combinations.

> Problems 5.14–5.27 deal with combinations—not the lock kind, the mathematical kind.

> A lock with the combination 10-20-30 will not open if you use the combination 30-20-10. You have to know the correct numbers and put them in the correct order.

5.13 Calculate the number of ways a poker player can arrange her five-card hand.

Note that you are *not* asked to calculate the number of possible poker hands (which is equal to the combination $_{52}C_5$). Instead, the problem asks you to calculate the number of ways all five of the cards can be permuted. Calculate $_nP_r$ given $n = r = 5$.

$$_5P_5 = \frac{5!}{(5-5)!} = \frac{5!}{0!} = \frac{5!}{1} = 5 \cdot 4 \cdot 3 \cdot 2 \cdot 1 = 120$$

> $0! = 1$

When all of the objects in a set are permuted (when $n = r$), the number of possible permutations is $n!$

$$_nP_n = \frac{n!}{(n-n)!} = \frac{n!}{0!} = \frac{n!}{1} = n!$$

This problem is very similar to Problem 5.6, in which you are asked to calculate the number of ways eight people can be seated at an eight-person table. In both problems, each time an item (a card or dinner guest) is assigned a position, the number of possibilities for the next position is reduced by one.

Card 1	Card 2	Card 3	Card 4	Card 5
5	4	3	2	1

Combinations
When the order of objects is not important

5.14 Contrast combinations and permutations and give an example of the former.

Combinations are similar to permutations, in terms of their role in combinatorics and their notation. Both calculate the number of ways elements from a group can be selected, but combinations do not differentiate between sequences that contain the same elements.

For instance, if you are asked to calculate the number of ways you can select two letters from the set {A, B, and C}, sequences BC and CB are considered the same combination. They contain the same letters, and the order of the letters in the sequence does not matter.

A combination of n elements, choosing r at a time, is written $_nC_r$ and is calculated according to the formula below.

$$_nC_r = \frac{n!}{(n-r)!r!}$$

To calculate the number of ways you can choose two letters from the set {A, B, C}, evaluate $_3C_2$.

> You can read this "n choose r." You can also use the notation $\binom{n}{r}$ instead of $_nC_r$.

$$_nC_r = \frac{n!}{(n-r)!r!}$$
$$_3C_2 = \frac{3!}{(3-2)!2!}$$
$$_3C_2 = \frac{3!}{2!}$$
$$_3C_2 = \frac{3\cdot2\cdot1}{2\cdot1}$$
$$_3C_2 = \frac{6}{2}$$
$$_3C_2 = 3$$

The three combinations are AB, AC, and BC.

5.15 A young woman bought seven books to read on vacation but only has time to read three of them. How many ways can she choose three of the seven books to bring with her?

You are asked to identify unique combinations of books; the order in which you choose them is irrelevant. Evaluate $_7C_3$.

> Choosing books 1, 3, and 7 is no different from choosing books 7, 3, and 1. Calculate $_7C_3$ instead of $_7P_3$, because order doesn't matter.

$$_nC_r = \frac{n!}{(n-r)!r!}$$
$$_7C_3 = \frac{7!}{(7-3)!3!}$$
$$_7C_3 = \frac{7!}{4!3!}$$
$$_7C_3 = \frac{7\cdot6\cdot5\cdot\cancel{4!}}{\cancel{4!}\cdot3!}$$
$$_7C_3 = \frac{7\cdot6\cdot5}{3\cdot2\cdot1}$$
$$_7C_3 = \frac{210}{6}$$
$$_7C_3 = 35$$

There are 35 unique ways to choose three out of seven books.

5.16 Calculate the number of unique four-person groups that can be formed by selecting from eight eligible candidates.

The order in which people are selected is not important. Evaluate $_8C_4$.

$$_8C_4 = \frac{8!}{(8-4)!4!} = \frac{8!}{4!4!} = \frac{8 \cdot 7 \cdot 6 \cdot 5 \cdot 4\!\!\!/}{4\!\!\!/ \cdot 4!} = \frac{1,680}{4 \cdot 3 \cdot 2 \cdot 1} = \frac{1,680}{24} = 70$$

> If you used a permutation, two groups with the same members would be considered different, and the problem asks for unique groups.

5.17 How many ways can you organize a class of 25 students into groups of five?

There is no indication that the order in which students are selected has any bearing on their role in the group, so evaluate $_{25}C_5$.

$$_{25}C_5 = \frac{25!}{(25-5)!5!} = \frac{25!}{20!5!} = \frac{25 \cdot 24 \cdot 23 \cdot 22 \cdot 21 \cdot 20\!\!\!/}{20\!\!\!/ \cdot 5!} = \frac{6,375,600}{5 \cdot 4 \cdot 3 \cdot 2 \cdot 1} = \frac{6,375,600}{120} = 53,130$$

5.18 An executive needs to select 3 stores from a total of 11 to participate in a customer service program. How many ways are there to select 3 of the 11 stores?

Evaluate $_{11}C_3$.

$$_{11}C_3 = \frac{11!}{(11-3)!3!} = \frac{11!}{8!3!} = \frac{11 \cdot 10 \cdot 9 \cdot 8\!\!\!/}{8\!\!\!/ \cdot 3!} = \frac{990}{3 \cdot 2} = \frac{990}{6} = 165$$

> By the way, you can't multiply the two factorials x!y! and get (xy)! The product 8!3! does not equal ~~11!~~ 24!

5.19 In a 6/49 state lottery, a participant picks 6 numbers from a field of 49 choices. Calculate the odds of selecting the correct combination of numbers.

The order in which the numbers are selected is not important in a 6/49 lottery. There are $_{49}C_6$ unique six-number combinations.

$$_{49}C_6 = \frac{49!}{(49-6)!6!} = \frac{49!}{43!6!} = \frac{49 \cdot 48 \cdot 47 \cdot 46 \cdot 45 \cdot 44 \cdot 43\!\!\!/}{43\!\!\!/ \cdot 6!} = \frac{10,068,347,520}{720} = 13,983,816$$

Only one of the 13,983,816 combinations wins the lottery, so the probability of winning is $\frac{1}{13,983,816} = 0.000000072$.

5.20 Determine the number of ways a jury of 6 men and 6 women can be chosen from an eligible pool of 12 men and 14 women.

First, calculate the number of unique ways 6 men can be selected from a group of 12.

$$_{12}C_6 = \frac{12!}{(12-6)!6!} = \frac{12!}{6!6!} = \frac{12 \cdot 11 \cdot 10 \cdot 9 \cdot 8 \cdot 7 \cdot 6\!\!\!/}{6\!\!\!/ \cdot 6!} = \frac{665,280}{6 \cdot 5 \cdot 4 \cdot 3 \cdot 2 \cdot 1} = \frac{665,280}{720} = 924$$

Second, calculate the number of unique ways 6 women can be selected from a group of 14.

$$_{14}C_6 = \frac{14!}{(14-6)!6!} = \frac{14!}{8!6!} = \frac{14 \cdot 13 \cdot 12 \cdot 11 \cdot 10 \cdot 9 \cdot 8!}{8! \cdot 6!} = \frac{2,162,160}{6 \cdot 5 \cdot 4 \cdot 3 \cdot 2 \cdot 1} = \frac{2,162,160}{720} = 3,003$$

Multiply the probabilities of each event occurring separately to calculate the probability of both events occurring together.

$$\left(_{12}C_6\right)\left(_{14}C_6\right) = (924)(3,003) = 2,774,772$$

There are 2,774,772 ways the jury can be chosen.

Note: Problems 5.21–5.22 refer to a jar that contains four blue marbles and six yellow marbles. In each problem, three marbles are randomly selected.

5.21 Calculate the probability of selecting exactly two blue marbles (without replacement).

> They are replaced at the end of the problem, though. When you start Problem 5.22, you can assume that all ten marbles are back in the jar for your marble-selecting pleasure.

If exactly two marbles are blue, then one marble must be yellow. Calculate the number of ways you can choose two of the four blue marbles in the jar.

$$_4C_2 = \frac{4!}{(4-2)!2!} = \frac{4!}{2!2!} = \frac{4 \cdot 3 \cdot 2!}{2! \cdot 2!} = \frac{12}{2} = 6$$

Calculate the number of ways you can choose one of six yellow marbles.

$$_6C_1 = \frac{6!}{(6-1)!1!} = \frac{6!}{5!1!} = \frac{6 \cdot 5!}{5! \cdot 1!} = \frac{6}{1} = 6$$

Apply the Fundamental Counting Principle to calculate the number of ways to draw two blue marbles and one yellow marble.

$$\left(_4C_2\right)\left(_6C_1\right) = (6)(6) = 36$$

The jar contains a total of ten marbles. Calculate the number of ways three marbles can be chosen (regardless of color).

> There are 36 ways to draw the marbles the way the problem describes and 120 ways to randomly draw 3 out of 10 marbles.

$$_{10}C_3 = \frac{10!}{(10-3)!3!} = \frac{10!}{7!3!} = \frac{10 \cdot 9 \cdot 8 \cdot 7!}{7! \cdot 3!} = \frac{720}{3 \cdot 2} = \frac{720}{6} = 120$$

There are 120 ways to choose three marbles. The probability of choosing exactly two blue marbles is $\frac{36}{120} = 0.30$.

Note: Problems 5.21–5.22 refer to a jar that contains four blue marbles and six yellow marbles. In each problem, three marbles are randomly selected.

5.22 Calculate the probability that at least two marbles are blue.

If three marbles are selected and at least two of them are blue, then either two or three blue marbles were drawn. According to Problem 5.21, the probability that exactly two blue marbles are drawn is 0.30.

Calculate the number of ways to choose three of four blue marbles.

$$_4C_3 = \frac{4!}{(4-3)!3!} = \frac{4!}{1!3!} = \frac{4 \cdot 3 \cdot 2 \cdot 1}{(1)(3 \cdot 2 \cdot 1)} = \frac{24}{6} = 4$$

Calculate the number of ways you can select zero of six yellow marbles.

$$_6C_0 = \frac{6!}{(6-0)!0!} = \frac{6!}{6!0!} = \frac{6!}{6!(1)} = 1$$

Apply the Fundamental Counting Principle to calculate the number of ways to select three blue and zero yellow marbles.

$$(_4C_3)(_6C_0) = (4)(1) = 4$$

Recall that there are 120 ways to select three of ten marbles. Thus, the probability of selecting three blue marbles is $\frac{4}{120} = 0.033$. To calculate the probability that *at least* two marbles are blue, add the probabilities that exactly two are blue and exactly three are blue.

$$P(\text{at least two blue}) = P(\text{two are blue}) + P(\text{three are blue})$$
$$= 0.30 + 0.033$$
$$= 0.333$$

There is a 33.3 percent chance of selecting at least two blue marbles (without replacement).

Note: In Problems 5.23–5.27, five cards are randomly selected from a standard 52-card deck.

5.23 Calculate the number of unique five-card poker hands.

The value of a poker hand is based on the suits and ranks of the cards, not the order in which the cards are received. Thus, the total number of possible hands is $_{52}C_5$.

$$_{52}C_5 = \frac{52!}{(52-5)!5!} = \frac{52!}{47!5!} = \frac{52 \cdot 51 \cdot 50 \cdot 49 \cdot 48 \cdot \cancel{47!}}{\cancel{47!} \cdot 5!} = \frac{311,875,200}{5 \cdot 4 \cdot 3 \cdot 2 \cdot 1} = \frac{311,875,200}{120} = 2,598,960$$

Note: In Problems 5.23–5.27, five cards are randomly selected from a standard 52-card deck.

5.24 Calculate the probability that a player is dealt a royal flush.

A royal flush is the highest-ranking poker hand, consisting of the 10, jack, queen, king, and ace of one suit. There are only four ways to receive a royal flush, one from each suit in the deck. Recall that there are 2,598,960 possible poker hands. Thus, the probability of being dealt a royal flush is $\dfrac{4}{2,598,960} = 0.00000154$.

Technically, a royal flush is a straight, a flush, and a straight flush. This problem wants to focus on straight flushes that aren't actually better hands in disguise.

Note: In Problems 5.23–5.27, five cards are randomly selected from a standard 52-card deck.

5.25 Determine the probability that a player is dealt a straight flush, but not a royal flush.

A straight flush consists of five cards of the same suit with consecutive ranks. For instance, the 5, 6, 7, 8, and 9 of spades constitute a straight flush. The highest card of a straight could be a 5, 6, 7, 8, 9, 10, jack, queen, or king. (The highest value could not be 4 or lower because K-A-2-3-4, termed by some a *wrap-around straight*, is not a straight according to the rules of poker.)

A-2-3-4-5 is a legitimate straight, called the wheel or bicycle straight. It's sort of odd, because you're treating the ace as a one, the lowest-ranking card instead of the highest-ranking card.

Notice that the ace is omitted as a possible straight flush high card. This is because an ace-high straight flush is a royal flush (as defined in Problem 5.24) and is disregarded in this problem. Therefore, there are a total of nine card combinations in a single suit that are straight flushes.

If each suit can form nine straight flushes, then a total of 4(9) = 36 straight flushes can be made from all four suits. Recall that there are 2,598,960 possible poker hands. Thus, the probability of being dealt a straight flush is $\dfrac{36}{2,598,960} = 0.0000139$.

Note: In Problems 5.23–5.27, five cards are randomly selected from a standard 52-card deck.

5.26 Calculate the probability that a player is dealt three of a kind.

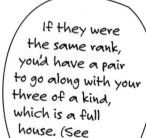

If a hand contains three poker cards of the same rank (and the remaining two cards are not a pair), the hand is classified as a three of a kind. There are 13 different ranks in a standard deck, so there are 13 ways to choose the rank that will appear three times in the hand. For the sake of illustration, assume the hand is 4-7-K-K-K.

A standard deck contains four cards of every rank. The poker hand 4-7-K-K-K contains three of the four kings. Calculate the number of ways three of the kings can be chosen.

$$_4C_3 = \frac{4!}{1!3!} = \frac{4 \cdot \cancel{3!}}{1 \cdot \cancel{3!}} = 4$$

If they were the same rank, you'd have a pair to go along with your three of a kind, which is a full house. (See Problem 5.27.)

The two remaining cards cannot be kings. Neither can they be the same rank. Thus, they must consist of two of the remaining twelve ranks. In the hand 4-7-K-K-K, the ranks are 4 and 7. Calculate the number of ways you can choose the two remaining ranks.

$$_{12}C_2 = \frac{12!}{(12-2)!2!} = \frac{12!}{10!2!} = \frac{12 \cdot 11 \cdot \cancel{10!}}{\cancel{10!} \cdot 2!} = \frac{132}{2} = 66$$

There are four ways to choose the suits of the two additional cards, one for each suit. For instance, there are four 4s and four 7s in the deck to complete the hand 4-7-K-K-K. Thus, these cards can be drawn (4)(4) = 16 different ways.

In order to form three of a kind, there are 13 possible ranks from which to pick the recurring card, four ways to select three of those cards, 66 ways to select the ranks of the two remaining cards, and 16 ways to select from among them. Thus, there are (13)(4)(66)(16) = 54,912 unique ways to accomplish the task.

$$P(\text{three of a kind}) = \frac{54,912}{2,598,960} = 0.0211$$

Note: In Problems 5.23–5.27, five cards are randomly selected from a standard 52-card deck.

5.27 Determine the probability that a player is dealt a full house.

A full house contains a pair and three of a kind. There are 13 ways to choose the card rank for the three of a kind and $_4C_3 = 4$ ways to select three of the four cards from that rank to place in the hand. There are 12 ways to select the card rank for the pair and $_4C_2 = 6$ ways to select two cards of that rank.

Therefore, there are (13)(4)(12)(6) = 3,744 ways to create a full house from a standard deck, and the probability of being dealt a full house is

$$\frac{3,744}{2,598,960} = 0.0014.$$

> There are 13 ranks of cards and you already used one for the three of a kind.

Probability Distributions

Probability using discrete data

5.28 Define the term *random variable*. Discuss the types of random variables that are used in statistics.

A random variable is an outcome that takes on a numerical value as a result of an experiment. The value is not known with certainty before the experiment. An example of a random variable would be tomorrow's high temperature. The experiment would be to measure and record the temperatures throughout the day and identify the maximum value.

There are two types of random variables. Continuous random variables are measured on a continuous number scale. Weight, for instance, is a continuous random variable, because it can be measured very precisely. An individual's weight may be 180 pounds one day and then 180.5 pounds the next.

Discrete random variables tally outcomes rather than measure them. Thus, their values are usually integers. A golf score, for instance, counts a golfer's total number of strokes and is therefore a discrete random variable. A golf score is an integer, because there are no fractional or partial strokes.

Note: Problems 5.29–5.31 refer to the following data set, the probability distribution for the number of students absent from a statistics class.

Students, x	Probability, P(x)
1	0.12
2	0.15
3	0.18
4	0.15
5	0.26
6	0.14
Total	**1.00**

5.29 Calculate the mean for this probability distribution.

The number of students absent is a discrete random variable because it is counted, or tallied, and is therefore an integer.

To compute the mean of a discrete probability distribution, multiply each event by its probability and then add the products.

$$\mu = \sum x \cdot P(x)$$
$$= (1)(0.12) + (2)(0.15) + (3)(0.18) + (4)(0.15) + (5)(0.26) + (6)(0.14)$$
$$= 0.12 + 0.30 + 0.48 + 0.60 + 1.05 + 0.84$$
$$= 3.7$$

An average of 3.7 students are absent per day.

Note: Problems 5.29–5.31 refer to the data set in Problem 5.29, the probability distribution for the number of students absent from a statistics class.

5.30 Calculate the variance and standard deviation of the data using the standard method.

Check out Problems 3.52–3.55, calculating the variance of grouped data. It's a very similar process.

The variance of a discrete probability distribution, according to the standard method, is $\sigma^2 = \sum (x - \mu)^2 \cdot P(x)$. Consider the table below. Columns A and B represent the probability distribution from the problem. Column C is the difference of column A and the mean $\mu = 3.7$, calculated in Problem 5.29. Column D is the square of column C. Column E is the product of columns B and D.

A	B	C	D	E
x	$P(x)$	$x - \mu$	$(x - \mu)^2$	$(x - \mu)^2 \cdot P(x)$
1	0.12	1 − 3.7 = −2.7	7.29	0.8748
2	0.15	2 − 3.7 = 1.7	2.89	0.4335
3	0.18	3 − 3.7 = −0.7	0.49	0.0882

4	0.15	4 − 3.7 = 0.3	0.09	0.0135
5	0.26	5 − 3.7 = 1.3	1.69	0.4394
6	0.14	6 − 3.7 = 2.3	5.29	0.7406
Total				**2.59**

$$\sigma^2 = \sum (x - \mu)^2 \cdot P(x) = 2.59$$

The variance of the distribution is the sum of Column E.

The standard deviation is the square root of the variance: $\sigma = \sqrt{2.59} = 1.61$.

Note: Problems 5.29–5.31 refer to the data set in Problem 5.29, the probability distribution for the number of students absent from a statistics class.

5.31 Verify the variance and standard deviation computed in Problem 5.30, using the shortcut method.

If you have the option, choose the shortcut method.

The shortcut method to calculate the variance of a discrete probability distribution is $\sigma^2 = \left[\sum x^2 \cdot P(x) \right] - \mu^2$. Use the table below to compute the bracketed expression in the formula.

A	B	C	D
x	*P(x)*	*x²*	*x² · P(x)*
1	0.12	1	0.12
2	0.15	4	0.60
3	0.18	9	1.62
4	0.15	16	2.40
5	0.26	25	6.50
6	0.14	36	5.04
Total			**16.28**

$$\sigma^2 = \left(\sum x^2 \cdot P[x] \right) - \mu^2$$
$$= 16.28 - (3.7)^2$$
$$= 16.28 - 13.69$$
$$= 2.59$$

And there was one less column of calculations using the shortcut method.

The standard deviation of this discrete probability distribution is the square root of the variance: $\sigma = \sqrt{2.59} = 1.61$. The variance and standard deviation values echo the values calculated using the standard method in Problem 5.30.

Note: Problems 5.32–5.34 refer to the data set below, the probability distribution for a survey conducted to determine the number of bedrooms in the respondent's household.

Number of Bedrooms	Probability
3	0.25
4	0.55
5	0.15
6	0.05
Total	**1.00**

5.32 Calculate the mean of this probability distribution.

Multiply each event by its probability and add the resulting products.

$$\mu = \sum x \cdot P(x)$$
$$= (3)(0.25) + (4)(0.55) + (5)(0.15) + (6)(0.05)$$
$$= 0.75 + 2.20 + 0.75 + 0.30$$
$$= 4.0$$

Note: Problems 5.32–5.34 refer to the data set in Problem 5.32, the probability distribution for a survey conducted to determine the number of bedrooms in the respondent's household.

5.33 Calculate the variance and standard deviation for this probability distribution, using the standard method.

Consider the table below.

x	$P(x)$	$x - \mu$	$(x - \mu)^2$	$(x - \mu)^2 \cdot P(x)$
3	0.25	$3 - 4 = -1.0$	1.0	0.25
4	0.55	$4 - 4 = 0$	0	0
5	0.15	$5 - 4 = 1.0$	1.0	0.15
6	0.05	$6 - 4 = 2.0$	4.0	0.20
Total				**0.60**

Compute the variance.

$$\sigma^2 = \sum (x - \mu)^2 \cdot P(x) = 0.60$$

The standard deviation of the probability distribution is the square root of the variance: $\sigma = \sqrt{0.60} = 0.775$.

Note: Problems 5.32–5.34 refer to the data set in Problem 5.32, the probability distribution for a survey conducted to determine the number of bedrooms in the respondent's household.

5.34 Verify the variance and standard deviation computed in Problem 5.33 using the shortcut method.

Recall that the shortcut method to calculate the variance of a discrete probability distribution is $\sigma^2 = \left[\sum x^2 \cdot P(x)\right] - \mu^2$. Use the table below to compute the bracketed expression.

x	$P(x)$	x^2	$x^2 \cdot P(x)$
3	0.25	9	2.25
4	0.55	16	8.80
5	0.15	25	3.75
6	0.05	36	1.80
Total			**16.6**

According to Problem 5.32, $\mu = 4.0$.

$$\sigma^2 = \left[\sum x^2 \cdot P(x)\right] - \mu^2$$
$$= 16.6 - (4.0)^2$$
$$= 16.6 - 16.00$$
$$= 0.60$$

The variance is 0.60, the same value calculated in Problem 5.33, so the standard deviation will be the same as well: $\sigma = 0.775$.

Chapter 6
DISCRETE PROBABILITY DISTRIBUTIONS

Binomial, Poisson, and hypergeometric

Chapter 5 introduced probability distributions of discrete random variables, which list the probabilities of discrete integer values, usually based on a tally. This chapter further investigates theoretical probability distributions, including binomial, Poisson, and hypergeometric distributions.

At the end of Chapter 5, general probability distributions were introduced, and (assuming you completed Chapter 5 already) you calculated the mean, variance, and standard deviation of those distributions. In this chapter, you'll be generating the probability values of the distribution yourself.

Binomial Probability Distribution
Using coefficients that are combinations

6.1   Define the characteristics of a binomial experiment and provide an example.

A binomial experiment has the following characteristics:

- The experiment consists of a fixed number of trials, n.
- Each trial has only two possible outcomes (for example, success or failure).
- The probabilities of both outcomes are constant throughout the experiment.
- Each trial in the experiment is independent.

Flipping a coin five times and recording the number of heads is one example of a binomial experiment. The number of trials is fixed ($n = 5$), the result of each coin flip is either heads or tails, the coin is just as likely to land heads on every toss, and each flip of the coin is unaffected by the other coin tosses.

Success does not necessarily imply a positive outcome, nor does failure imply a negative outcome. Success just means you met the defined objective. In the coin-flipping example below, success = heads.

Note: In Problems 6.2–6.5, a fair coin is flipped six times and the number of heads is counted.

6.2   Calculate the probability of exactly two heads.

Given n = the number of trials, r = the number of successes, p = the probability of a success, and q = the probability of a failure, the binomial probability distribution states that the probability of r successes in n trials is

$$\binom{n}{r} p^n q^{n-r} = \frac{n!}{(n-r)!\,r!} p^r q^{n-r}.$$ Note that the notation $\binom{n}{r}$ is used to represent $_nC_r$, the combination of n things, choosing r at a time.

In this problem, the coin is tossed $n = 6$ times, and you are asked to determine the probability of $r = 2$ heads. A fair coin is just as likely to land heads as it is to land tails, so $p = q = 0.5$. Apply the binomial probability formula.

It's based on the binomial theorem, an algebraic formula that calculates the coefficients for the terms of $(x + y)^n$ when the polynomial is expanded.

$$P(r) = \binom{n}{r} p^r q^{n-r}$$

$$P(2) = \binom{6}{2}(0.5)^2 (0.5)^{6-2}$$

$$P(2) = (15)(0.5)^2 (0.5)^4$$

$$P(2) = 0.2344$$

There is a 23.4% probability that the coin will land heads exactly twice.

There are only two possible outcomes—heads or tails—and the outcomes are equally likely. There's a 50-50 chance the coin lands heads, so p and q equal 50% (which is 0.5 in decimal form).

This is equal to $_6C_2$. The book assumes you've worked through Problems 5.14–5.27, which explain how to calculate combinations. Make sure you know how to turn $\binom{6}{2}$ into 15.

Note: In Problems 6.2–6.5, a fair coin is flipped six times and the number of heads is counted.

6.3 Calculate the probability that the coin will land heads fewer than three times.

The probability of fewer than three heads can be expressed as $P(0 \text{ or } 1 \text{ or } 2 \text{ heads})$. Apply the addition rule for mutually exclusive events.

$$P(0 \text{ or } 1 \text{ or } 2 \text{ heads}) = P(0) + P(1) + P(2)$$

Mutually exclusive events cannot occur at the same time. After flipping a coin six times, you can't have exactly one head and exactly two heads.

Substitute $n = 6$, $r = 0$, $p = 0.5$, and $q = 0.5$ into the binomial probability formula to calculate $P(0)$.

$$P(r) = \binom{n}{r} p^r q^{n-r}$$

$$P(0) = \binom{6}{0}(0.5)^0 (0.5)^{6-0}$$

$$P(0) = (1)(1)(0.5)^6$$

$$P(0) = 0.0156$$

Any nonzero value raised to the zero power is equal to one. Therefore, $(0.5)^0 = 1$.

Use the same method to determine the probabilities of $r = 1$ and $r = 2$.

$$P(1) = \binom{6}{1}(0.5)^1 (0.5)^{6-1} \qquad P(2) = \binom{6}{2}(0.5)^2 (0.5)^{6-2}$$

$$= (6)(0.5)^1 (0.5)^5 \qquad\qquad = (15)(0.5)^2 (0.5)^4$$

$$= 0.0938 \qquad\qquad\qquad = 0.2344$$

If you use more decimal places, you might get a slightly different answer.

The probability of landing heads fewer than three times is the sum of the probabilities of landing zero, one, or two heads.

$$P(0 \text{ or } 1 \text{ or } 2 \text{ heads}) = P(0) + P(1) + P(2)$$

$$= 0.0156 + 0.0938 + 0.2344$$

$$= 0.344$$

You flip the coin six times, so six heads is the most you can get.

6.4 Calculate the probability that the coin will land heads more than four times.

The probability of more than four heads can be expressed as $P(5 \text{ or } 6 \text{ heads})$. The events are mutually exclusive, so $P(5 \text{ or } 6 \text{ heads}) = P(5) + P(6)$. Calculate each of the probabilities individually.

$$P(5) = \binom{6}{5}(0.5)^5(0.5)^{6-5} \qquad P(6) = \binom{6}{6}(0.5)^6(0.5)^{6-6}$$
$$= (6)(0.5)^5(0.5)^1 \qquad\qquad = (1)(0.5)^6(0.5)^0$$
$$= 0.0938 \qquad\qquad\qquad = 0.0156$$

Evaluate $P(5 \text{ or } 6 \text{ heads})$ given $P(5) = 0.0938$ and $P(6) = 0.0156$.

$$P(5 \text{ or } 6 \text{ heads}) = P(5) + P(6)$$
$$= 0.0938 + 0.0156$$
$$= 0.109$$

6.5 Calculate the mean, variance, and standard deviation of the binomial distribution.

The mean of a binomial distribution is equal to $\mu = np$, where n is the number of trials and p is the probability that the coin will land heads.

$$\mu = np = (6)(0.5) = 3.0$$

This makes sense. The coin should land heads half of the time, and half of 6 is 3.

The coin will land heads an average of 3 times. The variance of a binomial distribution is equal to $\sigma^2 = npq$, where n is the number of trials, p is the probability that the coin lands heads, and q is the probability that the coin lands tails.

$$\sigma^2 = npq = (6)(0.5)(0.5) = 1.5$$

The standard deviation is the square root of the variance.

$$\sigma = \sqrt{1.5} = 1.225$$

There are only two outcomes, heads or tails. In all binomial distributions, p and q must have a sum of one.

6.6 Calculate the probability that a class of 10 students will have exactly four females.

This is a binomial experiment, because it has exactly two possible outcomes (male and female) and consists of a finite number of trials. The probability of choosing a female student is $p = 0.6$, and the probability of choosing a male student is $q = 1 - 0.6 = 0.4$.

Calculate the probability that a class of $n = 10$ students will contain exactly four females.

$$P(4) = \binom{10}{4}(0.6)^4 (0.4)^{10-4}$$
$$= (210)(0.6)^4 (0.4)^6$$
$$= 0.1115$$

Note: Problems 6.6–6.9 refer to a particular college at which 60% of the student population is female.

6.7 Calculate the probability that a class of 10 students will contain five, six, or seven females.

The probability that a class contains five, six, or seven females is expressed as $P(5 \text{ or } 6 \text{ or } 7)$. Each of the three events is mutually exclusive, so $P(5 \text{ or } 6 \text{ or } 7) = P(5) + P(6) + P(7)$. Calculate each probability individually.

$$P(5) = \binom{10}{5}(0.6)^5 (0.4)^{10-5} \quad P(6) = \binom{10}{6}(0.6)^6 (0.4)^{10-6} \quad P(7) = \binom{10}{7}(0.6)^7 (0.4)^{10-7}$$
$$= (252)(0.6)^5 (0.4)^5 \qquad = (210)(0.6)^6 (0.4)^4 \qquad = (120)(0.6)^7 (0.4)^3$$
$$= 0.2007 \qquad\qquad\quad = 0.2508 \qquad\qquad\quad = 0.2150$$

The probability that the class contains five, six, or seven female students is the sum of the probabilities calculated above.

$$P(5 \text{ or } 6 \text{ or } 7) = P(5) + P(6) + P(7)$$
$$= 0.2007 + 0.2508 + 0.2150$$
$$= 0.667$$

Note: Problems 6.6–6.9 refer to a particular college at which 60% of the student population is female.

6.8 Calculate the probability that at least five students in a class of 10 will be female.

Because the events are mutually exclusive, $P(\text{five or more females}) = P(5) + P(6) + P(7) + P(8) + P(9) + P(10)$. According to Problem 6.7, $P(5 \text{ or } 6 \text{ or } 7 \text{ females}) = 0.667$.

$$P(\text{five or more females}) = 0.667 + P(8) + P(9) + P(10)$$

"At least five" females includes the possibility of five females, P(5), rather than "more than five females," which excludes P(5).

Calculate the remaining probabilities that comprise the sum.

$$P(8) = \binom{10}{8}(0.6)^8(0.4)^{10-8} \quad P(8) = \binom{10}{9}(0.6)^9(0.4)^{10-9} \quad P(10) = \binom{10}{10}(0.6)^{10}(0.4)^{10-10}$$

$$= (45)(0.6)^8(0.4)^2 \qquad = (10)(0.6)^9(0.4)^1 \qquad = (1)(0.6)^{10}(1)$$

$$= 0.1209 \qquad\qquad = 0.0403 \qquad\qquad = 0.0060$$

Calculate the probability that at least five students will be female.

$$P(\text{five or more females}) = 0.667 + P(8) + P(9) + P(10)$$
$$= 0.667 + 0.1209 + 0.0403 + 0.0060$$
$$= 0.834$$

Note: Problems 6.6–6.9 refer to a particular college at which 60% of the student population is female.

6.9 Assume classes of 10 students are assigned randomly and the number of female students in each class is counted. Calculate the mean, variance, and standard deviation for this distribution.

The mean of a binomial distribution is $\mu = np$, where $n = 10$ is the number of students in each class and $p = 0.60$ is the probability that each student is female.

$$\mu = np = (10)(0.6) = 6.0$$

The variance of the binomial distribution is $\sigma^2 = npq$, where $q = 1 - 0.6 = 0.4$ is the probability that a randomly selected student is male.

$$\sigma^2 = npq = (10)(0.6)(0.4) = 2.4$$

The standard deviation is the square root of the variance.

$$\sigma = \sqrt{2.4} = 1.55$$

Note: In Problems 6.10–6.13, assume that NBA athlete LeBron James makes 73% of his free throw attempts.

6.10 Calculate the probability that LeBron James will make exactly six of his next eight free throw attempts.

Shooting a free throw has exactly two outcomes: a successful or an unsuccessful shot. Each player will take a finite number of throws in a game, so counting the number of free throws made by each player is a binomial experiment.

In this problem, James makes $r = 6$ of his $n = 8$ attempts; historically, the probability of James making a free throw is $p = 0.73$ and the probability of James missing a free throw is $q = 1 - 0.73 = 0.27$.

Notice that these exponents always add up to n.

$$P(6) = \binom{8}{6}(0.73)^6(0.27)^{8-6}$$

$$= (28)(0.73)^6(0.27)^2$$

$$= 0.309$$

6.11 Calculate the probability that LeBron James will make at least six of his next eight free throw attempts.

P(6 or 7 or 8) represents the probability that James will make at least six of his next eight free throw attempts. The three events are mutually exclusive, so P(6 or 7 or 8) = P(6) + P(7) + P(8). According to Problem 6.10, the probability of James making six out of eight free throws is $P(6) = 0.309$. Calculate $P(7)$ and $P(8)$.

> If this problem asked the probability of making 6 out of 9 free throws, you couldn't use $P(6)$ from Problem 6.10, because that value is based on 8 free throw attempts.

$$P(7) = \binom{8}{7}(0.73)^7(0.27)^{8-7} \qquad P(8) = \binom{8}{8}(0.73)^8(0.27)^{8-8}$$
$$= (8)(0.73)^7(0.27)^1 \qquad\qquad = (1)(0.73)^8(1)$$
$$= 0.2386 \qquad\qquad\qquad\quad = 0.0806$$

Evaluate P(6 or 7 or 8).

$$P(6 \text{ or } 7 \text{ or } 8) = P(6) + P(7) + P(8)$$
$$= 0.309 + 0.2386 + 0.0806$$
$$= 0.628$$

6.12 Use the complement rule to calculate the probability that LeBron James will make at least two of his next eight free throw attempts.

Rather than calculate $P(2) + P(3) + P(4) + P(5) + P(6) + P(7) + P(8)$, it is more expedient to calculate $P(0) + P(1)$, the complement of making at least two successful free throws. Calculate $P(0)$ and $P(1)$.

> The complement of an event is "anything other than the outcome that is described." In this case, the complement of making two or more free throws is making one or fewer.

$$P(0) = \binom{8}{0}(0.73)^0(0.27)^{8-0} \qquad P(1) = \binom{8}{1}(0.73)^1(0.27)^{8-1}$$
$$= (1)(1)(0.27)^8 \qquad\qquad\quad = (8)(0.73)(0.27)^7$$
$$= 0.0000282 \qquad\qquad\qquad = 0.000611$$

The probability that James will make one or fewer of his next eight free throws is $P(0) + P(1) = 0.0000282 + 0.000611 = 0.0006392$. Therefore, the probability that James will make at least two of his next eight free throws is $1 - 0.0006392 = 0.999$.

Note: In Problems 6.10–6.13, assume that NBA athlete LeBron James makes 73% of his free throw attempts.

6.13 Assume LeBron James shoots eight free throws per game for the first 20 games of the season, and the number of successful free throws per game is counted. Calculate the mean, variance, and standard deviation for this distribution.

Calculate the mean and variance of the binomial distribution.

$$\mu = np = (8)(0.73) = 5.84$$
$$\sigma^2 = npq = (8)(0.73)(0.27) = 1.5768$$

The standard deviation is the square root of the variance.

$$\sigma = \sqrt{1.5768} = 1.2557$$

Note: Problems 6.14–6.16 refer to a 10-question multiple choice test in which each question has four choices.

6.14 Calculate the probability that a student who guesses randomly will answer exactly two questions correctly.

This is a binomial experiment because there are two outcomes when you guess a multiple choice answer (right or wrong), there are a finite number of test questions, and guessing each answer is an independent event.

Each of the $n = 10$ questions has four possible choices, so the probability of randomly selecting the correct answer among those choices is $p = \dfrac{1}{4} = 0.25$.

Thus, the probability of randomly selecting an incorrect answer on each question is $q = 1 - 0.25 = 0.75$. Calculate the probability that a student who guesses randomly will answer exactly $r = 2$ questions correctly.

$$P(2) = \binom{10}{2}(0.25)^2 (0.75)^{10-2}$$
$$= (45)(0.25)^2 (0.75)^8$$
$$= 0.2816$$

Note: Problems 6.14–6.16 refer to a 10-question multiple choice test where each question has four choices.

6.15 Calculate the probability that a student who guesses randomly will answer fewer than three questions correctly.

To calculate $P(0 \text{ or } 1 \text{ or } 2)$, calculate the probability of each mutually exclusive event individually. Recall that $P(2) = 0.2816$, according to Problem 6.14.

$$P(0) = \binom{10}{0}(0.25)^0 (0.75)^{10-0} \qquad P(1) = \binom{10}{1}(0.25)^1 (0.75)^{10-1}$$
$$= (1)(1)(0.75)^{10} \qquad\qquad = (10)(0.25)(0.75)^9$$
$$= 0.0563 \qquad\qquad\qquad = 0.1877$$

Calculate the probability of answering two or fewer questions correctly.

$$P(0 \text{ or } 1 \text{ or } 2) = P(0) + P(1) + P(2)$$
$$= 0.0563 + 0.1877 + 0.2816$$
$$= 0.526$$

Note: Problems 6.14–6.16 refer to a 10-question multiple choice test where each question has four choices.

6.16 Assume each student in the class guesses randomly on each question of the test and the number of correct answers per student is tallied. Calculate the mean, variance, and standard deviation for this distribution.

Calculate the mean and variance of the binomial distribution.

$$\mu = np = (10)(0.25) = 2.5$$
$$\sigma^2 = npq = (10)(0.25)(0.75) = 1.875$$

The standard deviation is the square root of the variance.

$$\sigma = \sqrt{1.875} = 1.369$$

Poisson Probability Distribution

Determining probabilities over specific intervals

6.17 Define the characteristics of a Poisson process and provide an example.

A Poisson process has the following characteristics:

- The experiment counts the number of times an event occurs over a specific period of measurement (such as time, area, or distance).

- The mean of the Poisson distribution is the same for each interval of measurement.

- The number of occurrences in each interval is independent.

An example of a Poisson process would be the number of cars that pass through a tollbooth during one hour.

Named after Siméon Poisson, a French mathematician who described these processes during the early 1800s.

This is a Poisson process because the average number of cars that arrive at the tollbooth each hour does not change hour-to-hour and the number of cars that arrive during the first hour will have no impact on the number of cars that arrive any other hour.

Note: Problems 6.18–6.21 refer to a particular ice-cream stand, where the number of customers who arrive per hour averages seven and follows the Poisson probability distribution.

6.18 Calculate the probability that exactly four customers will arrive during the next hour.

If x = the number of occurrences per interval, λ = the average number of occurrences per interval, and e is Euler's number, then the probability of x occurrences per interval is $P(x) = \dfrac{\lambda^x e^{-\lambda}}{x!}$.

In this problem, $x = 4$ and $\lambda = 7$. Calculate the probability that exactly four customers will arrive during the next hour.

> A mathematical constant (like π) that's a nonrepeating, nonterminating decimal: 2.71828....

$$P(x) = \frac{\lambda^x e^{-\lambda}}{x!}$$

$$P(4) = \frac{(7^4)e^{-7}}{4!}$$

$$P(4) = 0.0912$$

There is a 9.12% chance that the ice-cream stand will have four visitors in the next hour.

> The more decimal places you use for e^{-7}, the more accurate your final result will be.

Note: Problems 6.18–6.21 refer to a particular ice-cream stand, where the number of customers who arrive per hour averages seven and follows the Poisson probability distribution.

6.19 Calculate the probability that fewer than two customers will arrive during the next hour.

The probability of fewer than two customers arriving can be expressed either as $P(x < 2)$, $P(0 \text{ or } 1)$, or $P(0 \cup 1)$. The events are mutually exclusive (you can either have zero customers *or* one customer in an hour, but not both zero and one total customers), so $P(0 \text{ or } 1) = P(0) + P(1)$.

> If the events weren't mutually exclusive, you'd have to subtract $P(0$ and $1)$.

Calculate the probabilities of $x = 0$ and $x = 1$ customers, assuming there are an average of $\lambda = 7$ customers per hour.

> $0!$ by definition $= 1$

$$P(0) = \frac{(7^0)e^{-7}}{0!} \qquad P(1) = \frac{(7^1)e^{-7}}{1!}$$
$$= \frac{(1)e^{-7}}{1} \qquad = \frac{(7)e^{-7}}{1}$$
$$= 0.000912 \qquad = 0.006383$$

Calculate $P(0 \text{ or } 1)$ by adding the probabilities calculated above.

$$P(0 \text{ or } 1) = P(0) + P(1)$$
$$= 0.000912 + 0.006383$$
$$= 0.0073$$

The probability of fewer than two customers arriving during the next hour is 0.73%.

Note: Problems 6.18–6.21 refer to a particular ice-cream stand, where the number of customers who arrive per hour averages seven and follows the Poisson probability distribution.

6.20 Calculate the probability that six, seven, or eight customers will arrive during the next hour.

The probability that six, seven, or eight customers will arrive during the next hour is equal to the sum of the mutually exclusive probabilities: $P(6) + P(7) + P(8)$. Let $\lambda = 7$ and calculate the values individually.

$$P(6) = \frac{(7^6)e^{-7}}{6!} \qquad P(7) = \frac{(7^7)e^{-7}}{7!} \qquad P(8) = \frac{(7^8)e^{-7}}{8!}$$

$$= \frac{(117,649)e^{-7}}{720} \qquad = \frac{(823,543)e^{-7}}{5,040} \qquad = \frac{(5,764,801)e^{-7}}{40,320}$$

$$= 0.149 \qquad\qquad = 0.149 \qquad\qquad = 0.130$$

Calculate the sum of the above probabilities.

$$P(6 \text{ or } 7 \text{ or } 8) = P(6) + P(7) + P(8)$$
$$= 0.149 + 0.149 + 0.130$$
$$= 0.428$$

The probability that six, seven, or eight customers will arrive during the next hour is 42.8%.

Note: Problems 6.18–6.21 refer to a particular ice-cream stand, where the number of customers who arrive per hour averages seven and follows the Poisson probability distribution.

6.21 Calculate the mean, variance, and standard deviation for this distribution.

Data that conforms to a Poisson distribution has a mean that is approximately equal to its variance. The problem states that the mean is $\lambda = 7$ customers per hour, so the variance is $\sigma^2 = 7$. The standard deviation is the square root of the variance: $\sigma = \sqrt{7} = 2.65$.

If the mean and variance of a data set are not nearly equal, then you shouldn't assume that the data has a Poisson distribution.

Note: Problems 6.22–6.24 refer to a particular intersection equipped with video surveillance cameras. The average number of tickets issued per month conforms to a Poisson distribution with an average of 3.7.

6.22 Calculate the probability that exactly five traffic tickets will be issued next month.

Substitute $\lambda = 3.7$ and $x = 5$ into the Poisson distribution formula.

$$P(x) = \frac{\lambda^x e^{-\lambda}}{x!}$$

$$P(5) = \frac{(3.7)^5 e^{-3.7}}{5!}$$

$$P(5) = \frac{(693.43957)e^{-3.7}}{120}$$

$$P(5) = 0.1429$$

Note: Problems 6.22–6.24 refer to a particular intersection equipped with video surveillance cameras. The average number of tickets issued per month conforms to a Poisson distribution with an average of 3.7.

"No more than two" means you include two as a possibility.

6.23 Calculate the probability that no more than two traffic tickets will be issued next month.

If no more than two traffic tickets are issued next month, then one of three mutually exclusive events will occur: one ticket will be issued, two tickets will be issued, or zero tickets will be issued. Calculate the probability of each.

$$P(0) = \frac{(3.7)^0 e^{-3.7}}{0!} \qquad P(1) = \frac{(3.7)^1 e^{-3.7}}{1!} \qquad P(2) = \frac{(3.7)^2 e^{-3.7}}{2!}$$

$$= \frac{(1)e^{-3.7}}{1} \qquad\qquad = \frac{(3.7)e^{-3.7}}{1} \qquad\qquad = \frac{(13.69)e^{-3.7}}{2}$$

$$= 0.0247 \qquad\qquad\quad = 0.0915 \qquad\qquad\quad = 0.1692$$

Add $P(0)$, $P(1)$, and $P(2)$ to calculate the probability that one of the three events will occur.

$$P(0 \text{ or } 1 \text{ or } 2) = P(0) + P(1) + P(2)$$

$$= 0.0247 + 0.0915 + 0.1692$$

$$= 0.285$$

There is a 28.5% chance that no more than two tickets will be issued next month.

Note: Problems 6.22–6.24 refer to a particular intersection equipped with video surveillance cameras. The average number of tickets issued per month conforms to a Poisson distribution with an average of 3.7.

6.24 Calculate the mean, variance, and standard deviation for this distribution.

The mean and variance of a Poisson distribution are approximately equal, so $\lambda = \sigma = 3.7$. The standard deviation is the square root of the variance: $\sigma = \sqrt{3.7} = 1.92$

Note: Problems 6.25–6.28 refer to an automated phone system that can answer three calls in a five-minute period. Assume that calls occur at an average rate of 1.2 every five minutes and follow the Poisson probability distribution.

6.25 Calculate the probability that no calls will occur during the next five minutes.

Substitute $\lambda = 1.2$ and $x = 0$ into the Poisson distribution formula.

$$P(0) = \frac{(1.2)^0 \, e^{-1.2}}{0!}$$
$$= \frac{(1)e^{-1.2}}{1}$$
$$= 0.3012$$

There is a 30.1% chance that no calls will occur during the next five minutes.

Note: Problems 6.25–6.28 refer to an automated phone system that can answer three calls in a five-minute period. Assume that calls occur at an average rate of 1.2 every five minutes and follow the Poisson probability distribution.

6.26 Calculate the probability that more calls will occur during the next five minutes than the system can handle.

The system can answer three calls during a five-minute span, so you are asked to calculate $P(x \geq 4)$. It is not possible to calculate infinitely many probabilities and add them, so compute the probability of the complement instead.

The complement of four or more calls per minute is three or fewer calls occurring. Hence, $P(x \geq 4) = 1 - P(0 \text{ or } 1 \text{ or } 2 \text{ or } 3)$. According to Problem 6.25, $P(0) = 0.3012$. Calculate $P(1)$, $P(2)$, and $P(3)$.

$$P(1) = \frac{(1.2)^1 \, e^{-1.2}}{1!} \qquad P(2) = \frac{(1.2)^2 \, e^{-1.2}}{2!} \qquad P(3) = \frac{(1.2)^3 \, e^{-1.2}}{3!}$$
$$= \frac{(1.2)e^{-1.2}}{1} \qquad\qquad = \frac{(1.44)e^{-1.2}}{2} \qquad\quad = \frac{(1.728)e^{-1.2}}{6}$$
$$= 0.3614 \qquad\qquad\quad = 0.2169 \qquad\qquad\quad = 0.0867$$

Calculate $P(0 \text{ or } 1 \text{ or } 2 \text{ or } 3)$.

$$P(0 \text{ or } 1 \text{ or } 2 \text{ or } 3) = P(0) + P(1) + P(2) + P(3)$$
$$= 0.3012 + 0.3614 + 0.2169 + 0.0867$$
$$= 0.966$$

Subtract this value from one to calculate the probability that more calls will be received than can be answered.

$$P(x \geq 4) = 1 - P(0 \text{ or } 1 \text{ or } 2 \text{ or } 3)$$
$$= 1 - 0.966$$
$$= 0.034$$

Note: Problems 6.25–6.28 refer to an automated phone system that can answer three calls in a five-minute period. Assume that calls occur at an average rate of 1.2 every five minutes and follow the Poisson probability distribution.

6.27 Calculate the probability that exactly seven calls will occur during the next 15 minutes.

If an average of $\lambda = 1.2$ calls occur over a five-minute interval, then an average of $3\lambda = 3.6$ calls occur over a 15-minute interval.

Substitute $\lambda = 3.6$ and $x = 7$ into the Poisson distribution formula.

You can also use a proportion:
$$\frac{\text{original average}}{\text{original interval}} = \frac{\text{new average}}{\text{new interval}}$$
In this case, you'd solve the proportion below to get the "new average":
$$\frac{1.2}{5} = \frac{\text{new average}}{15}$$

$$P(7) = \frac{(3.6)^7 \, e^{-3.6}}{7!}$$
$$= \frac{(7,836.41641) e^{-3.6}}{5,040}$$
$$= 0.0425$$

The probability that exactly seven calls will occur during the next 15 minutes is 4.25%.

Note: Problems 6.25–6.28 refer to an automated phone system that can answer three calls in a five-minute period. Assume that calls occur at an average rate of 1.2 every five minutes and follow the Poisson probability distribution.

6.28 Calculate the probability that exactly five calls will occur during the next 25 minutes.

If an average of 1.2 calls occur over a five-minute period, then solve the proportion below to determine the average number of calls during a 25-minute period.

$$\frac{1.2}{5} = \frac{\lambda}{25}$$

$$5\lambda = (1.2)(25)$$

$$5\lambda = 30$$

$$\lambda = \frac{30}{5}$$

$$\lambda = 6$$

You're answering the question, "1.2 is to 5 as what is to 25?"

Calculate the probability that exactly five calls will arrive during the next 25 minutes.

$$P(5) = \frac{(6^5)e^{-6}}{5!}$$

$$= \frac{(7,776)e^{-6}}{120}$$

$$= 0.1606$$

The probability that five calls will arrive during the next 25 minutes is 16.1%.

Note: Problems 6.29–6.31 refer to a service center that receives an average of 0.6 customer complaints per hour. Management's goal is to receive fewer than three complaints each hour. Assume the number of complaints follows the Poisson distribution.

6.29 Determine the probability that management's goal will be achieved during the next hour.

The probability that fewer than three complaints will be received is the sum of the probabilities that zero, one, or two complaints will be received: $P(x < 3) = P(0) + P(1) + P(2)$. Calculate each probability independently and add the results.

$$P(0) = \frac{(0.6)^0 e^{-0.6}}{0!} \quad P(1) = \frac{(0.6)^1 e^{-0.6}}{1!} \quad P(2) = \frac{(0.6)^2 e^{-0.6}}{2!}$$

$$= \frac{(1)e^{-0.6}}{1} \qquad\quad = \frac{(0.6)e^{-0.6}}{1} \qquad\quad = \frac{(0.36)e^{-0.6}}{2}$$

$$= 0.5488 \qquad\qquad = 0.3293 \qquad\qquad = 0.0988$$

This number is 0.5488 + 0.3293 + 0.0988.

The probability that the goal will be met is $P(x < 3) = 0.977.$

Note: Problems 6.29–6.31 refer to a service center that receives an average of 0.6 customer complaints per hour. Management's goal is to receive fewer than three complaints each hour. Assume the number of complaints follows the Poisson distribution.

6.30 Determine the probability that this goal will be achieved each hour for the next four hours.

According to Problem 6.29, the probability that the goal will be met during the next hour is approximately 0.977. In a Poisson distribution, the number of occurrences during each interval is independent, so the number of complaints received during one hour has no impact on the number of complaints received during any other hour.

Therefore, there is a 0.977 probability that the goal will be met during *any* hour. Raise 0.977 to the fourth power to determine the probability that the goal will be met every hour for the next four hours.

$$(0.977)^4 = 0.911$$

There is a 91.1% chance that management's goal will be met each hour for the next four hours.

Or every hour for four randomly selected hours. They don't have to be consecutive.

Note: Problems 6.29–6.31 refer to a service center that receives an average of 0.6 customer complaints per hour. Management's goal is to receive fewer than three complaints each hour. Assume the number of complaints follows the Poisson distribution.

6.31 Determine the probability that exactly four complaints will be received during the next eight hours.

If an average of 0.6 complaints are received every hour, then an average of $8(0.6) = 4.8$ complaints are received every eight hours. Substitute $x = 4$ and $\lambda = 4.8$ into the Poisson distribution formula.

$$P(4) = \frac{(4.8)^4 e^{-4.8}}{4!}$$

$$= 0.1820$$

The probability that four complaints will arrive during the next eight hours is 18.2%.

Remember, binomial experiments have only two outcomes. In the next problem, you're determining how many traffic lights are broken. In that case, finding a broken one is a "success."

The Poisson Distribution as an Approximation to the Binomial Distribution

A binomial shortcut

6.32 Describe the conditions under which the Poisson distribution can be used as an approximation to the binomial distribution.

The Poisson distribution can be used as an approximation to the binomial distribution when the number of trials n is greater than or equal to 20 and the probability of success p is less than or equal to 0.05. If these conditions are met, the probability is $P(x) = \dfrac{(np)^x e^{-(np)}}{x!}$.

In this example, n = 20 and p = .01, so both conditions have been met to use the Poisson approximation to the binomial distribution. You do that in Problem 6.34.

Note: Problems 6.33–6.34 refer to a town with 20 traffic lights. Each light has a 1% probability of not working properly on any given day.

6.33 Use the binomial distribution to calculate the probability that exactly 1 of the 20 lights will not work properly today.

Each of the $n = 20$ traffic lights has a $p = 0.01$ probability of malfunctioning and a $q = 1 - p = 0.99$ probability of functioning correctly. Apply the binomial distribution formula to determine the probability that exactly $r = 1$ of the traffic lights will malfunction.

$$P(r) = \binom{n}{r} p^n q^{n-r}$$
$$= \binom{20}{1}(0.01)^1 (0.99)^{20-1}$$
$$= (20)(0.01)^1 (0.99)^{19}$$
$$= 0.165$$

Note: Problems 6.33–6.34 refer to a town with 20 traffic lights. Each light has a 1% probability of not working properly on any given day.

6.34 Verify the solution to Problem 6.33 using the Poisson approximation to the binomial distribution.

Apply the formula stated in Problem 6.32.

$$P(x) = \frac{(np)^x e^{-(np)}}{x!}$$
$$= \frac{[(20)(0.01)]^1 e^{-[(20)(0.01)]}}{1!}$$
$$= \frac{(0.2) e^{-0.2}}{1}$$
$$= 0.164$$

The probability that exactly 1 of the 20 lights will not work properly today is 16.4%. This very closely approximates the probability of 16.5% calculated in Problem 6.33.

For an approximation, the answer was pretty accurate.

Note: Problems 6.35–6.36 refer to a particular website whose visitors have a 4% probability of making a purchase.

6.35 Use the binomial distribution to calculate the probability that exactly 2 of the next 25 people to visit the website will make a purchase.

Substitute $n = 25$, $r = 2$, $p = 0.04$, and $q = 1 - 0.04 = 0.96$ into the binomial distribution formula.

$$P(2) = \binom{25}{2}(0.04)^2 (0.96)^{25-2}$$
$$= (300)(0.04)^2 (0.96)^{23}$$
$$= 0.188$$

Note: Problems 6.35–6.36 refer to a particular website whose visitors have a 4% probability of making a purchase.

6.36 Verify the solution to Problem 6.35 using the Poisson approximation to the binomial distribution.

If $n = 25$ and $p = 0.04$, then $np = (25)(0.04) = 1$.

The e^{-1} in the numerator can be moved to the denominator and becomes e^{1}, or e. When you move a number with a negative exponent to the other side of the fraction, the exponent becomes positive.

$$P(x) = \frac{(np)^x \, e^{-(np)}}{x!}$$

$$P(2) = \frac{(1)^2 \, e^{-1}}{2!} = \frac{e^{-1}}{2}$$

$$P(2) = \frac{1}{2e}$$

$$P(2) = 0.184$$

According to the Poisson distribution, the probability that exactly 2 of the next 25 people will make a purchase is 18.4%.

Note: Problems 6.37–6.38 refer to a particular college that accepts 40% of the applications submitted.

6.37 Use the binomial distribution to determine the probability that exactly 2 of the next 10 applications will be accepted.

Notice that $p = 0.40$, which is not less than 0.05. That means you shouldn't use the Poisson approximation. Let's see what happens if you do anyway.

Calculate the probability that $r = 2$ of $n = 10$ applications will be accepted when $p = 0.4$ and $q = 0.6$.

$$P(2) = \binom{10}{2}(0.4)^2 (0.6)^{10-2}$$

$$= (45)(0.4)^2 (0.6)^8$$

$$= 0.121$$

Note: Problems 6.37–6.38 refer to a particular college that accepts 40% of the applications submitted.

6.38 Use the Poisson approximation to the binomial distribution to determine the probability that exactly 2 of the next 10 people will be accepted. Compare your answer to the result generated by Problem 6.37.

If $n = 10$ and $p = 0.4$, then $np = (10)(0.4) = 4$.

$$P(2) = \frac{(4)^2 \, e^{-4}}{2!}$$

$$= \frac{16e^{-4}}{2}$$

$$= 8e^{-4}$$

$$= 0.147$$

Because $p > \overset{0.05}{\cancel{0.5}}$, the Poisson distribution produces a probability that is significantly different than the binomial distribution: 14.7% rather than 12.1%.

Hypergeometric Probability Distribution

Determining probabilities when events are not independent

6.39 Define the characteristics of a hypergeometric probability distribution.

Unlike the binomial and Poisson distributions, the hypergeometric distribution does not require that events be independent of one another. Thus, the distribution is useful when samples are taken from small populations without replacement.

Consider an event that has only two possible outcomes, success or failure. Let N equal the population size and X equal the number of successes in the population; let n equal the sample size and x equal the number of successes in the sample. The formula below calculates the probability of x successes in a hypergeometric distribution.

$$P(x) = \frac{\binom{N-X}{n-x}\binom{X}{x}}{\binom{N}{n}}$$

The big parentheses signify combinations. For example, the denominator is $_NC_n$.

Note: Problems 6.40–6.42 refer to an experiment in which four balls are randomly selected from an urn containing five red balls and seven blue balls without replacement.

6.40 Four balls are randomly selected without replacement. Determine the probability that exactly one of them is red.

The phrase "without replacement" means that once the first ball is selected from the urn, it is not returned to the urn until the experiment is over. Choosing each ball affects the probability that the following ball will be a certain color, because the sample space has changed. Thus, the selection of each ball is not an independent event.

Hypergeometric distributions are useful in such cases. In this example, there are a total of $N = 12$ balls, of which $n = 4$ are selected at random. Drawing one of the $X = 5$ red balls counts as a success, and you are asked to calculate the probability of drawing exactly $x = 1$ red ball.

$$P(x) = \frac{\binom{N-X}{n-x}\binom{X}{x}}{\binom{N}{n}} = \frac{\binom{12-5}{4-1}\binom{5}{1}}{\binom{12}{4}} = \frac{\binom{7}{3}\binom{5}{1}}{\binom{12}{4}} = \frac{(35)(5)}{495} = \frac{175}{495} = 0.354$$

There is a 35.4% probability that exactly one of the four balls selected is red.

Note: Problems 6.40–6.42 refer to an experiment in which four balls are randomly selected from an urn containing five red balls and seven blue balls without replacement.

6.41 Calculate the probability that exactly three of the balls are blue.

Apply the hypergeometric distribution formula to calculate the probability of drawing $n = 4$ balls from a possible $N = 12$ and randomly selecting $x = 3$ of the $X = 7$ blue balls.

These are the same combinations as in Problem 6.40—they're just reversed.

$$P(x) = \frac{\binom{N-X}{n-x}\binom{X}{x}}{\binom{N}{n}} = \frac{\binom{12-7}{4-3}\binom{7}{3}}{\binom{12}{4}} = \frac{\binom{5}{1}\binom{7}{3}}{\binom{12}{4}} = \frac{(5)(35)}{495} = \frac{175}{495} = 0.354$$

There is a 35.4% probability that exactly three of the balls selected are blue, the same probability (according to Problem 6.40) that exactly one of the balls is red. The results are equal because selecting exactly one red ball means you also selected exactly three blue balls (and vice versa).

Note: Problems 6.40–6.42 refer to an experiment in which four balls are randomly selected from an urn containing five red balls and seven blue balls without replacement.

6.42 Calculate the probability that fewer than two of the balls are red.

The probability of selecting fewer than two red balls is equal to the sum of the probabilities of selecting zero red balls and exactly one red ball.

$$P(0 \text{ or } 1 \text{ red ball}) = P(0 \text{ red balls}) + P(1 \text{ red ball})$$

$$P(0 \text{ red balls}) = \frac{\binom{N-X}{n-x}\binom{X}{x}}{\binom{N}{n}} = \frac{\binom{12-5}{4-0}\binom{5}{0}}{\binom{12}{4}} = \frac{\binom{7}{4}\binom{5}{0}}{\binom{12}{4}} = \frac{(35)(1)}{495} = \frac{35}{495} = 0.071$$

According to Problem 6.41, the probability of selecting one red ball is 0.354. Thus, the probability of selecting fewer than two red balls is $0.071 + 0.354 = 42.5\%$.

Note: Problems 6.43–6.45 refer to a process that just produced 20 laptop computers, 3 of which have a defect. A school orders 5 of these laptops.

6.43 Determine the probability that none of the computers in the school's order have a defect.

Apply the hypergeometric distribution formula to calculate the probability of selecting $n = 5$ of the $N = 20$ computers and receiving $x = 0$ of the $X = 3$ defective laptops.

$$P(0) = \frac{\binom{N-X}{n-x}\binom{X}{x}}{\binom{N}{n}} = \frac{\binom{20-3}{5-0}\binom{3}{0}}{\binom{20}{5}} = \frac{\binom{17}{5}\binom{3}{0}}{\binom{20}{5}} = \frac{(6,188)(1)}{15,504} = 0.399$$

There is a 39.9% probability that none of the computers has a defect.

Note: Problems 6.43–6.45 refer to a process that just produced 20 laptop computers, 3 of which have a defect. A school orders 5 of these laptops.

6.44 Determine the probability that exactly two of the school's computers have a defect.

Apply the formula used in Problem 6.43, this time substituting $x = 2$ into it.

$$P(2) = \frac{\binom{N-X}{n-x}\binom{X}{x}}{\binom{N}{n}} = \frac{\binom{20-3}{5-2}\binom{3}{2}}{\binom{20}{5}} = \frac{\binom{17}{3}\binom{3}{2}}{\binom{20}{5}} = \frac{(680)(3)}{15,504} = \frac{2,040}{15,504} = 0.132$$

There is a 13.2% probability that exactly two of the school's laptops have a defect.

Note: Problems 6.43–6.45 refer to a process that just produced 20 laptop computers, 3 of which have a defect. A school orders 5 of these laptops.

6.45 Determine the probability that at least two of the computers in the school's order have a defect.

If *at least* two computers have a defect, then either two of them are defective or three of them are defective. ←

$$P(x \geq 2) = P(2) + P(3)$$

According to Problem 6.44, the probability of two defective computers in the order is 0.132. Calculate the probability that $x = 3$ of the $X = 3$ defective computers are among the $n = 5$ laptops ordered of the $N = 20$ manufactured.

Only 3 of the 20 computers are defective, so you can't have 4 or 5 defective computers in the order—3 is the max.

$$P(3) = \frac{\binom{N-X}{n-x}\binom{X}{x}}{\binom{N}{n}} = \frac{\binom{20-3}{5-3}\binom{3}{3}}{\binom{20}{5}} = \frac{\binom{17}{2}\binom{3}{3}}{\binom{20}{5}} = \frac{(136)(1)}{15,504} = 0.009$$

The probability that at least two defective computers are in the school's order is $0.132 + 0.009 = 0.141$.

This problem uses the hypergeometric distribution because you are selecting from a small population without replacement.

Note: *White boxers are dogs with a genetic predisposition for deafness within the first year of life. In Problems 6.46–6.48, assume 3 puppies from a litter of 10 will experience deafness before age one. A family has randomly selected two puppies from the litter to take home as pets.*

6.46 A family randomly selected two puppies from the litter. Calculate the probability that neither of the puppies will be deaf by age one.

Calculate the probability of selecting $x = 0$ of the $X = 3$ puppies that will experience deafness if $n = 2$ of the $N = 10$ puppies are chosen.

$$P(0) = \frac{\binom{N-X}{n-x}\binom{X}{x}}{\binom{N}{n}} = \frac{\binom{10-3}{2-0}\binom{3}{0}}{\binom{10}{2}} = \frac{\binom{7}{2}\binom{3}{0}}{\binom{10}{2}} = \frac{(21)(1)}{45} = 0.467$$

Note: *White boxers are dogs with a genetic predisposition for deafness within the first year of life. In Problems 6.46–6.48, assume 3 puppies from a litter of 10 will experience deafness before age one. A family has randomly selected two puppies from the litter to take home as pets.*

6.47 Determine the probability that exactly one of the selected puppies will experience deafness before age one.

Calculate the probability of selecting $x = 1$ of the $X = 3$ puppies that will experience deafness if $n = 2$ of the $N = 10$ puppies are chosen.

$$P(1) = \frac{\binom{N-X}{n-x}\binom{X}{x}}{\binom{N}{n}} = \frac{\binom{10-3}{2-1}\binom{3}{1}}{\binom{10}{2}} = \frac{\binom{7}{1}\binom{3}{1}}{\binom{10}{2}} = \frac{(7)(3)}{45} = 0.467$$

Add the results for Problems 6.46, 6.47, and 6.48:

$P(0) + P(1) + P(2)$

$= 0.4\overline{6} + 0.4\overline{6} + 0.0\overline{6}$

$= 1$

There are only three possible values of x—either zero, one, or two of the puppies will experience deafness, so all three probabilities add up to one.

Note: *White boxers are dogs with a genetic predisposition for deafness within the first year of life. In Problems 6.46–6.48, assume 3 puppies from a litter of 10 will experience deafness before age one. A family has randomly selected two puppies from the litter to take home as pets.*

6.48 Determine the probability that both of the selected pets will experience deafness before age one.

Calculate the probability of selecting $x = 2$ of the $X = 3$ puppies that will experience deafness if $n = 2$ of the $N = 10$ puppies are chosen.

$$P(2) = \frac{\binom{N-X}{n-x}\binom{X}{x}}{\binom{N}{n}} = \frac{\binom{10-3}{2-2}\binom{3}{2}}{\binom{10}{2}} = \frac{\binom{7}{0}\binom{3}{2}}{\binom{10}{2}} = \frac{(1)(3)}{45} = 0.067$$

Note: Problems 6.49–6.50 refer to a political committee that consists of seven Democrats, five Republicans, and two Independents. A randomly selected subcommittee of six people is formed from this group.

6.49 Determine the probability that the subcommittee will consist of two Democrats, three Republicans, and one Independent.

This problem is an extension of the hypergeometric distribution of events limited to two outcomes. Let *N* represent the population size and *n* represent the sample size, just as they did when only two outcomes were possible. In this case, there are *N* = 14 politicians on the committee and *n* = 6 on the subcommittee.

In this case there are three outcomes, because there are three political parties to choose from.

Let X_1 represent the population size of a subset and x_1 represent the sample size of that subset. For instance, there are $X_1 = 7$ Democrats on the committee and $x_1 = 2$ Democrats are selected for the subcommittee.

Similarly, X_2 and X_3 are other subsets of *N* (such that $X_1 + X_2 + X_3 = N$) and x_2 and x_3 are the sample sizes of those subsets. In this problem, there are $X_2 = 5$ Republicans and $X_3 = 2$ Independents on the committee, of which $x_2 = 3$ and $x_3 = 1$ are selected for the subcommittee. The probability of selecting x_1, x_2, and x_3 from groups of size X_1, X_2, and X_3 is calculated using the extended hypergeometric distribution formula below.

$$P(x_1, x_2, \cdots, x_k) = \frac{\binom{X_1}{x_1}\binom{X_2}{x_2}\cdots\binom{X_k}{x_k}}{\binom{N}{n}}$$

Substitute the values of *N*, *n*, X_1, X_2, X_3, x_1, x_2, and x_3 stated above into the formula.

$$P(2,3,1) = \frac{\binom{X_1}{x_1}\binom{X_2}{x_2}\binom{X_3}{x_3}}{\binom{N}{n}} = \frac{\binom{7}{2}\binom{5}{3}\binom{2}{1}}{\binom{14}{6}} = \frac{(21)(10)(2)}{3,003} = \frac{420}{3,003} = 0.140$$

P(2, 3, 1) is the probability of 2 Democrats, 3 Republicans, and 1 Independent.

The probability that the subcommittee of six will consist of two Democrats, three Republicans, and one Independent is 14.0%.

Note: Problems 6.49–6.50 refer to a political committee that consists of seven Democrats, five Republicans, and two Independents. A randomly selected subcommittee of six people is formed from this group.

6.50 Determine the probability that the subcommittee will consist of three Democrats and three Republicans.

Use the extended hypergeometric distribution formula to calculate the probability that $n = 6$ of $N = 14$ committee members will be chosen such that $x_1 = 3$ of the $X_1 = 7$ Democrats are selected, $x_2 = 3$ of the $X_2 = 5$ Republicans are selected, and $x_3 = 0$ of the $X_3 = 2$ Independents are selected.

$$P(3,3,0) = \frac{\binom{X_1}{x_1}\binom{X_2}{x_2}\binom{X_3}{x_3}}{\binom{N}{n}} = \frac{\binom{7}{3}\binom{5}{3}\binom{2}{0}}{\binom{14}{6}} = \frac{(35)(10)(1)}{3,003} = \frac{350}{3,003} = 0.117$$

6.51 A statistics class consisting of 16 students has the following grade distribution.

Grade	Number of Students
A	4
B	6
C	4
D	2
Total	**16**

Eight students are randomly selected from the class. Determine the probability that three students had an A, two students had a B, two students had a C, and one student had a D.

Set the capital X's equal to the total number of students with each grade ($X_1 = 4$, $X_2 = 6$, $X_3 = 4$, and $X_4 = 2$) and the lowercase x's equal to the number of selected students with each grade ($x_1 = 3$, $x_2 = 2$, $x_3 = 2$, and $x_4 = 1$).

Use the extended hypergeometric probability formula to determine the probability that the $n = 8$ selected students from the population of $N = 14$ have the indicated grades.

$$P(3,2,2,1) = \frac{\binom{X_1}{x_1}\binom{X_2}{x_2}\binom{X_3}{x_3}\binom{X_4}{x_4}}{\binom{N}{n}} = \frac{\binom{4}{3}\binom{6}{2}\binom{4}{2}\binom{2}{1}}{\binom{16}{8}} = \frac{(4)(15)(6)(2)}{12,870} = \frac{720}{12,870} = 0.056$$

Chapter 7

CONTINUOUS PROBABILITY DISTRIBUTIONS

Random variables that aren't whole numbers

The normal probability distribution is the most widely used distribution in statistics and is the major focus of this chapter. After the normal distribution is introduced, the empirical rule is explored, which establishes the amount of data that lies within one, two, or three standard deviations of the mean. The chapter concludes by exploring two additional continuous distributions: uniform and exponential.

Chapter 6 dealt with discrete probability distributions. This chapter brings you back to the land of continuity, where the data isn't always measured in integers. Weight, distance, and time are just a few examples of continuous random variables. This chapter focuses mostly on the normal distribution, which is also known as the bell curve. A normal distribution is not skewed left or right, and most of the data is clustered near the mean.

Normal Probability Distribution
Bell curves and z-scores

7.1 Identify the three defining characteristics of the normal probability distribution.

A continuous random variable is usually a measurement. Not only can it have integer values, it can also have all the decimal values that fall between integers. See Problem 5.28 for more details.

The normal probability distribution is a bell-shaped continuous distribution that fulfills the following conditions:

- The distribution is symmetrical around the mean.
- The mean, median, and mode are the same value.
- The total area under the curve is equal to one.

The shape of the normal probability distribution is shown below.

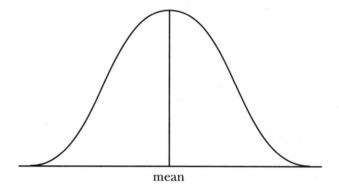

mean

Because the normal distribution is continuous, it represents infinitely many possible values, depending on the level of precision. Because there are an infinite number of possible values, the probability that a continuous random variable is equal to a specific single value is zero.

Instead of determining the probability of a single value occurring, when exploring normal distributions, you define two endpoints and calculate the probability that a chosen value will occur within the specified interval.

7.2 Describe the role that the mean, standard deviation, and z-score play in the normal probability distribution.

The mean μ is the center of a normal distribution. A higher mean shifts the position of the probability distribution to the right while a lower mean shifts its position to the left.

The standard deviation σ is a measure of dispersion—the higher the standard deviation, the wider the distribution. A smaller standard deviation results in a narrower bell-shaped curve.

The z-score measures the number of standard deviations between the mean and a specific value of x, according to the formula below.

$$z_x = \frac{x - \mu}{\sigma}$$

Note: Problems 7.3–7.6 refer to the speeds at which cars pass through a checkpoint. Assume the speeds are normally distributed such that $\mu = 61$ miles per hour and $\sigma = 4$ miles per hour.

7.3 Calculate the probability that the next car that passes through the checkpoint will be traveling slower than 65 miles per hour.

Calculate the z-score for $x = 65$ by substituting $\mu = 61$ and $\sigma = 4$ into the z-score formula.

$$z_x = \frac{x - \mu}{\sigma}$$

$$z_{65} = \frac{65 - 61}{4}$$

$$z_{65} = \frac{4}{4}$$

$$z_{65} = 1$$

> The value $x = 65$ is one standard deviation from the mean because 65 is four more than 61 and the standard deviation is four.

Use the standard normal table in Reference Table 1 (at the end of this book) to determine the area under the normal curve between the z-score of 1.00 and the mean. To better understand how to use the table, consider the excerpt below.

> At the mean of the distribution, $z = 0$.

Second digit of z

z	0.00	0.01	0.02	0.03	0.04	0.05	0.06	0.07	0.08	0.09
0.0	0.0000	0.0040	0.0080	0.0120	0.0160	0.0199	0.0239	0.0279	0.0319	0.0359
0.1	0.0398	0.0438	0.0478	0.0517	0.0557	0.0596	0.0636	0.0675	0.0714	0.0753
0.2	0.0793	0.0832	0.0871	0.0910	0.0948	0.0987	0.1026	0.1064	0.1103	0.1141
0.3	0.1179	0.1217	0.1255	0.1293	0.1331	0.1368	0.1406	0.1443	0.1480	0.1517
0.4	0.1554	0.1591	0.1628	0.1664	0.1700	0.1736	0.1772	0.1808	0.1844	0.1879
0.5	0.1915	0.1950	0.1985	0.2019	0.2054	0.2088	0.2123	0.2157	0.2190	0.2224
0.6	0.2257	0.2291	0.2324	0.2357	0.2389	0.2422	0.2454	0.2486	0.2517	0.2549
0.7	0.2580	0.2611	0.2642	0.2673	0.2704	0.2734	0.2764	0.2794	0.2823	0.2852
0.8	0.2881	0.2910	0.2939	0.2967	0.2995	0.3023	0.3051	0.3078	0.3106	0.3133
0.9	0.3159	0.3186	0.3212	0.3238	0.3264	0.3289	0.3315	0.3340	0.3365	0.3389
1.0	0.3413	0.3438	0.3461	0.3485	0.3508	0.3531	0.3554	0.3577	0.3599	0.3621

> If the z-score was 1.03, the area would be 0.3485.

Because $z = 1.00$, go to the 1.0 row and the 0.00 column; they intersect at the value 0.3413. Thus, the area under the normal curve between the mean and 1.00 standard deviations away from the mean is 0.3413.

The shaded area in the following figure represents the area to the left of $z = 1.0$, the portion of the distribution that traveled slower than 65 miles per hour, one standard deviation above the mean. Recall that the area beneath the curve is exactly one, so the shaded portion left of the mean has an area exactly half as large: 0.5.

> The normal distribution is symmetrical, so the mean splits the area beneath the curve in half. The total area is 1, so the area on either side of the mean is 0.5.

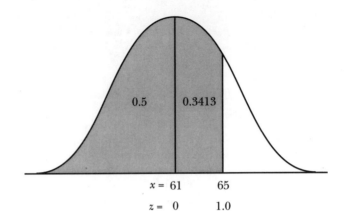

$x = 61$ 65

$z = $ 0   1.0

Add the shaded area left of the mean to the shaded area between the mean and $z = 1$ to calculate the probability that the car will be traveling less than 65 miles per hour.

$$P(x < 65) = P(z < 1) = 0.5 + 0.3413 = 0.8413$$

> *Note: Problems 7.3–7.6 refer to the speeds at which cars pass through a checkpoint. Assume the speeds are normally distributed such that $\mu = 61$ miles per hour and $\sigma = 4$ miles per hour.*

7.4   Calculate the probability that the next car passing will be traveling more than 66 miles per hour.

Calculate the z-score for $x = 66$.

> Think of the z-score and x as the same value in different units, like Fahrenheit and Celsius. If x is a raw data value (in this case speed), z represents how different that speed is when compared to the mean, measured in standard deviations.

$$z_x = \frac{x - \mu}{\sigma}$$

$$z_{66} = \frac{66 - 61}{4}$$

$$z_{66} = \frac{5}{4}$$

$$z_{66} = 1.25$$

According to Reference Table 1, the area corresponding to a z-score of 1.25 is 0.3944. This value represents the area between the mean (which has a z-score of 0) and 1.25 deviations either above or below the mean (in this case above, because $66 > 61$).

The probability that the next car will be traveling more than 66 miles per hour is the shaded area beneath the following normal curve.

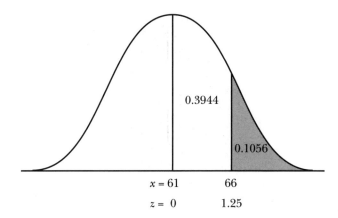

The area beneath the curve and right of the mean is 0.5. Recall that the area between the mean and $z = 1.25$ standard deviations above the mean is 0.3944. Thus the shaded area is $0.5 - 0.3944 = 0.1056$.

You could write this as $P(x > 66)$ or $P(x \geq 66)$. The notation doesn't matter. Remember, the probability that you get a single value on a continuous distribution is not defined, so $P(66) = 0$.

Note: Problems 7.3–7.6 refer to the speeds at which cars pass through a checkpoint. Assume the speeds are normally distributed such that $\mu = 61$ miles per hour and $\sigma = 4$ miles per hour.

7.5 Calculate the probability that the next car will be traveling less than 59 miles per hour.

Calculate the z-score for $x = 59$.

$$z_x = \frac{x - \mu}{\sigma}$$
$$z_{59} = \frac{59 - 61}{4}$$
$$z_{59} = -0.50$$

A z-score is negative when x is less than the mean.

Reference Table 1 can be used for negative z-scores as well as positive z-scores because the normal distribution is symmetrical. According to the table, the area corresponding to a z-score of 0.50 is 0.1915. This is the area between z-scores of 0 and 0.50 as well as the area between z-scores of −0.50 and 0.

The shaded region in the figure below represents the area of interest, the area beneath the curve left of the mean, excluding the area between $z = -0.50$ and $z = 0$.

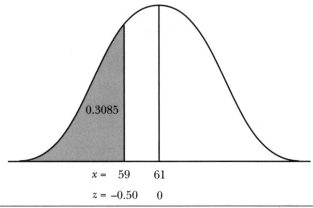

Calculate the probability that the next car to pass will be traveling less than 59 miles per hour.

$$P(x < 59) = P(z < 0.50) = 0.5 - 0.1915 = 0.3085$$

> *Note: Problems 7.3–7.6 refer to the speeds at which cars pass through a checkpoint. Assume the speeds are normally distributed such that $\mu = 61$ miles per hour and $\sigma = 4$ miles per hour.*

7.6 Calculate the probability that the next car to pass will be traveling more than 58 miles per hour.

Calculate the *z*-score for $x = 58$.

$$z_x = \frac{x - \mu}{\sigma}$$

$$z_{58} = \frac{58 - 61}{4}$$

$$z_{58} = -\frac{3}{4}$$

$$z_{58} = -0.75$$

If a car travels more than 58 miles per hour, then its speed is either greater than the mean ($x > 61$ and $z > 0$) or between the mean and 0.75 standard deviations below the mean ($-0.75 < z < 0$ and $58 < x < 61$). This probability corresponds to the area of the shaded region below.

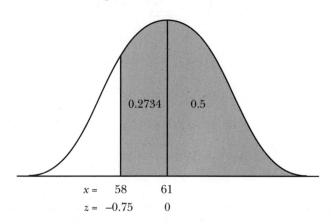

$x =$	58	61
$z =$	−0.75	0

Add the areas of the regions right and left of the mean to compute the probability that the next car will be traveling more than 58 miles per hour.

$$P(x > 58) = P(z > -0.75) = 0.2734 + 0.5 = 0.7734$$

This comes from the Reference Table 1 in the back of this book.

Note: Problems 7.7–7.10 refer to the selling prices of various homes in a community that follow the normal distribution with μ = $276,000 and σ = $32,000.

7.7 Calculate the probability that the next house in the community will sell for more than $206,000.

Calculate the z-score for $x = 206,000$.

> Round z to two decimal places because Reference Table 1 uses two decimal places.

$$z_x = \frac{x - \mu}{\sigma}$$

$$z_{206,000} = \frac{206,000 - 276,000}{32,000}$$

$$z_{206,000} = \frac{-70,000}{32,000}$$

$$z_{206,000} = -2.19$$

The probability that the next house in the community will sell for more than $206,000 corresponds to the area of the shaded region below.

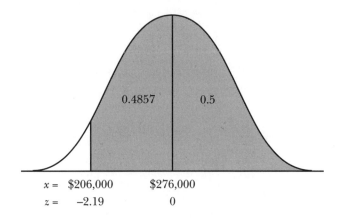

$x =$ $206,000	$276,000
$z =$ −2.19	0

The shaded area right of the mean is 0.5, and according to Reference Table 1, the area between $z = 0$ and $z = -2.19$ is 0.4857.

$$P(x > 206,000) = P(z > -2.19) = 0.4857 + 0.5 = 0.9857$$

Note: Problems 7.7–7.10 refer to the selling prices of various homes in a community that follow the normal distribution with μ = $276,000 and σ = $32,000.

7.8 Calculate the probability that the next house in the community will sell for less than $220,000.

Calculate the z-score for $x = 220,000$.

$$z_x = \frac{x - \mu}{\sigma}$$

$$z_{220,000} = \frac{220,000 - 276,000}{32,000}$$

$$z_{220,000} = \frac{-56,000}{32,000}$$

$$z_{220,000} = -1.75$$

The probability that the next house in the community will sell for less than $220,000 corresponds to the area of the shaded region below. According to Reference Table 1, the area between $x = 220,000$ and $x = 276,000$ is 0.4599.

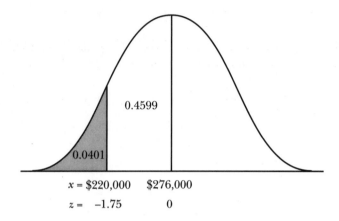

$$P(x < 220,000) = P(z < -1.75) = 0.5 - 0.4599 = 0.041$$

Note: Problems 7.7–7.10 refer to the selling prices of various homes in a community that follow the normal distribution with $\mu = \$276,000$ and $\sigma = \$32,000$.

7.9 Calculate the probability that the next house in the community will sell for more than $250,000 but less than $350,000.

Calculate the z-scores for $x = 250,000$ and $x = 350,000$.

$$z_{250,000} = \frac{250,000 - 276,000}{32,000} \qquad z_{350,000} = \frac{350,000 - 276,000}{32,000}$$

$$= \frac{-26,000}{32,000} \qquad = \frac{74,000}{32,000}$$

$$= -0.81 \qquad = 2.31$$

The probability that the selling price of the next house will be between $250,000 and $350,000 corresponds to the area of the shaded region below.

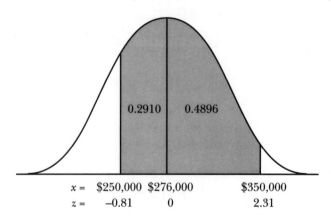

$$P(250,000 < x < 350,000) = P(-0.81 < z < 2.31)$$
$$= 0.2910 + 0.4896$$
$$= 0.7806$$

This is the area between 0.81 standard deviations below and 2.31 standard deviations above the mean for any normal distribution, not just this particular example.

Note: Problems 7.7–7.10 refer to the selling prices of various homes in a community that follow the normal distribution with μ = $276,000 and σ = $32,000.

7.10 Calculate the probability that the selling price of the next house in the community will be between $276,000 and $325,000.

Notice that $276,000 is the mean of the normal distribution, so $z_{276,000} = 0$. Calculate the z-score for x = 325,000.

$$z_{325,000} = \frac{325,000 - 276,000}{32,000} = \frac{49,000}{32,000} = 1.53$$

According to Reference Table 1, the area beneath the normal distribution curve between the mean and 1.53 standard deviations above the mean is 0.4370.

$$P(276,000 < x < 325,000) = P(0 < z < 1.53) = 0.4370$$

Note: In Problems 7.11–7.14, assume that a retail store has customers whose ages are normally distributed such that μ = 37.5 years and σ = 7.6 years.

7.11 Calculate the probability that a randomly chosen customer is more than 48 years old.

Calculate the z-score for x = 48.

$$z_{48} = \frac{48 - 37.5}{7.6} = \frac{10.5}{7.6} = 1.38$$

According to Reference Table 1

Note that $P(0 < z < 1.38) = 0.4162$. Calculate the probability of randomly selecting a customer older than 48.

$$P(x > 48) = P(z > 1.38)$$
$$= P(z > 0) - P(0 < z < 1.38)$$
$$= 0.5 - 0.4162$$
$$= 0.0838$$

There's a 50% chance the customer's age is over 37.5, the mean age. There's a 41.62% chance the customer's age is between 37.5 and 48. That means there's an 8.38% chance that the customer is older than 48.

Note: In Problems 7.11–7.14, assume that a retail store has customers whose ages are normally distributed such that $\mu = 37.5$ years and $\sigma = 7.6$ years.

7.12   Calculate the probability that a randomly chosen customer is younger than 44 years old.

Calculate the z-score for $x = 44$.

$$z_{44} = \frac{44 - 37.5}{7.6} = \frac{6.5}{7.6} = 0.86$$

There is a 0.5 probability that a randomly selected customer is younger than the mean age of 37.5. According to Reference Table 1, there is a 0.3051 probability that a customer is between 37.5 and 44 years of age. Thus, there is a $0.5 + 0.3051 = 0.8051$ probability that a customer is younger than 44 years of age.

Draw the shaded areas as this book did for Problems 7.3–7.10 if it helps you figure out what to add or subtract in this step.

Note: In Problems 7.11–7.14, assume that a retail store has customers whose ages are normally distributed such that $\mu = 37.5$ years and $\sigma = 7.6$ years.

7.13   Calculate the probability that a randomly chosen customer is between 46 and 54 years old.

Calculate the z-scores for $x = 46$ and $x = 52$.

$$z_{46} = \frac{46 - 37.5}{7.6} \qquad z_{54} = \frac{54 - 37.5}{7.6}$$
$$= \frac{8.5}{7.6} \qquad\qquad = \frac{16.5}{7.6}$$
$$= 1.12 \qquad\qquad = 2.17$$

The area between $z = 0$ and 2.17 is 0.4850, according to Reference Table 1. The area between $z = 0$ and 1.12 is 0.3686. Therefore, the area between $z = 1.12$ and 2.17 is $0.4850 - 0.3686 = 0.1164$, as illustrated below, and $P(46 < x < 54) = 0.1164$.

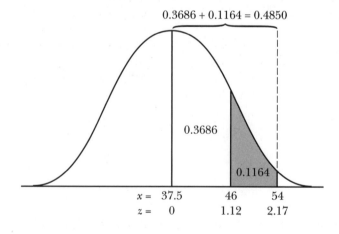

Note: In Problems 7.11–7.14, assume that a retail store has customers whose ages are normally distributed such that $\mu = 37.5$ years and $\sigma = 7.6$ years.

7.14 Calculate the probability that a randomly chosen customer is between 25 and 37.5 years old.

The upper age boundary is the mean, $x = 37.5$, for which $z = 0$. Calculate the z-score for $x = 25$.

$$z_{25} = \frac{25 - 37.5}{7.6} = \frac{-12.5}{7.6} = -1.64$$

According to Reference Table 1, $P(-1.64 < z < 0) = 0.4495$. Therefore, there is a 45.0% chance that a randomly chosen customer is between 25 and 37.5 years old.

Note: In Problems 7.15–7.18, assume that an individual's golf scores are normally distributed with a mean of 90.4 and a standard deviation of 5.3.

7.15 Calculate the probability that the golfer will shoot lower than a 76 during his next round.

Calculate the z-score for $x = 76$.

$$z_{76} = \frac{76 - 90.4}{5.3} = \frac{-14.4}{5.3} = -2.72$$

There is a 0.5 probability that the golfer will shoot below $\mu = 90.4$, his mean score. According to Reference Table 1, there is a 0.4967 probability that the golfer will shoot between $x = 76$ and $x = 90.4$. Therefore, $P(x < 76) = 0.5 - 0.4967 = 0.0033$. There is a 0.33% chance that the golfer will shoot lower than a 76 during his next round of golf.

> The z-score 2.72 is almost three standard deviations away from the mean. Almost none of the data in a normal distribution is that far away, which explains the slim chances of such a good golf score.

Note: In Problems 7.15–7.18, assume that an individual's golf scores are normally distributed with a mean of 90.4 and a standard deviation of 5.3.

7.16 Calculate the probability that the golfer will shoot between 87 and 95 during his next round. ←

Calculate the z-scores for $x = 87$ and $x = 95$.

$$z_{87} = \frac{87 - 90.4}{5.3} \qquad z_{95} = \frac{95 - 90.4}{5.3}$$

$$= \frac{-3.4}{5.3} \qquad\qquad = \frac{4.6}{5.3}$$

$$= -0.64 \qquad\qquad = 0.87$$

Note that $P(-0.64 < z < 0) = 0.2389$ and $P(0 < z < 0.87) = 0.3078$.

$$P(87 < x < 95) = P(-0.64 < z < 0.87)$$
$$= 0.2389 + 0.3078$$
$$= 0.5467$$

> One of these numbers is below the mean and one is above, so they represent two different regions on either side of the mean. Look up each of the z-scores and add the areas together.

Note: In Problems 7.15–7.18, assume that an individual's golf scores are normally distributed with a mean of 90.4 and a standard deviation of 5.3.

7.17 Calculate the probability that the golfer will shoot between 94 and 100 during his next round.

> Both of these values are greater than the mean, so the area between them is the larger area minus the smaller area.

Calculate the z-scores for $x = 94$ and $x = 100$.

$$z_{94} = \frac{94 - 90.4}{5.3} \qquad z_{100} = \frac{100 - 90.4}{5.3}$$

$$= \frac{3.6}{5.3} \qquad\qquad = \frac{9.6}{5.3}$$

$$= 0.68 \qquad\qquad = 1.81$$

Note that $P(0 < z < 0.68) = 0.2517$ and $P(0 < z < 1.81) = 0.4649$.

$$P(94 < x < 100) = P(0.68 < z < 1.81)$$
$$= 0.4649 - 0.2517$$
$$= 0.2132$$

Note: In Problems 7.15–7.18, assume that an individual's golf scores are normally distributed with a mean of 90.4 and a standard deviation of 5.3.

7.18 Calculate the probability that the golfer will shoot between 80 and 85 during his next round.

> These numbers are both below the mean of 90.4. The area between them is the area of the region that extends farther from the mean minus the smaller region. (Just like in Problem 7.17.)

Calculate the z-scores for $x = 80$ and $x = 85$.

$$z_{80} = \frac{80 - 90.4}{5.3} \qquad z_{85} = \frac{85 - 90.4}{5.3}$$

$$= \frac{-10.4}{5.3} \qquad\qquad = \frac{-5.4}{5.3}$$

$$= -1.96 \qquad\qquad = -1.02$$

Note that $P(-1.96 < z < 0) = 0.4750$ and $P(-1.02 < z < 0) = 0.3461$.

$$P(80 < x < 85) = P(-1.96 < z < -1.02)$$
$$= 0.4750 - 0.3461$$
$$= 0.1289$$

Note: In Problems 7.19–7.22, assume that the number of days it takes a homebuilder to complete a house is normally distributed with an average completion time of 176.7 days and a standard deviation of 24.8 days.

7.19 Calculate the probability that it will take between 185 and 225 days to complete the next home.

Calculate the z-scores for $x = 185$ and $x = 225$.

$$z_{185} = \frac{185 - 176.7}{24.8} \qquad z_{225} = \frac{225 - 176.7}{24.8}$$
$$= \frac{8.3}{24.8} \qquad\qquad = \frac{48.3}{24.8}$$
$$= 0.33 \qquad\qquad = 1.95$$

Note that $P(0 < z < 0.33) = 0.1293$ and $P(0 < z < 1.95) = 0.4744$.

$$P(185 < x < 225) = P(0.33 < z < 1.95)$$
$$= 0.4744 - 0.1293$$
$$= 0.3451$$

Note: In Problems 7.19–7.22, assume that the number of days it takes a homebuilder to complete a house is normally distributed with an average completion time of 176.7 days and a standard deviation of 24.8 days.

7.20 Calculate the probability that the next home built will be completed in 150 to 170 days.

Calculate the z-scores for $x = 150$ and $x = 170$.

$$z_{150} = \frac{150 - 176.7}{24.8} \qquad z_{170} = \frac{170 - 176.7}{24.8}$$
$$= \frac{-26.7}{24.8} \qquad\qquad = \frac{-6.7}{24.8}$$
$$= -1.08 \qquad\qquad = -0.27$$

According to Reference Table 1, $P(-1.08 < z < 0) = 0.3599$ and $P(-0.27 < z < 0) = 0.1064$. Therefore, the probability that it will take between $x = 150$ and $x = 170$ days to complete the house is $0.3599 - 0.1064 = 0.2535$.

Note: In Problems 7.19–7.22, assume that the number of days it takes a homebuilder to complete a house is normally distributed with an average completion time of 176.7 days and a standard deviation of 24.8 days.

7.21 Determine the completion time that the builder has a 95% probability of achieving.

Your goal is to find the value of x greater than the mean with a z-score z_x such that $P(0 < z < z_x) = 0.45$, as illustrated in the following figure.

> You want to find the z-score z_x on the right side of the mean that splits that side 45%/5%. You want a 95% probability, which means you'll add the entire region left of the mean (0.5) and most of the region right of the mean (0.45).

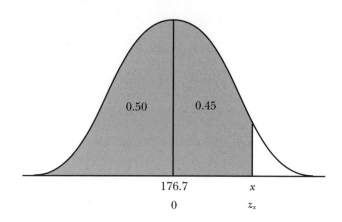

Search Reference Table 1 for the value closest to 0.4500. The closest approximations correspond to $z = 1.64$ and $z = 1.65$. Use either of these z-scores and calculate the corresponding x value.

You're calculating the number of days x it takes to complete a house that is $z = 1.64$ standard deviations greater than the mean.

$$z_x = \frac{x - \mu}{\sigma}$$

$$1.65 = \frac{x - 176.7}{24.8}$$

Cross-multiply and solve for x.

$$(1.65)(24.8) = x - 176.7$$
$$40.92 = x - 176.7$$
$$217.62 = x$$

There is a 95% probability that the builder can complete the home before 217.6 days have elapsed.

Note: In Problems 7.19–7.22, assume that the number of days it takes a homebuilder to complete a house is normally distributed with an average completion time of 176.7 days and a standard deviation of 24.8 days.

7.22 Determine the completion time that the builder has a 40% probability of achieving.

Your goal is to identify the value of x with a z-score z_x approximately equal to 0.1000. Therefore, $P(z_x < z < 0) = 0.10$. Subtracting this region from the area under the normal curve left of the mean results in the shaded region in the following figure.

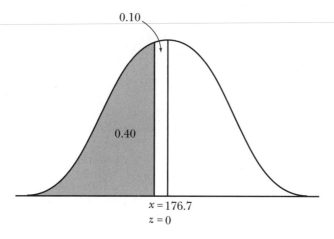

The closest approximation to 0.1000 in Reference Table 1 is 0.0987, which corresponds to $z = -0.25$. Calculate the corresponding value of x.

> Z has to be negative because it's less than the mean.

$$z_x = \frac{x - \mu}{\sigma}$$

$$-0.25 = \frac{x - 176.7}{24.8}$$

$$(-0.25)(24.8) = x - 176.7$$

$$-6.2 = x - 176.7$$

$$170.5 = x$$

There is a 40% probability that the builder can complete a home within 170.5 days.

The Empirical Rule

One, two, and three standard deviations from the mean

7.23 According to the empirical rule, how much of a normally distributed data set lies within one, two, and three standard deviations of the mean?

According to the empirical rule, 68% of the data lies within one standard deviation of the mean, 95% of the data lies within two standard deviations, and 99.7% of the data lies within three standard deviations.

> That's one standard deviation above and one below the mean.

7.24 Demonstrate that one standard deviation around the mean includes 68% of the area under the normal distribution curve.

According to Reference Table 1, the area between the mean and $z = 1.0$ standard deviation is 0.3413. The normal curve is symmetrical, so $P(-1 < z < 0)$ and $P(0 < z < 1)$ both equal 0.3413.

$$P(-1 < z < 1) = P(-1 < z < 0) + P(0 < z < 1)$$

$$= 0.3413 + 0.3413$$

$$= 0.6826$$

7.25 Grades for a statistics exam in a particular class follow the normal distribution with a mean of 84 and a standard deviation of 4. Using the empirical rule, identify the range of grades around the mean that includes 68% of the class.

The empirical rule states that 68% of the observations from a normal distribution fall within one standard deviation of the mean. One standard deviation is 4 in this example, so one standard deviation above the mean is $\mu + \sigma = 84 + 4 = 88$ and one standard deviation below the mean is $\mu - \sigma = 84 - 4 = 80$. Thus, 68% of the exam grades fall between 80 and 88.

7.26 The number of hot dogs sold by a street vendor each day during the same hour-long period is normally distributed, with a mean of 31.6 hot dogs and a standard deviation of 7.5. Using the empirical rule, identify the range of values around the mean that includes 95% of sales numbers.

The empirical rule states that 95% of the observations from a normal distribution fall within two standard deviations of the mean. Calculate the corresponding sales numbers.

Two standard deviations above the mean = how many hot dogs sales? How about below the mean?

$$\text{upper range} = \mu + z\sigma \qquad\qquad \text{lower range} = \mu - z\sigma$$
$$= 31.6 + 2(7.5) \qquad\qquad = 31.6 - 2(7.5)$$
$$= 31.6 + 15 \qquad\qquad\quad = 31.6 - 15$$
$$= 46.6 \qquad\qquad\qquad = 16.6$$

The expected range for 95% of the hot dog demand d is $16.6 < d < 46.6$.

Note: Problems 7.27–7.29 refer to the data set below, the double occupancy room rates (in euros) at 30 three-star Paris hotels. Assume the data is normally distributed, with a mean of 152.8 euros and a standard deviation of 20.5 euros.

Sorted Room Rates in Paris									
113	120	123	126	128	129	136	139	142	143
145	146	147	152	153	153	159	161	163	165
166	166	167	169	169	170	172	180	183	199

7.27 Verify that the empirical rule holds true for one standard deviation around the mean.

Identify the interval representing one standard deviation around the mean.

$$\text{upper range} = \mu + z\sigma \qquad\qquad \text{lower range} = \mu - z\sigma$$
$$= 152.8 + 1(20.5) \qquad\qquad = 152.8 - 1(20.5)$$
$$= 173.3 \qquad\qquad\qquad = 132.3$$

Of the 30 rates listed in the table, 21 are between 132.3 and 173.3 euros. Thus, $\frac{21}{30}$, or 70%, of the rates lie within one standard deviation of the mean. According to the empirical rule, approximately 68% of the observations should fall within that interval.

Real-life data (like hotel rates) rarely fit the normal distribution exactly, but this is pretty close.

Note: Problems 7.27–7.29 refer to the data set in Problem 7.27, the double occupancy room rates (in euros) at 30 three-star Paris hotels. Assume the data is normally distributed, with a mean of 152.8 euros and a standard deviation of 20.5 euros.

7.28 Verify that the empirical rule holds true for two standard deviations around the mean.

Identify the interval representing two standard deviations around the mean.

$$\text{upper range} = \mu + z\sigma \qquad\qquad \text{lower range} = \mu - z\sigma$$
$$= 152.8 + 2(20.5) \qquad\qquad = 152.8 - 2(20.5)$$
$$= 152.8 + 41 \qquad\qquad\quad = 152.8 - 41$$
$$= 193.8 \qquad\qquad\qquad = 111.8$$

Of the 30 rates listed in the table, 29 are between 193.8 and 111.8 euros. Thus, $\frac{29}{30}$, or 96.7%, of the rates lie within two standard deviations of the mean. According to the empirical rule, approximately 95% of the observations should fall within that interval.

Note: Problems 7.27–7.29 refer to the data set in Problem 7.27, the double occupancy room rates (in euros) at 30 three-star Paris hotels. Assume the data is normally distributed, with a mean of 152.8 euros and a standard deviation of 20.5 euros.

7.29 Verify that the empirical rule holds true for three standard deviations around the mean.

Identify the interval representing three standard deviations around the mean.

$$\text{upper range} = \mu + z\sigma \qquad\qquad \text{lower range} = \mu - z\sigma$$
$$= 152.8 + 3(20.5) \qquad\qquad = 152.8 - 3(20.5)$$
$$= 152.8 + 61.5 \qquad\qquad\quad = 152.8 - 61.5$$
$$= 214.3 \qquad\qquad\qquad = 91.3$$

All 30 rates listed in the table fall between 214.3 and 91.3 euros. The empirical rule states that approximately 99.7% of the observations should fall within that interval; in this case, 100% of the observations do.

Using the Normal Distribution to Approximate the Binomial Distribution

Another binomial probability shortcut

> **7.30** Describe the conditions under which the normal distribution can be used to approximate the binomial distribution.

When it comes to binomial distributions, p is the probability of a success and q is the probability of a failure. "Success" means you got the one result out of the two you were looking for.

If n represents the number of trials in which only outcomes p and q may occur, the normal distribution can be used to approximate the binomial distribution as long as $np \geq 5$ and $nq \geq 5$.

> **7.31** Describe the continuity correction that is applied when the normal distribution approximates the binomial distribution.

Continuity correction is used when a continuous distribution (such as the normal distribution) is used to approximate a discrete distribution (such as the binomial distribution). To correct for continuity, add 0.5 to a boundary of x or subtract 0.5 from a boundary of x as directed below:

- Subtract 0.5 from the x-value representing the left boundary under the normal curve.

- Add 0.5 to the x-value representing the right boundary under the normal curve.

Note that continuity correction is unnecessary when $n > 100$.

Note: Problems 7.32–7.33 refer to a statistics class in which 60% of the students are female; 15 students from the class are randomly selected.

The problem doesn't say how big the class is, but it's got to be bigger than 15 students if you're selecting that many of them randomly.

> **7.32** Use the normal approximation to the binomial distribution to calculate the probability that this randomly selected group will contain either seven or eight female students.

Determine whether conditions have been met to use the normal distribution to approximate the binomial distribution.

These formulas come from Problem 6.5.

$$np = (15)(0.6) = 9; \ 9 \geq 5$$
$$nq = (15)(0.4) = 6; \ 6 \geq 5$$

Calculate the mean and standard deviation of the binomial distribution.

$$\mu = np = (15)(0.6) = 9$$
$$\sigma = \sqrt{npq} = \sqrt{(15)(0.6)(0.4)} = \sqrt{3.6} = 1.90$$

Subtract 0.5 from the left boundary x = 7 and add 0.5 to the right boundary x = 8.

The problem asks you to calculate $P(7 \leq x \leq 8)$. Apply the continuity correction to adjust the boundaries: $P(6.5 \leq x \leq 8.5)$.

Calculate the z-scores for endpoints $x = 6.5$ and 8.5.

$$z_{6.5} = \frac{6.5 - 9}{1.90} \qquad z_{8.5} = \frac{8.5 - 9}{1.90}$$
$$= \frac{-2.5}{1.90} \qquad\quad = \frac{-0.5}{1.90}$$
$$= -1.32 \qquad\qquad = -0.26$$

According to Reference Table 1, $P(-1.32 < z < 0) = 0.4066$ and $P(-0.26 < z < 0) = 0.1026$.

$$P(6.5 \le x \le 8.5) = P(-1.32 \le z \le -0.26)$$
$$= 0.4066 - 0.1026$$
$$= 0.3040$$

> When both x values are on the same side of the mean (in this case, they're both below 9) and you're calculating the area of the region between them, subtract the smaller probability from the larger probability.

There is a 30.4% chance that the group of 15 students will contain either seven or eight females.

Note: Problems 7.32–7.33 refer to a statistics class in which 60% of the students are female; 15 students from the class are randomly selected.

7.33 Use the binomial distribution to calculate the probability that this randomly selected group will contain either seven or eight female students and compare this to the result in Problem 7.32.

> If you don't know how to work with binomial distributions, this problem is very similar to Problems 6.6 and 6.7.

You are selecting $n = 15$ students; there is a $p = 0.6$ probability of a success (selecting a female student) and a $q = 1 - p = 0.4$ probability of a failure (selecting a male student). Calculate the probability of exactly seven or exactly eight female students in the group.

$$P(7) = \binom{n}{r} p^r q^{n-r} \qquad P(8) = \binom{n}{r} p^r q^{n-r}$$
$$= \binom{15}{7}(0.6)^7 (0.4)^{15-7} \qquad = \binom{15}{8}(0.6)^8 (0.4)^{15-8}$$
$$= (6,435)(0.6)^7 (0.4)^8 \qquad = (6,435)(0.6)^8 (0.4)^7$$
$$= 0.1181 \qquad\qquad\qquad = 0.1771$$

The probability of selecting mutually exclusive events is equal to the sum of the probabilities of the individual events.

$$P(7 \text{ or } 8) = P(7) + P(8)$$
$$= 0.1181 + 0.1771$$
$$= 0.295$$

According to the binomial distribution, there is a 29.5% probability that the group will contain either seven or eight female students. This is very close to the 30.4% probability calculated in Problem 7.32.

7.34 Use the normal approximation to the binomial distribution to calculate the probability that exactly 3 of 16 randomly selected households have at least one high-definition television.

Ensure that you can use the normal distribution to approximate the binomial distribution by verifying that np and nq are greater than or equal to five. In this problem, selecting a house with a high-definition television is a success: $p = 0.35$. Choosing a house without a high-definition television is a failure: $q = 0.65$.

$$np = (16)(0.35) = 5.6 \geq 5$$
$$nq = (16)(1 - 0.35) = 16(0.65) = 10.4 \geq 5$$

Calculate the mean and standard deviation of the binomial distribution.

$$\mu = np = (16)(0.35) = 5.6$$
$$\sigma = \sqrt{npq} = \sqrt{(16)(0.35)(0.65)} = \sqrt{3.64} = 1.91$$

The problem asks you to calculate $P(3)$. Apply the continuity correction to get $P(2.5 \leq x \leq 3.5)$. Calculate the z-scores for $x = 2.5$ and $x = 3.5$.

You can only calculate the probability of a range of values using the normal distribution, not a single number like 3. So add and subtract 0.5 to get the interval 2.5 to 3.5.

$$z_{2.5} = \frac{2.5 - 5.6}{1.91} \qquad z_{3.5} = \frac{3.5 - 5.6}{1.91}$$
$$= \frac{-3.1}{1.91} \qquad\qquad = \frac{-2.1}{1.91}$$
$$= -1.62 \qquad\qquad = -1.10$$

According to Reference Table 1, $P(-1.62 \leq z \leq 0) = 0.4474$ and $P(-1.10 \leq z \leq 0) = 0.3643$.

$$P(2.5 \leq x \leq 3.5) = P(-1.62 \leq z \leq -1.10)$$
$$= 0.4474 - 0.3643$$
$$= 0.0831$$

7.35 Use the binomial distribution to calculate the probability that exactly 3 of 16 randomly selected households have at least one high-definition television.

Recall that $p = 0.35$ and $q = 1 - 0.35 = 0.65$. Apply the binomial distribution formula to determine the probability that $r = 3$ households out of $n = 16$ have at least one high-definition television.

$$P(r) = \binom{n}{r} p^r q^{n-r}$$

$$P(3) = \binom{16}{3} (0.35)^3 (0.65)^{16-3}$$

$$P(3) = (560)(0.35)^3 (0.65)^{13}$$

$$P(3) = 0.089$$

The binomial distribution reports an 8.9% probability; recall that Problem 7.34 estimated the probability at 8.3%.

Note: Problems 7.36–7.37 refer to a process that produces strings of holiday lights. Assume 96% of the strings produced are free of defects and a customer places an order for 20 strings of lights.

7.36 Use the normal approximation to the binomial distribution to calculate the probability that exactly one or exactly two of the ordered strings will be defective.

Only nq is greater than 5. I bet you this approximation is not going to be great.
$np = (20)(0.04) = 0.8$
$nq = (20)(0.96) = 19.2$

In this problem, you are calculating the probability that a faulty string will be received. Thus, receiving a functioning string of lights is defined as a failure ($q = 0.96$) and receiving faulty lights is a success ($p = 1 - q = 0.04$). Calculate the mean and standard deviation of the distribution.

$$\mu = np = (20)(0.04) = 0.8$$

$$\sigma = \sqrt{npq} = \sqrt{(20)(0.04)(0.96)} = \sqrt{0.768} = 0.876$$

You are asked to calculate $P(1 \leq x \leq 2)$; apply the continuity correction to get $P(0.5 \leq x \leq 2.5)$. Calculate the z-scores for the boundaries of the corrected interval.

$$z_{0.5} = \frac{0.5 - 0.80}{0.876} \qquad z_{2.5} = \frac{2.5 - 0.80}{0.876}$$

$$= \frac{-0.3}{0.876} \qquad = \frac{1.7}{0.876}$$

$$= -0.34 \qquad = 1.94$$

Calculate $P(-0.34 \leq z \leq 1.94)$ by adding the values in Reference Table 1 that correspond to $z_{0.5}$ and $z_{2.5}$.

$$P(0.5 \leq x \leq 2.5) = P(-0.34 \leq z \leq 1.94)$$

$$= 0.1331 + 0.4738$$

$$= 0.607$$

Note: Problems 7.36–7.37 refer to a process that produces strings of holiday lights. Assume 96% of the strings produced are free of defects and a customer places an order for 20 strings of lights.

7.37 Use the binomial distribution to determine how closely the normal distribution approximates the probability that either one or two of the strings of lights in the order will be defective.

Apply the binomial distribution to calculate the probability that exactly $r = 1$ or exactly $r = 2$ of the $n = 20$ strings of lights are defective.

$$P(1) = \binom{20}{1}(0.04)^1 (0.96)^{20-1} \qquad P(2) = \binom{20}{2}(0.04)^2 (0.96)^{20-2}$$

$$= (20)(0.04)^1 (0.96)^{19} \qquad\qquad = (190)(0.04)^2 (0.96)^{18}$$

$$= 0.3683 \qquad\qquad\qquad\qquad = 0.1458$$

Calculate the probability that either one or two of the strings of lights ordered will be defective.

$$P(1 \text{ or } 2) = P(1) + P(2)$$
$$= 0.3683 + 0.1458$$
$$= 0.514$$

The 60.7% probability calculated in Problem 7.36 does not accurately approximate the actual probability of 51.4%.

Continuous Uniform Distribution
Going from a bell-shaped to a box-shaped distribution

7.38 Identify the defining characteristics of the continuous uniform distribution.

The continuous probability distribution is defined by two values, a and b, that represent the minimum and maximum values respectively. All intervals of the same length between a and b are equally probable.

To calculate the probability that a random variable will lie between x_1 and x_2 in the distribution, apply the following equation.

$$P\left(x_1 \le x \le x_2\right) = \frac{x_2 - x_1}{b - a}$$

Note: In Problems 7.39–7.41, assume that a vending machine dispenses coffee in amounts that vary uniformly between 7.4 and 8.2 ounces per cup.

7.39 Calculate the probability that the next cup of coffee purchased will contain between 7.5 and 7.8 ounces.

The vending machine produces a minimum of $a = 7.4$ ounces and a maximum of $b = 8.2$ ounces. Let $x_1 = 7.5$ and $x_2 = 7.8$ represent boundaries of the interval specified by the problem, and apply the uniform probability equation.

> Make sure $x_2 > x_1$.

$$P\left(x_1 \le x \le x_2\right) = \frac{x_2 - x_1}{b - a}$$

$$P(7.5 \le x \le 7.8) = \frac{7.8 - 7.5}{8.2 - 7.4}$$

$$P(7.5 \le x \le 7.8) = \frac{0.3}{0.8}$$

$$P(7.5 \le x \le 7.8) = 0.375$$

There is a 37.5% probability that the next cup will contain between 7.5 and 7.8 ounces of coffee.

Note: In Problems 7.39–7.41, assume that a vending machine dispenses coffee in amounts that vary uniformly between 7.4 and 8.2 ounces per cup.

7.40 Calculate the probability that the next cup of coffee purchased will contain between 7.6 and 8.0 ounces.

Substitute $a = 7.4$, $b = 8.2$, $x_1 = 7.6$, and $x_2 = 8.0$ into the uniform probability equation.

$$P\left(x_1 \le x \le x_2\right) = \frac{x_2 - x_1}{b - a}$$

$$P(7.6 \le x \le 8.0) = \frac{8.0 - 7.6}{8.2 - 7.4}$$

$$P(7.6 \le x \le 8.0) = \frac{0.4}{0.8}$$

$$P(7.6 \le x \le 8.0) = 0.5$$

The interval bounded by the maximum and minimum volumes of coffee dispensed has a length of $8.2 - 7.4 = 0.8$. This problem specifies a subinterval that is half as long: $8.0 - 7.6 = 0.4$.

Intervals of equal length are equally probable, so if the subinterval is half the length of the total interval, then the probability that a randomly selected value will fall on the subinterval is 0.5, as illustrated in the following figure.

> If you'd picked a subinterval that was 25% as long as the original interval, there'd be a 25% chance that a randomly selected value would occur in the subinterval you chose.

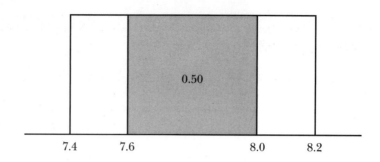

Note: In Problems 7.39–7.41, assume that a vending machine dispenses coffee in amounts that vary uniformly between 7.4 and 8.2 ounces per cup.

7.41   Calculate the mean and standard deviation of the distribution.

The mean of a continuous uniform distribution is the average of the minimum value a and the maximum value b.

$$\mu = \frac{a+b}{2} = \frac{7.4+8.2}{2} = \frac{15.6}{2} = 7.8$$

The standard deviation of the distribution is $\sigma = \frac{b-a}{\sqrt{12}}$.

The units of the mean and standard deviation match the units of the distribution. In this case, the mean is 7.8 ounces and the standard deviation is 0.231 ounces.

$$\sigma = \frac{b-a}{\sqrt{12}} = \frac{8.2-7.4}{\sqrt{12}} = \frac{0.8}{\sqrt{12}} = 0.231$$

Note: In Problems 7.42–7.44, assume that the time it takes the housekeeping crew of a local hotel to clean a room varies uniformly between 28 and 45 minutes.

7.42   Calculate the probability that the next room will require more than 35 minutes to clean.

It takes between $a = 28$ and $b = 45$ minutes to clean the room. Apply the uniform probability equation, setting $x_1 = 35$ and $x_2 = 45$.

45 minutes is the maximum value, so the subinterval you're testing can't get any bigger than 45.

$$P(x_1 \le x \le x_2) = \frac{x_2 - x_1}{b-a}$$
$$P(35 \le x \le 45) = \frac{45-35}{45-28}$$
$$P(35 \le x \le 45) = \frac{10}{17}$$
$$P(35 \le x \le 45) = 0.588$$

Note: In Problems 7.42–7.44, assume that the time it takes the housekeeping crew of a local hotel to clean a room varies uniformly between 28 and 45 minutes.

7.43 Calculate the probability that the next room will require less than 32 minutes to clean.

If it takes housekeeping less than 32 minutes to clean a room, then it took them between 28 and 32 minutes.

The problem says that 28 minutes is as fast as housekeeping can go.

$$P\left(x_1 \le x \le x_2\right) = \frac{x_2 - x_1}{b - a}$$

$$P\left(28 \le x \le 32\right) = \frac{32 - 28}{45 - 28}$$

$$P\left(28 \le x \le 32\right) = \frac{4}{17}$$

$$P\left(28 \le x \le 32\right) = 0.235$$

Note: In Problems 7.42–7.44, assume that the time it takes the housekeeping crew of a local hotel to clean a room varies uniformly between 28 and 45 minutes.

7.44 Calculate the mean and standard deviation for the distribution.

Substitute $a = 28$ and $b = 45$ into the formulas presented in Problem 7.41.

$$\mu = \frac{a + b}{2} = \frac{28 + 45}{2} = \frac{73}{2} = 36.5$$

$$\sigma = \frac{b - a}{\sqrt{12}} = \frac{45 - 28}{\sqrt{12}} = \frac{17}{\sqrt{12}} = 4.907$$

Exponential Distribution
Like the Poisson distribution, but continuous

7.45 Identify the defining characteristics of the exponential distribution.

The exponential probability distribution is a continuous distribution commonly used to measure the time between events of interest, such as the time between customer arrivals at a retail store or the time between failures in a process.

In Chapter 6, the variable λ was used to represent the mean of the Poisson distribution, a discrete distribution that counted the number of times an event occurred during a specific time period. The mean and the standard deviation of the exponential distribution are both $\frac{1}{\lambda}$.

To review the Poisson distribution, check out Problems 6.17–6.31.

The exponential distribution is the continuous counterpart of the discrete Poisson distribution. For example, if a random variable follows the Poisson distribution with an average occurrence of two times per minute ($\lambda = 2$), then the same random variable also follows the exponential distribution with a mean and standard deviation of $\frac{1}{\lambda} = \frac{1}{2} = 0.5$.

The constant e is Euler's number, the nonrepeating, nonterminating decimal 2.71828....

If x is a random variable that follows the exponential distribution, then the probability that $x \geq t$ is $e^{-\lambda t}$.

$$P(x \geq t) = e^{-\lambda t}$$

Note: In Problems 7.46–7.48, assume that the average elapsed time between customers entering a retail store is exponentially distributed and averages 12 minutes.

7.46 Calculate the probability that the elapsed time between two customers will be 10 minutes or more.

So, on average, a new person comes into the store every 12 minutes.

The mean of the exponential distribution is $\dfrac{1}{\lambda} = 12$ minutes per customer. Thus, there are $\lambda = \dfrac{1}{12} = 0.083$ customers per minute over a $t = 10$ minute period. Apply the exponential probability formula.

$$P(x \geq t) = e^{-\lambda t}$$
$$P(x \geq 10) = e^{-(0.083)(10)}$$
$$P(x \geq 10) = e^{-0.83}$$
$$P(x \geq 10) = 0.436$$

It's easy to confuse $\dfrac{1}{\lambda}$ and λ, but the units help. The units for $\dfrac{1}{\lambda}$ are always a continuous measurement, minutes in this problem. The units for λ are always a discrete measurement; in this problem, it's the number of customers.

There is a 43.6% probability the elapsed time between two customers will be 10 minutes or more.

Note: In Problems 7.46–7.48, assume that the average elapsed time between customers entering a retail store is exponentially distributed and averages 12 minutes.

7.47 Calculate the probability that the next customer will arrive less than four minutes after the previous customer.

Substituting $t = 4$ into the exponential probability formula calculates the complement, the probability of the next customer arriving *more than* four minutes after the previous customer. Subtract that probability from 1 to determine the probability that the next customer will arrive *less than* four minutes later.

$$P(x < t) = 1 - e^{-\lambda t}$$
$$P(x < 4) = 1 - e^{-(0.083)(4)}$$
$$P(x < 4) = 1 - e^{-0.332}$$
$$P(x < 4) = 0.283$$

Note: In Problems 7.46–7.48, assume that the average elapsed time between customers entering a retail store is exponentially distributed and averages 12 minutes.

7.48 Calculate the standard deviation of the distribution.

The standard deviation of an exponential distribution is equal to the mean: $\sigma = 12$.

Note: In Problems 7.49–7.51, assume that the tread life of a particular brand of tire is exponentially distributed and averages 32,000 miles.

7.49 Calculate the probability that a set of these tires will have a tread life of at least 38,000 miles.

The mean of the distribution is $\dfrac{1}{\lambda} = 32$ miles per set of tires (in thousands), so $\lambda = \dfrac{1}{32} = 0.03125$ sets of tires per thousand miles. Substitute $t = 38$ into the exponential probability formula to calculate the probability that a particular set of tires will have a tread life of more than 38,000 miles.

$$P(x \geq t) = e^{-\lambda t}$$
$$P(x \geq 38) = e^{-(0.03125)(38)}$$
$$P(x \geq 38) = e^{-1.1875}$$
$$P(x \geq 38) = 0.305$$

Note: In Problems 7.49–7.51, assume that the tread life of a particular brand of tire is exponentially distributed and averages 32,000 miles.

7.50 Calculate the probability that a set of these tires has a tread life of less than 22,000 miles.

In order to calculate the probability that a randomly selected value from an exponentially distributed population is less than $t = 22,000$ miles, apply the complement of the exponential probability formula.

$$P(x < 22) = 1 - e^{-(0.03125)(22)}$$
$$= 1 - e^{-0.6875}$$
$$= 0.497$$

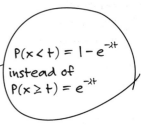

$P(x < t) = 1 - e^{-\lambda t}$
instead of
$P(x \geq t) = e^{-\lambda t}$

Note: In Problems 7.49–7.51, assume that the tread life of a particular brand of tire is exponentially distributed and averages 32,000 miles.

7.51 Calculate the probability that a set of these tires will have a tread life of between 33,000 and 40,000 miles.

In order to calculate $P(33 < x < 40)$, determine the probability that the tires will last less than 40,000 miles and subtract the probability that the tires will last less than 33,000 miles.

$$P(33 < x < 40) = P(x < 40) - P(x < 33)$$
$$= \left[1 - e^{-(0.03125)(40)}\right] - \left[1 - e^{-(0.03125)(33)}\right]$$
$$= \left[1 - e^{-1.25}\right] - \left[1 - e^{-1.03125}\right]$$

Distribute the negative sign through the second quantity.

$$= 1 - e^{-1.25} - 1 + e^{-1.03125}$$
$$= (1-1) + \left(-e^{-1.25} + e^{-1.03125}\right)$$
$$= 0.070$$

In this problem, you're given a discrete measurement (trucks) instead of a continuous measurement (hours). That means you're given λ instead of $\frac{1}{\lambda}$.

There is a 7.0% probability that a set of these tires will have a tread life of between 33,000 and 40,000 miles.

Note: In Problems 7.52–7.54, assume that an average of 3.5 trucks per hour arrive at a loading dock and that the elapsed time between arrivals is exponentially distributed.

7.52 Calculate the probability that the next truck will arrive at least 30 minutes after the previous truck.

Note that λ is 3.5 trucks per hour but t is expressed in minutes. Convert λ into trucks per minute so that the units are consistent.

You could make $t = 0.5$ hours instead, but Problems 7.53 and 7.54 are in minutes, too, so you might as well use minutes.

$$\lambda = \frac{3.5 \text{ trucks}}{1 \text{ hour}} = \frac{3.5 \text{ trucks}}{60 \text{ minutes}} = 0.0583 \text{ trucks per minute}$$

Apply the exponential probability formula.

$$P(x \geq 30) = e^{-(0.0583)(30)}$$
$$= e^{-1.749}$$
$$= 0.174$$

Note: In Problems 7.52–7.54, assume that an average of 3.5 trucks per hour arrive at a loading dock and that the elapsed time between arrivals is exponentially distributed.

7.53 Calculate the probability that the next truck will arrive no more than six minutes after the previous truck.

Calculate the complement of the exponential probability $P(x \geq 6)$.

$$P(x < 6) = 1 - e^{-(0.0583)(6)}$$
$$= 1 - e^{-0.3498}$$
$$= 0.295$$

Note: In Problems 7.52–7.54, assume that an average of 3.5 trucks per hour arrive at a loading dock and that the elapsed time between arrivals is exponentially distributed.

7.54 Calculate the probability that the next truck will arrive between three and ten minutes after the previous truck.

The probability that a truck will arrive between three and ten minutes after the previous truck is equal to the difference of those exponential probabilities.

$$P(3 < x < 10) = P(x < 10) - P(x < 3)$$
$$= \left[1 - e^{-(0.0583)(10)}\right] - \left[1 - e^{-(0.0583)(3)}\right]$$
$$= 1 - e^{-0.583} - 1 + e^{-0.175}$$
$$= (1 - 1) + \left(-e^{-0.583} + e^{-0.175}\right)$$
$$= 0.281$$

$P(x < 10)$ includes all values of x between 0 and 10. You want the probability that x will be between 3 and 10, so you need to remove the probability that x is less than 3: $P(x < 3)$.

Chapter 8
SAMPLING AND SAMPLING DISTRIBUTIONS

Working with a subset of a population

A population is defined as all possible outcomes or measurements of interest, whereas a sample is a subset of a population. Many populations are infinitely large; thus, virtually all statistical analyses are conducted on samples drawn from a population. In order to interpret the results of these analyses correctly, you must first understand the behavior of samples. In this chapter, you will do just that through the exploration of sampling distributions.

This chapter relies heavily on the normal probability distribution concepts introduced in Chapter 7. The two major topics are the sampling distribution of the mean and the sampling distribution of the proportion. Also make sure you understand binomial distributions, as they make a guest appearance late in the chapter.

Probability Sampling

So many ways to gather a sample

8.1 Describe how to select a simple random sample from a population.

A simple random sample is a sample that is randomly selected so that every combination has an equal chance of being chosen. If an urn contains six balls of different colors, selecting three of the balls without looking inside the urn is an example of a simple random sample.

8.2 Describe how to select a systematic sample from a population.

Systematic sampling includes every kth member of the population in the sample; the value of k will depend on the size of the population and the size of the sample that is desired. For instance, if a sample size of 50 is needed from a population of 1,000, then $k = \dfrac{1{,}000}{50} = 20$. Systematically, every twentieth person from the population is selected and included in the sample.

8.3 Describe how to select a cluster sample from a population.

Cluster sampling first divides the population into groups (or clusters) and then randomly selects clusters to include in the sample. The entire cluster or just a randomly selected portion of it may be selected. For example, if a researcher wishes to poll a sample of shoppers at a shopping mall, she might choose a few stores randomly, and then interview the customers inside those stores only. In this example, the stores are the clusters.

In order for cluster sampling to be effective, each cluster selected for the sample needs to be representative of the population at large.

> Cluster sampling is cost-effective because it requires minimal research about the population. In the mall example, you didn't have to know anything about the shoppers ahead of time—you just needed to pick a few stores from the map.

8.4 Describe how to select a stratified sample from a population.

Stratified sampling first divides the population into mutually exclusive groups (or strata) and then selects a random sample from each of those groups. It differs from cluster sampling in that strata are defined in terms of specific characteristics of the population, whereas clusters produce less homogeneous samples.

Consider the example presented in Problem 8.3, in which clusters are assigned based upon the stores in a mall. A stratified sample would be chosen in terms of a specific customer characteristic, such as gender. Stratified sampling is helpful when it is important that the sample have certain characteristics of the overall population. Usually the sample sizes are proportional to their known relationship in the population.

> If cluster sampling had been used at the mall to ask how male teenagers respond to a new product, there's no guarantee that the cluster sample would have included male teenagers at all.

Sampling Distribution of the Mean
Predicting the behavior of sample means

8.5 Identify the implications of the central limit theorem on the sampling distribution of the mean.

> It's called the CENTRAL limit theorem because it's the most important theorem in statistics.

According to the central limit theorem, as a sample size n gets larger, the distribution of the sample means more closely approximates a normal distribution, regardless of the distribution of the population from which the sample was drawn. As a general rule of thumb, the assertions of the central limit theorem are valid when $n \geq 30$. If the population itself is normally distributed, the sampling distribution of the mean is normal for any sample size.

As the sample size increases, the distribution of sample means converges toward the center of the distribution. Thus, as the sample size increases, the standard deviation of the sample means decreases. According to the central limit theorem, the standard deviation of the sample means $\sigma_{\bar{x}}$ is equal to $\dfrac{\sigma}{\sqrt{n}}$, where σ is the standard deviation of the population and n is the sample size.

The standard deviation of the sample mean is formally known as the standard error of the mean. The z–score for sample means is calculated based on the formula below.

$$z_{\bar{x}} = \frac{\bar{x} - \mu}{\sigma_{\bar{x}}}$$

> The variable $\bar{x}$ represents the mean of the sample.

Note: In Problems 8.6–8.8, assume that the systolic blood pressure of 30-year-old males is normally distributed, with an average of 122 mmHg and a standard deviation of 10 mmHg.

8.6 A random sample of 16 men from this age group is selected. Calculate the probability that the average blood pressure of the sample will be greater than 125 mmHg.

> The unit mmHg stands for "millimeters of mercury."

The population is normally distributed, so sample means are also normally distributed for any sample size. Calculate the standard error of the mean.

$$\sigma_{\bar{x}} = \frac{\sigma}{\sqrt{n}}$$

$$\sigma_{125} = \frac{10}{\sqrt{16}}$$

$$\sigma_{125} = \frac{10}{4}$$

$$\sigma_{125} = 2.5$$

Calculate the z-score for the sample mean, $\bar{x} = 125$.

$$z_{\bar{x}} = \frac{\bar{x} - \mu}{\sigma_{\bar{x}}}$$

$$z_{125} = \frac{125 - 122}{2.5}$$

$$z_{125} = \frac{3}{2.5}$$

$$z_{125} = 1.2$$

> See Problem 7.3 if you're not sure how to use Reference Table 1.

According to Reference Table 1, the normal probability associated with $z = 1.20$ is 0.3849. The probability that the sample mean will be greater than 125 is the area of the shaded region beneath the normal curve in the figure below. The area below the curve on each side of the mean is 0.5, and the area between the mean and the z-score 1.20 is 0.3849.

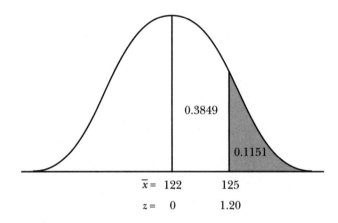

Calculate the probability that the average blood pressure of the sample will be greater than 125 mmHg.

$$P(\bar{x} > 125) = P(z_{\bar{x}} > 1.20)$$
$$= 0.5 - 0.3849$$
$$= 0.1151$$

Note: In Problems 8.6–8.8, assume that the systolic blood pressure of 30-year-old males is normally distributed, with an average of 122 mmHg and a standard deviation of 10 mmHg.

8.7 A random sample of 16 men from this age group had their blood pressure measured. Calculate the probability that the average blood pressure of this sample will be between 118 and 124 mmHg.

According to Problem 8.6, the standard error of the mean for a sample consisting of $n = 16$ members of the population is $\sigma_{\bar{x}} = 2.5$. Calculate the z-scores for $\bar{x} = 118$ and $\bar{x} = 124$.

$$z_{118} = \frac{118 - 122}{2.5} \qquad z_{124} = \frac{124 - 122}{2.5}$$

$$= \frac{-4}{2.5} \qquad\qquad = \frac{2}{2.5}$$

$$= -1.60 \qquad\qquad = 0.80$$

Identify the probabilities associated with these z-scores in Reference Table 1 and calculate the probability that the average blood pressure of the sample is between 118 and 124 mmHg. ←

One of the interval boundaries (118) is below the mean and one is above (124). When the boundaries are on different sides of the mean, add the probabilities together.

$$P\left(118 < \bar{x} < 124\right) = P\left(-1.60 < z_{\bar{x}} < 0.80\right)$$

$$= 0.4452 + 0.2881$$

$$= 0.7333$$

Note: In Problems 8.6–8.8, assume that the systolic blood pressure of 30-year-old males is normally distributed, with an average of 122 mmHg and a standard deviation of 10 mmHg.

8.8 Calculate the probability that the blood pressure of an individual male from this population will be between 118 and 124 mmHg.

Calculate the z-scores for $x = 118$ and $x = 124$.

$$z_x = \frac{x - \mu}{\sigma} \qquad\qquad z_x = \frac{x - \mu}{\sigma}$$

$$z_{118} = \frac{118 - 122}{10} \qquad z_{124} = \frac{124 - 122}{10}$$

$$z_{118} = \frac{-4}{10} \qquad\qquad z_{124} = \frac{2}{10}$$

$$z_{118} = -0.40 \qquad\qquad z_{124} = 0.20$$

This time, you're dealing with a single data point, not a sample mean, so you divide by the standard deviation of the population (10) instead of the standard error of the mean.

According to Reference Table 1, $P(-0.40 < z < 0) = 0.1554$ and $P(0 < z < 0.20) = 0.0793$.

$$P\left(118 < x < 124\right) = P\left(-0.40 < z < 0.20\right)$$

$$= 0.1554 + 0.0793$$

$$= 0.2347$$

This probability that a single value x lies in the interval $118 < x < 124$ is lower than the probability that a sample of $n = 16$ individuals has a mean that lies in the same interval ($23.5\% < 73.3\%$). Sample means more closely approximate the population mean than individual observations.

You calculate 0.7333 in Problem 8.7.

Note: In Problems 8.9–8.12, assume that the average weight of an NFL player is 245.7 pounds with a standard deviation of 34.5 pounds, but the probability distribution of the population is unknown.

8.9 If a random sample of 32 NFL players is selected, what is the probability that the average weight of the sample will be less than 234 pounds?

Because the probability distribution is unknown, you need a sample size of 30 or more to apply the central limit theorem's assertion that the sample means are normally distributed. In this problem, $n = 32 > 30$. Calculate the standard error of the mean.

$$\sigma_{\bar{x}} = \frac{\sigma}{\sqrt{n}} = \frac{34.5}{\sqrt{32}} = 6.099$$

Calculate the z-score for the sample mean, $\bar{x} = 234$.

$$z_{\bar{x}} = \frac{\bar{x} - \mu}{\sigma_{\bar{x}}} = \frac{234 - 245.7}{6.099} = \frac{-11.7}{6.099} = -1.92$$

The z-score for $z = -1.92$ is 0.4726.

Because the sample means are normally distributed, there is a 0.5 probability that the sample mean is less than the mean of the population, 245.7 pounds. According to Reference Table 1, there is a 0.4726 probability that the sample mean will be between 234 and 245.7 pounds. Thus, there is a $0.5 - 0.4726 = 0.0274$ probability that the sample mean will be less than 234 pounds.

Note: In Problems 8.9–8.12, assume that the average weight of an NFL player is 245.7 pounds with a standard deviation of 34.5 pounds, but the probability distribution of the population is unknown.

8.10 If a random sample of 32 NFL players is selected, what is the probability that the average weight of the sample is between 248 and 254 pounds?

According to Problem 8.9, the standard error of a sample consisting of $n = 32$ NFL players is $\sigma_{\bar{x}} = 6.099$. Calculate the z-scores for $\bar{x} = 248$ and $\bar{x} = 254$.

$$z_{248} = \frac{248 - 245.7}{6.099} \qquad z_{254} = \frac{254 - 245.7}{6.099}$$

$$= \frac{2.3}{6.099} \qquad\qquad = \frac{8.3}{6.099}$$

$$= 0.38 \qquad\qquad\quad = 1.36$$

Both z-scores are on the same side of the mean, so subtract the probability of the z-score closer to the mean from the probability of the z-score farther from the mean.

Calculate $P\left(248 < \bar{x} < 254\right)$.

$$P\left(248 < \bar{x} < 254\right) = P\left(0.38 < z_{\bar{x}} < 1.36\right)$$

$$= 0.4131 - 0.1480$$

$$= 0.2651$$

Note: In Problems 8.9–8.12, assume that the average weight of an NFL player is 245.7 pounds with a standard deviation of 34.5 pounds, but the probability distribution of the population is unknown.

8.11 If a random sample of 32 NFL players is selected, what is the probability that the average weight of the sample is between 242 and 251 pounds?

> You're going to compare this to the probability of another mean occurring in the same interval in Problem 8.12, but that sample will be almost four times as large.

Recall that $\sigma_{\bar{x}} = 6.099$. Calculate the z-scores for $\bar{x} = 242$ and $\bar{x} = 251$.

$$z_{242} = \frac{242 - 245.7}{6.099} \qquad z_{251} = \frac{251 - 245.7}{6.099}$$

$$= \frac{-3.7}{6.099} \qquad\qquad = \frac{5.3}{6.099}$$

$$= -0.61 \qquad\qquad = 0.87$$

Calculate $P(242 < \bar{x} < 251)$.

$$P(242 < \bar{x} < 251) = P(-0.61 < z_{\bar{x}} < 0.87)$$

$$= 0.2291 + 0.3078$$

$$= 0.5369$$

Note: In Problems 8.9–8.12, assume that the average weight of an NFL player is 245.7 pounds with a standard deviation of 34.5 pounds, but the probability distribution of the population is unknown.

8.12 Calculate the probability that the average weight of a sample is between 242 and 251 pounds if the sample consists of $n = 120$ players.

Calculate the standard error of the mean, given $\sigma = 34.5$ and $n = 120$.

$$\sigma_{\bar{x}} = \frac{\sigma}{\sqrt{n}} = \frac{34.5}{\sqrt{120}} = 3.149$$

Increasing the sample size from 32 to 120 has reduced the standard error of the mean from 6.099 to 3.149. The larger the sample, the more closely its mean will approximate the mean of the population.

Calculate the z-scores for $\bar{x} = 242$ and $\bar{x} = 251$.

> You have to recalculate z_{242} and z_{251} because the standard error is different than it was in Problem 8.11.

$$z_{242} = \frac{242 - 245.7}{3.149} \qquad z_{251} = \frac{251 - 245.7}{3.149}$$

$$= \frac{-3.7}{3.149} \qquad\qquad = \frac{5.3}{3.149}$$

$$= -1.17 \qquad\qquad = 1.68$$

Calculate $P(242 < \bar{x} < 251)$.

$$P\left(242 < \bar{x} < 251\right) = P\left(-1.17 < z_{\bar{x}} < 1.68\right)$$
$$= 0.3790 + 0.4535$$
$$= 0.8325$$

> The mean of the larger sample has an 83.3% chance of being between 252 and 241, whereas the mean of the smaller sample has only a 53.7% chance (according to Problem 8.11).

Larger samples have less variability, so it is more probable that the sample mean of a larger population will occur in an interval containing the population mean.

Note: Problems 8.13–8.16 refer to a 2001 report that claimed the average annual consumption of milk in the United States was 23.4 gallons per person with a standard deviation of 7.1 gallons per person.

8.13 If a random sample of 40 American citizens is selected, what is the probability that their average milk consumption is less than 25 gallons per person annually?

> In this example, $n > 40$, so you can assume the sampling distribution of the mean is approximately normal.

Calculate the standard error of the mean.

$$\sigma_{\bar{x}} = \frac{\sigma}{\sqrt{n}} = \frac{7.1}{\sqrt{40}} = 1.123$$

Calculate the z-score for $\bar{x} = 25$.

$$z_{25} = \frac{25 - 23.4}{1.123} = \frac{1.6}{1.123} = 1.42$$

There is a 0.5 probability that the sample mean is less than the population mean of 23.4. According to Reference Table 1, there is a 0.4222 probability that the sample mean is between the population mean of 23.4 gallons and 25 gallons (which has a z-score of 1.42).

$$P\left(\bar{x} < 25\right) = P\left(z_{\bar{x}} < 1.42\right)$$
$$= P\left(z_{\bar{x}} < 0\right) + P\left(0 < z_{\bar{x}} < 1.42\right)$$
$$= 0.5 + 0.4222$$
$$= 0.9222$$

Note: Problems 8.13–8.16 refer to a 2001 report that claimed the average annual consumption of milk in the United States was 23.4 gallons per person with a standard deviation of 7.1 gallons per person.

8.14 If a random sample of 40 American citizens is selected, what is the probability that their average milk consumption is between 21 and 22 gallons of milk per person annually?

According to Problem 8.13, $\sigma_{\bar{x}} = 1.123$. Calculate the z-scores for $\bar{x} = 21$ and $\bar{x} = 22$.

$$z_{21} = \frac{21 - 23.4}{1.123} \qquad z_{22} = \frac{22 - 23.4}{1.123}$$

$$= \frac{-2.4}{1.123} \qquad = \frac{-1.4}{1.123}$$

$$= -2.14 \qquad = -1.25$$

Calculate $P(21 < \bar{x} < 22)$.

$$P(21 < \bar{x} < 22) = P(-2.14 < z_{\bar{x}} < -1.25)$$

$$= 0.4838 - 0.3944$$

$$= 0.0894$$

Note: Problems 8.13–8.16 refer to a 2001 report that claimed the average annual consumption of milk in the United States was 23.4 gallons per person with a standard deviation of 7.1 gallons per person.

8.15 If a random sample of 60 American citizens is selected, what is the probability that their average milk consumption is more than 23 gallons of milk per person annually?

> Increasing the sample size from 40 to 60 decreases the standard error of the mean from 1.123 to 0.917.

Calculate the standard error of the mean.

$$\sigma_{\bar{x}} = \frac{\sigma}{\sqrt{n}} = \frac{7.1}{\sqrt{60}} = 0.917$$

Calculate the z-score for $\bar{x} = 23$.

$$z_{23} = \frac{23 - 23.4}{0.917} = \frac{-0.4}{0.917} = -0.44$$

There is a 0.5 probability that the sample mean is greater than the population mean of 23.4. According to Reference Table 1, there is a 0.1700 probability that the sample mean is between 23 and 23.4.

$$P(\bar{x} > 23) = P(-0.44 < z_{\bar{x}} < 0) + P(z_{\bar{x}} > 0)$$

$$= 0.1700 + 0.50$$

$$= .6700$$

Note: Problems 8.13–8.16 refer to a 2001 report that claimed the average annual consumption of milk in the United States was 23.4 gallons per person with a standard deviation of 7.1 gallons per person.

8.16 If a recent random sample of 60 American citizens is selected and the sample mean is 20.6 gallons per person, how likely is it that the true population mean is still 23.4 gallons per person?

Recall that $\sigma_{\bar{x}} = 0.917$. Calculate the z-score for $\bar{x} = 20.6$.

$$z_{20.6} = \frac{20.6 - 23.4}{0.917} = \frac{-2.8}{0.917} = -3.05$$

Calculate the probability that the average annual consumption of milk from this sample is 20.6 gallons or less per person.

$$
\begin{aligned}
P(\bar{x} \le 20.6) &= P(z_{\bar{x}} \le -3.05) \\
&= P(z_{\bar{x}} < 0) - P(-3.05 \le z_{\bar{x}} \le 0) \\
&= 0.5 - 0.4989 \\
&= 0.0011
\end{aligned}
$$

If the true population mean is 23.4 gallons per person, then there is only a 0.11% chance that a sample of size $n = 60$ will have a sample mean of 20.6 gallons or less. Therefore, it is highly unlikely that the actual average annual consumption of milk in the United States is still 23.4 gallons per person, assuming the sample is representative of the population.

8.17 Quality control programs often establish control limits that are three standard deviations from the target mean of a process. If the mean of a sample taken from the process is within the control limits, the process is deemed satisfactory.

A process is designed to fill bottles with 16 ounces of soda with a standard deviation of 0.5 ounces. Determine the control limits above and below the mean for this process using a sample size of $n = 30$.

This is the standard deviation of the sample mean. In order for the process to be satisfactory, it's got to be within three of these standard deviations above or below the mean.

Calculate the standard error of the mean.

$$\sigma_{\bar{x}} = \frac{\sigma}{\sqrt{n}} = \frac{0.5}{\sqrt{30}} = 0.091$$

The lower control limit is three standard errors below the mean, whereas the upper control limit is three standard errors above the mean.

$$\text{lower control limit} = \mu - 3(\sigma_{\bar{x}}) = 16 - 3(0.091) = 16 - 0.273 = 15.727$$

$$\text{upper control limit} = \mu + 3(\sigma_{\bar{x}}) = 16 + 3(0.091) = 16 + 0.273 = 16.273$$

If a 30-bottle sample is collected, the process is considered satisfactory if the sample mean is between 15.727 ounces and 16.273 ounces.

Finite Population Correction Factor
Sampling distribution of the mean with a small population

8.18 Describe the finite population correction factor for the sampling distribution for the mean and the conditions under which it should be applied.

When a population is very large, selecting something as part of a sample has a negligible impact on the population. For instance, if you randomly chose individuals from the continent of Europe and recorded the gender of the individuals you chose, selecting a finite number of men would not significantly change the probability that the next individual you chose would also be male.

However, when the sample size n is more than 5 percent of the population size N, the finite population correction factor below should be applied. Under this condition, the population size is small enough that the sampling events are no longer independent of one another. The selection of one item from the population impacts the probability of future items being selected.

In other words, when $\frac{n}{N} > 0.05$. Some textbooks say 10% instead of 5%.

$$\sigma_{\bar{x}} = \frac{\sigma}{\sqrt{n}} \sqrt{\frac{N-n}{N-1}}$$

Note: Problems 8.19–8.20 refer to a process that fills boxes with a mean of 340 grams of cereal, with a standard deviation of 20 grams. Assume the probability distribution for this population is unknown.

8.19 If a store purchases 600 boxes of cereal, what is the probability that a sample of 50 boxes from the order will average less than 336 grams?

Note that the sample is more than 5% of the total population: $\frac{n}{N} = \frac{50}{600} = 0.083 > 0.05$. Thus, you must apply the finite population correction factor when calculating the standard error of the mean.

Without the correction, the standard error of the mean is 2.83.

$$\sigma_{\bar{x}} = \frac{\sigma}{\sqrt{n}} \sqrt{\frac{N-n}{N-1}} = \frac{20}{\sqrt{50}} \sqrt{\frac{600-50}{600-1}} = \frac{20}{\sqrt{50}} \sqrt{\frac{550}{599}} = 2.710$$

Calculate the z-score for $\bar{x} = 336$.

$$z_{\bar{x}} = \frac{\bar{x} - \mu}{\sigma_{\bar{x}}} = \frac{336 - 340}{2.710} = \frac{-4}{2.710} = -1.48$$

There is a 0.5 probability that the sample mean will be less than the population mean of 340. According to Reference Table 1, there is a 0.4306 probability that the sample mean will be between 336 and 340.

$$P(\bar{x} < 336) = P(z_{\bar{x}} < -1.48)$$
$$= P(z_{\bar{x}} < 0) - P(-1.48 < z_{\bar{x}} < 0)$$
$$= 0.50 - 0.4306$$
$$= 0.0694$$

Note: Problems 8.19–8.20 refer to a process that fills boxes with a mean of 340 grams of cereal, with a standard deviation of 20 grams. Assume the probability distribution for this population is unknown.

8.20 If a store purchases 600 boxes of cereal, what is the probability that a sample of 100 boxes from the order will average between 342 and 345 grams?

The $n = 100$ box sample is more than 5 percent of the $N = 600$ box order $\left(\dfrac{n}{N} = \dfrac{100}{600} = 0.167 > 0.05\right)$, so apply the finite population correction factor to calculate the standard error of the mean.

> As the sample size gets closer to the population size, the finite population correction factor has a bigger effect on the standard error.

$$\sigma_{\bar{x}} = \frac{\sigma}{\sqrt{n}}\sqrt{\frac{N-n}{N-1}} = \frac{20}{\sqrt{100}}\sqrt{\frac{600-100}{600-1}} = \frac{20}{10}\sqrt{\frac{500}{599}} = 2\sqrt{\frac{500}{599}} = 1.827$$

Calculate the z-scores for $\bar{x} = 342$ and $\bar{x} = 345$.

$$z_{342} = \frac{342-340}{1.827} \qquad z_{345} = \frac{345-340}{1.827}$$

$$= \frac{2}{1.827} \qquad\qquad = \frac{5}{1.827}$$

$$= 1.09 \qquad\qquad\quad = 2.74$$

There is a 0.4969 probability that the sample mean is between 340 and 345; there is a 0.3631 probability that the sample mean is between 340 and 342. Thus, there is a 0.4969 – 0.3631 = 0.1338 probability that the sample mean is between 342 and 345 grams.

Note: In Problems 8.21–8.22, assume that a teacher needs to grade 155 exams and the amount of time it takes to grade each of those exams is a normally distributed population, with an average of 12 minutes per exam and a standard deviation of 4 minutes per exam.

8.21 Calculate the probability that it will take an average of more than 10 minutes per exam to grade a random sample of 20 exams.

A sample of $n = 20$ exams constitutes nearly 13% of the $N = 155$ exam population, so apply the finite population correction factor to calculate the standard error of the mean.

$$\sigma_{\bar{x}} = \frac{\sigma}{\sqrt{n}}\sqrt{\frac{N-n}{N-1}} = \frac{4}{\sqrt{20}}\sqrt{\frac{155-20}{155-1}} = \frac{4}{\sqrt{20}}\sqrt{\frac{135}{154}} = 0.837$$

Calculate the z-score for $\bar{x} = 10$.

$$z_{\bar{x}} = \frac{\bar{x}-\mu}{\sigma_{\bar{x}}} = \frac{10-12}{0.837} = \frac{-2}{0.837} = -2.39$$

There is a 0.4916 probability that the sample mean will be between 10 and 12 minutes per exam; there is a 0.5 probability that the sample mean is greater than the population mean of 12 minutes per exam. Thus, the probability that the sample mean is 10 minutes per exam or greater is 0.4916 + 0.5 = 0.9916.

Note: In Problems 8.21–8.22, assume that a teacher needs to grade 155 exams and the amount of time it takes to grade each of those exams is a normally distributed population, with an average of 12 minutes per exam and a standard deviation of 4 minutes per exam.

8.22 A sample of 16 exams requires an average of 11 minutes each to grade. How likely is it that the teacher actually grades each exam in 11 minutes or less?

> The teacher graded these 16 exams a little faster than he anticipated. Is the teacher actually overestimating the average time it takes him to grade an exam?

A sample of $n = 16$ exams constitutes more than 10% of the $N = 155$ exam population, so apply the finite population correction factor to calculate the standard error of the mean.

$$\sigma_{\bar{x}} = \frac{\sigma}{\sqrt{n}} \sqrt{\frac{N-n}{N-1}} = \frac{4}{\sqrt{16}} \sqrt{\frac{155-16}{155-1}} = \frac{4}{4} \sqrt{\frac{139}{154}} = \sqrt{\frac{139}{154}} = 0.950$$

Calculate the z-score for $\bar{x} = 11$.

$$z_{11} = \frac{11-12}{0.950} = \frac{-1}{0.950} = -1.05$$

If the population average truly is 12, then there is a 0.50 probability that the sample mean is less than 12; there is a 0.3531 probability that the sample mean is between 11 and 12. Thus, there is a $0.50 - 0.3531 = 0.1469$ probability that the teacher could grade 16 exams in an average of 11 minutes or less each.

Although a 14.7% probability is low, it is not low enough to assert that the population mean is inaccurate. Conventionally, a probability of less than 5% is required to reject a hypothesis. Therefore, it is reasonable for the teacher to claim that each exam takes an average of 12 minutes to grade.

> More on this in Chapter 10.

Sampling Distribution of the Proportion
Predicting the behavior of discrete random variables

8.23 Describe the sampling distribution of the proportion and the circumstances under which it is used.

> Make sure that p is between zero and one. If p is greater than one, 1 – p will be negative and your calculator will explode when you try to take the square root of a negative number.

The sampling distribution of the proportion is applied when the random variable is binomially distributed. Divide the number of successes s by the sample size n to calculate p_s, the proportion of successes in the sample.

$$p_s = \frac{s}{n}$$

Calculate the standard error of the proportion σ_p by substituting the population proportion p (not the sample proportion p_s) into the formula below.

$$\sigma_p = \sqrt{\frac{p(1-p)}{n}}$$

The z–score for the sampling distribution of the proportion is equal to the difference of the sample proportion p_s and the population proportion p divided by the standard error of the proportion σ_p.

$$z_p = \frac{p_s - p}{\sigma_p}$$

> Even if a population has a very precise proportion of success p, that doesn't mean each sample will have the exact proportion of success p_s as well, especially if the sample is small.

One final word of caution: you can only use the normal distribution to approximate the binomial distribution if two specific requirements are met. As explained in Problems 7.30–7.37, two products must be greater than or equal to five: $np \geq 5$ and $n(1 - p) \geq 5$.

Note: Problems 8.24–8.27 refer to a report that claims 15% of men are left-handed.

8.24 Calculate the probability that more than 12% of a random sample of 100 men is left-handed.

The probability of success for the population is $p = 0.15$. Given a sample size of $n = 100$, both np and $n(1 - p)$ are greater than or equal to 5. Thus, it is appropriate to use the normal distribution to approximate the binomial distribution.

$$np = (100)(0.15) = 15 \geq 5$$
$$n(1 - p) = (100)(1 - 0.15) = (100)(0.85) = 85 \geq 5$$

> Left-handedness is binomially distributed, because there are only two outcomes to the experiment: either you are left-handed or you're not. Sure, there are ambidextrous people, but let's say we made them choose which hand they found more dominant.

Calculate the standard error of the proportion.

$$\sigma_p = \sqrt{\frac{p(1-p)}{n}} = \sqrt{\frac{(0.15)(1-0.15)}{100}} = \sqrt{\frac{(0.15)(0.85)}{100}} = \sqrt{\frac{0.1275}{100}} = 0.03571$$

Calculate the z-score for $p_s = 0.12$.

$$z_p = \frac{p_s - p}{\sigma_p} = \frac{0.12 - 0.15}{0.03571} = \frac{-0.03}{0.03571} = -0.84$$

There is a 0.2995 probability that the proportion of left-handed men in the sample is between 12% and 15%; there is a 0.5 probability that the proportion in the sample is greater than the population proportion of 15%. Thus, there is a 0.2995 + 0.5 = 0.7995 probability that more than 12% of the sample is left-handed.

Note: Problems 8.24–8.27 refer to a report that claims 15% of men are left-handed.

8.25 Calculate the probability that more than 16% of a random sample of 150 men is left-handed.

> $np = (150)(0.15) = 22.5$
>
> $n(1 - p) = (150)(1 - 0.15) = 127.5$

Because np and $n(1 - p)$ are both greater than or equal to 5, you can approximate the binomial distribution using the normal distribution. Calculate the standard error of the proportion.

$$\sigma_p = \sqrt{\frac{p(1-p)}{n}} = \sqrt{\frac{(0.15)(1-0.15)}{150}} = \sqrt{\frac{(0.15)(0.85)}{150}} = \sqrt{\frac{0.1275}{150}} = 0.0292$$

Calculate the z-score for $p_s = 0.16$.

$$z_p = \frac{p_s - p}{\sigma_p} = \frac{0.16 - 0.15}{0.0292} = \frac{0.01}{0.0292} = 0.34$$

There is a 0.5 probability that the sample proportion is greater than the population proportion of 15%. There is a 0.1331 probability that the sample proportion lies between 15% and 16%. Thus, there is a $0.5 - 0.1331 = 0.3669$ probability that more than 16% of the sample is left-handed.

Note: Problems 8.24–8.27 refer to a report that claims 15% of men are left-handed.

8.26 Calculate the probability that 11% to 16% of a 60-man random sample is left-handed.

Note that $np = 9$ and $n(1 - p) = 51$. Both products are greater than 5, so the normal approximation to the binomial distribution can be used. Calculate the standard error of the proportion.

$$\sigma_p = \sqrt{\frac{p(1-p)}{n}} = \sqrt{\frac{(0.15)(1-0.15)}{60}} = \sqrt{\frac{(0.15)(0.85)}{60}} = \sqrt{\frac{0.1275}{60}} = 0.0461$$

Calculate the z-scores for $p_s = 0.11$ and $p_s = 0.16$. $\longleftarrow$

$$z_{0.11} = \frac{0.11 - 0.15}{0.0461} \qquad\qquad z_{0.16} = \frac{0.16 - 0.15}{0.0461}$$

$$= \frac{-0.04}{0.0461} \qquad\qquad\qquad = \frac{0.01}{0.0461}$$

$$= -0.87 \qquad\qquad\qquad\quad = 0.22$$

Even though Problem 8.25 also used a sample proportion of 0.16, you have to recalculate $z_{0.16}$ because the sample size is 60 this time, not 150.

There is a 0.3078 probability that the proportion of left-handers in the sample is between 11% and 15%; there is a 0.0871 probability that the proportion is between 15% and 16%. Thus, there is a $0.3078 + 0.0871 = 0.3949$ probability that 11% to 16% of the sample is left-handed.

Note: Problems 8.24–8.27 refer to a report that claims 15% of men are left-handed.

8.27 If a random sample of 125 men contains only 10 who are left-handed, is it reasonable to assert that 15% of all males are left-handed?

Note that the products np and $n(1 - p)$ are sufficiently large to proceed using the normal approximation to the binomial distribution. Calculate the standard error of the proportion.

$$\sigma_p = \sqrt{\frac{p(1-p)}{n}} = \sqrt{\frac{(0.15)(1-0.15)}{125}} = \sqrt{\frac{(0.15)(0.85)}{125}} = \sqrt{\frac{0.1275}{125}} = 0.0319$$

> Problems 8.24–8.26 wanted you to calculate probabilities based on possible values of p_s. This problem makes you calculate p_s yourself.

Calculate the sample proportion p_s.

$$p_s = \frac{s}{n} = \frac{10}{125} = 0.08$$

Calculate the z-score for $p_s = 0.08$.

$$z_{0.08} = \frac{0.08 - 0.15}{0.0319} = \frac{-0.07}{0.0319} = -2.19$$

If 15% of all men are truly left-handed, then there is a 0.5 probability that the proportion of left-handed men in a sample is less than 15%; there is a 0.4857 probability that the proportion of the sample is between 8% and 15%. Thus, there is only a 0.5 – 0.4857 = 0.0143 probability that 8% or less of the sample is left-handed.

The proportion of left-handers in the sample ($p_s = 0.08$) is significantly lower than the assumed population proportion ($p = 0.15$). In fact, if 15% of men truly are left-handed, then there is only a 1.43% chance of selecting a random sample of 125 men and finding that 10 are left-handed. Because 1.43% < 5%, this is a statistically significant result, and the sample provides little, if any, support that the reported proportion of 15% is correct.

Note: Problems 8.28–8.30 refer to a poll that reported 42% of voters favor the Republican candidate in an upcoming election.

8.28 Calculate the probability that less than 45% of a sample of 40 voters will vote for the Republican candidate.

Calculate the standard error of the proportion.

> Even though this book doesn't always start these problems by ensuring $np \geq 5$ and $n(1-p) \geq 5$, it's an important prerequisite. If those conditions aren't met, the answer you end up with will be pretty inaccurate.

$$\sigma_p = \sqrt{\frac{p(1-p)}{n}} = \sqrt{\frac{(0.42)(1-0.42)}{40}} = \sqrt{\frac{(0.42)(0.58)}{40}} = \sqrt{\frac{0.2436}{40}} = 0.0780$$

Calculate the z-score for $p_s = 0.45$.

$$z_{0.45} = \frac{0.45 - 0.42}{0.0780} = \frac{0.03}{0.0780} = 0.38$$

There is a 0.5 probability that less than 42% of the sample will vote Republican; there is a 0.1480 probability that between 42% and 45% of the sample will vote Republican. Thus, there is a 0.5 + 0.1480 = 0.6480 probability that less than 45% percent of the sample will vote Republican.

Note: Problems 8.28–8.30 refer to a poll that reported 42% of voters favor the Republican candidate in an upcoming election.

8.29 If a random sample of 60 voters is selected, what is the probability that between 28 and 32 of them favor the Republican candidate?

Calculate the standard error of the proportion.

$$\sigma_p = \sqrt{\frac{p(1-p)}{n}} = \sqrt{\frac{(0.42)(1-0.42)}{60}} = \sqrt{\frac{(0.42)(0.58)}{60}} = \sqrt{\frac{0.2436}{60}} = 0.0637$$

Calculate the two proposed sample proportions.

$$p_{28} = \frac{28}{60} = 0.4667 \qquad p_{32} = \frac{32}{60} = 0.5333$$

Calculate the z-scores for $p_{28} = 0.4667$ and $p_{32} = 0.5333$.

$$z_{0.4667} = \frac{0.4667 - 0.42}{0.0637} \qquad z_{0.5333} = \frac{0.5333 - 0.42}{0.0637}$$
$$= \frac{0.0467}{0.0637} \qquad\qquad = \frac{0.1133}{0.0637}$$
$$= 0.73 \qquad\qquad\quad = 1.78$$

Calculate $(0.0467 < p_s < 0.5333)$.

$$P(0.4667 < p_s < 0.5333) = P(0.73 < z_p < 1.78)$$
$$= P(0 < z_p < 1.78) - P(0 < z_p < 0.73)$$
$$= 0.4625 - 0.2673$$
$$= 0.1952$$

Note: Problems 8.28–8.30 refer to a poll that reported 42% of voters favor the Republican candidate in an upcoming election.

8.30 If a random sample of 120 people contains only 47 that favor the Republican candidate, does the sample support the results of the poll?

Calculate the standard error of the proportion.

$$\sigma_p = \sqrt{\frac{p(1-p)}{n}} = \sqrt{\frac{(0.42)(1-0.42)}{120}} = \sqrt{\frac{(0.42)(0.58)}{120}} = \sqrt{\frac{0.2436}{120}} = 0.0451$$

Calculate the sample proportion p_s.

$$p_s = \frac{47}{120} = 0.3917$$

Calculate the z-score for $p_s = 0.3917$.

$$z_{0.3917} = \frac{0.3917 - 0.42}{0.0451} = \frac{-0.0283}{0.0451} = -0.63$$

This is a rounded percentage version of $P_s = 0.3917$.

Assuming 42% of the voters prefer the Republican candidate, there is a 0.5 probability that less than 42% of the sample will vote Republican; there is a 0.2357 probability that between 39.2% and 42% of the sample will vote Republican. Thus, there is a 0.5 – 0.2357 = 0.2643 probability that 39.2% of the sample or less will vote Republican.

If 42% of the population will actually vote for the Republican candidate, then the probability of selecting a sample containing 39.2% Republican voters is 0.2643. Because this probability is greater than 0.05, it is large enough to support the validity of the poll.

Note: Problems 8.31–8.35 refer to a study conducted in 2000 that reported 71.3% of men between the ages of 45 and 54 are considered overweight.

8.31 If a random sample of 90 men in this age group is selected, what is the probability that more than 70% of them will be overweight?

Calculate the standard error of the proportion.

$$\sigma_p = \sqrt{\frac{p(1-p)}{n}} = \sqrt{\frac{(0.713)(1-0.713)}{90}} = \sqrt{\frac{(0.713)(0.287)}{90}} = \sqrt{\frac{0.204631}{90}} = 0.0477$$

Calculate the z-score for $p_s = 0.70$.

$$z_{0.70} = \frac{0.70 - 0.713}{0.0477} = \frac{-0.013}{0.0477} = -0.27$$

There is a 0.1064 probability that between 70% and 71.3% of the sample will be overweight; there is a 0.5 probability that more than 71.3% of the sample will be overweight. Thus, there is a 0.1064 + 0.5 = 0.6064 probability that more than 70% of the sample will be overweight.

Note: Problems 8.31–8.35 refer to a study conducted in 2000 that reported 71.3% of men between the ages of 45 and 54 are considered overweight.

8.32 If a random sample of 60 men in this age group is selected, what is the probability that between 66% and 75% of them are overweight?

If the problem gives you percentages, you don't have to calculate the sample proportions p_s—the percentages are the proportions.

Calculate the standard error of the proportion.

$$\sigma_p = \sqrt{\frac{(0.713)(1-0.713)}{60}} = \sqrt{\frac{(0.713)(0.287)}{60}} = \sqrt{\frac{0.204631}{60}} = 0.0584$$

Calculate the z-scores for $p_s = 0.66$ and $p_s = 0.75$.

$$z_{0.66} = \frac{0.66 - 0.713}{0.0584} \qquad z_{0.75} = \frac{0.75 - 0.713}{0.0584}$$

$$= \frac{-0.053}{0.0584} \qquad = \frac{0.037}{0.0584}$$

$$= -0.91 \qquad = 0.63$$

Calculate $P(0.66 < p_s < 0.75)$.

$$P\left(0.66 < p_s < 0.75\right) = P\left(-0.91 < z_p < 0.63\right)$$

$$= P\left(-0.91 < z_p < 0\right) + P\left(0 < z_p < 0.63\right)$$

$$= 0.3186 + 0.2357$$

$$= 0.5543$$

Note: Problems 8.31–8.35 refer to a study conducted in 2000 that reported 71.3% of men between the ages of 45 and 54 are considered overweight.

8.33 If a random sample of 150 men in this age group is selected, what is the probability that between 66% and 75% of them are overweight?

> New sample size, but with the same boundaries as Problem 8.32: 66% and 75%. Unfortunately, a new sample size means a new standard error and therefore new z-scores.

Calculate the standard error of the proportion.

$$\sigma_p = \sqrt{\frac{(0.713)(1 - 0.713)}{150}} = \sqrt{\frac{(0.713)(0.287)}{150}} = \sqrt{\frac{0.204631}{150}} = 0.0369$$

Calculate the z-scores for $p_s = 0.66$ and $p_s = 0.75$.

$$z_{0.66} = \frac{0.66 - 0.713}{0.0369} \qquad z_{0.75} = \frac{0.75 - 0.713}{0.0369}$$

$$= \frac{-0.053}{0.0369} \qquad = \frac{0.037}{0.0369}$$

$$= -1.44 \qquad = 1.00$$

Calculate $P(0.66 < p_s < 0.75)$.

$$P\left(0.66 < p_s < 0.75\right) = P\left(-1.44 < z_p < 1.00\right)$$

$$= P\left(-1.44 < z_p < 0\right) + P\left(0 < z_p < 1.00\right)$$

$$= 0.4251 + 0.3413$$

$$= 0.7664$$

> It jumps from a 55% chance to a 77% chance that the sample proportion will be between $p_s = 0.66$ and $p_s = 0.75$, just a few digits off the population proportion $p = 0.713$.

Compare $P(0.66 < p_s < 0.75)$ with sample size 60 in Problem 8.32 and sample size 150 in this problem. The larger the sample size, the more likely it is that the sample proportion will better approximate the population proportion.

Note: Problems 8.31–8.35 refer to a study conducted in 2000 that reported 71.3% of men between the ages of 45 and 54 are considered overweight.

8.34 A recent sample of 22 men from this age group included 18 who were considered overweight. Is this sufficient evidence to conclude that the proportion of overweight men from this age group is still 71.3%?

Calculate the standard error of the proportion.

$$\sigma_p = \sqrt{\frac{(0.713)(1-0.713)}{22}} = \sqrt{\frac{(0.713)(0.287)}{22}} = \sqrt{\frac{0.204631}{22}} = 0.0964$$

Calculate the sample proportion p_s.

$$p_s = \frac{18}{22} = 0.8182$$

Calculate the z-score for $p_s = 0.8182$.

$$z_{0.8182} = \frac{0.8182 - 0.713}{0.0964} = \frac{0.1052}{0.0964} = 1.09$$

Assuming the population proportion truly is 71.3%, determine the probability that 81.82% or more of the sample could be overweight, as found in the recent sample.

$$P(p_s \geq 0.818) = P(z_p \geq 1.09)$$
$$= P(z_p \geq 0) - P(0 \leq z_p \leq 1.09)$$
$$= 0.5 - 0.3621$$
$$= 0.1379$$

Although there is only a 13.79% probability that 81.82% of the men in the sample were overweight given a population proportion of 71.3%, the probability would have to be less than 5% to reject the hypothesis that the population proportion is false.

If the sample size had been larger than 22, the results might have been more convincing. Check out the next problem.

Note: Problems 8.31–8.35 refer to a study conducted in 2000 that reported 71.3% of men between the ages of 45 and 54 are considered overweight.

8.35 A recent sample of 154 men from this age group included 126 who were considered overweight. Is there sufficient evidence to conclude that the proportion of overweight men from this age group is still 71.3%?

Just a wild hunch, but I am guessing that this time there is.

Calculate the standard error of the proportion.

$$\sigma_p = \sqrt{\frac{(0.713)(1-0.713)}{154}} = \sqrt{\frac{(0.713)(0.287)}{154}} = \sqrt{\frac{0.204631}{154}} = 0.0365$$

Calculate the sample proportion p_s.

$$p_s = \frac{126}{154} = 0.8182$$

This is the same sample proportion as in Problem 8.34: 81.8% of the sample is overweight. However, this time the sample is much larger.

Calculate the z-score for $p_s = 0.8182$.

$$z_{0.818} = \frac{0.8182 - 0.713}{0.0365} = \frac{0.1052}{0.0365} = 2.88$$

Assuming the population proportion truly is 71.3 percent, determine the probability that 81.82 percent or more of the sample could be overweight, as found in the recent sample.

$$P(p_s \geq 0.8182) = P(z_p \geq 2.88)$$
$$= P(z_p \geq 0) - P(0 \leq z_p \leq 2.88)$$
$$= 0.5 - 0.4980$$
$$= 0.0020$$

There is only a 0.20% chance that 81.82% of the men in the sample would be overweight given a population proportion of 71.3%. Because 0.20% < 5%, this sample lends no support to the claim that the population proportion is still 71.3%.

Finite Population Correction Factor for the Sampling Distribution of the Proportion

Predicting proportions from a small population

8.36 Describe the finite population correction factor for the sampling distribution for the proportion and the conditions under which it should be applied.

As explained in Problem 8.18, when a population is very large, selecting something as part of a sample has a negligible impact on the population. However, when the population N is small relative to the sample size n, a finite population correction factor is applied when calculating the standard error of the proportion.

$$\sigma_p = \sqrt{\frac{p(1-p)}{n}} \sqrt{\frac{N-n}{N-1}}$$

The radical on the right is also part of the correction factor defined in Problem 8.18.

As with the sampling distribution for the mean, the finite population correction factor for the proportion should be applied when the sample size is more than 5% of the population size, when $\frac{n}{N} > 0.05$.

The only difference between Problems 8.37–8.40 and the problems in the last section is the way you calculate σ_p. The old rules apply, so np and $n(1-p)$ still have to be greater than or equal to 5.

Note: Problems 8.37–8.40 refer to a 2001 study that reported 27.7% of high school students smoke. Random samples are selected from a high school that has 632 students.

8.37 If a random sample of 60 students is selected, what is the probability that fewer than 19 of the students smoke?

Verify that you can use the normal approximation to the binomial distribution.

$$np = (60)(0.277) = 16.62 \geq 5$$
$$n(1-p) = (60)(1-0.277) = (60)(0.723) = 43.38 \geq 5$$

Notice that the sample size $n = 60$ is more than 5% of the population size: $\dfrac{n}{N} = \dfrac{60}{632} = 0.095 > 0.05$. Thus, you should calculate the standard error of the proportion using the finite population correction factor.

$$\sigma_p = \sqrt{\frac{p(1-p)}{n}}\sqrt{\frac{N-n}{N-1}} = \sqrt{\frac{(0.277)(0.723)}{60}}\sqrt{\frac{(632-60)}{(632-1)}} = \sqrt{\frac{0.200271}{60}}\sqrt{\frac{572}{631}} = 0.0550$$

Calculate the sample proportion p_s and the corresponding z-score.

$$p_s = \frac{19}{60} = 0.3167$$

$$z_{0.3167} = \frac{p_s - p}{\sigma_p} = \frac{0.3167 - 0.277}{0.0550} = \frac{0.0397}{0.0550} = 0.72$$

There is a 0.5 probability that the sample proportion is less than the population proportion of 27.7%; there is a 0.2642 probability that the sample proportion is between 27.7% and 31.67%. Thus, there is a 0.5 + 0.2642 = 0.7642 probability that fewer than 19 of the high school students in the sample (31.67% of the 60 students) were smokers.

Note: Problems 8.37–8.40 refer to a 2001 study that reported 27.7% of high school students smoke. Random samples are selected from a high school that has 632 students.

8.38 If a random sample of 75 students is selected, what is the probability that more than 17 of the students smoke?

A random sample of $n = 75$ students constitutes 11.9% of the $N = 632$ population. Calculate the standard error of the proportion using the finite population correction factor.

$$\sigma_p = \sqrt{\frac{p(1-p)}{n}}\sqrt{\frac{N-n}{N-1}} = \sqrt{\frac{(0.277)(0.723)}{75}}\sqrt{\frac{(632-75)}{(632-1)}} = \sqrt{\frac{0.200271}{75}}\sqrt{\frac{557}{631}} = 0.0486$$

Calculate the sample proportion p_s and the corresponding z-score.

$$p_s = \frac{17}{75} = 0.2267$$

$$z_{0.2267} = \frac{p_s - p}{\sigma_p} = \frac{0.2267 - 0.277}{0.0486} = \frac{-0.0503}{0.0486} = -1.03$$

There is a 0.3485 probability that the sample proportion is between 22.67% and 27.7%; there is a 0.5 probability that the sample proportion is greater than the population proportion of 27.7%. Thus, there is a $0.3485 + 0.5 = 0.8485$ probability that more than 17 of the students in the sample are smokers.

Note: Problems 8.37–8.40 refer to a 2001 study that reported 27.7% of high school students smoke. Random samples are selected from a high school that has 632 students.

8.39 If a random sample of 90 students is selected, what is the probability that between 31 and 37 of the students smoke?

Because a sample of $n = 90$ students constitutes 14.2% of the $N = 632$ student population, calculate the standard error of the proportion using the finite population correction factor.

$$\sigma_p = \sqrt{\frac{p(1-p)}{n}} \sqrt{\frac{N-n}{N-1}} = \sqrt{\frac{(0.277)(0.723)}{90}} \sqrt{\frac{(632-90)}{(632-1)}} = \sqrt{\frac{0.200271}{90}} \sqrt{\frac{542}{631}} = 0.0437$$

Calculate the proposed sample proportions.

$$p_s = \frac{27}{90} = 0.3 \qquad p_s = \frac{31}{90} = 0.3444$$

Identify the z-scores for $p_s = 0.3$ and $p_s = 0.3444$.

$$z_{0.3} = \frac{0.3 - 0.277}{0.0437} \qquad z_{0.3444} = \frac{0.3444 - 0.277}{0.0437}$$

$$= \frac{0.023}{0.0437} \qquad\qquad = \frac{0.0674}{0.0437}$$

$$= 0.53 \qquad\qquad\qquad = 1.54$$

Calculate $P(0.3 < p_s < 0.3444)$.

$$P(0.3 < p_s < 0.3444) = P(0.53 < z_p < 1.54)$$

$$= P(0 < z_p < 1.54) - P(0 < z_p < 0.53)$$

$$= 0.4382 - 0.2019$$

$$= 0.2363$$

Note: Problems 8.37–8.40 refer to a 2001 study that reported 27.7% of high school students smoke. Random samples are selected from a high school that has 632 students.

8.40 If a random sample of 110 students contains 20 smokers, does this result support the 2001 study?

Because a sample of $n = 110$ students is 17.4% of the total student population, you should calculate the standard error of the proportion using the finite population correction factor.

$$\sigma_p = \sqrt{\frac{p(1-p)}{n}}\sqrt{\frac{N-n}{N-1}} = \sqrt{\frac{(0.277)(0.723)}{110}}\sqrt{\frac{(632-110)}{(632-1)}} = \sqrt{\frac{0.200271}{110}}\sqrt{\frac{522}{631}} = 0.0388$$

Calculate the sample proportion p_s and the corresponding z-score.

$$p_s = \frac{20}{110} = 0.1818$$

$$z_{0.1818} = \frac{0.1818 - 0.277}{0.0388} = \frac{-0.0952}{0.0388} = -2.45$$

Assuming the population proportion truly is 0.277, calculate the probability that the sample proportion is less than or equal to $p_s = 0.1818$.

$$P(p_s \le 0.1818) = P(z_p \le -2.45)$$
$$= P(z_p \le 0) - P(-2.45 \le z_p \le 0)$$
$$= 0.50 - 0.4929$$
$$= 0.0071$$

If the actual proportion of high school smokers is 27.7%, then there is only a 0.71% chance that a sample of 110 students will include 20 or fewer smokers. These results do not support the 2001 study.

Chapter 9
CONFIDENCE INTERVALS

Putting samples to work

One of the most important roles of statistics is to draw conclusions about a population based on information garnered from a sample of that population. Thus, it is important to contextualize the calculations performed on the samples, and confidence intervals play a key role by quantifying the accuracy of population estimates.

At the risk of sounding like a broken record, you'll only understand this chapter if you understand the chapters before it, especially Chapter 8. Confidence intervals for the mean are affected by the size of the sample, just as the sample means and sample proportions were in Chapter 8. (Remember, this led to the finite population correction factors for the sampling distributions of the mean and the proportion.) The techniques you use to calculate confidence intervals also vary based on whether or not you know the standard deviation of the population.

Introduction to Confidence Intervals for the Mean
How unrepresentative could a sample be?

9.1 Define sampling error.

It's unlikely that a sample will have the exact same mean as the population from which it's drawn. However, the larger the sample, the more likely its mean will be close to the population mean.

A population is often too large or too inaccessible for every element to be measured. In these situations, a sample from the population is randomly selected and the sample mean is used to estimate the population mean. Sampling error accounts for the difference between the sample mean and the population mean. Whenever populations are sampled to estimate the population mean, sampling error will most likely be present.

Note: Problems 9.2–9.3 refer to the data set below, the ages of 10 customers in a retail store.

Customer Age									
36	29	55	22	34	67	30	41	35	21

9.2 If the first three customers in the table are chosen to estimate the average age of all 10 customers, what is the sampling error?

Calculate the population mean.

$$\mu = \frac{\sum x}{N} = \frac{36+29+55+22+34+67+30+41+35+21}{10} = \frac{370}{10} = 37$$

Now calculate the sample mean, the average age of the first three ages in the table.

$$\bar{x} = \frac{36+29+55}{3} = \frac{120}{3} = 40$$

The average age of the first three customers is three years greater than the average age of all 10 customers.

Calculate the difference between the sample mean and the population mean.

$$\bar{x} - \mu = 40 - 37 = 3$$

The sampling error is 3 years.

Note: Problems 9.2–9.3 refer to the data set in Problem 9.2, giving the ages of 10 customers in a retail store.

9.3 If the first seven customers in the table are chosen to estimate the average age of all 10 customers, what is the sampling error?

According to Problem 9.2, $\mu = 37$. Calculate the mean of the seven-person sample.

$$\bar{x} = \frac{36+29+55+22+34+67+30}{7} = \frac{273}{7} = 39$$

The sampling error is the difference between the sample mean and the population mean: $\bar{x} - \mu = 39 - 37 = 2$ years. Notice that the sampling error decreases (from three to two years) when the sample size increases from three to seven.

9.4 Describe the difference between a point estimate and a confidence interval for the mean.

A point estimate for the mean is a sample mean used to estimate the population mean. A confidence interval represents a range of values around the point estimate within which the true population mean most likely lies.

9.5 Describe the role that confidence levels play in the confidence interval.

Confidence intervals are stated in terms of confidence levels. Typical confidence levels range from, but are not limited to, 90% to 98%. For example, a 95% confidence interval represents a range of values around the sample mean that is 95% certain to contain the true population mean. Given two 95% confidence intervals of different sizes, the smaller confidence interval is a more precise estimate of the true population mean. ←

> Two things determine how accurate the prediction is: (1) the confidence level percentage and (2) the size of the sample.

Confidence Intervals for the Mean with Large Samples and Sigma Known

Central limit theorem to the rescue

Note: Problems 9.6–9.9 refer to a random sample of customer order totals with an average of $78.25 and a population standard deviation of $22.50.

9.6 Calculate a 90% confidence interval for the mean, given a sample size of 40 orders. ←

> The sample size has to be greater than $n = 30$ to be large enough to use the technique described in this section. If it's smaller, skip ahead to Problem 9.15.

Calculate the standard error of the mean.

$$\sigma_{\bar{x}} = \frac{\sigma}{\sqrt{n}} = \frac{22.50}{\sqrt{40}} = 3.558$$

The sample mean is the center of a confidence interval, so half of the interval (in this case, 45%) is directly to the right of the sample mean and half is directly to the left. Refer to the standard normal table in Reference Table 1, locate the area that most closely approximates 0.45, and set z_c equal to the corresponding z-score: $z_c = 1.64$. ←

Substitute $\bar{x} = 78.25$, $\sigma_{\bar{x}} = 3.558$, and $z_c = 1.64$ into the confidence interval boundary formulas below. Note that the term $z_c \sigma_{\bar{x}}$ is commonly called the margin of error, E. In this problem, $E = (1.64)(3.558) = 5.835$.

> The z-score $z = 1.64$ has an area of 0.4495. You could use $z = 1.65$ instead, because 0.4505 is just as close to 0.45 as 0.4495.

lower limit $= \bar{x} - z_c \sigma_{\bar{x}}$

    $= 78.25 - (1.64)(3.558)$

    $= 72.41$

upper limit $= \bar{x} + z_c \sigma_{\bar{x}}$

    $= 78.25 + (1.64)(3.558)$

    $= 84.09$

Based on the sample, you can be 90% confident that the true population mean of the order totals lies on the interval bounded below by $72.41 and above by $84.09.

9.7   Calculate a 90% confidence interval for the mean, given a sample of 75 orders.

Calculate the standard error of the mean.

You don't have to recalculate this every time. When you're dealing with a 90% confidence level in any problem, set $z_c = 1.64$.

$$\sigma_{\bar{x}} = \frac{\sigma}{\sqrt{n}} = \frac{22.50}{\sqrt{75}} = 2.598$$

The sample mean and confidence level are the same as in Problem 9.6: $\bar{x} = 78.25$ and $z_c = 1.64$. Apply the confidence interval boundary formulas.

$\text{lower limit} = \bar{x} - z_c \sigma_{\bar{x}}$ $\text{upper limit} = \bar{x} + z_c \sigma_{\bar{x}}$

$\quad = 78.25 - (1.64)(2.598)$ $\quad = 78.25 + (1.64)(2.598)$

$\quad = 73.99$ $\quad = 82.51$

Based on the sample, you can be 90% confident that the true population mean for the order totals is between $73.99 and $82.51.

9.8   Explain the difference in the 90% confidence intervals calculated in Problems 9.6 and 9.7.

Although both problems developed a 90% confidence interval from the same population, the size of the intervals differed due to the different sample sizes. The larger sample size in Problem 9.7 resulted in a smaller confidence interval than in Problem 9.6. Specifically, increasing the sample size from $n = 40$ to $n = 75$ reduced the margin of error from $5.84 to $4.26. If two intervals of the same confidence level are different sizes, then the smaller interval provides a more precise estimate of the population mean.

9.9   Calculate the minimum sample size needed to identify a 90% confidence interval for the mean, assuming a $5.00 margin of error.

Recall that the margin of error E is the product of the z-score representing the correct confidence level and the standard error of the mean: $E = z_c \sigma_{\bar{x}}$. Substitute the standard error of the mean formula $\left(\sigma_{\bar{x}} = \dfrac{\sigma}{\sqrt{n}} \right)$ into the margin of error equation.

$$E = z_c \sigma_{\bar{x}}$$

$$E = z_c \left(\frac{\sigma}{\sqrt{n}} \right)$$

$$E = \frac{z_c \sigma}{\sqrt{n}}$$

> Once you derive the formula here, you won't have to do it over and over again to calculate the minimum sample size n. In Problem 9.14, for example, the book skips right to the formula.

Cross multiply and solve for n.

$$\sqrt{n}\,(E) = z_c \sigma$$

$$\sqrt{n} = \frac{z_c \sigma}{E}$$

$$\left(\sqrt{n} \right)^2 = \left(\frac{z_c \sigma}{E} \right)^2$$

$$n = \left(\frac{z_c \sigma}{E} \right)^2$$

Evaluate the expression for n given $z_c = 1.64$, $\sigma = 22.50$, and $E = 5.00$.

$$n = \left[\frac{(1.64)(22.50)}{5.00} \right]^2 = \left(\frac{36.9}{5} \right)^2 = (7.38)^2 = 54.4644 \approx 55$$

Sample size needs to be an integer value, so 54.4644 is rounded up to 55, as rounding it down produces a sample size that is not sufficiently large: $54 < 54.4644$.

Note: Problems 9.10–9.14 refer to a random sample of 35 teenagers who averaged 7.3 hours of sleep per night. Assume the population standard deviation is 1.8 hours.

9.10 Calculate a 95% confidence interval for the mean.

Use the z-score $z_c = 1.96$ when calculating 95% confidence intervals, because Reference Table 1 states that its area is equal to $\frac{0.95}{2} = 0.475$. Calculate the standard error of the mean.

$$\sigma_{\bar{x}} = \frac{\sigma}{\sqrt{n}} = \frac{1.8}{\sqrt{35}} = 0.304$$

Recall that the sample mean is $\bar{x} = 7.3$; apply the confidence interval boundary formulas.

lower limit $= \bar{x} - z_c \sigma_{\bar{x}}$ upper limit $= \bar{x} + z_c \sigma_{\bar{x}}$

$\qquad = 7.3 - (1.96)(0.304)$ $\qquad = 7.3 + (1.96)(0.304)$

$\qquad = 6.70$ $\qquad = 7.90$

Note: Problems 9.10–9.14 refer to a random sample of 35 teenagers who averaged 7.3 hours of sleep per night. Assume the population standard deviation is 1.8 hours.

9.11 Calculate a 98% confidence interval for the mean.

To identify the appropriate value of z_c, divide the decimal form of the confidence level by two and locate the z-score in Reference Table 1 whose area most closely approximates that quotient.

> So $0.98 \div 2 = 0.49$ and the z-score $z_c = 2.33$ has an area of 0.4901.

Apply the confidence interval boundary formulas. This problem has the same sample size as Problem 9.10, so the standard error of the mean is unchanged: $\sigma_{\bar{x}} = 0.304$.

$$\text{lower limit} = \bar{x} - z_c\sigma_{\bar{x}} \qquad\qquad \text{upper limit} = \bar{x} + z_c\sigma_{\bar{x}}$$
$$= 7.3 - (2.33)(0.304) \qquad\qquad = 7.3 + (2.33)(0.304)$$
$$= 6.59 \qquad\qquad\qquad\qquad = 8.01$$

Increasing the confidence level from 95% to 98% increases the margin of error from 0.60 to 0.71 hours.

Note: Problems 9.10–9.14 refer to a random sample of 35 teenagers who averaged 7.3 hours of sleep per night. Assume the population standard deviation is 1.8 hours.

9.12 A recent report claims that teenagers sleep an average of 7.8 hours per night. Discuss the validity of the claim using the 98% confidence interval calculated in Problem 9.11.

According to Problem 9.11, you can be 98% confident that the actual average is between 6.59 and 8.01 hours of sleep. The study's reported average of 7.3 hours falls within this confidence interval, so the sample in Problem 9.11 supports the validity of this claim.

Note: Problems 9.10–9.14 refer to a random sample of 35 teenagers who averaged 7.3 hours of sleep per night. Assume the population standard deviation is 1.8 hours.

9.13 Explain the difference in the confidence intervals calculated in Problems 9.10 and 9.11.

> If you want to be more certain that you create an interval that contains the actual answer, you'll need to give yourself a little more room.

Both problems selected a sample size of $n = 35$ from the same population, but the confidence levels were different. In order to be more confident that the interval includes the true population mean, the interval itself needs to be wider. Thus, the larger confidence level (98%) required a wider confidence interval.

Note: Problems 9.10–9.14 refer to a random sample of teenagers who averaged 7.3 hours of sleep per night. Assume the population standard deviation is 1.8 hours.

9.14 Calculate the minimum sample size needed to identify a 95% confidence interval for the mean, assuming a 0.40 hour margin of error.

Apply the formula for the minimum sample size generated in Problem 9.9.

$$n = \left(\frac{z_c \sigma}{E}\right)^2 = \left[\frac{(1.96)(1.8)}{0.40}\right]^2 = \left(\frac{3.528}{0.4}\right)^2 = (8.82)^2 = 77.7924 \approx 78$$

A minimum sample size of 78 teenagers is required to provide a 95% confidence interval with a margin of error of 0.40 hours.

The value $z_c = 1.96$ comes from Problem 9.10. It's always used for a 95% confidence interval.

Confidence Intervals for the Mean with Small Samples and Sigma Known

Life without the central limit theorem

Small means a sample size of less than 30.

Note: Problems 9.15–9.20 refer to a random sample of 15 cars of the same model. Assume that the gas mileage for the population is normally distributed with a standard deviation of 5.2 miles per gallon.

9.15 Identify the bounds for a 90% confidence interval for the mean, given a sample mean of 26.7 miles per gallon.

Because the sample size is less than 30, you cannot rely on the central limit theorem to ensure that the sample means will also be normally distributed. However, the problem states that the population is normally distributed, so you can assume that samples of any size are normally distributed as well.

Calculate the standard error of the mean.

$$\sigma_{\bar{x}} = \frac{\sigma}{\sqrt{n}} = \frac{5.2}{\sqrt{15}} = 1.343$$

Substitute $\bar{x} = 26.7$, $z_c = 1.64$, and $\sigma_{\bar{x}} = 1.343$ into the confidence interval boundary formulas.

$$\text{lower limit} = \bar{x} - z_c \sigma_{\bar{x}} \qquad\qquad \text{upper limit} = \bar{x} + z_c \sigma_{\bar{x}}$$
$$= 26.7 - (1.64)(1.343) \qquad\qquad = 26.7 + (1.64)(1.343)$$
$$= 24.50 \qquad\qquad\qquad\qquad = 28.90$$

Note: Problems 9.15–9.20 refer to a random sample of 15 cars of the same model. Assume that the gas mileage for the population is normally distributed with a standard deviation of 5.2 miles per gallon.

9.16 Identify the bounds for a 90% confidence interval for the mean, given a sample mean of 22.8 miles per gallon.

Substitute $\bar{x} = 22.8$, as well as the standard error for the mean and the appropriate value of z_c identified in Problem 9.15, into the confidence interval boundary formulas.

$$\text{lower limit} = \bar{x} - z_c \sigma_{\bar{x}} \qquad\qquad \text{upper limit} = \bar{x} + z_c \sigma_{\bar{x}}$$
$$= 22.8 - (1.64)(1.343) \qquad\qquad = 22.8 + (1.64)(1.343)$$
$$= 20.60 \qquad\qquad\qquad\qquad = 25.00$$

Note: Problems 9.15–9.20 refer to a random sample of 15 cars of the same model. Assume that the gas mileage for the population is normally distributed with a standard deviation of 5.2 miles per gallon.

9.17 The car manufacturer of this particular model claims that the average gas mileage is 26 miles per gallon. Discuss the validity of this claim using the 90% confidence interval calculated in Problem 9.16.

Because the manufacturer's claim of 26 miles per gallon is greater than the upper boundary of the confidence interval (25.00), this sample does not validate the claim of the manufacturer.

Note: Problems 9.15–9.20 refer to a random sample of 15 cars of the same model. Assume that the gas mileage for the population is normally distributed with a standard deviation of 5.2 miles per gallon.

9.18 Explain the difference in the confidence intervals calculated in Problems 9.15 and 9.16.

Both problems select a sample of the same size (15) from the same population and use the same confidence level (90%). However, the sample means were different. Because confidence intervals are built around sample means, changing the sample mean changes the corresponding confidence interval as well.

As long as the sample size and confidence level remain constant, the width of the confidence interval will remain constant as well. The interval will merely shift right or left, depending on the location of the sample mean. Because the width remains constant under these conditions, the level of precision for the approximate population mean also remains constant from sample to sample.

> Even if the confidence level is unchanged.

Note: Problems 9.15–9.20 refer to a random sample of 15 cars of the same model. Assume that the gas mileage for the population is normally distributed with a standard deviation of 5.2 miles per gallon.

9.19 Let a and b represent the lower and upper boundaries of the 90% confidence interval for the mean of the population. Is it correct to conclude that there is a 90% probability the true population mean lies between a and b? Explain your answer.

A confidence interval does not describe the probability that any particular interval constructed around the mean of a single sample will contain the actual population mean. In this problem, it would be inaccurate to state that there is a 90% probability the interval bounded below by a and above by b contains the population mean.

If you were to collect 10 different samples from the population, calculate the sample mean for each, and then construct the 10 corresponding confidence intervals, a 90% confidence level implies that 9 of the 10 intervals will include the true population mean. Consider the illustration below, which represents 10 different confidence intervals calculated around the sample means of 10 different samples.

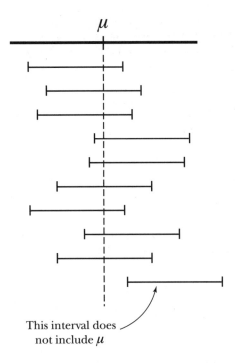

This interval does
not include μ

Because 9 of the 10 samples have confidence intervals that include the population mean, the samples exhibit a 90% confidence level.

Note: Problems 9.15–9.20 refer to a random sample of 15 cars of the same model. Assume that the gas mileage for the population is normally distributed with a standard deviation of 5.2 miles per gallon.

9.20 Calculate the minimum sample size needed to identify a 95% confidence interval for the mean, assuming a 2.0 miles per gallon margin of error.

Substitute $E = 2$, $\sigma = 5.2$, and $z_c = 1.96$ into the formula generated in Problem 9.9.

$$n = \left(\frac{z_c \sigma}{E}\right)^2 = \left[\frac{(1.96)(5.2)}{2}\right]^2 = (5.096)^2 = 25.969$$

A minimum sample size of 26 cars is required.

Note: Problems 9.21–9.23 refer to a random sample of 20 paperback novels that average 425.1 pages in length. Assume that the page count for all paperback novels is normally distributed with a standard deviation of 92.8 pages.

9.21 Identify the bounds of a 95% confidence interval for the mean.

Calculate the standard error of the mean.

$$\sigma_{\bar{x}} = \frac{\sigma}{\sqrt{n}} = \frac{92.8}{\sqrt{20}} = 20.751$$

Substitute $\bar{x} = 425.1$, $z_c = 1.96$, and $\sigma_{\bar{x}} = 20.751$ into the confidence interval boundary formulas.

lower limit $= \bar{x} - z_c \sigma_{\bar{x}}$

$= 425.1 - (1.96)(20.751)$

$= 384.43$

upper limit $= \bar{x} + z_c \sigma_{\bar{x}}$

$= 425.1 + (1.96)(20.751)$

$= 465.77$

Based on this sample, you can be 95% confident that the true population mean for the page count of paperback novels is between 384.43 and 465.77.

> *Note: Problems 9.21–9.23 refer to a random sample of 20 paperback novels that average 425.1 pages in length. Assume that the page count for all paperback novels is normally distributed with a standard deviation of 92.8 pages.*

9.22 Identify the bounds of a 99% confidence interval for the mean.

0.99 ÷ 2 = 0.4950, and the closest value in Reference Table 1 is 0.4949. It's in the 2.5 row and 0.07 column of the table, so z_c = 2.57. Actually, you could use 2.58 as well—both work.

A 99% confidence interval has a corresponding z_c value of 2.57. The sample size is $n = 20$, as it was in Problem 9.21, so there is no need to recalculate the standard error of the mean: $\sigma_{\bar{x}} = 20.751$.

lower limit $= \bar{x} - z_c \sigma_{\bar{x}}$

$= 425.1 - (2.57)(20.751)$

$= 371.77$

upper limit $= \bar{x} + z_c \sigma_{\bar{x}}$

$= 425.1 + (2.57)(20.751)$

$= 478.43$

Based on the sample, you can be 99% confident that the true population mean is between 371.8 and 478.5.

Increasing the confidence level from 95% to 99% increases the margin of error, makes the confidence interval wider, and produces a less precise estimate.

> *Note: Problems 9.21–9.23 refer to a random sample of 20 paperback novels that average 425.1 pages in length. Assume that the page count for all paperback novels is normally distributed with a standard deviation of 92.8 pages.*

9.23 Calculate the minimum sample size needed to identify a 98% confidence interval for the mean assuming a margin of error of 52 pages.

Substitute $E = 52$, $\sigma = 92.8$, and $z_c = 2.33$ into the formula generated in Problem 9.9 to calculate the minimum sample size n.

$$n = \left(\frac{z_c \sigma}{E}\right)^2 = \left[\frac{(2.33)(92.8)}{52}\right]^2 = (4.158154)^2 = 17.290$$

Even though 17.3 is closer to 17 than 18, always round up when calculating minimum sample size.

A minimum sample size of 18 books is required to provide a 98% confidence interval with a margin of error of 52 pages.

Confidence Intervals for the Mean with Small Samples and Sigma Unknown

Introducing the Student's t-distribution

9.24 Describe how to construct confidence intervals when the population standard deviation σ is unknown.

When the population standard deviation σ is unknown, the sample standard deviation s is used in its place as an approximation. When you substitute s for σ, the Student's t-distribution (or, more simply, the t-distribution) is used in lieu of the normal distribution.

When the sample size is less than 30, the population needs to be normally distributed when using the t-distribution. When the sample size is 30 or more, the normal distribution can be used as an approximation to the t-distribution, even if the population is not normally distributed.

> The Student's t-distribution was developed by William Gosset, an employee of the Guinness Brewing Company in Ireland, in 1908. He published his findings using the pseudonym "Student."

Note: Problems 9.25–9.27 refer to the data set below, the amount of trash generated by ten households (in pounds per day). Assume that the population is normally distributed.

Pounds of Trash									
3.9	4.6	15.6	10.5	16.0	6.7	12.0	9.2	13.8	16.8

9.25 Construct a 95% confidence interval for the mean based on the sample.

Calculate the sample mean.

$$\bar{x} = \frac{\sum x}{n} = \frac{3.9+4.6+15.6+10.5+16+6.7+12+9.2+13.8+16.8}{10} = \frac{109.1}{10} = 10.91$$

> See Problem 3.49 if you need to review this.

The standard deviation of the population is not known. In order to calculate the sample standard deviation s using the shortcut formula below, you need to calculate the sum of the squares of the data values and the square of the sum of the data values.

$$s = \sqrt{\frac{\sum x^2 - \frac{(\sum x)^2}{n}}{n-1}}$$

The table that follows lists each data value and its square. Compute the sums of both columns.

x	x^2
3.9	15.21
4.6	21.16
15.6	243.36
10.5	110.25
16.0	256.00
6.7	44.89
12.0	144.00
9.2	84.64
13.8	190.44
16.8	282.24
Total	**109.1**	**1,392.19**

Substitute $\sum x = 109.1$ and $\sum x^2 = 1{,}392.19$ into the sample standard deviation shortcut formula.

$$s = \sqrt{\dfrac{1{,}392.19 - \dfrac{(109.1)^2}{10}}{10-1}} = \sqrt{\dfrac{1{,}392.19 - \dfrac{11{,}902.81}{10}}{9}} = \sqrt{\dfrac{1{,}392.19 - 1{,}190.281}{9}} = \sqrt{\dfrac{201.909}{9}} = 4.736$$

Use s to calculate the approximate standard error of the mean. (Note that $\hat{\sigma}_{\bar{x}}$ is used in place of $\sigma_{\bar{x}}$ because s is used in place of σ.)

$$\hat{\sigma}_{\bar{x}} = \dfrac{s}{\sqrt{n}} = \dfrac{4.736}{\sqrt{10}} = 1.50$$

Rather than use z-scores from Reference Table 1 to calculate the boundaries of the confidence interval, you will use critical t-scores from Reference Table 2. To identify the appropriate t-score, first locate the column for the confidence level indicated by the problem (in this case, 95%).

To locate the appropriate row within that column, calculate the degrees of freedom (df) according to the following formula: df = $n - 1$. In this problem, $df = 10 - 1 = 9$. Consider the following excerpt of Reference Table 2. The correct value of t_c is underlined: $t_c = 2.262$.

Probabilities Under the *t*-Distribution Curve

1 Tail	0.2000	0.1500	0.1000	0.0500	0.0250	0.0100	0.0050	0.0010	0.0005
2 Tail	0.4000	0.3000	0.2000	0.1000	0.0500	0.0200	0.0100	0.0020	0.0010
Conf Lev	0.6000	0.7000	0.8000	0.9000	0.9500	0.9800	0.9900	0.9980	0.9990
df									
1	1.376	1.963	3.078	6.314	12.706	31.821	63.657	318.31	636.62
2	1.061	1.386	1.886	2.920	4.303	6.965	9.925	22.327	31.599
3	0.978	1.250	1.638	2.353	3.182	4.541	5.841	10.215	12.924
4	0.941	1.190	1.533	2.132	2.776	3.747	4.604	7.173	8.610
5	0.920	1.156	1.476	2.015	2.571	3.365	4.032	5.893	6.869
6	0.906	1.134	1.440	1.943	2.447	3.143	3.707	5.208	5.959
7	0.896	1.119	1.415	1.895	2.365	2.998	3.499	4.785	5.408
8	0.889	1.108	1.397	1.860	2.306	2.896	3.355	4.501	5.041
9	0.883	1.100	1.383	1.833	<u>2.262</u>	2.821	3.250	4.297	4.781
10	0.879	1.093	1.372	1.812	2.228	2.764	3.169	4.144	4.587

Use the modified versions of the confidence interval boundary formulas below to construct a 95% confidence interval for the mean.

$$\text{lower limit} = \bar{x} - t_c \hat{\sigma}_{\bar{x}} \qquad \text{upper limit} = \bar{x} + t_c \hat{\sigma}_{\bar{x}}$$
$$= 10.91 - (2.262)(1.50) \qquad = 10.91 + (2.262)(1.50)$$
$$= 7.52 \qquad = 14.30$$

Based on this sample, you are 95% confident that the true population mean is between 7.52 and 14.30 pounds per household.

Note: Problems 9.25–9.27 refer to the data set in Problem 9.25, the amount of trash generated by ten households (in pounds per day). Assume that the population is normally distributed.

9.26 Construct a 90% confidence interval for the mean.

Given a sample size of 10, there are 10 – 1 = 9 degrees of freedom. The critical *t*-score in the 90% confidence level column and the df = 9 row of Reference Table 2 is $t_c = 1.833$. According to Problem 9.25, $\hat{\sigma}_{\bar{x}} = 1.50$. Compute the boundaries of the confidence interval.

$$\text{lower limit} = \bar{x} - t_c \hat{\sigma}_{\bar{x}} \qquad \text{upper limit} = \bar{x} + t_c \hat{\sigma}_{\bar{x}}$$
$$= 10.91 - (1.833)(1.50) \qquad = 10.91 + (1.833)(1.50)$$
$$= 8.16 \qquad = 13.66$$

Based on this sample, you are 90% confident that the true population mean for the amount of trash generated per household is between 8.16 and 13.66 pounds. Decreasing the confidence level from 95% (in Problem 9.25) to 90% results in a narrower confidence interval. ⟵

> The interval is narrower (and more precise), but you're less confident that the actual mean falls in the smaller interval.

Note: Problems 9.25–9.27 refer to the data set in Problem 9.25, the amount of trash generated by ten households (in pounds per day). Assume that the population is normally distributed.

9.27 Construct an 80% confidence interval for the mean.

Once again, $\bar{x} = 10.91$, $\hat{\sigma}_{\bar{x}} = 1.50$, and df = 9; according to Reference Table 2, $t_c = 1.383$. Calculate the bounds of the confidence interval.

Need to calculate the margin of error E?

$$E = t_c \hat{\sigma}_{\bar{x}}$$
$$= 1.383(1.5)$$
$$= 2.0745$$

$$\text{lower limit} = \bar{x} - t_c \hat{\sigma}_{\bar{x}} \qquad \text{upper limit} = \bar{x} + t_c \hat{\sigma}_{\bar{x}}$$
$$= 10.91 - (1.383)(1.50) \qquad\qquad = 10.91 + (1.383)(1.50)$$
$$= 8.84 \qquad\qquad\qquad\qquad = 12.98$$

Note: Problems 9.28–9.30 refer to the data below, the current ages of eight randomly selected aircraft passengers (in years). Assume that the population is normally distributed.

Age in Years							
7.0	3.5	8.6	15.8	14.7	7.8	0.5	4.2

9.28 Construct a 95% confidence interval for the mean based on the sample.

Calculate the sample mean.

$$\bar{x} = \frac{\sum x}{n} = \frac{7.0 + 3.5 + 8.6 + 15.8 + 14.7 + 7.8 + 0.5 + 4.2}{8} = \frac{62.1}{8} = 7.7625$$

In order to determine the sample standard deviation s using the shortcut method, you need to calculate the sum of the squares of the data values and the square of the sum. The table below contains each data value and its square.

x	x^2
7.0	49.00
3.5	12.25
8.6	73.96
15.8	249.64
14.7	216.09
7.8	60.84
0.5	0.25
4.2	17.64
Total 62.1	**679.67**

Substitute $\sum x = 62.1$ and $\sum x^2 = 679.67$ into the sample standard deviation shortcut formula.

$$s = \sqrt{\frac{\sum x^2 - \frac{\left(\sum x\right)^2}{n}}{n-1}} = \sqrt{\frac{679.67 - \frac{(62.1)^2}{8}}{8-1}} = \sqrt{\frac{679.67 - \frac{3,856.41}{8}}{7}} = \sqrt{\frac{679.67 - 482.05125}{7}} = 5.313$$

Use $s = 5.313$ to calculate the approximate standard error of the mean.

$$\hat{\sigma}_{\bar{x}} = \frac{s}{\sqrt{n}} = \frac{5.313}{\sqrt{8}} = 1.88$$

The sample size is $n = 8$, so df $= 8 - 1 = 7$. According to Reference Table 2, $t_c = 2.365$. Calculate the boundaries of the confidence interval.

lower limit $= \bar{x} - t_c \hat{\sigma}_{\bar{x}}$ upper limit $= \bar{x} + t_c \hat{\sigma}_{\bar{x}}$

      $= 7.7625 - (2.365)(1.88)$ $= 7.7625 + (2.365)(1.88)$

      $= 3.32$ $= 12.21$

Based on this sample, you can be 95% confident that the average age of an aircraft passenger is between 3.31 and 12.21 years. This interval is fairly wide for two reasons: the small sample size and the comparatively high standard deviation.

> Were there no adults on this plane?

Note: Problems 9.28–9.30 refer to the data set in Problem 9.28, the current ages of eight randomly selected aircraft passengers (in years). Assume that the population is normally distributed.

9.29 Construct a 98% confidence interval for the mean.

According to Reference Table 2, $t_c = 2.998$. Recall that $\bar{x} = 7.7625$ and $\hat{\sigma}_{\bar{x}} = 1.88$.

lower limit $= \bar{x} - t_c \hat{\sigma}_{\bar{x}}$ upper limit $= \bar{x} + t_c \hat{\sigma}_{\bar{x}}$

      $= 7.7625 - (2.998)(1.88)$ $= 7.7625 + (2.998)(1.88)$

      $= 2.13$ $= 13.40$

Note: Problems 9.28–9.30 refer to the data set in Problem 9.28, the current ages of eight randomly selected aircraft passengers (in years). Assume that the population is normally distributed.

9.30 Construct a 90% confidence interval for the mean.

According to Reference Table 2, $t_c = 1.895$. Recall that $\bar{x} = 7.7625$ and $\hat{\sigma}_{\bar{x}} = 1.88$.

lower limit $= \bar{x} - t_c \hat{\sigma}_{\bar{x}}$ upper limit $= \bar{x} + t_c \hat{\sigma}_{\bar{x}}$

      $= 7.7625 - (1.895)(1.88)$ $= 7.7625 + (1.895)(1.88)$

      $= 4.20$ $= 11.33$

This is the standard deviation of the sample: s, not σ.

Note: Problems 9.31–9.33 refer to a sample of 22 households that used an average of 346.2 gallons of water per day, with a standard deviation of 50.5 gallons. Assume the population of household water usage is normally distributed.

9.31 Construct an 80% confidence interval for the mean.

A sample size of $n = 22$ has $22 - 1 = 21$ degrees of freedom. The value at which 80% confidence and df = 21 intersects in Reference Table 2 is $t_c = 1.323$. Calculate the approximate standard error of the mean, given $s = 50.5$.

$$\hat{\sigma}_{\bar{x}} = \frac{s}{\sqrt{n}} = \frac{50.5}{\sqrt{22}} = 10.767$$

Identify the bounds of the confidence interval.

lower limit $= \bar{x} - t_c \hat{\sigma}_{\bar{x}}$ upper limit $= \bar{x} + t_c \hat{\sigma}_{\bar{x}}$

$= 346.2 - (1.323)(10.767)$ $= 346.2 + (1.323)(10.767)$

$= 331.96$ $= 360.44$

Note: Problems 9.31–9.33 refer to a sample of 22 households that used an average of 346.2 gallons of water per day, with a standard deviation of 50.5 gallons. Assume the population of household water usage is normally distributed.

9.32 Construct a 99% confidence interval for the mean.

According to Reference Table 2, $t_c = 2.831$. Recall that $\bar{x} = 346.2$ and $\hat{\sigma}_{\bar{x}} = 10.767$.

lower limit $= \bar{x} - t_c \hat{\sigma}_{\bar{x}}$ upper limit $= \bar{x} + t_c \hat{\sigma}_{\bar{x}}$

$= 346.2 - (2.831)(10.767)$ $= 346.2 + (2.831)(10.767)$

$= 315.72$ $= 376.68$

Note: Problems 9.31–9.33 refer to a sample of 22 households that used an average of 346.2 gallons of water per day, with a standard deviation of 50.5 gallons. Assume the population of household water usage is normally distributed.

9.33 Construct a 95% confidence interval for the mean.

According to Reference Table 2, $t_c = 2.080$. Recall that $\bar{x} = 346.2$ and $\hat{\sigma}_{\bar{x}} = 10.767$.

lower limit $= \bar{x} - t_c \hat{\sigma}_{\bar{x}}$ upper limit $= \bar{x} + t_c \hat{\sigma}_{\bar{x}}$

$= 346.2 - (2.080)(10.767)$ $= 346.2 + (2.080)(10.767)$

$= 323.80$ $= 368.60$

Confidence Intervals for the Mean with Large Samples and Sigma Unknown

Welcome back, central limit theorem!

9.34 Describe the impact of a large sample size on the construction of a confidence interval when the population standard deviation is unknown.

The *t*-distribution should be used whenever the sample standard deviation *s* is used in place of the population standard deviation σ. However, when the sample size reaches 30 or more, the *t*-score values become very close to the *z*-score values from the normal distribution. Thus, the normal distribution becomes a good approximation to the *t*-distribution when $n \geq 30$.

Most published tables for the *t*-distribution only show probabilities up to a sample size of 30 and then only in increments beyond this value. Because the normal distribution table does not depend on sample size, it is more convenient to substitute the *z*-score for the *t*-score when $n \geq 30$. ←

> The population doesn't need to be normally distributed when the sample size is greater than 30.

9.35 How do you identify the minimum sample size needed to construct a confidence interval for the mean with a specific margin of error when the population standard deviation is unknown?

Consider the minimum sample size formula generated in Problem 9.9.

$$n = \left(\frac{z_c \sigma}{E} \right)^2$$

> The range is the highest data value minus the lowest data value.

This equation requires you to know the population standard deviation σ. If σ is unknown, estimate it using the range *R* of the population, according to the formula below.

$$\hat{\sigma} = \frac{R}{6}$$

This estimate is based on the empirical rule, which states that 99.7% of a population's values lie within three standard deviations of the mean—three standard deviations less than the mean and three standard deviations greater than the mean, for a total of six.

> See Problem 7.23 for a more detailed explanation of the empirical rule.

Note: Problems 9.36–9.37 refer to a sample of 36 mechanical engineers averaging 41.7 years of age, with a sample standard deviation of 6.9 years.

9.36 Construct a 95% confidence interval to estimate the average age of a mechanical engineer.

Approximate the standard error of the mean.

$$\hat{\sigma}_{\bar{x}} = \frac{s}{\sqrt{n}} = \frac{6.9}{\sqrt{36}} = \frac{6.9}{6} = 1.15$$

See Problem 9.10 for more information.

According to Reference Table 1, $z_c = 1.96$ for a 95% confidence interval. Apply the confidence interval boundary formulas.

$$\text{lower limit} = \bar{x} - z_c \hat{\sigma}_{\bar{x}} \qquad \text{upper limit} = \bar{x} + z_c \hat{\sigma}_{\bar{x}}$$
$$= 41.7 - (1.96)(1.15) \qquad\qquad = 41.7 + (1.96)(1.15)$$
$$= 39.45 \qquad\qquad\qquad = 43.95$$

Note: Problems 9.36–9.37 refer to a sample of 36 mechanical engineers averaging 41.7 years of age, with a sample standard deviation of 6.9 years.

9.37 Construct a 99% confidence interval to estimate the average age of a mechanical engineer.

The critical z-score that corresponds to a 99% confidence interval is $z_c = 2.58$. Recall that $\hat{\sigma}_{\bar{x}} = 1.15$.

$$\text{lower limit} = \bar{x} - z_c \hat{\sigma}_{\bar{x}} \qquad \text{upper limit} = \bar{x} + z_c \hat{\sigma}_{\bar{x}}$$
$$= 41.7 - (2.58)(1.15) \qquad\qquad = 41.7 + (2.58)(1.15)$$
$$= 38.73 \qquad\qquad\qquad = 44.67$$

9.38 A formal wear store determines that its customer population has a minimum age of 27 years and a maximum age of 62 years. What is the minimum sample size needed to construct a 90% confidence interval for the mean age of its customers, assuming a margin of error of one year?

The range R of the sample is the difference between the maximum and minimum values: $R = 62 - 27 = 35$. Estimate the standard deviation of the population.

$$\hat{\sigma} = \frac{R}{6} = \frac{35}{6} = 5.833$$

Calculate the minimum sample size using $\hat{\sigma}$.

$$n = \left(\frac{z_c \hat{\sigma}}{E}\right)^2 = \left[\frac{(1.64)(5.833)}{1}\right]^2 = (9.56612)^2 = 91.51$$

A minimum sample size of 92 customers is required for a 90% confidence level.

$0.92 \div 2 = 0.4600$, and the closest value in Reference Table 1 is 0.4599, which is $z_c = 1.75$.

Note: Problems 9.39–9.40 refer to a sample of 60 high school teachers with an average annual salary of $52,113 and a sample standard deviation of $7,804.

9.39 Construct a 92% confidence interval to estimate the mean salary of a high school teacher.

The critical z-score that corresponds to a 92% confidence interval is $z_c = 1.75$. Approximate the standard error of the mean.

$$\hat{\sigma}_{\bar{x}} = \frac{s}{\sqrt{n}} = \frac{7,804}{\sqrt{60}} = 1,007.492$$

Calculate the boundaries of the confidence interval.

lower limit $= \bar{x} - z_c \hat{\sigma}_{\bar{x}}$ upper limit $= \bar{x} + z_c \hat{\sigma}_{\bar{x}}$

$\quad = 52,113 - (1.75)(1,007.492)$ $\quad = 52,113 + (1.75)(1,007.492)$

$\quad = 50,349.89$ $\quad = 53,876.11$

Note: Problems 9.39–9.40 refer to a sample of 60 high school teachers with an average annual salary of $52,113 and a sample standard deviation of $7,804.

9.40 Construct a 97% confidence interval to estimate the mean salary of a high school teacher.

The critical z-score that corresponds to a 97% confidence interval is $z_c = 2.17$. Recall that $\hat{\sigma}_{\bar{x}} = 1,007.492$.

> This z-score has a corresponding area of 0.485, which is equal to 0.97 ÷ 2.

lower limit $= \bar{x} - z_c \hat{\sigma}_{\bar{x}}$ upper limit $= \bar{x} + z_c \hat{\sigma}_{\bar{x}}$

$\quad = 52,113 - (2.17)(1,007.492)$ $\quad = 52,113 + (2.17)(1,007.492)$

$\quad = \$49,926.74$ $\quad = \$54,299.26$

9.41 A nationwide organization examines the salaries of its administrative assistants and determines that the minimum and maximum annual salaries are $29,500 and $68,300, respectively. What is the minimum sample size needed to construct a 95% confidence interval for the mean salary, assuming a $1,500 margin of error?

The critical z-score that corresponds to a 95% confidence interval is $z_c = 1.96$. Estimate the population standard deviation.

> 90%: $z_c = 1.64$
> 91%: $z_c = 1.70$
> 92%: $z_c = 1.75$
> 93%: $z_c = 1.81$
> 94%: $z_c = 1.88$
> 95%: $z_c = 1.96$
> 96%: $z_c = 2.05$
> 97%: $z_c = 2.17$
> 98%: $z_c = 2.33$
> 99%: $z_c = 2.58$

$$\hat{\sigma} = \frac{R}{6} = \frac{68,300 - 29,500}{6} = \frac{38,800}{6} = 6,466.667$$

Calculate the minimum sample size using $\hat{\sigma}$.

$$n = \left(\frac{z_c \hat{\sigma}}{E}\right)^2 = \left[\frac{(1.96)(6,466.667)}{1,500}\right]^2 = (8.449778)^2 = 71.40$$

A minimum sample size of 72 administrative assistant salaries is required.

Note: Problems 9.42–9.43 refer to a sample of 45 golfers at a particular golf course with an average golf score of 94.5 and a sample standard deviation of 9.3.

9.42 Construct a 90% confidence interval to estimate the mean score at the golf course.

Approximate the standard error of the mean.

$$\hat{\sigma}_{\bar{x}} = \frac{s}{\sqrt{n}} = \frac{9.3}{\sqrt{45}} = 1.386$$

Calculate the boundaries of the confidence interval.

$$\text{lower limit} = \bar{x} - z_c\hat{\sigma}_{\bar{x}} \qquad\qquad \text{upper limit} = \bar{x} + z_c\hat{\sigma}_{\bar{x}}$$
$$= 94.5 - (1.64)(1.386) \qquad\qquad = 94.5 + (1.64)(1.386)$$
$$= 92.23 \qquad\qquad\qquad\qquad = 96.77$$

Note: Problems 9.42–9.43 refer to a sample of 45 golfers at a particular golf course with an average golf score of 94.5 and a sample standard deviation of 9.3.

9.43 Construct a 96% confidence interval to estimate the mean score at the golf course.

According to Problem 9.42, if a 45-golfer sample has a standard deviation of 9.3, then $\hat{\sigma}_{\bar{x}} = 1.386$. Apply the confidence interval boundary formulas.

$$\text{lower limit} = \bar{x} - z_c\hat{\sigma}_{\bar{x}} \qquad\qquad \text{upper limit} = \bar{x} + z_c\hat{\sigma}_{\bar{x}}$$
$$= 94.5 - (2.05)(1.386) \qquad\qquad = 94.5 + (2.05)(1.386)$$
$$= 91.66 \qquad\qquad\qquad\qquad = 97.34$$

9.44 A local country club has a population of golfers that shoot a minimum score of 76 and a maximum score of 117. What is the minimum sample size needed to construct a 98% confidence interval for the mean score, assuming a margin of error of two strokes?

Estimate the population standard deviation.

$$\hat{\sigma} = \frac{R}{6} = \frac{117 - 76}{6} = \frac{41}{6} = 6.833$$

Calculate the minimum sample size using $\hat{\sigma}$.

$$n = \left(\frac{z_c\hat{\sigma}}{E}\right)^2 = \left[\frac{(2.33)(6.833)}{2}\right]^2 = (7.960445)^2 = 63.37$$

A minimum sample size of 64 golfers is required.

Confidence Intervals for the Proportion
Estimating percentages from a population

9.45 Describe the procedure used to construct a confidence interval for a proportion.

The confidence interval for the proportion is used when the random variable is binomially distributed. This interval identifies boundaries around a sample proportion within which the true population proportion lies.

You may need to calculate the sample proportion p_s by dividing the number of successes in the sample by the sample size: $p_s = \dfrac{s}{n}$. The standard error of the proportion σ_p is calculated using the formula below, in which p is the population proportion.

$$\sigma_p = \sqrt{\frac{p(1-p)}{n}}$$

If you know the population proportion, why try to estimate it with an interval?

When the population proportion is not known, the standard error of the proportion is approximated using the formula below, in which the sample proportion p_s replaces the population proportion.

$$\hat{\sigma}_p = \sqrt{\frac{p_s(1-p_s)}{n}}$$

Once the standard error of the proportion is estimated, the limits of the confidence interval are calculated by adding the margin of error to and subtracting it from the sample proportion.

$$\text{lower limit} = p_s - z_c\hat{\sigma}_p \qquad \text{upper limit} = p_s - z_c\hat{\sigma}_p$$

9.46 Explain how to identify the minimum sample size needed to construct a confidence interval for the proportion with a specific margin of error when the population proportion is unknown.

The minimum sample size n needed to construct a confidence interval for population proportion p with margin of error E is calculated using the formula below.

$$n = \frac{p(1-p)z_c^2}{E^2}$$

That's because $p(1-p)$ reaches its maximum value when $p = 0.5$.

The formula requires you to know the population proportion p. There are, however, two common ways to address a missing p value. One option is to use a sample proportion p_s to approximate p. The second option is to choose a conservative value for p.

The closer p is to 0.5, the larger the required sample size. Setting $p = 0.5$ guarantees a sample size large enough to provide the desired margin of error.

9.47 A random sample of 175 registered voters revealed that 54% of them voted in the last election. Construct a 95% confidence interval to estimate the true proportion of voter turnout.

Of the $n = 175$ registered voters, $p_s = 0.54$ voted. You are constructing a 95% confidence interval; the corresponding critical z-score is $z_c = 1.96$. Approximate the standard error of the proportion using the sample proportion p_s.

$$\hat{\sigma}_p = \sqrt{\frac{p_s(1-p_s)}{n}} = \sqrt{\frac{(0.54)(1-0.54)}{175}} = \sqrt{\frac{(0.54)(0.46)}{175}} = \sqrt{\frac{0.2484}{175}} = 0.0377$$

Calculate the boundaries of the confidence interval.

$$\text{lower limit} = p_s - z_c\hat{\sigma}_p \qquad\qquad \text{upper limit} = p_s - z_c\hat{\sigma}_p$$
$$= 0.54 - (1.96)(0.0377) \qquad\qquad = 0.54 + (1.96)(0.0377)$$
$$= 0.466 \qquad\qquad\qquad\qquad = 0.614$$

Based on this sample, you can be 95% confident that the true proportion of voter turnout was between 46.6% and 61.4%.

9.48 A pilot sample of 50 voters found that 30 of them voted in the last election. How many more voters must be sampled to construct a 95% interval with a margin of error equal to 0.04?

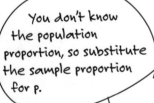

You don't know the population proportion, so substitute the sample proportion for p.

Calculate the sample proportion for the pilot sample.

$$p_s = \frac{s}{n} = \frac{30}{50} = 0.60$$

Substitute $p_s = 0.60$, $z_c = 1.96$, and $E = 0.04$ into the minimum sample size formula presented in Problem 9.46.

$$n = \frac{p_s(1-p_s)z_c^2}{E^2} = \frac{(0.6)(1-0.6)(1.96)^2}{(0.04)^2} = \frac{0.921984}{0.0016} = 576.24$$

A sample size of 577 voters is required. So far, 50 voters have been polled, so an additional $577 - 50 = 527$ voters need to be sampled.

9.49 In a random sample of 220 men (18 or older), 139 are married. Construct a 98% confidence interval to estimate the true proportion of married men (18 or older).

Calculate the sample proportion.

$$p_s = \frac{s}{n} = \frac{139}{220} = 0.632$$

Approximate the standard error of the proportion.

$$\hat{\sigma}_p = \sqrt{\frac{p_s(1-p_s)}{n}} = \sqrt{\frac{(0.632)(1-0.632)}{220}}\sqrt{\frac{(0.632)(0.368)}{220}} = \sqrt{\frac{0.232576}{220}} = 0.0325$$

Calculate the boundaries of the confidence interval.

lower limit = $p_s - z_c\hat{\sigma}_p$ upper limit = $p_s - z_c\hat{\sigma}_p$

$= 0.632 - (2.33)(0.0325)$ $= 0.632 + (2.33)(0.0325)$

$= 0.556$ $= 0.708$

Based on this sample, you are 98% confident that the true proportion of married men (18 or older) is between 55.6% and 70.8%.

9.50 Determine the minimum sample size of men (21 or older) required to construct a 90% confidence interval to estimate the proportion of married men (21 or older), assuming a 0.05 margin of error.

Because there is no information available to estimate the population proportion, use $p_s = 0.5$ to provide a conservative estimate of the sample size.

$$n = \frac{p(1-p)z_c^2}{E^2} = \frac{(0.5)(1-0.5)(1.64)^2}{(0.05)^2} = \frac{(0.5)^2(1.64)^2}{(0.05)^2} = \left[\frac{(0.5)(1.64)}{0.05}\right]^2 = 268.96$$

Each of these factors is squared, so you can multiply 0.5 and 1.64, divide by 0.05, and then square the answer (instead of squaring each number first).

A sample size of 269 men is required.

9.51 A random sample of 500 American citizens (who were at least five years old) contains 96 people who speak a language other than English at home. Construct a 90% confidence interval to estimate the true proportion of people (five years old or older) who speak a language other than English at home.

Calculate the sample proportion.

$$p_s = \frac{s}{n} = \frac{96}{500} = 0.192$$

Approximate the standard error of the proportion.

$$\hat{\sigma}_p = \sqrt{\frac{p_s(1-p_s)}{n}} = \sqrt{\frac{(0.192)(1-0.192)}{500}}\sqrt{\frac{(0.192)(0.808)}{500}} = \sqrt{\frac{0.155136}{500}} = 0.0176$$

Calculate the boundaries of the confidence interval.

lower limit = $p_s - z_c\hat{\sigma}_p$ upper limit = $p_s - z_c\hat{\sigma}_p$

$= 0.192 - (1.64)(0.0176)$ $= 0.192 + (1.64)(0.0176)$

$= 0.163$ $= 0.221$

9.52 A pilot sample of 75 American citizens (who were at least five years old) found that 11 of the people spoke a language other than English at home. Determine the additional number of people that need to be sampled to construct a 90% confidence interval for the proportion, assuming a margin of error of 0.02.

Calculate the sample proportion for the pilot sample.

$$p_s = \frac{11}{75} = 0.1467$$

Determine the minimum required sample size.

$$n = \frac{p_s(1-p_s)z_c^2}{E^2} = \frac{(0.1467)(1-0.1467)(1.64)^2}{(0.02)^2} = \frac{(0.1467)(0.8533)(1.64)^2}{(0.02)^2} = 841.70$$

An additional 842 – 75 = 767 people need to be sampled.

9.53 A random sample of 415 women between the ages of 40 and 45 contains 330 mothers. Construct a 92% confidence interval to estimate the true proportion of women in that age range who are mothers.

Calculate the sample proportion.

$$p_s = \frac{330}{415} = 0.7952$$

Approximate the standard error of the proportion.

$$\hat{\sigma}_p = \sqrt{\frac{p_s(1-p_s)}{n}} = \sqrt{\frac{(0.7952)(1-.7952)}{415}} = \sqrt{\frac{(0.7952)(0.2048)}{415}} = 0.020$$

Calculate the boundaries of the confidence interval.

lower limit = $p_s - z_c\hat{\sigma}_p$ upper limit = $p_s - z_c\hat{\sigma}_p$
= 0.7952 – (1.75)(0.020)   = 0.7952 + (1.75)(0.020)
= 0.760   = 0.830

9.54 Determine the minimum sample size required to construct a 92% confidence interval to estimate the proportion of women between the ages of 35 and 45 who are mothers, assuming a margin of error equal to 0.03.

Because there is no information available to estimate the population proportion, use $p_s = 0.50$ to provide a conservative estimate of sample size. Calculate the minimum sample size.

$$n = \frac{p_s(1-p_s)z_c^2}{E^2} = \frac{(0.5)(1-0.5)(1.75)^2}{(0.03)^2} = \frac{0.765625}{0.0009} = 850.69$$

A sample size of 851 women between the ages of 40 and 45 will provide a 92% confidence interval with a margin of error equal to 0.03.

Chapter 10

HYPOTHESIS TESTING FOR A SINGLE POPULATION

Time to reject some null hypotheses

Hypothesis testing is the cornerstone of inferential statistics. At its heart, it involves making a statement about a population parameter, such as the mean or proportion, and then using a sample from the population to conclude whether that statement is true.

The challenge for this chapter is to choose the appropriate hypothesis test for a given problem. One important factor is the size of the sample. For large samples, you can rely on the central limit theorem and the normal distribution. When the sample size is small and the population standard deviation is unknown, switch to the t-distribution (introduced in Chapter 9).

Introduction to Hypothesis Testing for the Mean

What are null and alternative hypotheses?

10.1 Describe the null and alternative hypotheses used in hypothesis testing.

> The null hypothesis says, "Unless you tell me otherwise, this is what I think about the population." It may, for instance, insist that the population mean is greater than 250.

A hypothesis is a statement about a population that may or may not be true. The purpose of hypothesis testing is to make a statistical conclusion about whether or not to accept such a statement. Every hypothesis test has both a null hypothesis and an alternative hypothesis.

The null hypothesis, denoted H_0, represents the status quo, comparing the mean of a population to a specific value. The null hypothesis is believed to be true unless there is overwhelming evidence to the contrary. The alternative hypothesis, denoted H_1, represents the opposite of the null hypothesis; it is true if the null hypothesis is false. (Some texts denote the alternative hypothesis H_A.)

The alternative hypothesis always states that the mean of the population is less than, greater than, or not equal to a specific value. It is also known as the *research hypothesis* because it states the position a researcher is attempting to establish.

10.2 A lightbulb manufacturer has developed a new lightbulb that it claims has an average life of more than 1,000 hours. State the null and alternative hypotheses that would be used to verify this claim.

> If something is not greater than 1,000, then it is either less than or equal to 1,000.

The alternative, or research, hypothesis represents the claim the company is attempting to establish: $\mu > 1,000$ hours. The null hypothesis is the opposite of the research hypothesis: $\mu \le 1,000$ hours. Given an alternative hypothesis that contains an inequality, convention states that the null hypothesis will include the possibility of equality.

> In other words, the null hypothesis includes the "or equal to" possibility.

10.3 A pizza delivery company claims that its average delivery time is less than 45 minutes. State the null and alternative hypotheses that would be used to prove this claim.

The alternative hypothesis H_1 represents the claim the company is attempting to verify (in this case, that the population mean of the delivery time is less than 45 minutes). The null hypothesis H_0 is the opposite of H_1 and includes the possibility of equality.

$$H_0: \mu \ge 45 \text{ minutes}$$
$$H_1: \mu < 45 \text{ minutes}$$

> Some textbooks state the null hypothesis using = instead of ≤ or ≥.

10.4 A cereal manufacturer uses a filling process designed to add 18 ounces of cereal to each box. State the null and alternative hypotheses the manufacturer would use to verify the accuracy of this process.

The alternative hypothesis states the goal of the process: to fill each box with 18 ounces of cereal. The null hypothesis is the opposite: a population mean that is *not* equal to 18 ounces.

$$H_0 : \mu = 18 \text{ ounces}$$
$$H_1 : \mu \ne 18 \text{ ounces}$$

10.5 Describe Type I and Type II errors in hypothesis testing.

The purpose of a hypothesis test is to verify the validity of a claim about a population based on a single sample. However, a sample may not be representative of the population as a whole, which would invalidate the claims made based on the sample.

Consider Problem 10.4, in which a process is used to fill cereal boxes. If the sample mean was 16 ounces, the hypothesis test might reject the null hypothesis, which states that the population mean equals 18 ounces. If the population mean actually is 18 ounces, the conclusion is wrong. This is known as a Type I error. The probability of making a Type I error is known as α, the level of significance. The value for α is determined before the population is sampled; typical values of α range from 0.01 to 0.10.

> *This occurs because of sampling error. See Problems 9.1–9.3.*

If a sample from the filling process had a mean of 18 ounces, the hypothesis test would fail to reject the null hypothesis. If the filling process is actually not operating accurately and the population mean is 16 ounces, a Type II error has occurred. The probability of making a Type II error is known as β, and the power of the hypothesis test is $1-\beta$.

> *You never have enough evidence to accept the null hypothesis unless you sample the entire population. When using a sample, you can only "fail to reject" the null hypothesis.*

10.6 Explain how to perform a two-tailed hypothesis test.

A two-tailed hypothesis test is used when the alternative hypothesis is expressed as "not equal to" a specific value. The cereal box problem (Problem 10.4) is one such example, because the alternative hypothesis is $\mu \neq 18$. To better understand the two-tailed hypothesis test, consider the normal distribution curve below.

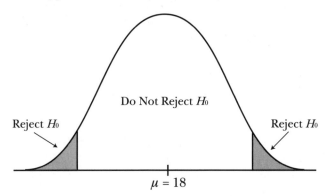

The bell curve in the figure represents the sampling distribution for the average weight of a box of cereal. The mean of the population, $\mu = 18$ ounces, according to the null hypothesis, is the mean of the sampling distribution. The area of the shaded regions is α, the level of significance.

To conduct a two-tailed hypothesis test, complete the following steps:

- Collect a sample of size n and calculate the test statistic: in this case, the sample mean.

- Plot the sample mean on the x-axis of the sampling distribution curve.

- If the sample mean lies within the unshaded region, do not reject H_0; you do not have enough evidence to support H_1, the alternative hypothesis.

- If the sample mean lies within either of the shaded regions (known as the rejection region), reject H_0; you have sufficient evidence to support H_1.

Because there are two rejection regions in the preceding figure, this procedure is called a two-tailed hypothesis test.

10.7 Explain how to perform a one-tailed hypothesis test.

A one-tailed hypothesis test is used when the alternative hypothesis is expressed as "greater than" or "less than" a specific value. The pizza delivery problem (Problem 10.3) is one such example, because the alternative hypothesis is $\mu < 45$. Consider the figure below.

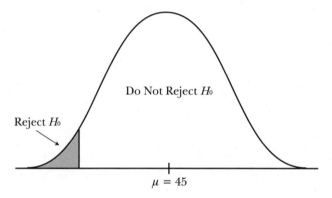

A one-tailed test has only one rejection region: in this case, the shaded area on the left side the distribution. The area of this shaded region is based on α. Follow the same procedure you used for the two-tailed test (outlined in Problem 10.6) and plot the sample mean. (In Problem 10.3, the mean is 45 minutes.) Two possible outcomes may occur:

- If the sample mean lies in the unshaded region, do not reject H_0; you do not have enough evidence to support the alternative hypothesis.

- If the sample mean lies in the shaded rejection region, reject H_0; you have enough evidence to support H_1.

In the pizza delivery time example, the company can only reject the null hypothesis (the delivery time is 45 minutes or longer) if the sample mean is low enough to fall within the shaded region.

Hypothesis Testing for the Mean with $n \geq 30$ and Sigma Known

Calling on the central limit theorem once again

Note: Problems 10.8–10.10 refer to a company that claims the average time a customer waits on hold is less than 5 minutes. A sample of 35 customers has an average wait time of 4.78 minutes. Assume the population standard deviation for wait time is 1.8 minutes.

10.8 Test the company's claim at the $\alpha = 0.05$ significance level by comparing the calculated z-score to the critical z-score.

Identify the null and alternative hypotheses.

$$H_0 : \mu \geq 5 \text{ minutes}$$
$$H_1 : \mu < 5 \text{ minutes}$$

The hypotheses are written in terms of "less than" or "greater than," so a one-tailed test is used. You are attempting to verify that the population mean is less than 5 minutes, so the rejection region is 5% of the total area beneath the normal curve less than the sample mean. ←

> This area is at the far left of the curve. Look at the diagram in Problem 10.7.

If the shaded rejection region has an area of $\alpha = 0.5$, the area between the mean of the distribution and the rejection region is $0.50 - 0.05 = 0.45$. According to Reference Table 1, the corresponding critical z-score is $z_c = -1.64$. Note that z_c is negative because it is on the left side of the mean.

Calculate the standard error of the mean.

$$\sigma_{\bar{x}} = \frac{\sigma}{\sqrt{n}} = \frac{1.8}{\sqrt{35}} = 0.3043$$

Now calculate $z_{\bar{x}}$, the z-score of the sample mean, if the population mean is $\mu = 5$ minutes.

$$z_{\bar{x}} = \frac{\bar{x} - \mu}{\sigma_{\bar{x}}}$$
$$z_{4.78} = \frac{4.78 - 5}{0.3043}$$
$$z_{4.78} = -0.72$$

Plot both z-scores, as illustrated below.

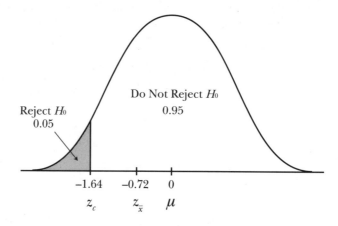

Even though the sample average is 4.78 minutes, it's not far enough below 5 minutes to support the claim that the entire population averages less than 5 minutes.

Because $z_{\bar{x}}$ does not lie in the shaded rejection region (-0.72 is not less than -1.64), there is not enough evidence to support the alternative hypothesis. Thus, you cannot conclude, based on this sample, that the average wait time is less than 5 minutes.

Note: Problems 10.8–10.10 refer to a company that claims the average time a customer waits on hold is less than 5 minutes. A sample of 35 customers has an average wait time of 4.78 minutes. Assume the population standard deviation for wait time is 1.8 minutes.

10.9 Verify your answer to Problem 10.8 by comparing the sample mean $\bar{x} = 4.78$ to the critical sample mean $\bar{x}_c$.

The critical sample mean $\bar{x}_c$ is the sum of the population mean $\mu = 5$ and the product of the critical z-score z_c and the standard error of the mean $\sigma_{\bar{x}}$.

$$\bar{x}_c = \mu + z_c \sigma_{\bar{x}} = 5 + (-1.64)(0.3043) = 5.0 - 0.499 = 4.5$$

Problem 10.8 showed that 4.78 minutes was not far enough below 5 minutes to prove the alternative hypothesis. Turns out that the sample mean needed to be below 4.5 minutes.

In order to reject the null hypothesis that the population mean is less than 5 minutes, the sample mean needs to be less than 4.5 minutes. However, the sample mean is 4.78 minutes; there is insufficient evidence to support the alternative hypothesis.

Note: Problems 10.8–10.10 refer to a company that claims the average time a customer waits on hold is less than 5 minutes. A sample of 35 customers has an average wait time of 4.78 minutes. Assume the population standard deviation for wait time is 1.8 minutes.

10.10 Verify your answer to Problem 10.8 by comparing the p-value to the level of significance $\alpha = 0.05$.

The p-value is the observed level of significance, the smallest level of significance at which the null hypothesis can be rejected. When the p-value is less than the level of significance α, you reject the null hypothesis; otherwise, you fail to reject the null hypothesis.

Recall that the z-score of the sample mean is $z_{4.78} = -0.72$. Calculate the probability that the sample mean lies in the shaded region of the distribution illustrated in Problem 10.8.

$$p\text{-value} = P\left(z_{\bar{x}} < -0.72\right)$$
$$= P\left(z_{\bar{x}} < 0\right) - P\left(-0.72 < z_{\bar{x}} < 0\right)$$
$$= 0.50 - 0.2642$$
$$= 0.2358$$

This comes from Reference Table 1.

Because the p-value is greater than $\alpha = 0.05$, you fail to reject the null hypothesis and must conclude that there is not enough evidence to support the company's claim.

Note: Problems 10.11–10.14 refer to a computer company that claims its laptop batteries average more than 3.5 hours of use per charge. A sample of 45 batteries last an average of 3.72 hours. Assume the population standard deviation is 0.7 hours.

10.11 Test the company's claim at the $\alpha = 0.10$ significance level by comparing the calculated z-score to the critical z-score.

Identify the null and alternative hypotheses.

$$H_0 : \mu \le 3.5 \text{ hours}$$
$$H_1 : \mu > 3.5 \text{ hours}$$

Use a one-tailed test to identify the rejection region, which is—like the sample mean—greater than the proposed population mean $\mu = 3.5$ hours. If the shaded rejection region has an area of $\alpha = 0.10$, the area between the mean of the distribution and the rejection region is $0.50 - 0.10 = 0.40$. According to Reference Table 1, the corresponding critical z-score is $z_c = 1.28$.

In this example, z_c is positive because the rejection region is right of the mean.

Calculate the standard error of the mean.

$$\sigma_{\bar{x}} = \frac{\sigma}{\sqrt{n}} = \frac{0.7}{\sqrt{45}} = 0.1043$$

Calculate the z-score for the sample mean.

$$z_{\bar{x}} = \frac{\bar{x} - \mu}{\sigma_{\bar{x}}}$$
$$z_{3.72} = \frac{3.72 - 3.5}{0.1043}$$
$$z_{3.72} = 2.11$$

Consider the figure below, which illustrates both z-scores.

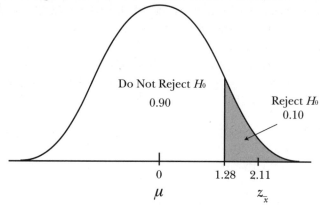

Do Not Reject H_0
0.90

Reject H_0
0.10

0   1.28 2.11
μ $z_{\bar{x}}$

Because 2.11 > 1.28, there is sufficient evidence to support the company's claim that its laptop batteries will average more than 3.5 hours of use per charge.

Use the same procedure as in Problem 10.11, but change the significance level from 0.10 to 0.01.

Note: Problems 10.11–10.14 refer to a computer company that claims its laptop batteries average more than 3.5 hours of use per charge. A sample of 45 batteries last an average of 3.72 hours. Assume the population standard deviation is 0.7 hours.

10.12 Test the company's claim at the $\alpha = 0.01$ significance level by comparing the calculated z-score to the critical z-score.

If the shaded rejection region has an area of $\alpha = 0.01$, the area between the mean of the distribution and the rejection region is $0.50 - 0.01 = 0.49$. The corresponding critical z-score is $z_c = 2.33$.

The values of the sample mean, the population mean, and the standard error of the mean are unaffected by the change in the significance level. According to Problem 10.11, $z_{\bar{x}} = 2.11$. In Problem 10.11, you rejected the null hypothesis because $z_{\bar{x}}$ was greater than the critical z-score $z_c = 1.28$.

2.11 < 2.33

In this problem, however, the critical z-score is higher, because of the change in the level of significance: $z_{\bar{x}} < z_c$, so you fail to reject the null hypothesis and cannot support the company's claim. Lowering α from 0.10 to 0.01 makes rejecting the null hypothesis a more formidable challenge.

Note: Problems 10.11–10.14 refer to a computer company that claims its laptop batteries average more than 3.5 hours of use per charge. A sample of 45 batteries last an average of 3.72 hours. Assume the population standard deviation is 0.7 hours.

10.13 Verify your answer to Problem 10.11 by comparing the sample mean to the critical sample mean $\bar{x}_c$ at an $\alpha = 0.10$ level of significance.

According to Problem 10.11, $z_c = 1.28$ and $\sigma_{\bar{x}} = 0.1043$. Calculate the critical sample mean $\bar{x}_c$.

$$\bar{x}_c = \mu + z_c \sigma_{\bar{x}} = 3.5 + (1.28)(0.1043) = 3.634$$

You can reject the null hypothesis if the batteries in the sample have an average charge of at least 3.634 hours. The sample mean is $\bar{x} = 3.72$, so there is enough evidence to reject H_0 and support the company's claim.

Note: Problems 10.11–10.14 refer to a computer company that claims its laptop batteries average more than 3.5 hours of use per charge. A sample of 45 batteries last an average of 3.72 hours. Assume the population standard deviation is 0.7 hours.

10.14 Verify your answer to Problem 10.11 by comparing the *p*-value to the level of significance $\alpha = 0.10$.

According to Problem 10.11, if you assume that 3.5 is the population mean, then the *z*-score of the sample mean is $z_{3.72} = 2.11$. The *p*-value is the probability that a randomly selected sample could have a mean greater than 2.11.

$$
\begin{aligned}
p\text{-value} &= P\left(z_{\bar{x}} > 2.11\right) \\
&= P\left(z_{\bar{x}} > 0\right) - P\left(0 < z_{\bar{x}} < 2.11\right) \\
&= 0.50 - 0.4826 \\
&= 0.0174
\end{aligned}
$$

Because the *p*-value 0.0174 is less than the level of significance $\alpha = 0.10$, you reject the null hypothesis and conclude that there is enough evidence to support the company's claim.

> You can support the company's claim at any level of significance 0.0174 or higher. Because 0.01 < 0.0174, you couldn't reject the null hypothesis in Problem 10.12.

Note: In Problems 10.15–10.18, a researcher is testing the claim that the average adult consumes 1.7 cups of coffee per day. Assume the population standard deviation is 0.5 cups per day.

10.15 A sample of 30 adults averaged 1.85 cups of coffee per day. Test the researcher's claim at the $\alpha = 0.05$ significance level by comparing the calculated *z*-score to the critical *z*-score.

Identify the null and alternative hypotheses.

$$H_0 : \mu = 1.7 \text{ cups per day}$$
$$H_1 : \mu \neq 1.7 \text{ cups per day}$$

> The alternative hypothesis can only be less than, greater than, or not equal to something. It can never be equal to a value.

The hypotheses are written in terms of "equal to" and "not equal to," so a two-tailed test is used. Half of $\alpha = 0.05$ is placed on the left side of the distribution and half is placed on the right, as illustrated in the figure below.

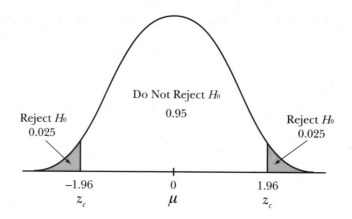

Each shaded region has an area of 0.025 (half of 0.05), so the areas between the mean and each shaded region are $0.50 - 0.025 = 0.475$. According to Reference Table 1, the corresponding critical z-scores are $z_c = -1.96$ and $z_c = 1.96$. In order to reject the null hypothesis, the z-score of the sample mean must be less than -1.96 or greater than 1.96.

Calculate the standard error of the mean.

$$\sigma_{\bar{x}} = \frac{\sigma}{\sqrt{n}} = \frac{0.5}{\sqrt{30}} = 0.0913$$

Calculate $z_{\bar{x}}$, the z-score of the sample mean.

$$z_{1.85} = \frac{\bar{x} - \mu}{\sigma_{\bar{x}}} = \frac{1.85 - 1.7}{0.0912} = \frac{0.15}{0.0913} = 1.64$$

Because $z_{1.85}$ is neither less than -1.96 nor greater than 1.96, the amount of coffee consumed by the average adult is not significantly greater or less than 1.7 cups per day. The claim appears to be valid.

> So the average adult probably does drink about 1.7 cups of coffee per day. You can't reject H_o.

Note: In Problems 10.15–10.18, a researcher is testing the claim that the average adult consumes 1.7 cups of coffee per day. Assume the population standard deviation is 0.5 cups per day.

10.16 A sample of 60 adults averaged 1.85 cups of coffee per day. Test the researcher's claim at the $\alpha = 0.05$ level of significance by comparing the calculated z-score to the critical z-score.

> The samples have the same mean: 1.85.

The null and alternative hypotheses are the same as in Problem 10.15, as are the critical z-scores $z_c = \pm 1.96$. However, the standard error of the mean is different because of the change in sample size.

$$\sigma_{\bar{x}} = \frac{\sigma}{\sqrt{n}} = \frac{0.5}{\sqrt{60}} = 0.0645$$

Calculate z-score of the sample mean $\bar{x} = 1.85$.

$$z_{1.85} = \frac{1.85 - 1.7}{0.0645} = \frac{0.15}{0.0645} = 2.33$$

Because 2.33 > 1.96, the z-score of the mean lies within the rejection region, and there is sufficient evidence to support the researcher's claim. There is a statistically significant difference between 1.7 and 1.85 daily cups of coffee at the $\alpha = 0.05$ level of significance, when $n = 60$.

> Upping the sample size from 30 to 60 was all it took to reject the null hypothesis.

Note: In Problems 10.15–10.18, a researcher is testing the claim that the average adult consumes 1.7 cups of coffee per day. Assume the population standard deviation is 0.5 cups per day.

10.17 Verify your answer to Problem 10.15 by comparing the sample mean to the critical sample mean.

According to Problem 10.15, the critical z-scores at the $\alpha = 0.05$ significance level are $z_c = \pm 1.96$ and $\sigma_{\bar{x}} = 0.0913$. Because you are applying a two-tailed test, calculate two critical sample means by adding $z_c \sigma_{\bar{x}}$ to the population mean and subtracting it from the population mean.

$$\bar{x}_c = \mu + z_c \sigma_{\bar{x}} \qquad\qquad \bar{x}_c = \mu + z_c \sigma_{\bar{x}}$$
$$= 1.7 + (-1.96)(0.0913) \qquad = 1.7 + (1.96)(0.0913)$$
$$= 1.52 \qquad\qquad\qquad\quad = 1.88$$

In order to reject the null hypothesis, the sample mean must be less than 1.52 or greater than 1.88. The sample in Problem 10.15 had a mean of 1.85, which is not large enough to reject the null hypothesis.

Note: In Problems 10.15–10.18, a researcher is testing the claim that the average adult consumes 1.7 cups of coffee per day. Assume the population standard deviation is 0.5 cups per day.

10.18 Verify your answer to Problem 10.15 by comparing the p-value to the level of significance.

You are applying a two-tailed test, so multiply the p-value you would calculate with a one-tailed test by two. According to Problem 10.15, the z-score of the sample mean is $z_{1.85} = 1.64$.

$$p\text{-value} = 2 \cdot P\left(z_{\bar{x}} > 1.64\right)$$
$$= 2\left[P\left(z_{\bar{x}} > 0\right) - P\left(0 < z_{\bar{x}} < 1.64\right)\right]$$
$$= 2(0.50 - 0.4495)$$
$$= (2)(0.0505)$$
$$= 0.101$$

Because the p-value is greater than $\alpha = 0.05$, you fail to reject the null hypothesis.

Note: Problems 10.19–10.21 refer to a claim that the average SAT math score for graduating high school students in the state of Virginia has recently exceeded 500. A sample of 70 students from Virginia had an average SAT math score of 530. Assume the population standard deviation for Virginia students' math SAT scores is 125.

10.19 Test the claim at the $\alpha = 0.05$ significance level by comparing the calculated z-score to the critical z-score.

Identify the null and alternative hypotheses.

$$H_0 : \mu \le 500$$
$$H_1 : \mu > 500$$

> The claim is that average math scores are over 500, so that is the alternative hypothesis.

The hypotheses are stated in terms of "greater than" and "less than," so a one-tailed test is used. The alternative hypothesis contains "greater than," so the rejection region has a critical z-score boundary that is greater than the population mean of 500.

Recall that $z_c = 1.64$. In order to reject the null hypothesis, the z-score of the sample mean will need to be more than 1.64 standard deviations above the mean.

> Significance level 0.05 has different critical z-scores for one- and two-tailed tests. Problem 10.8 says $z_c = 1.64$ for one-tailed tests and Problem 10.15 says $z_c = \pm 1.96$ for two-tailed tests.

Calculate the standard error of the mean.

$$\sigma_{\bar{x}} = \frac{\sigma}{\sqrt{n}} = \frac{125}{\sqrt{70}} = 14.940$$

Calculate the z-score of the sample mean.

$$z_{530} = \frac{530 - 500}{14.940} = \frac{30}{14.940} = 2.01$$

Because $z_{530} = 2.01$ is greater than $z_c = 1.64$, you reject H_0 and conclude that there is sufficient evidence to support the claim that the average SAT math score of Virginia students has recently exceeded 500.

Note: Problems 10.19–10.21 refer to a claim that the average SAT math score for graduating high school students in the state of Virginia has recently exceeded 500. A sample of 70 students from Virginia had an average SAT math score of 530. Assume the population standard deviation for Virginia students' math SAT scores is 125.

10.20 Verify your answer to Problem 10.19 by comparing the sample mean to the critical sample mean.

According to Problem 10.19, $z_c = 1.64$ and $\sigma_{\bar{x}} = 14.940$. Calculate the critical sample mean.

$$\bar{x}_c = \mu + z_c \sigma_{\bar{x}} = 500 + (1.64)(14.940) = 524.50$$

Because the sample mean $\bar{x} = 530$ is greater than the critical sample mean $\bar{x}_c = 524.50$, you reject the null hypothesis.

Note: Problems 10.19–10.21 refer to a claim that the average SAT math score for graduating high school students in the state of Virginia has recently exceeded 500. A sample of 70 students from Virginia had an average SAT math score of 530. Assume the population standard deviation for Virginia students' math SAT scores is 125.

10.21 Verify your answer to Problem 10.19 by comparing the *p*-value to the level of significance $\alpha = 0.05$.

According to Problem 10.19, $z_{530} = 2.01$. Calculate the probability that a random sample has a mean that is 2.01 standard deviations or more above the population mean $\mu = 500$.

$$
\begin{aligned}
p\text{-value} &= P\left(z_{\bar{x}} > 2.01\right) \\
&= P\left(z_{\bar{x}} > 0\right) - P\left(0 < z_{\bar{x}} < 2.01\right) \\
&= 0.50 - 0.4778 \\
&= 0.0222
\end{aligned}
$$

The null hypothesis is rejected when the level of significance is greater than or equal to $\alpha = 0.0222$. Here, $0.05 > 0.0222$, so you reject the null hypothesis.

> The p-value is the lowest possible significance level at which the alternative hypothesis may be rejected.

Hypothesis Testing for the Mean with *n* < 30 and Sigma Known

They need to be normally distributed

Note: Problems 10.22–10.24 refer to a random sample of 20 undergraduate students who worked an average of 13.5 hours per week for a university. Assume the population is normally distributed with a standard deviation of 5 hours per week.

10.22 Test the claim that the average student works less than 15 hours per week at the $\alpha = 0.02$ significance level by comparing the calculated *z*-score to the critical *z*-score.

> The population size is small (less than 30), so the population needs to be normally distributed to use z-scores.

Identify the null and alternative hypotheses.

$$H_0 : \mu \geq 15 \text{ hours}$$
$$H_1 : \mu < 15 \text{ hours}$$

> The sample mean will have to have a z-score of less than −2.05 in order to reject the null hypothesis.

Apply a one-tailed test on the left side of the sampling distribution, to determine whether the sample mean of 13.5 is significantly less than the proposed population mean of 15. If the rejection region has an area of $\alpha = 0.02$, the area between the mean and the rejection region is $0.50 - 0.02 = 0.48$, which has a corresponding *z*-score of $z_c = -2.05$. Note that z_c is negative because the rejection region is on the left side of the distribution.

Calculate the standard error of the mean.

$$\sigma_{\bar{x}} = \frac{\sigma}{\sqrt{n}} = \frac{5}{\sqrt{20}} = 1.118$$

Now calculate the z-score of the sample mean.

$$z_{13.5} = \frac{13.5 - 15.0}{1.118} = \frac{-1.5}{1.118} = -1.34$$

Because $z_{13.5} = -1.34$ is *not* less than the critical z-score $z_c = -2.05$, the sample mean does not lie within the rejection region and you fail to reject H_0. You cannot conclude that the average student works less than 15 hours per week.

Note: Problems 10.22–10.24 refer to a random sample of 20 undergraduate students who worked an average of 13.5 hours per week for a university. Assume the population is normally distributed with a standard deviation of 5 hours per week.

10.23 Verify your answer to Problem 10.22 by comparing the sample mean to the critical sample mean.

According to Problem 10.22, $z_c = -2.05$ and $\sigma_{\bar{x}} = 1.118$. Calculate the critical sample mean.

> Even though the rejection region is left of the mean, you still use a plus sign right here. The negative sign will come from the z-score, in this case $z_c = -2.05$.

$$\bar{x}_c = \mu + z_c \sigma_{\bar{x}} = 15 + (-2.05)(1.118) = 12.71$$

The sample mean is $\bar{x} = 13.5$, which is not low enough to reject the sample mean, because it is not less than 12.71. Thus, you fail to reject the null hypothesis.

Note: Problems 10.22–10.24 refer to a random sample of 20 undergraduate students who worked an average of 13.5 hours per week for a university. Assume the population is normally distributed with a standard deviation of 5 hours per week.

10.24 Verify your answer to Problem 10.22 by comparing the p-value and the significance level $\alpha = 0.02$.

According to Problem 10.22, $z_{13.5} = -1.34$. Calculate the corresponding p-value.

$$
\begin{aligned}
p\text{-value} &= P\left(z_{\bar{x}} < -1.34\right) \\
&= P\left(z_{\bar{x}} < 0\right) - P\left(-1.34 < z_{\bar{x}} < 0\right) \\
&= 0.50 - 0.4099 \\
&= 0.0901
\end{aligned}
$$

The p-value is greater than $\alpha = 0.02$, so you fail to reject the null hypothesis.

Note: Problems 10.25–10.27 refer to a random sample of 25 cars that passed a specific interstate milepost at an average speed of 67.4 miles per hour. Assume the speed of cars passing that milepost is normally distributed with a standard deviation of 6 miles per hour. A researcher claims that the average speed of the population is not 65 miles per hour.

10.25 Test the claim at the $\alpha = 0.10$ significance level by comparing the calculated z-score to the critical z-score.

Identify the null and alternative hypotheses.

$$H_0 : \mu = 65 \text{ mph}$$
$$H_1 : \mu \neq 65 \text{ mph}$$

Apply a two-tailed test, dividing $\alpha = 0.10$ in half and defining rejection regions of area 0.05 on the left and right sides of the distribution. The area between the mean and each rejection region is $0.50 - 0.05 = 0.45$; according to Reference Table 1, the corresponding critical z-scores are $z_c = \pm 1.64$.

Calculate the standard error of the mean.

$$\sigma_{\bar{x}} = \frac{\sigma}{\sqrt{n}} = \frac{6}{\sqrt{25}} = \frac{6}{5} = 1.2$$

Calculate the z-score of the sample mean.

$$z_{67.4} = \frac{67.4 - 65}{1.2} = \frac{2.4}{1.2} = 2$$

Because $z_{67.4} = 2$ is greater than $z_c = 1.64$, you reject H_0 and conclude that the average speed is not 65 miles per hour.

> Two-tailed critical z-scores (and critical sample means) are the same distance from the mean: in this case, 1.64 standard deviations above and below the mean.

> In a two-tailed test, the sample z-score (in this case, 2) has to be higher or lower than the critical z-scores (in this case, lower than −1.64 or higher than 1.64).

Note: Problems 10.25–10.27 refer to a random sample of 25 cars that passed a specific interstate milepost at an average speed of 67.4 miles per hour. Assume the speed of cars passing that milepost is normally distributed with a standard deviation of 6 miles per hour. A researcher claims that the average speed of the population is not 65 miles per hour.

10.26 Verify your answer to Problem 10.25 by comparing the sample mean to the critical sample mean.

According to Problem 10.25, $z_c = \pm 1.64$ and $\sigma_{\bar{x}} = 1.2$. Calculate the critical sample means, both greater and less than the population mean $\mu = 65$.

$$\bar{x}_c = \mu + z_c \sigma_{\bar{x}} \qquad\qquad \bar{x}_c = \mu + z_c \sigma_{\bar{x}}$$
$$= 65 + (-1.64)(1.2) \qquad = 65 + (1.64)(1.2)$$
$$= 63.03 \qquad\qquad\qquad = 66.97$$

A sample mean that is less than 63.03 or greater than 66.97 allows you to reject the null hypothesis. In this problem, $\bar{x} = 67.4$ is greater than 66.97, so you reject the null hypothesis.

Note: Problems 10.25–10.27 refer to a random sample of 25 cars that passed a specific interstate milepost at an average speed of 67.4 miles per hour. Assume the speed of cars passing that milepost is normally distributed with a standard deviation of 6 miles per hour. A researcher claims that the average speed of the population is not 65 miles per hour.

10.27 Verify your answer to Problem 10.25 by comparing the *p*-value to the $\alpha = 0.10$ significance level.

According to Problem 10.25, $z_{67.4} = 2$. You are applying a two-tailed test, so multiply the *p*-value for a one-tailed test by two.

$$p\text{-value} = 2 \cdot P\left(z_{\bar{x}} > 2\right)$$
$$= 2\left[P\left(z_{\bar{x}} > 0\right) + P\left(0 < z_{\bar{x}} < 2\right)\right]$$
$$= 2(0.50 - 0.4772)$$
$$= 2(0.0228)$$
$$= 0.0456$$

You can reject the null hypothesis when the significance level is greater than or equal to 0.0456. In this problem, the significance level is 0.10, so you reject the null hypothesis.

Note: In Problems 10.28–10.30, a professor claims the average class size at a university is greater than 35 students because a random sample of 18 classes contained an average of 38.1 students. Assume the class size distribution is normal with a population standard deviation of 7.6 students.

10.28 Test the claim at the $\alpha = 0.01$ significance level by comparing the calculated *z*-score to the critical *z*-score.

Identify the null and alternative hypotheses.

$$H_0 : \mu \le 35 \text{ students}$$
$$H_1 : \mu > 35 \text{ students}$$

See Problem 10.12.

The critical *z*-score for a one-tailed test on the right side of the distribution is $z_c = 2.33$ when $\alpha = 0.01$. Calculate the standard error of the mean.

$$\sigma_{\bar{x}} = \frac{\sigma}{\sqrt{n}} = \frac{7.6}{\sqrt{18}} = \frac{7.6}{4.24} = 1.791$$

The rejection region is on the right side of the mean, so the sample mean has to be bigger than z_c to reject the null hypothesis.

Calculate the *z*-score for the sample mean.

$$z_{38.1} = \frac{38.1 - 35}{1.791} = \frac{3.1}{1.791} = 1.73$$

Because $z_{38.1} = 1.73$ is less than $z_c = 2.33$, you fail to reject H_0 and conclude there is insufficient evidence to support the claim.

Note: In Problems 10.28–10.30, a professor claims the average class size at a university is greater than 35 students because a random sample of 18 classes contained an average of 38.1 students. Assume the class size distribution is normal with a population standard deviation of 7.6 students.

10.29 Verify your answer to Problem 10.28 by comparing the sample mean to the critical sample mean.

According to Problem 10.28, $z_c = 2.33$ and $\sigma_{\bar{x}} = 1.791$. Calculate the critical sample mean.

$$\bar{x}_c = \mu + z_c \sigma_{\bar{x}} = 35 + (2.33)(1.791) = 39.17$$

The sample mean $\bar{x} = 38.1$ is less than the critical sample mean $\bar{x}_c = 39.17$, so you fail to reject the null hypothesis.

Note: In Problems 10.28–10.30, a professor claims the average class size at a university is greater than 35 students because a random sample of 18 classes contained an average of 38.1 students. Assume the class size distribution is normal with a population standard deviation of 7.6 students.

10.30 Verify your answer to Problem 10.28 by comparing the p-value to the $\alpha = 0.01$ significance level.

According to Problem 10.28, $z_{38.1} = 1.73$. The p-value is the probability that the sample mean could be greater than 38.1, assuming the population mean is $\mu = 35$.

$$
\begin{aligned}
p\text{-value} &= P\left(z_{\bar{x}} > 1.73\right) \\
&= P\left(z_{\bar{x}} > 0\right) - P\left(0 < z_{\bar{x}} < 1.73\right) \\
&= 0.50 - 0.4582 \\
&= 0.0418
\end{aligned}
$$

The p-value 0.0418 is greater than the significance level $\alpha = 0.01$, so you fail to reject H_0.

> You could have rejected the null hypothesis at a significance level of 0.05, because 0.0418 is less than 0.05.

Hypothesis Testing for the Mean with $n < 30$ and Sigma Unknown

Bringing back the t-distribution

10.31 Describe the hypothesis testing procedure used when sample sizes are less than 30 and the population standard deviation is unknown.

When the population standard deviation is unknown, use the sample standard deviation to approximate it. Additionally, you should apply the t-distribution in place of the normal distribution. When the sample size is less than 30, it is important that the population be normally distributed, because you cannot apply the central limit theorem.

> See Problems 9.24–9.33 for a review of Student's t-distribution.

Recall that Reference Table 2 lists probabilities for the *t*-distribution. Note that it is not possible to discern *p*-values using the table, as is possible for normally distributed data and the corresponding *z*-scores. Statistical software, however, can provide these *p*-values if absolutely required.

Note: Problems 10.32–10.33 refer to a claim that houses in a particular community average less than 90 days on the market. A random sample of 9 homes averaged 77.4 days on the market with a sample standard deviation of 29.6 days. Assume the population is normally distributed.

10.32 Test the claim at the $\alpha = 0.05$ significance level by comparing the calculated *t*-score to the critical *t*-score.

Identify the null and alternative hypotheses.

$$H_0 : \mu \geq 90 \text{ days}$$
$$H_1 : \mu < 90 \text{ days}$$

The sample size is $n = 9$; calculate the corresponding degrees of freedom.

$$df = n - 1 = 9 - 1 = 8$$

You are applying a one-tailed test with a significance level of $\alpha = 0.05$ and df = 8. Consider the excerpt from Reference Table 2 below. The critical *t*-value, t_c, is the intersection of row df = 8 and 1-Tailed significance level 0.0500, underlined below.

Don't use the "Conf Lev" values in Table 2; use the "1-Tailed" values instead. If you were doing a two-tailed test at the same significance level, you'd get $t_c = 2.306$, which is right of the underlined number.

Probabilities Under the *t*-Distribution Curve

1-Tailed	0.2000	0.1500	0.1000	0.0500	0.0250	0.0100	0.0050	0.0010	0.0005
2-Tailed	0.4000	0.3000	0.2000	0.1000	0.0500	0.0200	0.0100	0.0020	0.0010
Conf Lev	0.6000	0.7000	0.8000	0.9000	0.9500	0.9800	0.9900	0.9980	0.9990
df									
1	1.376	1.963	3.078	6.314	12.706	31.821	63.657	318.31	636.62
2	1.061	1.386	1.886	2.920	4.303	6.965	9.925	22.327	31.599
3	0.978	1.250	1.638	2.353	3.182	4.541	5.841	10.215	12.924
4	0.941	1.190	1.533	2.132	2.776	3.747	4.604	7.173	8.610
5	0.920	1.156	1.476	2.015	2.571	3.365	4.032	5.893	6.869
6	0.906	1.134	1.440	1.943	2.447	3.143	3.707	5.208	5.959
7	0.896	1.119	1.415	1.895	2.365	2.998	3.499	4.785	5.408
8	0.889	1.108	1.397	<u>1.860</u>	2.306	2.896	3.355	4.501	5.041
9	0.883	1.100	1.383	1.833	2.262	2.821	3.250	4.297	4.781
10	0.879	1.093	1.372	1.812	2.228	2.764	3.169	4.144	4.587

The alternative hypothesis contains "less than," so the rejection region is left of the mean and t_c must be a negative number: $t_c = -1.860$. In order to reject the null hypothesis, the *t*-score of the sample mean will have to be less than −1.860, farther than 1.860 standard deviations left of the population mean.

Calculate the approximate standard error of the mean.

$$\hat{\sigma}_{\bar{x}} = \frac{s}{\sqrt{n}} = \frac{29.6}{\sqrt{9}} = \frac{29.6}{3} = 9.8667$$

Calculate the t-score of the sample mean.

$$t_{\bar{x}} = \frac{\bar{x} - \mu}{\hat{\sigma}_{\bar{x}}} = \frac{77.4 - 90}{9.87} = \frac{-12.6}{9.8667} = -1.28$$

Because $t_{77.4} = -1.28$ is greater than $t_c = -1.860$, you fail to reject H_0 and conclude that there is not sufficient evidence to support the claim.

Note: Problems 10.32–10.33 refer to a claim that houses in a particular community average less than 90 days on the market. A random sample of 9 homes averaged 77.4 days on the market with a sample standard deviation of 29.6 days. Assume the population is normally distributed.

10.33 Verify your answer to problem 10.32 by comparing the sample mean to the critical sample mean.

According to Problem 10.32, $t_c = -1.860$ and $\hat{\sigma}_{\bar{x}} = 9.8667$. Calculate the critical sample mean.

$$\bar{x}_c = \mu + t_c \hat{\sigma}_{\bar{x}} = 90 + (-1.860)(9.8667) = 71.65$$

In order to lie in the rejection region, which is left of the population mean, the sample mean must be less than 71.65. However, $77.4 > 71.65$ so you fail to reject the null hypothesis.

> If the rejection region is right of the mean, a sample mean that's greater than the critical sample mean leads to a rejection of the null hypothesis.

Note: In Problems 10.34–10.35, an auditor claims that the average annual salary of a project manager at a construction company exceeds $82,000. A random sample of 20 project managers had an average salary of $89,600, with a sample standard deviation of $12,700. Assume the salaries of the managers are normally distributed.

10.34 Test the claim at the $\alpha = 0.01$ significance level by comparing the calculated t-score to the critical t-score.

Identify the null and alternative hypotheses.

$$H_0 : \mu \leq \$82,000$$
$$H_1 : \mu > \$82,000$$

> H_1 contains "greater than," so the rejection region is right of the mean and t_c is positive.

A sample of 20 salaries has df $= 20 - 1 = 19$ degrees of freedom. According to Reference Table 2, the corresponding critical t-score for a one-tailed test is $t_c = 2.539$.

Calculate the approximate standard error of the mean.

$$\hat{\sigma}_{\bar{x}} = \frac{s}{\sqrt{n}} = \frac{12,700}{\sqrt{20}} = 2,839.806$$

Calculate the *t*-score of the sample mean.

$$t_x = \frac{\bar{x} - \mu}{\hat{\sigma}_{\bar{x}}} = \frac{89,600 - 82,000}{2,839.806} = \frac{7,600}{2,839.806} = 2.68$$

Because $t_{89,600} = 2.68$ is greater than $t_c = 2.539$, you reject H_0 and conclude that there is sufficient evidence to support the claim.

Note: In Problems 10.34–10.35, an auditor claims that the average annual salary of a project manager at a construction company exceeds $82,000. A random sample of 20 project managers had an average salary of $89,600 with a sample standard deviation of $12,700. Assume the salaries of the managers are normally distributed.

10.35 Verify your answer to Problem 10.34 by comparing the sample mean to the critical sample mean.

According to Problem 10.34, $t_c = 2.539$ and $\hat{\sigma}_{\bar{x}} = 2,839.806$. Calculate the critical sample mean.

$$\bar{x}_c = \mu + t_c \hat{\sigma}_{\bar{x}} = 82,000 + (2.539)(2,839.806) = 89,210.27$$

Because the sample mean $\bar{x} = \$89,600$ is more than the critical sample mean $\bar{x}_c = \$89,210.27$, you reject the null hypothesis.

Note: Problems 10.36–10.37 refer to a claim that the average cost for a family of four to attend a Major League Baseball game is not equal to $172. A random sample of 22 families reported an average cost of $189.34, with a sample standard deviation of $33.65. Assume the population is normally distributed.

10.36 Test the claim at the 0.10 level of significance by comparing the calculated *t*-score to the critical *t*-score.

Identify the null and alternative hypotheses.

$$H_0 : \mu = \$172$$
$$H_1 : \mu \neq \$172$$

This problem requires a two-tailed test with df = 22 − 1 = 21 and $\alpha = 0.10$. The corresponding critical *t*-score in Reference Table 2 is $t_c = 1.721$. A two-tailed test has two rejection regions—one on each side of the mean—so $t_c = \pm 1.721$.

In order to reject the null hypothesis, the sample mean must have a *t*-score of less than −1.721 or more than 1.721. Calculate the approximate standard error of the mean.

$$\hat{\sigma}_{\bar{x}} = \frac{s}{\sqrt{n}} = \frac{33.65}{\sqrt{22}} = 7.174$$

Calculate the *t*-score of the sample mean.

$$t_{189.34} = \frac{189.34 - 172}{7.174} = \frac{17.37}{7.174} = 2.42$$

Because $t_{189.34} = 2.42$ is greater than $t_c = 1.721$, you reject H_0 and conclude that there is sufficient evidence to support the claim.

Note: Problems 10.36–10.37 refer to a claim that the average cost for a family of four to attend a Major League Baseball game is not equal to $172. A random sample of 22 families reported an average cost of $189.34, with a sample standard deviation of $33.65. Assume the population is normally distributed.

10.37 Verify your answer to Problem 10.36 by comparing the sample mean to the critical sample mean.

According to Problem 10.36, $t_c = \pm 1.721$ and $\hat{\sigma}_{\bar{x}} = 7.174$. Calculate the critical sample means that bound the rejection regions left and right of the population mean $\mu = 172$.

$$\bar{x}_c = \mu + t_c \hat{\sigma}_{\bar{x}} \qquad\qquad \bar{x}_c = \mu + t_c \hat{\sigma}_{\bar{x}}$$
$$= 172 + (-1.721)(7.174) \qquad = 172 + (1.721)(7.174)$$
$$= 159.65 \qquad\qquad\qquad = 184.35$$

Because the sample mean $\bar{x} = \$189.34$ is more than $\bar{x}_c = \$184.34$, you reject the null hypothesis and conclude that there is sufficient evidence to support the claim.

> You could reject H_0 for a sample mean less than $159.65.

Note: In Problems 10.38–10.39, an insurance company claims that the average automobile on the road today is less than 6 years old. A random sample of 15 cars had an average age of 5.4 years with a sample standard deviation of 1.1 years. Assume the population is normally distributed.

10.38 Test the claim at the $\alpha = 0.05$ significance level by comparing the calculated t-score to the critical t-score.

> This repetitive disclaimer means the population you took your sample from needs to be normally distributed (or you can't actually calculate t-scores).

Identify the null and alternative hypotheses.

$$H_0 : \mu \geq 6 \text{ years}$$
$$H_1 : \mu < 6 \text{ years}$$

This problem uses a one-tailed test with df = 14 degrees of freedom and a significance level of $\alpha = 0.05$. The rejection region is left of the mean, so $t_c = -1.761$.

Calculate the approximate standard error of the mean. You can reject the null hypothesis only if the t-score of the sample mean is less than -1.761.

$$\hat{\sigma}_{\bar{x}} = \frac{s}{\sqrt{n}} = \frac{1.1}{\sqrt{15}} = 0.284$$

Calculate the t-score of the sample mean.

$$t_{5.4} = \frac{5.4 - 6}{0.284} = \frac{-0.6}{0.284} = -2.11$$

Because $t_{5.4} = -2.11$ is less than $t_c = -1.761$, you reject H_0 and conclude that there is sufficient evidence to support the claim.

Note: In Problems 10.38–10.39, an insurance company claims that the average automobile on the road today is less than 6 years old. A random sample of 15 cars had an average age of 5.4 years with a sample standard deviation of 1.1 years. Assume the population is normally distributed.

10.39 Verify your answer to Problem 10.38 by comparing the sample mean to the critical sample mean.

According to Problem 10.38, $t_c = -1.761$ and $\hat{\sigma}_{\bar{x}} = 0.284$. Calculate the critical sample mean.

$$\bar{x}_c = \mu + t_c \hat{\sigma}_{\bar{x}} = 6 + (-1.761)(0.284) = 5.50$$

The sample mean $\bar{x} = 5.4$ is less than the critical sample mean $\bar{x}_c = 5.5$, so you reject the null hypothesis.

Note: In Problems 10.40–10.41, a golfer claims that the average score at a particular course is not equal to 96. A random sample of 18 golfers shot an average score of 93.7, with a sample standard deviation of 22.8. Assume the population of golf scores is normally distributed.

10.40 Test the claim at the $\alpha = 0.02$ significance level by comparing the calculated t-score to the critical t-score.

Identify the null and alternative hypotheses.

$$H_0 : \mu = 96$$
$$H_1 : \mu \neq 96$$

Apply a two-tailed test with df = 18 − 1 = 17 and $\alpha = 0.02$. According to Reference Table 2, $t_c = \pm 2.567$. Calculate the approximate standard error of the mean.

$$\hat{\sigma}_{\bar{x}} = \frac{s}{\sqrt{n}} = \frac{22.8}{\sqrt{18}} = 5.374$$

Calculate the t-score of the sample mean.

$$t_{93.7} = \frac{93.7 - 96}{5.374} = \frac{-2.3}{5.374} = -0.43$$

Because $t_{93.7} = -0.43$ is neither less than $t_c = -2.567$ nor greater than $t_c = 2.567$, you fail to reject H_0 and conclude that there is insufficient evidence to support the claim.

Note: In Problems 10.40–10.41, a golfer claims that the average score at a particular course is not equal to 96. A random sample of 18 golfers shot an average score of 93.7, with a sample standard deviation of 22.8. Assume the population of golf scores is normally distributed.

10.41 Verify your answer to Problem 10.40 by comparing the sample mean to the critical sample mean.

According to Problem 10.40, $t_c = \pm 2.567$ and $\hat{\sigma}_{\bar{x}} = 5.374$. Calculate the critical sample means left and right of the population mean $\mu = 96$.

$$\bar{x}_c = \mu + t_c \hat{\sigma}_{\bar{x}} \qquad\qquad \bar{x}_c = \mu + t_c \hat{\sigma}_{\bar{x}}$$
$$= 96 + (-2.567)(5.374) \qquad = 96 + (2.567)(5.374)$$
$$= 82.20 \qquad\qquad\qquad = 109.80$$

Because the sample mean $\bar{x} = 93.7$ is neither less than 82.20 nor greater than 109.80, you fail to reject the null hypothesis.

Hypothesis Testing for the Mean with $n > 30$ and Sigma Unknown

Like the last section, but with z-scores

10.42 What impact does a large sample size have on hypothesis testing for the mean when the population standard deviation is unknown?

The *t*-distribution should be used whenever the sample standard deviation s is used in place of the population standard deviation σ. However, when the sample size reaches 30 or more, *t*-scores approximate *z*-scores from the normal distribution. Thus, the normal distribution can be used to approximate the *t*-distribution when $n \geq 30$.

Statistical software programs will continue to use *t*-values rather than approximate them, because they are not limited to a finite number of values in a table. Therefore, these programs will yield slightly different results. ←

So don't panic if you use a computer (or even an advanced calculator) to check your answers and end up getting a slightly different result.

Note: In Problems 10.43–10.45, a wireless phone company claims that its customers' cell phone bills average less than $100 per month. A random sample of 75 customers reported an average monthly bill of $94.25 with a sample standard deviation of $17.38.

10.43 Test the claim at the $\alpha = 0.05$ significance level by comparing the calculated *z*-score to the critical *z*-score.

Identify the null and alternative hypotheses.

$$H_0 : \mu \geq \$100$$
$$H_1 : \mu < \$100$$

See
Problem 10.8.

Use the normal distribution to approximate the *t*-distribution, as $n = 75$ is greater than 30. A one-tailed test on the left side of the distribution at the $\alpha = 0.05$ significance level has a critical z-score of $z_c = -1.64$; the null hypothesis is rejected only if the z-score of the sample mean is less than -1.64.

Approximate the standard error of the mean.

$$\hat{\sigma}_{\bar{x}} = \frac{s}{\sqrt{n}} = \frac{17.38}{\sqrt{75}} = 2.007$$

Calculate the z-score of the sample mean.

$$z_{\bar{x}} = \frac{\bar{x} - \mu}{\hat{\sigma}_{\bar{x}}} = \frac{94.25 - 100}{2.007} = \frac{-5.75}{2.007} = -2.86$$

Because $z_{94.25} = -2.86$ is less than $z_c = -1.64$, you reject H_0 and conclude that there is sufficient evidence to support the claim that the average cell phone bill is less than \$100 per month.

Note: In Problems 10.43–10.45, a wireless phone company claims that its customers' cell phone bills average less than \$100 per month. A random sample of 75 customers reported an average monthly bill of \$94.25 with a sample standard deviation of \$17.38.

10.44 Verify your answer to Problem 10.43 by comparing the sample mean to the critical sample mean.

According to Problem 10.43, $z_c = -1.64$ and $\hat{\sigma}_{\bar{x}} = 2.007$. Calculate the critical sample mean.

$$\bar{x}_c = \mu + z_c \hat{\sigma}_{\bar{x}} = 100 + (-1.64)(2.007) = 96.71$$

Because the sample mean $\bar{x} = \$94.25$ is less than $\bar{x}_c = \$96.71$, you reject the null hypothesis.

Note: In Problems 10.43–10.45, a wireless phone company claims that its customers' cell phone bills average less than \$100 per month. A random sample of 75 customers reported an average monthly bill of \$94.25 with a sample standard deviation of \$17.38.

10.45 Verify your answer to Problem 10.43 by comparing the *p*-value to the level of significance $\alpha = 0.05$.

It's also the smallest level of significance that lets you reject the null hypothesis.

According to Problem 10.43, $z_{\bar{x}} = -2.86$. The *p*-value is the probability that a randomly chosen sample could have a mean more than 2.86 standard deviations below the population mean of $\mu = 100$.

$$
\begin{aligned}
p\text{-value} &= P\left(z_{\bar{x}} < -2.86\right) \\
&= P\left(z_{\bar{x}} < 0\right) - P\left(-2.86 < z_{\bar{x}} < 0\right) \\
&= 0.50 - 0.4979 \\
&= 0.0021
\end{aligned}
$$

As long as $\alpha > 0.0021$, you are able to reject the null hypothesis. In this problem, $\alpha = 0.05$, so you reject the null hypothesis.

Note: In Problems 10.46–10.48, a researcher claims that the average college student spends more than 16 hours on the Internet per month. A random sample of 60 college students spent an average of 17.3 hours online per month, with a sample standard deviation of 5.3 hours.

10.46 Test the claim at the $\alpha = 0.02$ significance level by comparing the calculated z-score to the critical z-score.

Identify the null and alternative hypotheses.

$$H_0 : \mu \leq 16 \text{ hours}$$
$$H_1 : \mu > 16 \text{ hours}$$

A one-tailed test at the $\alpha = 0.02$ significance level with a rejection region on the right side of the distribution has a critical z-score of $z_c = 2.05$. In order to reject the null hypothesis, the z-score of the sample mean must be greater than or equal to 2.05.

See Problem 10.22.

Approximate the standard error of the mean.

$$\hat{\sigma}_{\bar{x}} = \frac{s}{\sqrt{n}} = \frac{5.3}{\sqrt{60}} = \frac{5.3}{7.75} = 0.6842$$

Calculate the z-score of the sample mean.

$$z_{17.3} = \frac{17.3 - 16}{0.6842} = \frac{1.3}{0.6842} = 1.90$$

You fail to reject H_0 because $z_{17.3} < z_c$; the sample does not provide sufficient evidence to support the claim.

Note: In Problems 10.46–10.48, a researcher claims that the average college student spends more than 16 hours on the Internet per month. A random sample of 60 college students spent an average of 17.3 hours online per month, with a sample standard deviation of 5.3 hours.

10.47 Verify your answer to Problem 10.46 by comparing the sample mean to the critical sample mean.

According to Problem 10.46, $z_c = 2.05$ and $\hat{\sigma}_{\bar{x}} = 0.6842$. Calculate the critical sample mean.

$$\bar{x}_c = \mu + z_c \hat{\sigma}_{\bar{x}} = 16 + (2.05)(0.6842) = 17.40$$

The sample mean $\bar{x} = 17.3$ is less than $\bar{x}_c = 17.40$, so you fail to reject the null hypothesis.

When H_1 contains >, the sample mean and its z-score have to be larger than the critical sample mean and the critical z-score.

> *Note: In Problems 10.46–10.48, a researcher claims that the average college student spends more than 16 hours on the Internet per month. A random sample of 60 college students spent an average of 17.3 hours online per month, with a sample standard deviation of 5.3 hours.*

10.48 Verify your answer to Problem 10.46 by comparing the *p*-value to the level of significance $\alpha = 0.02$.

According to Problem 10.46, $z_c = 1.90$. Calculate the probability that a random sample will have a mean more than 1.90 standard deviations above the population mean of $\mu = 17.3$.

$$
\begin{aligned}
p\text{-value} &= P\left(z_{\bar{x}} > 1.90\right) \\
&= P\left(z_{\bar{x}} > 0\right) - P\left(0 < z_{\bar{x}} < 1.90\right) \\
&= 0.50 - 0.4713 \\
&= 0.0287
\end{aligned}
$$

The level of significance $\alpha = 0.02$ does not exceed the *p*-value 0.0287, so you fail to reject the null hypothesis.

> *Note: In Problems 10.49–10.51, a study claims that the average annual tuition for private high schools is more than $7,000. A random sample of 55 private high schools had an average annual tuition of $7,225 and a sample standard deviation of $1,206.*

10.49 Test the claim at the $\alpha = 0.10$ significance level by comparing the calculated *z*-score to the critical *z*-score.

Identify the null and alternative hypotheses.

$$
H_0 : \mu \leq \$7,000
$$
$$
H_1 : \mu > \$7,000
$$

See Problem 10.11.

A one-tailed test at the $\alpha = 0.10$ level of significance with a rejection region right of the distribution has a critical *z*-score of $z_c = 1.28$

Calculate the approximate standard error of the mean.

$$
\hat{\sigma}_{\bar{x}} = \frac{s}{\sqrt{n}} = \frac{1,206}{\sqrt{55}} = 162.62
$$

Calculate the *z*-score of the sample mean.

$$
z_{7,225} = \frac{7,225 - 7,000}{162.62} = \frac{225}{162.62} = 1.38
$$

Because $z_{7,225} = 1.38$ is greater than $z_c = 1.28$, you reject H_0 and conclude there is sufficient evidence to support the claim.

Note: In Problems 10.49–10.51, a study claims that the average annual tuition for private high schools is more than $7,000. A random sample of 55 private high schools had an average annual tuition of $7,225 and a sample standard deviation of $1,206.

10.50 Verify your answer to Problem 10.49 by comparing the sample mean to the critical sample mean.

According to Problem 10.49, $z_c = 1.28$ and $\hat{\sigma}_{\bar{x}} = 162.62$. Calculate the critical sample mean.

$$\bar{x}_c = \mu + z_c \hat{\sigma}_{\bar{x}} = 7,000 + (1.28)(\$162.62) = 7,208.15$$

The sample mean $\bar{x} = \$7,225$ is greater than the critical sample mean $\bar{x}_c = \$7,208.15$, so you reject the null hypothesis.

Note: In Problems 10.49–10.51, a study claims that the average annual tuition for private high schools is more than $7,000. A random sample of 55 private high schools had an average annual tuition of $7,225 and a sample standard deviation of $1,206.

10.51 Verify your answer to Problem 10.49 by comparing the p-value to the level of significance $\alpha = 0.10$.

According to Problem 10.49, $z_{7,225} = 1.38$. Calculate the probability that a random sample has a mean 1.38 standard deviations above the population mean $\mu = 7,000$.

$$p\text{-value} = P\left(z_{\bar{x}} > 1.38\right)$$
$$= P\left(z_{\bar{x}} > 0\right) - P\left(0 < z_{\bar{x}} < 1.38\right)$$
$$= 0.50 - 0.4162$$
$$= 0.0838$$

The p-value 0.0838 is less than the significance level $\alpha = 0.10$, so you reject the null hypothesis.

Note: In Problems 10.52–10.54, a breeder claims that the average weight of an adult male Labrador retriever is not equal to 70 pounds. A random sample of 45 male Labradors weighed an average of 72.6 pounds, with a sample standard deviation of 14.1 pounds.

10.52 Test the claim at the $\alpha = 0.01$ confidence level by comparing the calculated z-score to the critical z-score.

Identify the null and alternative hypotheses.

$$H_0 : \mu = 70 \text{ pounds}$$
$$H_1 : \mu \neq 70 \text{ pounds}$$

Apply a two-tailed test, with rejection regions of area $0.01 \div 2 = 0.005$ at both ends of the distribution. The area beneath the normal curve between $\mu = 70$ and the rejection region is $0.5 - 0.005 = 0.495$, which has corresponding critical z-scores $z_c = \pm 2.57$. Thus, you can reject H_0 if the z-score of the sample mean is either less than -2.57 or greater than 2.57.

Calculate the approximate standard error of the mean.

$$\hat{\sigma}_{\bar{x}} = \frac{s}{\sqrt{n}} = \frac{14.1}{\sqrt{45}} = \frac{14.1}{6.71} = 2.1019$$

Calculate the z-score for the sample mean.

$$z_{72.6} = \frac{72.6 - 70}{2.1019} = \frac{2.6}{2.1019} = 1.24$$

Because $z_{72.6} = 1.24$ is neither less than –2.57 nor greater than 2.57, you fail to reject H_0 and conclude that there is insufficient evidence to support the claim.

Note: In Problems 10.52–10.54, a breeder claims that the average weight of an adult male Labrador retriever is not equal to 70 pounds. A random sample of 45 male Labradors weighed an average of 72.6 pounds, with a sample standard deviation of 14.1 pounds.

10.53 Verify your answer to Problem 10.52 by comparing the sample mean to the critical sample mean.

According to Problem 10.52, $z_c = \pm 2.57$ and $\hat{\sigma}_{\bar{x}} = 2.1019$. Calculate the critical sample means that define the boundaries of the rejection regions.

$$\bar{x}_c = \mu + z_c \hat{\sigma}_{\bar{x}} \qquad \bar{x}_c = \mu + z_c \hat{\sigma}_{\bar{x}}$$
$$= 70 + (-2.57)(2.1019) \qquad = 70 + (2.57)(2.1019)$$
$$= 64.60 \qquad = 75.40$$

The sample mean $\bar{x} = 72.6$ is neither less than 64.60 nor greater than 75.40, so you fail to reject the null hypothesis.

Note: In Problems 10.52–10.54, a breeder claims that the average weight of an adult male Labrador retriever is not equal to 70 pounds. A random sample of 45 male Labradors weighed an average of 72.6 pounds, with a sample standard deviation of 14.1 pounds.

10.54 Verify your answer to Problem 10.52 by comparing the p-value to the level of significance $\alpha = 0.01$.

According to Problem 10.52, $z_{72.6} = 1.24$. Calculate the probability that the mean of a random sample will be greater than 1.24 standard deviations above the mean and double it because you are applying a two-tailed test.

$$p\text{-value} = 2 \cdot P(z_{\bar{x}} > 1.24)$$
$$= 2\left[P(z_{\bar{x}} > 0) - P(0 < z_{\bar{x}} < 1.24) \right]$$
$$= 2(0.50 - 0.3925)$$
$$= (2)(0.1075)$$
$$= 0.215$$

The significance level $\alpha = 0.01$ does not exceed the p-value 0.215, so you fail to reject the null hypothesis.

Hypothesis Testing for the Proportion
Testing percentages instead of means

10.55 Explain how to conduct a hypothesis test for a proportion.

Hypothesis testing for the proportion investigates claims about a population proportion based on a sample proportion. Recall that p_s, the proportion of successes in a sample, is equal to the number of successes s divided by the sample size n: $p_s = \dfrac{s}{n}$.

The standard error of the proportion σ_p, in which p is the population proportion, and the calculated z-score z_p for the sample proportion are evaluated using the formulas below.

$$\sigma_p = \sqrt{\frac{p(1-p)}{n}} \qquad z_p = \frac{p_s - p}{\sigma_p}$$

In order to reject the null hypothesis, z_p will be compared to a critical z-score z_c, the value of which will depend on the level of significance α stated in the problem.

Note: In Problems 10.56–10.58, a government bureau claims that more than 50% of U.S. tax returns were filed electronically last year. A random sample of 150 tax returns for last year contained 86 that were filed electronically.

10.56 Test the claim at the $\alpha = 0.05$ significance level by comparing the calculated z-score to the critical z-score.

Identify the null and alternative hypotheses.

$$H_0 : p \le 0.50$$
$$H_1 : p > 0.50$$

The alternative hypothesis is stated in terms of "greater than," so a one-tailed test at a $\alpha = 0.05$ significance level is applied; the corresponding critical z-score is $z_c = 1.64$. The z-score of the sample proportion will need to be greater than 1.64 to reject the null hypothesis.

Calculate the sample proportion.

$$p_s = \frac{s}{n} = \frac{86}{150} = 0.573$$

Calculate the standard error of the proportion.

$$\sigma_p = \sqrt{\frac{p(1-p)}{n}} = \sqrt{\frac{0.50(1-0.50)}{150}} = \sqrt{\frac{(0.50)^2}{150}} = 0.0408$$

Make sure to substitute the population proportion in for p, not the sample proportion.

Calculate z_p, the z-score for the sample proportion.

$$z_p = \frac{p_s - p}{\sigma_p} = \frac{0.573 - 0.50}{0.0408} = \frac{0.073}{0.0408} = 1.79$$

Because $z_{0.50} = 1.79$ is greater than $z_c = 1.64$, you reject H_0 and conclude that there is sufficient evidence to support the claim.

Note: In Problems 10.56–10.58, a government bureau claims that more than 50% of U.S. tax returns were filed electronically last year. A random sample of 150 tax returns for last year contained 86 that were filed electronically.

10.57 Verify your answer to Problem 10.56 by comparing the sample proportion to the critical sample proportion.

According to Problem 10.56, $z_c = 1.64$ and $\sigma_p = 0.0408$. Calculate the critical sample proportion p_c.

$$p_c = p + z_c\sigma_p = 0.50 + (1.64)(0.0408) = 0.567$$

The sample proportion $p_s = 0.57333$ is greater than the critical sample proportion $p_c = 0.567$, so you reject the null hypothesis.

Note: In Problems 10.56–10.58, a government bureau claims that more than 50% of U.S. tax returns were filed electronically last year. A random sample of 150 tax returns for last year contained 86 that were filed electronically.

10.58 Verify your answer to Problem 10.56 by comparing the p-value to the level of significance $\alpha = 0.05$.

According to Problem 10.56, $z_{0.573} = 1.79$. Calculate the probability that the mean of a random sample is more than 1.79 standard deviations above the population mean.

$$\begin{aligned} p\text{-value} &= P(z_p > 1.79) \\ &= P(z_p > 0) - P(0 < z_p < 1.79) \\ &= 0.50 - 0.4632 \\ &= 0.0368 \end{aligned}$$

Because $\alpha = 0.05$ exceeds the p-value 0.0368, you reject the null hypothesis.

Note: In Problems 10.59–10.61, a nationwide poll claims that the president of the United States has less than a 64% approval rating. In a random sample of 125 people, 74 people gave the president a positive approval rating.

10.59 Test the claim at the $\alpha = 0.02$ significance level by comparing the calculated z-score to the critical z-score.

Identify the null and alternative hypotheses.

$$H_0 : p \geq 0.64$$
$$H_1 : p < 0.64$$

The alternative hypothesis claims that $p < 0.64$, so a one-tailed test is applied, with a rejection region on the left side of the distribution with a critical z-score of –2.05. In order to reject the null hypothesis, z_p will have to be less than –2.05.

See Problem 10.22.

The sample proportion is $p_s = \dfrac{74}{125} = 0.592$. Calculate the standard error of the proportion.

$$\sigma_p = \sqrt{\frac{p(1-p)}{n}} = \sqrt{\frac{0.64(1-0.64)}{125}} = \sqrt{\frac{(0.64)(0.36)}{125}} = \sqrt{\frac{0.2304}{125}} = 0.04293$$

Calculate the z-score of the sample proportion.

$$z_p = \frac{p_s - p}{\sigma_p} = \frac{0.592 - 0.64}{0.04293} = \frac{-0.048}{0.04293} = -1.12$$

Because $z_{0.592} = -1.12$ is not less than $z_c = -2.05$, you fail to reject H_0 and conclude that the sample provides insufficient evidence to support the claim.

Note: In Problems 10.59–10.61, a nationwide poll claims that the president of the United States has less than a 64% approval rating. In a random sample of 125 people, 74 people gave the president a positive approval rating.

10.60 Verify your answer to Problem 10.59 by comparing the sample proportion to the critical sample proportion.

According to Problem 10.59, $z_c = -2.05$ and $\sigma_p = 0.04293$. Calculate the critical sample proportion p_c.

$$p_c = p + z_c \sigma_p = 0.64 + (-2.05)(0.04293) = 0.552$$

The sample proportion $p_s = 0.592$ is not less than the critical sample proportion $p_c = 0.552$, so you fail to reject the null hypothesis.

Just as the calculated z-score had to be less than z_c in Problem 10.59, the sample proportion has to be less than the critical sample proportion z_c here in order to reject the null hypothesis.

Note: In Problems 10.59–10.61, a nationwide poll claims that the president of the United States has less than a 64% approval rating. In a random sample of 125 people, 74 people gave the president a positive approval rating.

10.61 Verify your answer to Problem 10.59 by comparing the p-value to the level of significance $\alpha = 0.02$.

According to Problem 10.59, $z_{0.592} = -1.12$. Calculate the probability that the proportion of a random sample is more than 1.12 standard deviations below the population proportion $p = 0.64$.

$$p\text{-value} = P\left(z_p < -1.12\right)$$
$$= P\left(z_p < 0\right) - P\left(-1.12 < z_p < 0\right)$$
$$= 0.50 - 0.3686$$
$$= 0.1314$$

The significance level $\alpha = 0.02$ does not exceed the p-value 0.1314, so you fail to reject the null hypothesis.

Note: Problems 10.62–10.64 refer to a claim that the proportion of U.S. households that watches the Super Bowl on television is not 40%. In a random sample, 72 of 140 households had watched the most recent Super Bowl.

10.62 Test the claim at the $\alpha = 0.05$ significance level by comparing the calculated z-score to the critical z-score.

Identify the null and alternative hypotheses.

$$H_0 : p = 0.40$$
$$H_1 : p \neq 0.40$$

See Problem 10.15.

A two-tailed test at the $\alpha = 0.05$ significance level has critical z-scores $z_c = \pm 1.96$. The sample proportion is $p_s = \dfrac{72}{140} = 0.514$. Calculate the standard error of the proportion.

$$\sigma_p = \sqrt{\frac{p(1-p)}{n}} = \sqrt{\frac{0.40(1-0.40)}{140}} = \sqrt{\frac{(0.40)(0.60)}{140}} = \sqrt{\frac{0.24}{140}} = 0.0414$$

Calculate the z-score for the sample proportion.

$$z_{0.514} = \frac{0.514 - 0.40}{0.0414} = 2.75$$

Because $z_{0.514} = 2.75$ is greater than $z_c = 1.96$, you reject H_0 and conclude that there is sufficient evidence to support the claim.

Note: Problems 10.62–10.64 refer to a claim that the proportion of U.S. households that watches the Super Bowl on television is not 40%. In a random sample, 72 of 140 households had watched the most recent Super Bowl.

10.63 Verify your answer to Problem 10.62 by comparing the calculated sample proportion to the critical sample proportion.

According to Problem 10.62, $z_c = \pm 1.96$ and $\sigma_p = 0.0414$. Calculate the critical sample proportions that bound the left and right rejection regions.

$$p_c = p + z_c \sigma_p \qquad\qquad p_c = p + z_c \sigma_p$$
$$= 0.40 + (-1.96)(0.0414) \qquad = 0.40 + (1.96)(0.0414)$$
$$= 0.319 \qquad\qquad\qquad = 0.481$$

Because the sample proportion $p_s = 0.514$ is greater than $p_c = 0.481$, you reject the null hypothesis.

Note: Problems 10.62–10.64 refer to a claim that the proportion of U.S. households that watches the Super Bowl on television is not 40%. In a random sample, 72 of 140 households had watched the most recent Super Bowl.

10.64 Verify your answer to Problem 10.62 by comparing the p-value to the level of significance $\alpha = 0.05$.

According to Problem 10.62, $z_{0.514} = 2.75$. You are performing a two-tailed test, so multiply the p-value from the one-tailed test by two.

$$p\text{-value} = 2 \cdot P\left(z_p > 2.75\right)$$
$$= 2\left[P\left(z_p > 0\right) - P\left(0 < z_p < 2.75\right)\right]$$
$$= 2(0.50 - 0.4970)$$
$$= 2(0.003)$$
$$= 0.006$$

0.05 > 0.006

Because the significance level is greater than the confidence level, you reject the null hypothesis.

Note: In Problems 10.65–10.67, a union claims that less than 12% of the current U.S. workforce are union members. A random sample of 160 workers included 12 union members.

10.65 Test the claim at the $\alpha = 0.10$ significance level by comparing the calculated z-score to the critical z-score.

Identify the null and alternative hypotheses.

$$H_0 : p \geq 0.12$$
$$H_1 : p < 0.12$$

A one-tailed test on the left side of the distribution at the $\alpha = 0.10$ significance level has a critical z-score of $z_c = -1.28$. The sample proportion is $p_s = \dfrac{12}{160} = 0.075$.
Calculate the standard error of the proportion.

See Problem 10.11.

$$\sigma_p = \sqrt{\frac{p(1-p)}{n}} = \sqrt{\frac{0.12(1-0.12)}{160}} = \sqrt{\frac{(0.12)(0.88)}{160}} = \sqrt{\frac{0.1056}{160}} = 0.0257$$

Calculate the z-score for the sample proportion.

$$z_{0.075} = \frac{0.075 - 0.12}{0.0257} = \frac{-0.045}{0.0257} = -1.75$$

Because $z_{0.075} = -1.75$ is less than $z_c = -1.28$, you reject H_0 and conclude that there is sufficient evidence to support the claim.

Note: In Problems 10.65–10.67, a union claims that less than 12% of the current U.S. workforce are union members. A random sample of 160 workers included 12 union members.

10.66 Verify your answer to Problem 10.65 by comparing the calculated sample proportion to the critical sample proportion.

According to Problem 10.65, $z_c = -1.28$ and $\sigma_p = 0.0257$. Calculate the critical sample proportion.

$$p_c = p + z_c\sigma_p = 0.12 + (-1.28)(0.0257) = 0.087$$

The sample proportion $p_s = 0.075$ is less than the critical sample proportion $p_c = 0.087$, so you reject the null hypothesis.

Note: In Problems 10.65–10.67, a union claims that less than 12% of the current U.S. workforce are union members. A random sample of 160 workers included 12 union members.

10.67 Verify your answer to Problem 10.65 by comparing the *p*-value to the level of significance $\alpha = 0.10$.

According to Problem 10.65, $z_{0.075} = -1.75$. You are applying a one-tailed test on the left side of the distribution, so the *p*-value is the area beneath the normal curve that is more than 1.75 standard deviations below the mean.

$$
\begin{aligned}
p\text{-value} &= P\left(z_p < -1.75\right) \\
&= P\left(z_p < 0\right) + P\left(-1.75 < z_p < 0\right) \\
&= 0.50 - 0.4599 \\
&= 0.0401
\end{aligned}
$$

The *p*-value 0.0401 is less than the level of significance $\alpha = 0.10$, so you reject the null hypothesis.

Note: In Problems 10.68–10.70, a researcher claims that the proportion of U.S. households with at least one pet is not equal to 70%. A random sample of 120 households contained 90 that owned at least one pet.

10.68 Test the claim at the $\alpha = 0.10$ level of significance by comparing the calculated *z*-score to the critical *z*-score.

Identify the null and alternative hypotheses.

$$H_0 : p = 0.70$$
$$H_1 : p \neq 0.70$$

See Problem 10.25.

A two-tailed test at the $\alpha = 0.10$ level of significance has critical *z*-scores $z_c = \pm 1.64$.

The sample proportion is $p_s = \dfrac{90}{120} = 0.75$. Calculate the standard error of the proportion.

$$\sigma_p = \sqrt{\frac{p(1-p)}{n}} = \sqrt{\frac{0.70(1-0.70)}{120}} = \sqrt{\frac{(0.70)(0.30)}{120}} = \sqrt{\frac{0.21}{120}} = 0.0418$$

Calculate the z-score of the sample proportion.

$$z_{0.75} = \frac{0.75 - 0.70}{0.0418} = \frac{0.05}{0.0418} = 1.20$$

Because $z_{0.75} = 1.20$ is neither less than $z_c = -1.64$ nor greater than $z_c = 1.64$, you fail to reject H_0 and conclude that there is insufficient evidence to support the claim.

Note: In Problems 10.68–10.70, a researcher claims that the proportion of U.S. households with at least one pet is not equal to 70%. A random sample of 120 households contained 90 that owned at least one pet.

10.69 Verify your answer to Problem 10.68 by comparing the sample mean to the critical sample mean.

According to Problem 10.68, $z_c = \pm 1.64$ and $\sigma_p = 0.0418$. Calculate the critical sample proportions that bound the left and right rejection regions.

$$
\begin{aligned}
p_c &= p + z_c \sigma_p & p_c &= p + z_c \sigma_p \\
&= 0.70 + (-1.64)(0.0418) & &= 0.70 + (1.64)(0.0418) \\
&= 0.631 & &= 0.769
\end{aligned}
$$

Because the sample proportion p_s is neither less than 0.631 nor greater than 0.769, you fail to reject the null hypothesis.

Note: In Problems 10.68–10.70, a researcher claims that the proportion of U.S. households with at least one pet is not equal to 70%. A random sample of 120 households contained 90 that owned at least one pet.

10.70 Verify your answer to Problem 10.68 by comparing the p-value to the level of significance $\alpha = 0.10$.

According to Problem 10.68, $z_c = 1.20$. Calculate the p-value for the two-tailed test.

$$
\begin{aligned}
p\text{-value} &= 2 \cdot P\left(z_p > 1.20\right) \\
&= 2\left[P\left(z_p > 0\right) - P\left(0 < z_p < 1.20\right) \right] \\
&= 2(0.50 - 0.3849) \\
&= 2(0.1151) \\
&= 0.2302
\end{aligned}
$$

The p-value 0.2302 exceeds the significance level $\alpha = 0.10$, so you fail to reject the null hypothesis.

Chapter 11
HYPOTHESIS TESTING FOR TWO POPULATIONS

Hypothesizing

This chapter expands the hypothesis testing procedures outlined in Chapter 10 from one population to two. Because each procedure depends upon the sample size and whether or not the population standard deviation is known, the structure of the chapter closely mirrors the structure of Chapter 10. Additionally, the case of dependent samples is investigated.

In Chapter 10, you proved that a population mean (or proportion) was either less than, greater than, or not equal to a specific value. In this chapter, you'll examine two data sets at once and compare their means and proportions, trying to prove that one is bigger than the other or that they're not equal.

There's one thing to worry about when you've got two populations that wasn't a concern in Chapter 10: what if the two populations are actually related? That's covered in the fifth section of the chapter, starting with Problem 11.38.

Hypothesis Testing for Two Means with *n* > 30 and Sigma Known

Comparing two population means

11.1 Explain the hypothesis testing procedure for two population means, including examples of hypotheses and identifying the standard error and z-score formulas. Assume the sample sizes of both populations are greater than 30 and the population standard deviation is known.

As Chapter 10 demonstrated, hypothesis testing requires the creation of a null hypothesis and an alternative hypothesis. Given a population with mean μ_1 and a population with mean μ_2, you can test claims that one mean is larger than the other using either of the following pairs of hypotheses.

$$H_0 : \mu_1 \le \mu_2 \qquad H_0 : \mu_1 - \mu_2 \le 0$$
$$H_1 : \mu_1 > \mu_2 \qquad H_1 : \mu_1 - \mu_2 > 0$$

> According to H_1, mean 1 is bigger than mean 2. Subtracting a smaller number from a bigger number gives you a positive number (which is greater than zero).

Instead of proving that one mean is larger than the other, you may wish to prove that the means are simply not equal.

$$H_0 : \mu_1 - \mu_2 = 0$$
$$H_1 : \mu_1 - \mu_2 \ne 0$$

You may choose to be more specific. Rather than claim that the means are unequal, you could prove that the difference of the means is larger than a fixed value. For instance, the hypotheses below claim that μ_1 and μ_2 differ by more than 100.

$$H_0 : \mu_1 - \mu_2 \le 100$$
$$H_1 : \mu_1 - \mu_2 > 100$$

If σ_1 and σ_2 are the standard deviations of the populations and n_1 and n_2 are sample sizes, then the standard error for the difference between the means is $\sigma_{\bar{x}_1 - \bar{x}_2}$, as calculated below.

$$\sigma_{\bar{x}_1 - \bar{x}_2} = \sqrt{\frac{\sigma_1^2}{n_1} + \frac{\sigma_2^2}{n_2}}$$

> The values observed in sample 1 are not affected by the values observed in sample 2. Later in this chapter you'll deal with samples that aren't independent.

The calculated z-score $z_{\bar{x}_1 - \bar{x}_2}$ for the hypothesis test (assuming $n > 30$) with population standard deviation σ is $z_{\bar{x}_1 - \bar{x}_2}$, which is calculated according to the following formula. This test assumes that the two samples are independent of each other.

$$z_{\bar{x}_1 - \bar{x}_2} = \frac{\left(\bar{x}_1 - \bar{x}_2 \right) - \left(\mu_1 - \mu_2 \right)}{\sigma_{\bar{x}_1 - \bar{x}_2}}$$

The term $(\mu_1 - \mu_2)$ is the hypothesized difference between the two population means. If you are testing a claim that there is no difference between population means, then $\mu_1 - \mu_2 = 0$. If you are testing a claim that the difference between population means is greater than some value *k*, then the term $\mu_1 - \mu_2 = k$.

Note: Problems 11.2–11.4 refer to the table below, salary data from two samples of high school teachers from New Jersey and Delaware.

	New Jersey	Delaware
Sample mean	$52,378	$48,773
Sample size	40	42
Population standard deviation	$6,812	$7,514

11.2 Test the hypothesis that the average teacher salary in New Jersey is more than the average teacher salary in Delaware by comparing the calculated z-score to the critical z-score at the $\alpha = 0.05$ significance level.

State the null and alternative hypotheses, using New Jersey as population 1 and Delaware as population 2. If salaries in New Jersey are greater than salaries in Delaware, then $\mu_1 - \mu_2 > 0$ is the alternative hypothesis. The null hypothesis is the opposite statement: $\mu_1 - \mu_2 \leq 0$.

$$H_0 : \mu_1 - \mu_2 \leq 0$$
$$H_1 : \mu_1 - \mu_2 > 0$$

It's always a good idea to state which is which, so you can keep it consistent for the entire problem.

The critical z-score for a one-tailed test on the right side of the distribution with $\alpha = 0.05$ is $z_c = 1.64$. If $z_{\bar{x}_1 - \bar{x}_2}$ is greater than 1.64, you will reject the null hypothesis. Calculate the standard error for the difference between the means.

See Problem 10.8.

$$\sigma_{\bar{x}_1 - \bar{x}_2} = \sqrt{\frac{\sigma_1^2}{n_1} + \frac{\sigma_2^2}{n_2}}$$

$$= \sqrt{\frac{(6,812)^2}{40} + \frac{(7,514)^2}{42}}$$

$$= \sqrt{\frac{46,403,344}{40} + \frac{56,460,196}{42}}$$

$$= \sqrt{1,160,083.60 + 1,344,290.38}$$

$$= \sqrt{2,504,373.98}$$

$$= 1,582.52$$

The difference between the sample means is $52,378 - 48,773 = 3,605$. Calculate the corresponding z-score.

You're only proving the New Jersey salaries are greater, so use 0 here. If you were proving that the New Jersey salaries had a population mean that was $4,000 more, then this number would be 4,000 instead.

$$z_{\bar{x}_1 - \bar{x}_2} = \frac{(\bar{x}_1 - \bar{x}_2) - (\mu_1 - \mu_2)}{\sigma_{\bar{x}_1 - \bar{x}_2}}$$

$$= \frac{3,605 - 0}{1,582.52}$$

$$= 2.28$$

Because $z_{\bar{x}_1-\bar{x}_2} = 2.28$ is greater than $z_c = 1.64$, you reject H_0 and conclude that there is sufficient evidence to support the claim.

Note: Problems 11.2–11.4 refer to the table in Problem 11.2, salary data from two samples of high school teachers from New Jersey and Delaware.

11.3 Verify your answer to Problem 11.2 by comparing the p-value to the level of significance $\alpha = 0.05$.

> See Problem 10.10 for an explanation of the p-value.

According to Problem 11.2, $z_{\bar{x}_1-\bar{x}_2} = 2.28$. You are applying a one-tailed test on the right side of the distribution, so subtract the area between the population mean and $z_{\bar{x}_1-\bar{x}_2}$ (from Reference Table 1) from the area right of the mean (0.5).

$$p\text{-value} = P\left(z_{\bar{x}_1-\bar{x}_2} > 2.28\right) = 0.50 - 0.4887 = 0.0113$$

Because the p-value 0.0113 is less than $\alpha = 0.05$, you can reject the null hypothesis.

Note: Problems 11.2–11.4 refer to the table in Problem 11.2, salary data from two samples of high school teachers from New Jersey and Delaware.

11.4 Construct a 95% confidence interval for the difference between average salaries of New Jersey and Delaware teachers.

The following formulas are used to construct a confidence interval around the difference between sample means.

$$\text{lower limit} = \left(\bar{x}_1 - \bar{x}_2\right) - z_c\sigma_{\bar{x}_1-\bar{x}_2} \qquad \text{upper limit} = \left(\bar{x}_1 - \bar{x}_2\right) + z_c\sigma_{\bar{x}_1-\bar{x}_2}$$

The critical z-score z_c is determined using the same approach that was discussed in Chapter 9. For a 95% confidence interval, $z_c = 1.96$. According to Problem 11.2, $\bar{x}_1 - \bar{x}_2 = \$3,605$ and $\sigma_{\bar{x}_1-\bar{x}_2} = \$1,582.52$.

> See Problem 9.10 for more details on $z_c = 1.96$.

$$
\begin{aligned}
\text{lower limit} &= \left(\bar{x}_1 - \bar{x}_2\right) - z_c\sigma_{\bar{x}_1-\bar{x}_2} \\
&= 3,605 - (1.96)(1,582.52) \\
&= 503
\end{aligned}
\qquad
\begin{aligned}
\text{upper limit} &= \left(\bar{x}_1 - \bar{x}_2\right) + z_c\sigma_{\bar{x}_1-\bar{x}_2} \\
&= 3,605 + (1.96)(1,582.52) \\
&= 6,707
\end{aligned}
$$

Based on this sample, you are 95% confident that New Jersey teacher salaries average between $503 and $6,707 more than Delaware teacher salaries.

Note: Problems 11.5–11.7 refer to the table below, average hotel room rates in Buffalo and Cleveland based on two random samples.

	Buffalo	**Cleveland**
Sample mean	$126.15	$135.60
Sample size	35	46
Population standard deviation	$42.00	$39.00

11.5 Test the hypothesis that the average hotel room rate in Buffalo is not equal to the average room rate in Cleveland by comparing the calculated z-score to the critical z-score with $\alpha = 0.05$.

State the null and alternative hypotheses, using Buffalo as population 1 and Cleveland as population 2.

$$H_0 : \mu_1 - \mu_2 = 0$$
$$H_1 : \mu_1 - \mu_2 \neq 0$$

See Problem 10.15.

The critical z-score for a two-tailed test with $\alpha = 0.05$ is $z_c = \pm 1.96$. In order to reject the null hypothesis, $z_{\bar{x}_1 - \bar{x}_2}$ will need to be less than -1.96 or greater than 1.96. Calculate the standard error for the difference between the means.

$$\sigma_{\bar{x}_1 - \bar{x}_2} = \sqrt{\frac{\sigma_1^2}{n_1} + \frac{\sigma_2^2}{n_2}} = \sqrt{\frac{(42)^2}{35} + \frac{(39)^2}{46}} = \sqrt{\frac{1{,}764}{35} + \frac{1{,}521}{46}} = \sqrt{83.4652} = 9.14$$

Subtract in the correct order: sample mean 1 minus sample mean 2. In this case, you end up with a negative number.

Calculate the difference between the sample means.

$$\bar{x}_1 - \bar{x}_2 = \$126.15 - \$135.60 = -\$9.45$$

Calculate the z-score for the difference between the means.

$$z_{\bar{x}_1 - \bar{x}_2} = \frac{(\bar{x}_1 - \bar{x}_2) - (\mu_1 - \mu_2)}{\sigma_{\bar{x}_1 - \bar{x}_2}} = \frac{-9.45 - 0}{9.14} = -1.03$$

Because $z_{\bar{x}_1 - \bar{x}_2} = -1.03$ is between $z_c = -1.96$ and $z_c = 1.96$, you fail to reject H_0 and conclude that the evidence is insufficient to support the claim.

Note: Problems 11.5–11.7 refer to the table in Problem 11.5, showing average hotel room rates in Buffalo and Cleveland based on two random samples.

11.6 Verify your answer to Problem 11.5 by comparing the p-value to the level of significance $\alpha = 0.05$.

According to Problem 11.5, $z_{\bar{x}_1 - \bar{x}_2} = -1.03$. Because this is a two-tailed test, the p-value from the one-tailed test is doubled.

$$p\text{-value} = 2 \cdot P\left(z_{\bar{x}_1 - \bar{x}_2} < -1.03\right) = 2(0.50 - 0.3485) = 0.303$$

Because the p-value 0.303 is greater than $\alpha = 0.05$, you fail to reject the null hypothesis.

Note: Problems 11.5–11.7 refer to the table in Problem 11.5, average hotel room rates in Buffalo and Cleveland based on two random samples.

11.7 Construct a 90% confidence interval for the difference between average hotel room rates in Buffalo and Cleveland.

See Problem 9.6.

A 90% confidence interval has a corresponding critical z-score of $z_c = 1.64$.

According to Problem 11.5, $\bar{x}_1 - \bar{x}_2 = -9.45$ and $\sigma_{\bar{x}_1 - \bar{x}_2} = 9.14$. Construct a 90% confidence interval around the difference in sample means.

$$\text{lower limit} = \left(\bar{x}_1 - \bar{x}_2\right) - z_c \sigma_{\bar{x}_1 - \bar{x}_2} \qquad \text{upper limit} = \left(\bar{x}_1 - \bar{x}_2\right) + z_c \sigma_{\bar{x}_1 - \bar{x}_2}$$
$$= -9.45 + (-1.64)(9.14) \qquad\qquad\qquad = -9.45 + (1.64)(9.14)$$
$$= -24.44 \qquad\qquad\qquad\qquad\qquad = 5.54$$

Positive difference = population 1 has a bigger mean.

Negative difference = population 2 has a bigger mean.

Zero difference = the means are the same.

Based on this sample, you are 90% confident that the difference between hotel room rates in Buffalo and Cleveland is between –$24.44 and $5.54. Because this confidence interval includes zero, you can support the hypothesis that there is no difference between the average room rates in these two cities.

Note: Problems 11.8–11.10 refer to the table below, the average hourly wages at day-care centers in the Northeast and Southeast, based on two random samples.

	Northeast	Southeast
Sample mean	$9.60	$8.40
Sample size	52	38
Population standard deviation	$1.25	$1.30

11.8 Test the hypothesis that the average hourly wage in the Northeast is at least $0.50 higher than the average hourly wage in the Southeast by comparing the calculated z-score to the critical z-score at the $\alpha = 0.02$ significance level.

State the null and alternative hypotheses, using the Northeast as population 1 and the Southeast as population 2.

$$H_0 : \mu_1 - \mu_2 \leq \$0.50$$
$$H_1 : \mu_1 - \mu_2 > \$0.50$$

See Problem 10.22.

The critical z-score for a one-tailed test with $\alpha = 0.02$ is $z_c = 2.05$.

Calculate the standard error for the difference between the means.

$$\sigma_{\bar{x}_1 - \bar{x}_2} = \sqrt{\frac{\sigma_1^2}{n_1} + \frac{\sigma_2^2}{n_2}} = \sqrt{\frac{(1.25)^2}{52} + \frac{(1.30)^2}{38}} = \sqrt{\frac{1.5625}{52} + \frac{1.69}{38}} = \sqrt{0.074} = 0.273$$

The difference between the means is $\bar{x}_1 - \bar{x}_2 = 9.60 - 8.40 = 1.2$. Calculate the corresponding z-score.

$$z_{\bar{x}_1 - \bar{x}_2} = \frac{(\bar{x}_1 - \bar{x}_2) - (\mu_1 - \mu_2)}{\sigma_{\bar{x}_1 - \bar{x}_2}} = \frac{1.2 - 0.50}{0.273} = \frac{0.7}{0.273} = 2.56$$

> You're trying to show that population 1 is $0.50 bigger than population 2, so $\mu_1 - \mu_2 = 0.50$.

Because $z_{\bar{x}_1 - \bar{x}_2} = 2.56$ is greater than $z_c = 2.05$, you reject H_0 and conclude that there is sufficient evidence to support the claim that the average hourly wage in the Northeast is at least $0.50 higher than the average hourly wage in the Southeast.

> **Note: Problems 11.8–11.10 refer to the table in Problem 11.8, the average hourly wages at day-care centers in the Northeast and Southeast, based on two random samples.**

11.9 Verify your answer to Problem 11.8 by comparing the p-value to the level of significance $\alpha = 0.02$.

According to Problem 11.8, $z_{\bar{x}_1 - \bar{x}_2} = 2.56$. Calculate the p-value for a one-tailed test on the right side of the distribution.

$$p\text{-value} = P\left(z_{\bar{x}_1 - \bar{x}_2} > 2.56\right) = 0.50 - 0.4948 = 0.0052$$

Because the significance level $\alpha = 0.02$ exceeds the p-value 0.0052, you reject the null hypothesis.

> **Note: Problems 11.8–11.10 refer to the table in Problem 11.8, the average hourly wages at day-care centers in the Northeast and Southeast, based on two random samples.**

11.10 Construct a 95% confidence interval for the difference between hourly wages in the Northeast and the Southeast.

A 95% confidence interval has a critical z-score of $z_c = 1.96$. According to Problem 11.8, $\bar{x}_1 - \bar{x}_2 = 1.2$ and $\sigma_{\bar{x}_1 - \bar{x}_2} = 0.273$. Apply the upper and lower boundary formulas to identify the confidence interval.

lower limit $= (\bar{x}_1 - \bar{x}_2) - z_c \sigma_{\bar{x}_1 - \bar{x}_2}$ upper limit $= (\bar{x}_1 - \bar{x}_2) + z_c \sigma_{\bar{x}_1 - \bar{x}_2}$

$= 1.2 - (1.96)(0.273)$ $= 1.2 + (1.96)(0.273)$

$= 0.66$ $= 1.74$

> This entire confidence interval exceeds $0.50, which supports the conclusion in Problem 11.9 that the difference in salaries exceeds $0.50.

Based on this sample, you are 95% confident that the difference between hourly wages in the Northeast and the Southeast is between $0.66 and $1.74.

Hypothesis Testing for Two Means with $n < 30$ and Sigma Known

When populations need to be normally distributed

Note: Problems 11.11–11.13 refer to the table below, the average bill per customer at a restaurant when different types of background music were played. The managers would like to determine the impact music has on the size of the bill. Assume the population is normally distributed.

	Fast Music	Slow Music
Sample mean	$39.65	$42.60
Sample size	18	23
Population standard deviation	$4.21	$5.67

11.11 Test the hypothesis that the average bill of customers exposed to fast music is different from the average bill of customers exposed to slow music by comparing the calculated z-score to the critical z-score at the $\alpha = 0.10$ significance level.

State the null and alternative hypotheses, using fast music customers as population 1 and slow music customers as population 2.

$$H_0 : \mu_1 - \mu_2 = 0$$
$$H_1 : \mu_1 - \mu_2 \neq 0$$

See Problem 10.25.

The critical z-score for a two-tailed test with $\alpha = 0.10$ is $z_c = \pm 1.64$. Calculate the standard error for the difference between the means.

$$\sigma_{\bar{x}_1 - \bar{x}_2} = \sqrt{\frac{\sigma_1^2}{n_1} + \frac{\sigma_2^2}{n_2}} = \sqrt{\frac{(4.21)^2}{18} + \frac{(5.67)^2}{23}} = \sqrt{\frac{17.7241}{18} + \frac{32.1489}{23}} = \sqrt{2.3825} = 1.544$$

Calculate the difference between the sample means.

$$\bar{x}_1 - \bar{x}_2 = 39.65 - 42.60 = -2.95$$

Calculate the z-score for the difference between the means.

$$z_{\bar{x}_1 - \bar{x}_2} = \frac{(\bar{x}_1 - \bar{x}_2) - (\mu_1 - \mu_2)}{\sigma_{\bar{x}_1 - \bar{x}_2}} = \frac{-2.95 - 0}{1.544} = -1.91$$

Because $z_{\bar{x}_1 - \bar{x}_2} = -1.91$ is less than $z_c = -1.64$, it lies within the left rejection region. Thus, you reject H_0 and conclude that there is sufficient evidence to support the claim.

Note: Problems 11.11–11.13 refer to the table in Problem 11.11, the average bill per customer at a restaurant when different types of background music were played. The managers would like to determine the impact music has on the size of the bill. Assume the population is normally distributed.

11.12 Verify your answer to Problem 11.11 by comparing the *p*-value to the level of significance $\alpha = 0.10$.

According to Problem 11.10, $z_{\bar{x}_1-\bar{x}_2} = -1.91$. Remember that the *p*-value of a two-tailed test is twice the *p*-value of a one-tailed test.

$$p\text{-value} = 2 \cdot P\left(z_{\bar{x}_1-\bar{x}_2} < -1.91\right) = 2(0.50 - 0.4719) = (2)(0.0281) = 0.0562$$

The significance level $\alpha = 0.10$ exceeds the *p*-value 0.0562, so you reject the null hypothesis.

Note: Problems 11.11–11.13 refer to the table in Problem 11.11, the average bill per customer at a restaurant when different types of background music were played. The managers would like to determine the impact music has on the size of the bill. Assume the population is normally distributed.

11.13 Construct a 90% confidence interval for the difference between the bills of customers exposed to fast or slow music.

A 90% confidence interval has a corresponding critical z-score of $z_c = 1.64$. According to Problem 11.11, $\bar{x}_1 - \bar{x}_2 = -2.95$ and $\sigma_{\bar{x}_1-\bar{x}_2} = 1.544$. Construct a 90% confidence interval around the difference in sample means.

See Problem 9.6.

$$\text{lower limit} = \left(\bar{x}_1 - \bar{x}_2\right) - z_c\sigma_{\bar{x}_1-\bar{x}_2} \qquad \text{upper limit} = \left(\bar{x}_1 - \bar{x}_2\right) + z_c\sigma_{\bar{x}_1-\bar{x}_2}$$
$$= -2.95 - (1.64)(1.544) \qquad\qquad = -2.95 - (1.64)(1.544)$$
$$= -5.48 \qquad\qquad\qquad\qquad = -0.42$$

Based on this sample, you are 90% confident that the difference between bills for customers exposed to fast or slow music is between – \$5.48 and – \$0.42. Because this confidence interval does not include zero, you can support the hypothesis that the average bills are different.

The limits of the confidence interval are negative because the bills for customers exposed to fast music are lower than the bills for customers exposed to slow music.

Note: Problems 11.14–11.16 refer to the table below, customer satisfaction data for two similar stores. The scores are averages of ratings on a scale of 1 to 10. Assume the populations of satisfaction scores are normally distributed.

	Store A	Store B
Sample mean	7.9	8.6
Sample size	25	27
Population standard deviation	1.4	1.9

11.14 Test the hypothesis that the average customer rating in Store A is lower than the average rating in Store B by comparing the calculated z-score to the critical z-score with $\alpha = 0.05$.

State the null and alternative hypotheses using Store A as population 1 and Store B as population 2.

$$H_0 : \mu_1 - \mu_2 \geq 0$$
$$H_1 : \mu_1 - \mu_2 < 0$$

The critical z-score for a one-tailed test with $\alpha = 0.05$ is $z_c = 1.64$. Calculate the standard error for the difference between the means.

$$\sigma_{\bar{x}_1 - \bar{x}_2} = \sqrt{\frac{\sigma_1^2}{n_1} + \frac{\sigma_2^2}{n_2}} = \sqrt{\frac{(1.4)^2}{25} + \frac{(1.9)^2}{27}} = \sqrt{\frac{1.96}{25} + \frac{3.61}{27}} = \sqrt{0.2121} = 0.461$$

The difference in sample means is $\bar{x}_1 - \bar{x}_2 = 7.9 - 8.6 = -0.7$. Calculate the corresponding z-score.

$$z_{\bar{x}_1 - \bar{x}_2} = \frac{(\bar{x}_1 - \bar{x}_2) - (\mu_1 - \mu_2)}{\sigma_{\bar{x}_1 - \bar{x}_2}} = \frac{-0.7 - 0}{0.461} = -1.52$$

Because $z_{\bar{x}_1 - \bar{x}_2} = -1.52$ is not less than $z_c = -1.64$, you fail to reject H_0 and conclude that there is insufficient evidence to support the claim.

Note: Problems 11.14–11.16 refer to the table in Problem 11.14, customer satisfaction data for two similar stores. The scores are averages of ratings on a scale of 1 to 10. Assume the populations of satisfaction scores are normally distributed.

11.15 Verify your answer to Problem 11.14 by comparing the p-value to the level of significance $\alpha = 0.05$.

According to Problem 11.14, $z_{\bar{x}_1 - \bar{x}_2} = -1.52$. Calculate the p-value for a one-tailed test on the left side of the distribution.

$$p\text{-value} = P\left(z_{\bar{x}_1 - \bar{x}_2} < -1.52\right) = 0.50 - 0.4357 = 0.0643$$

The p-value 0.0643 is greater than $\alpha = 0.05$, so you fail to reject the null hypothesis.

Note: Problems 11.14–11.16 refer to the table in Problem 11.14, customer satisfaction data for two similar stores. The scores are averages of ratings on a scale of 1 to 10. Assume the populations of satisfaction scores are normally distributed.

11.16 Construct a 95% confidence interval for the difference between customer satisfaction ratings for Stores A and B.

A 95% confidence interval has a corresponding critical z-score of $z_c = 1.96$. According to Problem 11.14, $\bar{x}_1 - \bar{x}_2 = -0.7$ and $\sigma_{\bar{x}_1 - \bar{x}_2} = 0.461$. Construct a 95% confidence interval around the difference in sample means.

$$\text{lower limit} = (\bar{x}_1 - \bar{x}_2) - z_c \sigma_{\bar{x}_1 - \bar{x}_2} \qquad \text{upper limit} = (\bar{x}_1 - \bar{x}_2) + z_c \sigma_{\bar{x}_1 - \bar{x}_2}$$
$$= -0.7 - (1.96)(0.461) \qquad\qquad\qquad = -0.7 + (1.96)(0.461)$$
$$= -1.60 \qquad\qquad\qquad\qquad\qquad = 0.20$$

Based on these samples, you are 95% confident that the difference between the customer ratings at Stores A and B is between –1.6 and 0.2. Because this confidence interval includes zero, the average ratings for Store A are most likely not lower than those for Store B.

Hypothesis Testing for Two Means with $n < 30$ and Sigma Unknown

No sigma + small samples = t-distribution

Note: In Problems 11.17–11.18, assume you are testing a claim about two populations for which the standard deviations are unknown. A sample is selected from each population, and both sample sizes are less than 30.

11.17 Explain how to calculate the t-score for the difference of two sample means.

When the population standard deviation is unknown, the sample standard deviation is used as an approximation. When this substitution is made, Student's t-distribution is used in place of the normal distribution. Hence, t-scores are substitutes for z-scores. Note that the population from which the samples are selected must be normally distributed.

See Problems 9.24–9.33 to review Student's t-distribution.

The formula for the t-score of the difference between the sample means is very similar to the corresponding z-score formula, presented in Problem 11.1.

$$t_{\bar{x}_1 - \bar{x}_2} = \frac{(\bar{x}_1 - \bar{x}_2) - (\mu_1 - \mu_2)}{\hat{\sigma}_{\bar{x}_1 - \bar{x}_2}}$$

This is the approximated standard error for the difference between two means. Look at Problem 11.22 for more details.

Chapter 12 includes a test that tells you whether or not population variances are equal.

n_1 = size of sample 1

n_2 = size of sample 2

s_1 = standard deviation of sample 1

s_2 = standard deviation of sample 2

Note: In Problems 11.17–11.18, assume you are testing a claim about two populations for which the standard deviations are unknown. A sample is selected from each population, and both sample sizes are less than 30.

11.18 Identify the formulas for $\hat{\sigma}_{\bar{x}_1 - \bar{x}_2}$ (the approximated standard error for the difference between the means) and df (the degrees of freedom), when the variances of the populations are equal. What formulas are used when the populations do not exhibit the same variance?

If the variances of two populations are equal, there are df = $n_1 + n_2 - 2$ degrees of freedom. In order to calculate $\hat{\sigma}_{\bar{x}_1 - \bar{x}_2}$, you must first calculate s_p, the pooled variance, using the formula below.

$$s_p = \sqrt{\frac{(n_1 - 1)s_1^2 + (n_2 - 1)s_2^2}{n_1 + n_2 - 2}}$$

Substitute s_p into the approximated standard error for the difference between the means.

$$\hat{\sigma}_{\bar{x}_1 - \bar{x}_2} = s_p \sqrt{\frac{1}{n_1} + \frac{1}{n_2}}$$

If the variances of the two populations are not equal, you calculate the approximated standard error for the difference between two means using a different formula, one that does not include pooled variance.

$$\hat{\sigma}_{\bar{x}_1 - \bar{x}_2} = \sqrt{\frac{s_1^2}{n_1} + \frac{s_2^2}{n_2}}$$

The formula used to calculate the degrees of freedom is also vastly different when the populations have different variances.

$$df = \frac{\left(\dfrac{s_1^2}{n_1} + \dfrac{s_2^2}{n_2}\right)^2}{\dfrac{\left(\dfrac{s_1^2}{n_1}\right)^2}{n_1 - 1} + \dfrac{\left(\dfrac{s_2^2}{n_2}\right)^2}{n_2 - 1}}$$

Note: Problems 11.19–11.20 refer to the table below, the average number of chocolate chips per cookie in two competing products. Assume the population of the number of chocolate chips per cookie is normally distributed and the population variances are equal.

	Brand A	Brand B
Sample mean	6.4	5.6
Sample size	10	11
Sample standard deviation	1.1	1.7

11.19 The makers of Brand A claim that their cookies average more chocolate chips than Brand B. Test this hypothesis at the $\alpha = 0.05$ level of significance.

State the null and alternative hypotheses, using Brand A as population 1 and Brand B as population 2.

$$H_0 : \mu_1 - \mu_2 \leq 0$$
$$H_1 : \mu_1 - \mu_2 > 0$$

If Brand A has more chips than Brand B, then the population mean of Brand A is larger than the population mean of Brand B. Subtracting a smaller number from a larger number gives you a positive result.

There are $n_1 + n_2 - 2 = 10 + 11 - 2 = 19$ degrees of freedom. A one-tailed test with df = 19 and $\alpha = 0.05$ has a critical t-score of $t_c - 1.729$. Calculate the pooled variance.

$$s_p = \sqrt{\frac{(n_1 - 1)s_1^2 + (n_2 - 1)s_2^2}{n_1 + n_2 - 2}} = \sqrt{\frac{(10-1)(1.1)^2 + (11-1)(1.7)^2}{10+11-2}} = \sqrt{\frac{10.89 + 28.9}{19}} = 1.447$$

Calculate the approximated standard error for the difference between the means.

$$\hat{\sigma}_{\bar{x}_1 - \bar{x}_2} = s_p \sqrt{\frac{1}{n_1} + \frac{1}{n_2}} = 1.447\sqrt{\frac{1}{10} + \frac{1}{11}} = 1.447\sqrt{0.19091} = 0.632$$

The difference between the sample means is $6.4 - 5.6 = 0.8$. Calculate the corresponding t-score.

$$t_{\bar{x}_1 - \bar{x}_2} = \frac{(\bar{x}_1 - \bar{x}_2) - (\mu_1 - \mu_2)}{\hat{\sigma}_{\bar{x}_1 - \bar{x}_2}} = \frac{0.8 - 0}{0.632} = 1.27$$

Because $t_{\bar{x}_1 - \bar{x}_2} = 1.27$ is less than $t_c = 1.729$, you fail to reject H_0 and conclude that the samples provide insufficient evidence to support the claim.

Note: Problems 11.19–11.20 refer to the table in Problem 11.19, the average number of chocolate chips per cookie in two competing products. Assume the population of the number of chocolate chips per cookie is normally distributed and the population variances are equal.

11.20 Verify your answer to Problem 11.19 by constructing a 95% confidence interval for the difference between the chocolate chip count averages.

This number comes from Reference Table 2.

According to Problem 11.19, $\bar{x}_1 - \bar{x}_2 = 0.80$, $\hat{\sigma}_{\bar{x}_1 - \bar{x}_2} = 0.632$, and df = 19. A 95% confidence interval has a corresponding critical t-score of $t_c = 2.093$.

$$\text{lower limit} = (\bar{x}_1 - \bar{x}_2) - t_c \hat{\sigma}_{\bar{x}_1 - \bar{x}_2}$$
$$= 0.80 + (2.093)(0.632)$$
$$= -0.52$$

$$\text{upper limit} = (\bar{x}_1 - \bar{x}_2) + t_c \hat{\sigma}_{\bar{x}_1 - \bar{x}_2}$$
$$= 0.80 + (2.093)(0.632)$$
$$= 2.12$$

Based on these samples, you are 95% confident that the difference between the average number of chocolate chips per cookie in Brands A and B is between −0.52 and 2.12. Because this interval includes zero, there may be no difference between the per-cookie chocolate chip average, so there is insufficient evidence to support the claim that Brand A has more chips per cookie.

Note: Problems 11.21–11.22 refer to the table below, the average number of minutes of battery life per charge for nickel-metal hydride (NiMH) batteries and lithium-ion (Li-ion) batteries, based on two random samples. Assume the populations from which the samples are taken have the same variance and are normally distributed.

	Li-ion	NiMH
Sample mean	90.5	68.4
Sample size	15	12
Sample standard deviation	16.2	14.0

11.21 A manufacturer claims that an average lithium-ion battery charge lasts 10 minutes longer than an average nickel-metal hydride battery charge. Test the claim at the $\alpha = 0.05$ significance level.

State the null and alternative hypotheses, using Li-ion as population 1 and NiMH as population 2.

$$H_0 : \mu_1 - \mu_2 \leq 10 \text{ minutes}$$
$$H_1 : \mu_1 - \mu_2 > 10 \text{ minutes}$$

If the test was on the left side of the distribution (if H_1 was < instead of >), t_c would be −1.708.

There are $n_1 + n_2 - 2 = 15 + 12 - 2 = 25$ degrees of freedom. A one-tailed test with $\alpha = 0.05$ and 25 degrees of freedom has a critical t-score of $t_c = 1.708$. Calculate the pooled variance.

$$s_p = \sqrt{\frac{(n_1 - 1)s_1^2 + (n_2 - 1)s_2^2}{n_1 + n_2 - 2}} = \sqrt{\frac{(15 - 1)(16.2)^2 + (12 - 1)(14)^2}{15 + 12 - 2}} = \sqrt{\frac{3674.16 + 2{,}156.0}{25}} = 15.271$$

Calculate the approximated standard error for the difference between the means.

$$\hat{\sigma}_{\bar{x}_1 - \bar{x}_2} = s_p \sqrt{\frac{1}{n_1} + \frac{1}{n_2}} = 15.271 \sqrt{\frac{1}{15} + \frac{1}{12}} = 15.271 \sqrt{0.15} = 5.914$$

The difference between the sample means is 90.5 − 68.4 = 22.1. Calculate the corresponding *t*-score.

$$t_{\bar{x}_1 - \bar{x}_2} = \frac{\left(\bar{x}_1 - \bar{x}_2\right) - \left(\mu_1 - \mu_2\right)}{\hat{\sigma}_{\bar{x}_1 - \bar{x}_2}} = \frac{22.1 - 10}{5.914} = 2.05$$

Because $t_{\bar{x}_1 - \bar{x}_2} = 2.05$ is greater than $t_c = 1.708$, you reject H_0; the samples provide sufficient data to support the claim.

> *Note: Problems 11.21–11.22 refer to the table in Problem 11.21, the average number of minutes of battery life per charge for nickel-metal hydride (NiMH) batteries and lithium-ion (Li-ion) batteries, based on two random samples. Assume the populations from which the samples are taken have the same variance and are normally distributed.*

11.22 Construct a 90% confidence interval for the difference between the average number of minutes of battery life per charge of NiMH and Li-ion batteries to verify your answer to Problem 11.21.

According to Problem 11.21, $t_c = 1.708$ and $\hat{\sigma}_{\bar{x}_1 - \bar{x}_2} = 5.914$.

lower limit $= \left(\bar{x}_1 - \bar{x}_2\right) - t_c \hat{\sigma}_{\bar{x}_1 - \bar{x}_2}$ upper limit $= \left(\bar{x}_1 - \bar{x}_2\right) + t_c \hat{\sigma}_{\bar{x}_1 - \bar{x}_2}$

$= 22.1 - (1.708)(5.914)$ $= 22.1 + (1.708)(5.914)$

$= 12.00$ $= 32.20$

Based on these samples, you are 90% confident that the difference between the average number of minutes of battery life per charge of NiMH and Li-ion batteries is between 12.00 and 32.20 minutes. This entire interval exceeds 10 minutes, so there is sufficient evidence to support the claim.

Note: Problems 11.23–11.24 refer to the table below, the average ages of men and women at a retirement community based on two random samples. Assume that age is normally distributed and population variances are equal.

	Men	Women
Sample mean	84.6	87.1
Sample size	17	14
Sample standard deviation	6.0	7.3

11.23 An employee claims that the average ages of men and women in the community are not equal. Test the claim at the $\alpha = 0.02$ significance level.

State the null and alternative hypotheses, such that population 1 represents the men and population 2 represents the women.

$$H_0 : \mu_1 - \mu_2 = 0$$
$$H_1 : \mu_1 - \mu_2 \neq 0$$

Proving that the populations have different average ages requires a different technique than proving that one population is older than the other. The first calls for a two-tailed test. The second calls for a one-tailed test.

A two-tailed test at the $\alpha = 0.02$ significance level with df $= n_1 + n_2 - 2 = 17 + 14 - 2 = 29$ degrees of freedom has critical t-scores $t_c = \pm 2.462$. Calculate the pooled variance.

$$s_p = \sqrt{\frac{(n_1 - 1)s_1^2 + (n_2 - 1)s_2^2}{n_1 + n_2 - 2}} = \sqrt{\frac{(17 - 1)(6)^2 + (14 - 1)(7.3)^2}{17 + 14 - 2}} = \sqrt{\frac{576 + 692.77}{29}} = 6.614$$

Calculate the approximated standard error for the difference between the means.

$$\hat{\sigma}_{\bar{x}_1 - \bar{x}_2} = s_p \sqrt{\frac{1}{n_1} + \frac{1}{n_2}} = 6.614 \sqrt{\frac{1}{17} + \frac{1}{14}} = 6.614\sqrt{0.130} = 2.385$$

The difference of the sample means is $84.6 - 87.1 = -2.5$. Calculate the corresponding t-score.

$$t_{\bar{x}_1 - \bar{x}_2} = \frac{(\bar{x}_1 - \bar{x}_2) - (\mu_1 - \mu_2)}{\hat{\sigma}_{\bar{x}_1 - \bar{x}_2}} = \frac{-2.5 - 0}{2.385} = -1.05$$

Because $t_{\bar{x}_1 - \bar{x}_2} = -1.05$ is neither less than $t_c = -2.462$ nor greater than $t_c = 2.462$, you fail to reject H_0; there is insufficient evidence to support the claim that the average ages are not equal.

Note: Problems 11.23–11.24 refer to the table in Problem 11.23, the average ages of men and women at a retirement community based on two random samples. Assume that age is normally distributed and population variances are equal.

11.24 Construct a 99% confidence interval for the difference between the average ages of men and women in the retirement community.

According to Problem 11.23, $\hat{\sigma}_{\bar{x}_1 - \bar{x}_2} = 2.385$ years. A 99% confidence interval with 29 degrees of freedom has a corresponding critical t-score of $t_c = 2.756$.

When you're using Reference Table 2 for this problem, ignore the one- and two-tailed column labels. The value 2.756 is where the 99% confidence column and df = 29 row intersect.

$$\text{lower limit} = (\bar{x}_1 - \bar{x}_2) - t_c \hat{\sigma}_{\bar{x}_1 - \bar{x}_2} \qquad \text{upper limit} = (\bar{x}_1 - \bar{x}_2) + t_c \hat{\sigma}_{\bar{x}_1 - \bar{x}_2}$$
$$= -2.5 - (2.756)(2.385) \qquad\qquad = -2.5 + (2.756)(2.385)$$
$$= -9.07 \qquad\qquad\qquad\qquad = 4.07$$

Based on these samples, you are 99% confident that the difference between the average ages of men and women is between –9.07 and 4.07 years.

The interval includes zero, so there could be no difference in the average ages.

Note: Problems 11.25–11.26 refer to the table below, samples of golf scores for two friends. Assume the golfers have normally distributed scores with the same variance.

	Brian	John
Sample mean	82.6	85.3
Sample size	10	10
Sample standard deviation	8.1	9.5

In golf, lower scores beat higher scores.

11.25 Brian claims that he is the better golfer because his scores are lower. Test his claim at the $\alpha = 0.10$ level of significance.

By choosing a relatively high value for alpha, Brian has improved his chances of finding support for his claim.

State the null and alternative hypotheses, using Brian's scores as population 1 and John's scores as population 2.

$$H_0 : \mu_1 - \mu_2 \geq 0$$
$$H_1 : \mu_1 - \mu_2 < 0$$

A one-tailed test on the left side of the distribution, with df $= n_1 + n_2 - 2 = 10 + 10 - 2 = 18$ degrees of freedom and $\alpha = 0.10$, has a critical t-score of $t_c = -1.330$. Calculate the pooled variance.

$$s_p = \sqrt{\frac{(n_1 - 1)s_1^2 + (n_2 - 1)s_2^2}{n_1 + n_2 - 2}} = \sqrt{\frac{(10-1)(8.1)^2 + (10-1)(9.5)^2}{10 + 10 - 2}} = \sqrt{\frac{590.49 + 812.25}{18}} = 8.828$$

Calculate the approximated standard error for the difference between the means.

$$\hat{\sigma}_{\bar{x}_1 - \bar{x}_2} = s_p \sqrt{\frac{1}{n_1} + \frac{1}{n_2}} = 8.828 \sqrt{\frac{1}{10} + \frac{1}{10}} = 8.828 \sqrt{0.2} = 3.948$$

The difference between the sample means is 82.6 – 85.3 = –2.7. Calculate the corresponding *t*-score.

$$t_{\bar{x}_1-\bar{x}_2} = \frac{(\bar{x}_1 - \bar{x}_2)-(\mu_1-\mu_2)}{\hat{\sigma}_{\bar{x}_1-\bar{x}_2}} = \frac{-2.7-0}{3.948} = -0.68$$

Because $t_{\bar{x}_1-\bar{x}_2} = -0.68$ is greater than $t_c = -1.330$, you fail to reject H_0 and conclude that the evidence is insufficient to support the claim that Brian is the better golfer.

You get –0.68 as the t-score for the difference of the means. It needed to be less than $t_c = -1.330$ to reject the null hypothesis.

Note: Problems 11.25–11.26 refer to the table in Problem 11.25, samples of golf scores for two friends. Assume the golfers have normally distributed scores with the same variance.

11.26 Construct a 95% confidence interval for the difference between Brian's and John's golf scores.

According to Problem 11.29, $\hat{\sigma}_{\bar{x}_1-\bar{x}_2} = 3.948$. A 95% confidence interval with 18 degrees of freedom has a critical *t*-score of $t_c = 2.101$.

lower limit $= (\bar{x}_1 - \bar{x}_2) - t_c\hat{\sigma}_{\bar{x}_1-\bar{x}_2}$ upper limit $= (\bar{x}_1 - \bar{x}_2) + t_c\hat{\sigma}_{\bar{x}_1-\bar{x}_2}$

$= -2.7 - (2.101)(3.948)$ $= -2.7 + (2.101)(3.948)$

$= -10.99$ $= 5.59$

Based on these samples, you are 95% confident that the difference between Brian's and John's average golf scores is between –10.99 and 5.59.

Don't miss this sentence! These populations have unequal variances, so you have to use the ugly df formula.

Note: Problems 11.27–11.28 refer to the table below, the average number of words a random sample of five-year-old girls and boys were able to recognize. Assume the populations from which the samples are taken are normally distributed but the variances of the populations are unequal.

	Girls	Boys
Sample mean	26.6	20.1
Sample size	11	12
Sample standard deviation	7.3	3.9

11.27 Is there a statistically significant difference between the average number of words recognized by five-year-old girls and five-year-old boys when $\alpha = 0.05$?

Proving two population means are different requires a two-tailed test and an alternative hypothesis that includes "not equal to."

Construct the null and alternative hypotheses, defining girls as population 1 and boys as population 2.

$$H_0 : \mu_1 - \mu_2 = 0$$
$$H_1 : \mu_1 - \mu_2 \neq 0$$

The populations have different variances, so apply the formula defined in Problem 11.18.

$$df = \frac{\left(\dfrac{s_1^2}{n_1}+\dfrac{s_2^2}{n_2}\right)^2}{\dfrac{\left(\dfrac{s_1^2}{n_1}\right)^2}{n_1-1}+\dfrac{\left(\dfrac{s_2^2}{n_2}\right)^2}{n_2-1}} = \frac{\left(\dfrac{(7.3)^2}{11}+\dfrac{(3.9)^2}{12}\right)^2}{\dfrac{\left(\dfrac{(7.3)^2}{11}\right)^2}{11-1}+\dfrac{\left(\dfrac{(3.9)^2}{12}\right)^2}{12-1}} = \frac{(4.84455+1.2675)^2}{\dfrac{(4.84455)^2}{10}+\dfrac{(1.2675)^2}{11}} = \frac{37.357}{2.493} = 14.98$$

A two-tailed test with $\alpha = 0.05$ and 15 degrees of freedom has critical t-scores of $t_c = \pm 2.131$. Apply the approximated standard error formula given unequal variances.

$$\hat{\sigma}_{\bar{x}_1-\bar{x}_2} = \sqrt{\frac{s_1^2}{n_1}+\frac{s_2^2}{n_2}} = \sqrt{\frac{(7.3)^2}{11}+\frac{(3.9)^2}{12}} = \sqrt{\frac{53.29}{11}+\frac{15.21}{12}} = \sqrt{4.84455+1.2675} = 2.472$$

The difference between the sample means is $26.6 - 20.1 = 6.5$. Calculate the corresponding t-score.

$$t_{\bar{x}_1-\bar{x}_2} = \frac{\left(\bar{x}_1-\bar{x}_2\right)-\left(\mu_1-\mu_2\right)}{\hat{\sigma}_{\bar{x}_1-\bar{x}_2}} = \frac{6.5-0}{2.472} = 2.63$$

Because $t_{\bar{x}_1-\bar{x}_2} = 2.63$ is greater than $t_c = 2.131$, you reject the null hypothesis. The difference between the number of words recognized by five-year-old girls and five-year-old boys is statistically significant when $\alpha = 0.05$.

Note: Problems 11.27–11.28 refer to the table in Problem 11.27, the average number of words a random sample of five-year-old girls and boys were able to recognize. Assume the populations from which the samples are taken are normally distributed but the variances of the populations are unequal.

11.28 Construct a 95% confidence interval for the difference between the average number of words recognized by five-year-old girls and five-year-old boys.

According to Problem 11.27, $\bar{x}_1 - \bar{x}_2 = 6.5$ and $\hat{\sigma}_{\bar{x}_1-\bar{x}_2} = 2.472$. The t-score for a 95% confidence interval with 15 degrees of freedom is $t_c = 2.131$.

$$\begin{aligned}\text{lower limit} &= \left(\bar{x}_1-\bar{x}_2\right)-t_c\hat{\sigma}_{\bar{x}_1-\bar{x}_2} & \text{upper limit} &= \left(\bar{x}_1-\bar{x}_2\right)+t_c\hat{\sigma}_{\bar{x}_1-\bar{x}_2}\\ &= 6.5-(2.131)(2.472) & &= 6.5-(2.131)(2.472)\\ &= 1.23 & &= 11.77\end{aligned}$$

Based on these samples, you are 95% confident that the difference between the number of words recognized by five-year-old girls and five-year-old boys is between 1.23 and 11.77.

You have to round df = 14.98 to the nearest whole number.

Note: Problems 11.29–11.30 refer to the table below, the average costs of seven-day cruises to Alaska and the Caribbean based on a random sample of various cruise lines. Assume the populations from which the samples are taken have equal variances and are normally distributed.

	Alaska	Caribbean
Sample mean	$884	$702
Sample size	8	7
Sample standard deviation	$135	$120

11.29 A travel agent claims the average seven-day cruise to Alaska is more expensive than the average seven-day cruise to the Caribbean. Test this claim at the $\alpha = 0.01$ significance level.

State the null and alternative hypotheses, using Alaska cruise costs as population 1 and Caribbean cruise costs as population 2.

$$H_0 : \mu_1 - \mu_2 \leq 0$$
$$H_1 : \mu_1 - \mu_2 > 0$$

When the population variances are equal, df is much easier to calculate. However, you do have to calculate pooled variance.

The critical t-score of a one-tailed test on the right side of the distribution with df = 8 + 7 − 2 = 13 degrees of freedom and $\alpha = 0.01$ is $t_c = 2.650$. Calculate the pooled variance.

$$s_p = \sqrt{\frac{(n_1 - 1)s_1^2 + (n_2 - 1)s_2^2}{n_1 + n_2 - 2}} = \sqrt{\frac{(8 - 1)(135)^2 + (7 - 1)(120)^2}{8 + 7 - 2}} = \sqrt{\frac{127{,}575 + 86{,}400}{13}} = 128.295$$

Calculate the approximated standard error for the difference between the means.

$$\hat{\sigma}_{\bar{x}_1 - \bar{x}_2} = s_p \sqrt{\frac{1}{n_1} + \frac{1}{n_2}} = 128.295 \sqrt{\frac{1}{8} + \frac{1}{7}} = 128.295 \sqrt{0.26786} = 66.399$$

The sample means have a difference of 884 − 702 = 182. Calculate the corresponding t-score.

$$t_{\bar{x}_1 - \bar{x}_2} = \frac{(\bar{x}_1 - \bar{x}_2) - (\mu_1 - \mu_2)}{\hat{\sigma}_{\bar{x}_1 - \bar{x}_2}} = \frac{182 - 0}{66.399} = 2.74$$

Because $t_{\bar{x}_1 - \bar{x}_2} = 2.74$ is greater than $t_c = 2.650$, there is sufficient evidence to reject the null hypothesis.

Note: Problems 11.29–11.30 refer to the table in Problem 11.29, the average costs of seven-day cruises to Alaska and the Caribbean based on a random sample of various cruise lines. Assume the populations from which the samples are taken have equal variances and are normally distributed.

11.30 Construct a 98% confidence interval for the difference between the average cruise fares to Alaska and the Caribbean.

According to Problem 11.29, $\bar{x}_1 - \bar{x}_2 = 182$ and $\hat{\sigma}_{\bar{x}_1 - \bar{x}_2} = 66.399$. A 98% confidence interval with 13 degrees of freedom has a critical t-score of $t_c = 2.650$.

$$\text{lower limit} = \left(\bar{x}_1 - \bar{x}_2\right) - t_c\,\hat{\sigma}_{\bar{x}_1 - \bar{x}_2} \qquad \text{upper limit} = \left(\bar{x}_1 - \bar{x}_2\right) + t_c\,\hat{\sigma}_{\bar{x}_1 - \bar{x}_2}$$
$$= 182 - (2.650)(66.399) \qquad\qquad = 182 + (2.650)(66.399)$$
$$= 6.04 \qquad\qquad\qquad\qquad = 357.96$$

Based on these samples, you are 98% confident that the difference between the average cruise fares to Alaska and the Caribbean is between $6.04 and $357.96.

Hypothesis Testing for Two Means with $n \geq 30$ and Sigma Unknown
Zs instead of Ts

11.31 What impact does a large sample size have on hypothesis testing for two means when the population standard deviation is unknown?

The t-distribution should be used whenever the sample standard deviation s is used in place of the population standard deviation σ. However, when sample sizes are greater than or equal to 30, the t-score values approximate z-score values from the normal distribution.

Note: Problems 11.32–11.34 refer to the table below, the results of a taste test between competing soda brands Cola A and Cola B. Two independent random samples were selected and the respondents rated the colas on a scale of 1 to 10. ←

	Cola A	Cola B
Sample mean	7.92	7.22
Sample size	38	45
Sample standard deviation	2.7	1.4

When $n \geq 30$, the populations don't have to be normally distributed. Another bonus: you don't have to check whether the populations have the same variance like you did in Problems 11.17–11.30.

11.32 Test the hypothesis that Cola A is preferred over Cola B by comparing the calculated z-score to the critical z-score at the $\alpha = 0.05$ significance level.

Identify the null and alternative hypotheses, using Cola A ratings as population 1 and Cola B ratings as population 2.

$$H_0 : \mu_1 - \mu_2 \leq 0$$
$$H_1 : \mu_1 - \mu_2 > 0$$

The critical z-score for a one-tailed test with $\alpha = 0.05$ is $z_c = 1.64$. Calculate the approximated standard error for the difference between the means.

$$\hat{\sigma}_{\bar{x}_1 - \bar{x}_2} = \sqrt{\frac{s_1^2}{n_1} + \frac{s_2^2}{n_2}} = \sqrt{\frac{(2.7)^2}{38} + \frac{(1.4)^2}{45}} = \sqrt{0.19184 + 0.04356} = 0.485$$

The sample means have a difference of $7.92 - 7.22 = 0.70$. Calculate the corresponding z-score.

$$z_{\bar{x}_1 - \bar{x}_2} = \frac{\left(\bar{x}_1 - \bar{x}_2\right) - \left(\mu_1 - \mu_2\right)}{\hat{\sigma}_{\bar{x}_1 - \bar{x}_2}} = \frac{0.70 - 0}{0.485} = 1.44$$

Because $z_{\bar{x}_1 - \bar{x}_2} = 1.44$ is less than $z_c = 1.64$, you fail to reject the null hypothesis.

Note: Problems 11.32–11.34 refer to the table in Problem 11.32, the results of a taste test between competing soda brands Cola A and Cola B. Two independent random samples were selected and the respondents rated the colas on a scale of 1 to 10.

11.33 Verify your answer to Problem 11.32 by comparing the p-value to the level of significance $\alpha = 0.05$.

According to Problem 11.32, $z_{\bar{x}_1 - \bar{x}_2} = 1.44$. Calculate the p-value for a one-tailed test on the right side of the distribution.

$$p\text{-value} = P\left(z_{\bar{x}_1 - \bar{x}_2} > 1.44\right) = 0.50 - 0.4251 = 0.0749$$

The p-value 0.0749 is greater than the confidence level $\alpha = 0.05$, so you fail to reject the null hypothesis.

Note: Problems 11.32–11.34 refer to the table in Problem 11.32, the results of a taste test between competing soda brands Cola A and Cola B. Two independent random samples were selected and the respondents rated the colas on a scale of 1 to 10.

11.34 Construct a 95% confidence interval for the difference between the average ratings for Cola A and Cola B.

A 95% confidence interval has a critical z-score of $z_c = 1.96$. According to Problem 11.32, $\bar{x}_1 - \bar{x}_2 = 0.70$ and $\sigma_{\bar{x}_1 - \bar{x}_2} = 0.485$.

$$\text{lower limit} = \left(\bar{x}_1 - \bar{x}_2\right) - z_c \hat{\sigma}_{\bar{x}_1 - \bar{x}_2} \qquad \text{upper limit} = \left(\bar{x}_1 - \bar{x}_2\right) + z_c \hat{\sigma}_{\bar{x}_1 - \bar{x}_2}$$
$$= 0.70 - (1.96)(0.485) \qquad\qquad = 0.70 + (1.96)(0.485)$$
$$= -0.25 \qquad\qquad\qquad\qquad\qquad = 1.65$$

Based on this sample, you are 95% confident that the difference between average customer ratings of Cola A and Cola B is between -0.25 and 1.65.

Note: Problems 11.35–11.37 refer to the table below, the average systolic blood pressure (in mmHg) of men ages 20–30 and 40–50, based on two random samples.

	20–30	40–50
Sample mean	128.1	133.5
Sample size	60	52
Sample standard deviation	10.7	12.0

11.35 Test the claim that the age groups have a different average systolic blood pressure by comparing the calculated z-score to the critical z-score at the $\alpha = 0.05$ significance level.

State the null and alternative hypotheses, using the 20–30 age group as population 1 and the 40–50 age group as population 2.

$$H_0 : \mu_1 - \mu_2 = 0$$
$$H_1 : \mu_1 - \mu_2 \neq 0$$

The critical z-scores for a two-tailed test with $\alpha = 0.05$ are $z_c = \pm 1.96$. Calculate the approximated standard error for the difference between the means.

$$\hat{\sigma}_{\bar{x}_1 - \bar{x}_2} = \sqrt{\frac{s_1^2}{n_1} + \frac{s_2^2}{n_2}} = \sqrt{\frac{(10.7)^2}{60} + \frac{(12)^2}{52}} = \sqrt{1.9082 + 2.7692} = 2.163$$

The difference between the sample means is $128.1 - 133.5 = -5.4$. Calculate the corresponding z-score.

$$z_{\bar{x}_1 - \bar{x}_2} = \frac{(\bar{x}_1 - \bar{x}_2) - (\mu_1 - \mu_2)}{\hat{\sigma}_{\bar{x}_1 - \bar{x}_2}} = \frac{-5.4 - 0}{2.163} = -2.50$$

Because $z_{\bar{x}_1 - \bar{x}_2} = -2.50$ is less than $z_c = -1.96$, there is sufficient evidence to reject H_0 and support the claim that there is a difference in average systolic blood pressure between the two age groups.

Note: Problems 11.35–11.37 refer to the table in Problem 11.35, the average systolic blood pressure (in mmHg) of men ages 20–30 and 40–50, based on two random samples.

11.36 Verify your answer to Problem 11.35 by comparing the p-value to the level of significance $\alpha = 0.05$.

According to Problem 11.35, $z_{\bar{x}_1 - \bar{x}_2} = -2.50$. Calculate the p-value for a two-tailed test.

$$p\text{-value} = 2 \cdot P\left(z_{\bar{x}_1 - \bar{x}_2} < -2.50\right) = 2(0.50 - 0.4938) = 0.0124$$

The significance level $\alpha = 0.05$ exceeds the p-value 0.0124, so you reject the null hypothesis.

This book calculates the probability that the sample mean difference could be 2.50 standard deviations below the population mean difference. Don't forget to multiply by two for a two-tailed test.

Note: Problems 11.35–11.37 refer to the table in Problem 11.35, the average systolic blood pressure (in mmHg) of men ages 20–30 and 40–50, based on two random samples.

11.37 Construct a 90% confidence interval for the difference between average systolic blood pressures of the different age groups.

A 90% confidence interval has a critical z-score of $z_c = 1.64$. According to Problem 11.35, $\bar{x}_1 - \bar{x}_2 = -5.4$ and $\sigma_{\bar{x}_1 - \bar{x}_2} = 2.163$. Apply the confidence interval boundary formulas.

$$\text{lower limit} = \left(\bar{x}_1 - \bar{x}_2\right) - z_c \hat{\sigma}_{\bar{x}_1 - \bar{x}_2} \qquad \text{upper limit} = \left(\bar{x}_1 - \bar{x}_2\right) + z_c \hat{\sigma}_{\bar{x}_1 - \bar{x}_2}$$
$$= -5.4 - (1.64)(2.163) \qquad\qquad\qquad = -5.4 + (1.64)(2.163)$$
$$= -8.95 \qquad\qquad\qquad\qquad\qquad = -1.85$$

Based on this sample, you are 90% confident that the difference between the average systolic blood pressures is between –8.95 mmHg and –1.85 mmHg.

Hypothesis Testing for Two Means with Dependent Samples

What happens when the two samples are related?

11.38 Describe the procedure for testing the difference between two means with dependent samples.

All of the preceding problems in this chapter assume their samples are independent—observations from one sample have no impact on observations in the other sample. Dependent samples, however, are related in some way, affecting the values in each sample.

Consider a weight-loss study in which each person is weighed at the beginning (population 1) and end (population 2) of the program. The change in weight of each person is calculated by subtracting the weights in population 2 from the corresponding weights in population 1. Every observation in population 1 is matched to an observation in population 2.

> That's why some books call this procedure a matched-pair test.

Dependent samples of two populations are tested differently than independent samples. The difference between the two samples is treated as a one-sample hypothesis test in which the variables are defined as follows:

- d = difference between a single pair of observations
- $\bar{d}$ = average difference of all the sample pairs
- μ_d = population mean paired difference stated in H_0
- s_d = the standard deviation of the differences
- $t_{\bar{d}}$ = the t-score of the average difference

The following four equations are used to perform the one-sample hypothesis test using the t-distribution with df = $n - 1$ degrees of freedom. Note that sample sizes less than 30 require normally distributed populations in order to apply this technique.

$$\bar{d} = \frac{\sum d}{n} \qquad s_d = \sqrt{\frac{\sum d^2 - \frac{(\sum d)^2}{n}}{n-1}} \qquad \mu_d = \mu_1 - \mu_2 \qquad t_{\bar{d}} = \frac{\bar{d} - \mu_d}{s_d / \sqrt{n}}$$

This is the shortcut standard deviation formula from Problem 3.38. (Standard deviation is the square root of variance.)

Note: Problems 11.39–11.40 refer to the table below, the before and after weights of nine individuals who completed a weight-loss program.

Person	1	2	3	4	5	6	7	8	9
Before	221	215	206	185	202	197	244	188	218
After	200	192	195	166	187	177	227	165	201

11.39 The company offering the weight-loss program claims that the average participant will have lost more than 15 pounds upon completion of the program. Test the claim at the $\alpha = 0.05$ significance level.

State the null and alternative hypotheses using the before weights as population 1 and the after weights as population 2.

$$H_0 : \mu_1 - \mu_2 \le 15 \text{ pounds}$$
$$H_1 : \mu_1 - \mu_2 > 15 \text{ pounds}$$

The hypotheses can also be written in terms of the difference of the means.

$$H_0 : \mu_d \le 15 \text{ pounds}$$
$$H_1 : \mu_d > 15 \text{ pounds}$$

Calculate the paired differences d = before − after and the square of the paired differences d^2.

Person	Before	After	d	d^2
1	221	200	21	441
2	215	192	23	529
3	206	195	11	121
4	185	166	19	361
5	202	187	15	225
6	197	177	20	400
7	244	227	17	289
8	188	165	23	529
9	218	201	17	289
Total			**166**	**3,184**

Calculate the standard deviation of the differences s_d of the $n = 9$ paired samples.

$$s_d = \sqrt{\frac{\sum d^2 - \frac{(\sum d)^2}{n}}{n-1}} = \sqrt{\frac{3,184 - \frac{(166)^2}{9}}{9-1}} = \sqrt{\frac{3,184 - 3,061.778}{8}} = \sqrt{15.27775} = 3.908$$

Calculate the average weight loss $\bar{d}$ and the corresponding t-score.

$$\bar{d} = \frac{\sum d}{n} = \frac{166}{9} = 18.444 \qquad t_{\bar{d}} = \frac{\bar{d} - \mu_d}{s_d / \sqrt{n}} = \frac{18.444 - 15}{3.908 / \sqrt{9}} = \frac{3.444}{1.303} = 2.64$$

The critical t-score for a one-tailed test on the right side of the distribution with $\alpha = 0.05$ and df $= n - 1 = 9 - 1 = 8$ degrees of freedom is $t_c = 1.860$. Because $t_{\bar{d}} = 2.64$ is greater than t_c, you reject H_0 and support the claim that the average weight loss is more than 15 pounds.

Note: Problems 11.39–11.40 refer to the table in Problem 11.39, the before and after weights of nine individuals who completed a weight-loss program.

11.40 Construct a 95% confidence interval for the population mean paired difference.

According to Problem 11.39, $\bar{d} = 18.444$ and $s_d = 3.908$. A 95% confidence interval with 8 degrees of freedom has a critical t-score of $t_c = 2.306$. Apply the confidence interval boundary formulas for a matched-pair test below.

This comes from Reference Table 2.

$$\text{lower limit} = \bar{d} - t_c \frac{s_d}{\sqrt{n}} \qquad\qquad \text{upper limit} = \bar{d} + t_c \frac{s_d}{\sqrt{n}}$$

$$= 18.444 - (2.306)\frac{3.908}{\sqrt{9}} \qquad\qquad = 18.444 + (2.306)\frac{3.908}{\sqrt{9}}$$

$$= 15.44 \qquad\qquad\qquad\qquad = 21.45$$

Based on these samples, you are 95% confident that the average weight loss of the population is between 15.43 and 21.45 pounds. The entire interval is greater than 15, so this interval supports the claim established in Problem 11.38.

Note: Problems 11.41–11.42 refer to the table below, the pretest and posttest scores of seven students who participated in an experimental instruction program for a standardized test.

Student	1	2	3	4	5	6	7
Pretest	85	72	79	75	84	89	90
Posttest	92	78	86	83	84	91	84

11.41 Test a claim that the experimental program increases student scores at the $\alpha = 0.05$ level of significance.

State the null and alternative hypotheses in terms of the population mean paired difference.

$$H_0 : \mu_d \le 0 \text{ points}$$
$$H_1 : \mu_d > 0 \text{ points}$$

Calculate the paired differences d = posttest – pretest and the square of the paired differences d^2.

Student	Posttest	Pretest	d	d^2
1	92	85	7	49
2	78	72	6	36
3	86	79	7	49
4	83	75	8	64
5	84	84	0	0
6	91	89	2	4
7	84	90	−6	36
Total			**24**	**238**

Calculate the standard deviation of the differences s_d.

$$s_d = \sqrt{\frac{\sum d^2 - \frac{\left(\sum d\right)^2}{n}}{n-1}} = \sqrt{\frac{238 - \frac{(24)^2}{7}}{7-1}} = \sqrt{\frac{238 - 82.28571}{6}} = \sqrt{25.95238} = 5.094$$

Calculate the average increase in test scores and the corresponding t-score.

$$\bar{d} = \frac{\sum d}{n} = \frac{24}{7} = 3.428 \qquad t_{\bar{d}} = \frac{\bar{d} - \mu_d}{s_d / \sqrt{n}} = \frac{3.428 - 0}{5.094 / \sqrt{7}} = \frac{3.428}{1.92535} = 1.78$$

A one-tailed test with $\alpha = 0.05$ and df = 7 – 1 = 6 degrees of freedom has a critical t-score of $t_c = 1.943$. Because $t_{\bar{d}} = 1.78$ is less than t_c, you fail to reject H_0 and conclude there is insufficient evidence to support the claim that the new instructional program increases student scores.

Note: Problems 11.41–11.42 refer to the table in Problem 11.41, the pretest and posttest scores of seven students who participated in an experimental instruction program for a standardized test.

11.42 Construct a 95% confidence interval for the population mean paired difference.

According to Problem 11.41, $\bar{d} = 3.428$ and $s_d = 5.094$. A 95% confidence interval with df = 7 – 1 = 6 degrees of freedom has a critical t-score of $t_c = 2.447$.

$$\text{lower limit} = \bar{d} - t_c \frac{s_d}{\sqrt{n}} \qquad \text{upper limit} = \bar{d} + t_c \frac{s_d}{\sqrt{n}}$$

$$= 3.428 - (2.447)\frac{5.094}{\sqrt{7}} \qquad\qquad = 3.428 + (2.447)\frac{5.094}{\sqrt{7}}$$

$$= -1.28 \qquad\qquad\qquad\qquad = 8.14$$

Based on these samples, you are 95% confident that the actual improvement in student scores is between −1.28 and 8.14 points.

Note: Problems 11.43–11.44 refer to the table below, the number of sales per week for an energy drink when the inventory was located in a middle aisle display and an end aisle display at eight different stores.

Store	1	2	3	4	5	6	7	8
End display	64	49	108	97	37	74	117	90
Middle display	72	41	100	62	40	60	122	62

11.43 Test the claim that the location of the display affects weekly sales at the $\alpha = 0.10$ level of significance.

State the null and alternative hypotheses.

$$H_0 : \mu_d = 0$$
$$H_1 : \mu_d \neq 0$$

Calculate the paired differences $d = \text{end} - \text{middle}$ and the squares of the differences.

Store	End	Middle	d	d^2
1	64	72	−8	64
2	49	41	8	64
3	108	100	8	64
4	97	62	35	1,225
5	37	40	−3	9
6	74	60	14	196
7	117	122	−5	25
8	90	62	28	784
Total			**77**	**2,431**

Calculate the standard deviation of the differences s_d.

$$s_d = \sqrt{\frac{\sum d^2 - \frac{\left(\sum d\right)^2}{n}}{n-1}} = \sqrt{\frac{2,431 - \frac{(77)^2}{8}}{8-1}} = \sqrt{\frac{2,431 - 741.125}{7}} = \sqrt{241.41071} = 15.537$$

Calculate the difference in sales of the samples and the corresponding t-score.

$$\bar{d} = \frac{\sum d}{n} = \frac{77}{8} = 9.625 \qquad t_{\bar{d}} = \frac{\bar{d} - \mu_d}{s_d / \sqrt{n}} = \frac{9.625 - 0}{15.537/\sqrt{8}} = \frac{9.625}{5.49316} = 1.75$$

A two-tailed test with $\alpha = 0.10$ and 7 degrees of freedom has critical t-scores $t_c = \pm 1.895$. However, $t_{\bar{d}} = 1.75$ is neither less than -1.895 nor greater than 1.895, so you fail to reject the null hypothesis and conclude that the evidence is insufficient to support the claim.

Note: Problems 11.43–11.44 refer to the table in Problem 11.43, showing the number of sales per week for an energy drink when the inventory was located in a middle aisle display and an end aisle display at eight different stores.

11.44 Construct a 90% confidence interval for the population mean paired difference.

According to Problem 11.43, $\bar{d} = 9.625$ and $s_d = 15.537$. A 90% confidence interval with df = 7 degrees of freedom has a critical t-score of $t_c = 1.895$.

$$\text{lower limit} = \bar{d} - t_c \frac{s_d}{\sqrt{n}} \qquad \text{upper limit} = \bar{d} + t_c \frac{s_d}{\sqrt{n}}$$

$$= 9.625 - (1.895)\frac{15.537}{\sqrt{8}} \qquad\qquad = 9.625 - (1.895)\frac{15.537}{\sqrt{8}}$$

$$= -0.78 \qquad\qquad\qquad = 20.03$$

Based on these samples, you are 90% confident that the true difference in sales is between -0.78 and 20.03.

Note: Problems 11.45–11.46 refer to the table below, the golf scores of eight people before and after a lesson with a golf professional.

Golfer	1	2	3	4	5	6	7	8
Before lesson	96	88	94	86	102	90	100	91
After lesson	88	81	95	79	96	90	103	86

11.45 The instructor claims that the average golfer will lower his score by more than 3 strokes after a single lesson. Test this claim at the $\alpha = 0.05$ significance level.

State the null and alternative hypotheses.

$$H_0 : \mu_d \le 3$$
$$H_1 : \mu_d > 3$$

Calculate the paired differences d = before – after and their squares d^2.

Golfer	Before	After	d	d^2
1	96	88	8	64
2	88	81	7	49
3	94	95	−1	1
4	86	79	7	49
5	102	96	6	36
6	90	90	0	0
7	100	103	−3	9
8	91	86	5	25
Total			**29**	**233**

Calculate the standard deviation of the differences.

$$s_d = \sqrt{\dfrac{\sum d^2 - \dfrac{\left(\sum d\right)^2}{n}}{n-1}} = \sqrt{\dfrac{233 - \dfrac{(29)^2}{8}}{8-1}} = \sqrt{\dfrac{233 - 105.125}{7}} = \sqrt{18.26786} = 4.274$$

Calculate the average paired difference $\bar{d}$ and the corresponding t-score.

$$\bar{d} = \dfrac{\sum d}{n} = \dfrac{29}{8} = 3.625 \qquad t_{\bar{d}} = \dfrac{\bar{d} - \mu_d}{s_d / \sqrt{n}} = \dfrac{3.625 - 3}{4.274 / \sqrt{8}} = \dfrac{0.625}{1.51109} = 0.41$$

The critical t-score of a one-tailed test on the right side of the distribution with $\alpha = 0.05$ and 7 degrees of freedom is $t_c = 1.895$. Because $t_{\bar{d}} = 0.41$ is less than $t_c = 1.895$, you fail to reject H_0 because of insufficient evidence.

Note: Problems 11.45–11.46 refer to the table in Problem 11.45, the golf scores of eight people before and after a lesson with a golf professional.

11.46 Construct a 98% confidence interval for the population mean paired difference between the golf scores before and after the lesson.

According to Problem 11.45, $\bar{d} = 3.625$ and $s_d = 4.274$. A 98% confidence interval with 7 degrees of freedom has a corresponding critical t-score of $t_c = 2.998$.

$$\text{lower limit} = \bar{d} - t_c \dfrac{s_d}{\sqrt{n}} \qquad\qquad \text{upper limit} = \bar{d} + t_c \dfrac{s_d}{\sqrt{n}}$$

$$= 3.625 - (2.998)\dfrac{4.274}{\sqrt{8}} \qquad\qquad = 3.625 + (2.998)\dfrac{4.274}{\sqrt{8}}$$

$$= -0.91 \qquad\qquad\qquad\qquad = 8.16$$

Based on these samples, you are 99% confident that the average improvement in golf scores is between −0.91 and 8.16 strokes.

Hypothesis Testing for Two Proportions

Comparing population percentages

11.47 Explain the hypothesis testing procedure for two proportions, identifying the formulas for estimated standard error $\hat{\sigma}_{\bar{p}_1-\bar{p}_2}$ and the calculated z-score $z_{\bar{p}_1-\bar{p}_2}$.

If they are not provided by the problem, calculate the proportions $\bar{p}_1$ and $\bar{p}_2$ of the samples: $\bar{p}_1 = \dfrac{x_1}{n_1}$ and $\bar{p}_2 = \dfrac{x_2}{n_2}$. In these formulas, x_1 and x_2 are the numbers of successes in the samples and n_1 and n_2 are the sample sizes. You also calculate the overall proportion of both populations $\hat{p}$ using the formulas below.

$$\hat{p} = \frac{x_1 + x_2}{n_1 + n_2}$$

Substitute $\hat{p}$ into the formula for the standard error of the difference between two proportions $\hat{\sigma}_{\bar{p}_1-\bar{p}_2}$.

$$\hat{\sigma}_{\bar{p}_1-\bar{p}_2} = \sqrt{(\hat{p})(1-\hat{p})\left(\frac{1}{n_1}+\frac{1}{n_2}\right)}$$

The calculated z-score can now be determined using the following formula, in which $p_1 - p_2$ represents the hypothesized difference between the population proportions.

$$z_{\bar{p}_1-\bar{p}_2} = \frac{(\bar{p}_1 - \bar{p}_2) - (p_1 - p_2)}{\hat{\sigma}_{\bar{p}_1-\bar{p}_2}}$$

> If you're predicting one proportion is bigger by a specific amount, $p_1 - p_2$ is that number. If you're just predicting that one is bigger (or that they're unequal), $p_1 - p_2 = 0$.

Note: Problems 11.48–11.50 refer to a sample of 400 Florida residents, of which 272 were home owners, and a sample of 600 Maryland residents, of which 390 were home owners.

11.48 A real estate agent claims that the proportion of home ownership in Florida exceeds the proportion of home ownership in Maryland. Test this claim by comparing the calculated z-score to the critical z-score at the $\alpha = 0.01$ significance level.

State the null and alternative hypotheses, using Florida home owners as population 1 and Maryland home owners as population 2.

$$H_0 : p_1 - p_2 \leq 0$$
$$H_1 : p_1 - p_2 > 0$$

A one-tailed test on the right side of the distribution with $\alpha = 0.01$ has a critical z-score of $z_c = 2.33$. In order to reject H_0, $z_{\bar{p}_1-\bar{p}_2}$ will need to be greater than 2.33.

> See Problem 10.12 for more details on determining this z_c.

Calculate the sample proportions and the estimated overall proportion.

$$\bar{p}_1 = \frac{x_1}{n_1} = \frac{272}{400} = 0.68$$

$$\bar{p}_2 = \frac{x_2}{n_2} = \frac{390}{600} = 0.65$$

$$\hat{p} = \frac{x_1 + x_2}{n_1 + n_2} = \frac{272 + 390}{400 + 600} = \frac{662}{1000} = 0.662$$

Determine the estimated standard error of the difference between the two proportions.

$$\hat{\sigma}_{\bar{p}_1 - \bar{p}_2} = \sqrt{(\hat{p})(1-\hat{p})\left(\frac{1}{n_1} + \frac{1}{n_2}\right)} = \sqrt{(0.662)(1-0.662)\left(\frac{1}{400} + \frac{1}{600}\right)} = \sqrt{(0.22376)(0.00417)} = 0.0305$$

The calculated z-score can now be determined using the following equation.

H₁ (P₁ − P₂ > 0) has a constant of zero. That constant goes here.

$$z_{\bar{p}_1 - \bar{p}_2} = \frac{(\bar{p}_1 - \bar{p}_2) - (p_1 - p_2)}{\hat{\sigma}_{\bar{p}_1 - \bar{p}_2}} = \frac{(0.68 - 0.65) - 0}{0.0305} = \frac{0.03}{0.0305} = 0.98$$

Because $z_{\bar{p}_1 - \bar{p}_2} = 0.98$ is less than $z_c = 2.33$, you fail to reject H_0; the evidence is not sufficient to support the claim that the proportion of home ownership in the state of Florida exceeds the proportion in Maryland.

Note: Problems 11.48–11.50 refer to a sample of 400 Florida residents, of which 272 were home owners, and a sample of 600 Maryland residents, of which 390 were home owners.

11.49 Verify your answer to Problem 11.48 by comparing the p-value to the level of significance $\alpha = 0.01$.

According to Problem 11.59, $z_{\bar{p}_1 - \bar{p}_2} = 0.98$. Calculate the p-value of a one-tailed test on the right side of the distribution.

$$p\text{-value} = P\left(z_{\bar{p}_1 - \bar{p}_2} > 0.98\right) = 0.50 - 0.3365 = 0.1635$$

The p-value 0.1635 is greater than $\alpha = 0.01$, so you fail to reject the null hypothesis.

See Problem 9.11. This is the same z-score as in Problems 11.48–11.49, but it wouldn't have been if this book had asked for, let's say, a 99% confidence level.

Note: Problems 11.48–11.50 refer to a sample of 400 Florida residents, of which 272 were home owners, and a sample of 600 Maryland residents, of which 390 were home owners.

11.50 Construct a 98% confidence interval for the difference between the proportion of home ownership in Florida and Maryland.

A 98% confidence interval has a critical z-score of $z_c = 2.33$. According to Problem 11.48, $\bar{p}_1 - \bar{p}_2 = 0.03$ and $\hat{\sigma}_{\bar{p}_1 - \bar{p}_2} = 0.0305$.

$$\text{lower limit} = \left(\bar{p}_1 - \bar{p}_2\right) - z_c \hat{\sigma}_{\bar{p}_1 - \bar{p}_2} \qquad \text{upper limit} = \left(\bar{p}_1 - \bar{p}_2\right) + z_c \hat{\sigma}_{\bar{p}_1 - \bar{p}_2}$$

$$= 0.03 - (2.33)(0.0305) \qquad\qquad = 0.03 + (2.33)(0.0305)$$

$$= -0.0411 \qquad\qquad\qquad\qquad = 0.1011$$

Based on these samples, you are 98% confident that the difference between the proportions is between −0.0411 and 0.1011. This interval includes zero, so the conclusion is consistent with Problems 11.48 and 11.49.

> The −0.0411 limit represents a higher Maryland population proportion; the 0.1011 limit represents a higher Florida proportion. Because either could happen, you cannot conclude that Florida has a higher proportion of home ownership.

Note: Problems 11.51–11.53 refer to a sample of Pittsburgh residents 25 years of age or older, in which 51 of 150 had at least a Bachelor's degree. A sample of 160 Phoenix residents of the same age contained 38 with at least a Bachelor's degree.

11.51 Test the claim that there is a difference in the proportion of adults in Pittsburgh and Phoenix who have at least a Bachelor's degree by comparing the calculated z-score to the critical z-score at the $\alpha = 0.05$ significance level.

State the null and alternative hypotheses, using Pittsburgh as population 1 and Phoenix as population 2.

$$H_0 : p_1 - p_2 = 0$$
$$H_1 : p_1 - p_2 \neq 0$$

The critical z-scores for a two-tailed test with $\alpha = 0.05$ are $z_c = \pm 1.96$.

> See Problem 10.15.

Calculate the sample and overall proportions.

$$\bar{p}_1 = \frac{x_1}{n_1} = \frac{51}{150} = 0.34$$

$$\bar{p}_2 = \frac{x_2}{n_2} = \frac{38}{160} = 0.2375$$

$$\hat{p} = \frac{x_1 + x_2}{n_1 + n_2} = \frac{51 + 38}{150 + 160} = \frac{89}{310} = 0.287$$

Determine the estimated standard error of the difference between the sample proportions.

$$\hat{\sigma}_{\bar{p}_1 - \bar{p}_2} = \sqrt{\left(\hat{p}\right)\left(1 - \hat{p}\right)\left(\frac{1}{n_1} + \frac{1}{n_2}\right)} = \sqrt{(0.287)(1 - 0.287)\left(\frac{1}{150} + \frac{1}{160}\right)} = \sqrt{(0.20463)(0.01292)} = 0.0514$$

Now calculate the z-score of the difference of the sample proportions.

$$z_{\bar{p}_1 - \bar{p}_2} = \frac{\left(\bar{p}_1 - \bar{p}_2\right) - \left(p_1 - p_2\right)}{\hat{\sigma}_{\bar{p}_1 - \bar{p}_2}} = \frac{(0.34 - 0.2375) - 0}{0.0514} = \frac{0.1025}{0.0514} = 1.99$$

Because $z_{\bar{p}_1 - \bar{p}_2} = 1.99$ is greater than $z_c = 1.96$, you reject H_0 and support the claim that there is a difference in the proportion of adults with degrees.

Note: Problems 11.51–11.53 refer to a sample of Pittsburgh residents 25 years of age or older, in which 51 of 150 had at least a Bachelor's degree. A sample of 160 Phoenix residents of the same age contained 38 with at least a Bachelor's degree.

11.52 Verify your answer to Problem 11.51 by comparing the *p*-value to the level of significance $\alpha = 0.05$.

According to Problem 11.51, $z_{\bar{p}_1-\bar{p}_2} = 1.99$. Calculate the *p*-value of the two-tailed test.

$$p\text{-value} = 2 \cdot P\left(z_{\bar{p}_1-\bar{p}_2} > 1.99\right) = 2\,(0.50 - 0.4767) = 0.0466$$

The significance level $\alpha = 0.05$ exceeds the *p*-value 0.0466, so you reject the null hypothesis.

Note: Problems 11.51–11.53 refer to a sample of Pittsburgh residents 25 years of age or older, in which 51 of 150 had at least a Bachelor's degree. A sample of 160 Phoenix residents of the same age contained 38 with at least a Bachelor's degree.

11.53 Construct a 95% confidence interval for the difference between the population proportions.

A 95% confidence interval has a critical *z*-score of $z_c = 1.96$. According to Problem 11.51, $\bar{p}_1 - \bar{p}_2 = 0.1025$ and $\hat{\sigma}_{\bar{p}_1-\bar{p}_2} = 0.0514$.

$$\begin{aligned}
\text{lower limit} &= \left(\bar{p}_1 - \bar{p}_2\right) - z_c\,\hat{\sigma}_{\bar{p}_1-\bar{p}_2} & \text{upper limit} &= \left(\bar{p}_1 - \bar{p}_2\right) + z_c\,\hat{\sigma}_{\bar{p}_1-\bar{p}_2} \\
&= 0.1025 - (1.96)(0.0514) & &= 0.1025 + (1.96)(0.0514) \\
&= 0.0018 & &= 0.2032
\end{aligned}$$

Based on these samples, you are 95% confident that the difference between the proportions is between 0.0018 and 0.2032.

Note: Problems 11.54–11.56 refer to a pair of samples in which 85 of 210 adult men and 60 of 225 adult women were overweight.

11.54 Test the claim that the proportion of overweight adult men exceeds the proportion of overweight adult women by more than 5% by comparing the calculated *z*-score to the critical *z*-score at the $\alpha = 0.05$ level of significance.

State the null and alternative hypotheses, defining adult men as population 1 and adult women as population 2.

$$H_0 : p_1 - p_2 \le 0.05$$
$$H_1 : p_1 - p_2 > 0.05$$

The critical *z*-score for a one-tailed test with $\alpha = 0.05$ is $z_c = 1.64$. Calculate the sample and overall proportions.

$$\bar{p}_1 = \frac{x_1}{n_1} = \frac{85}{210} = 0.405$$

$$\bar{p}_2 = \frac{x_2}{n_2} = \frac{60}{225} = 0.267$$

$$\hat{p} = \frac{x_1 + x_2}{n_1 + n_2} = \frac{85 + 60}{210 + 225} = \frac{145}{435} = 0.333$$

Determine the estimated standard error of the difference between the proportions.

$$\hat{\sigma}_{\bar{p}_1 - \bar{p}_2} = \sqrt{(\hat{p})(1-\hat{p})\left(\frac{1}{n_1} + \frac{1}{n_2}\right)} = \sqrt{(0.333)(1-0.333)\left(\frac{1}{210} + \frac{1}{225}\right)} = \sqrt{(0.22211)(0.00921)} = 0.0452$$

Calculate the z-score of the difference between the proportions.

This time, you're proving that P_1 is more than 5% larger than P_2, so $P_1 - P_2$ in the z-score formula is equal to 0.05.

$$z_{\bar{p}_1 - \bar{p}_2} = \frac{(\bar{p}_1 - \bar{p}_2) - (p_1 - p_2)}{\hat{\sigma}_{\bar{p}_1 - \bar{p}_2}} = \frac{(0.405 - 0.267) - 0.05}{0.0452} = \frac{0.088}{0.0452} = 1.95$$

Because $z_{\bar{p}_1 - \bar{p}_2} = 1.95$ is greater than $z_c = 1.64$, you reject H_0 and support the claim that the proportion of overweight men exceeds the proportion of overweight women by more than 5%.

Note: Problems 11.54–11.56 refer to a pair of samples in which 85 of 210 adult men and 60 of 225 adult women were overweight.

11.55 Verify your answer to Problem 11.54 by comparing the *p*-value to the level of significance $\alpha = 0.05$.

According to Problem 11.54, $z_{\bar{p}_1 - \bar{p}_2} = 1.95$. Calculate the *p*-value for the one-tailed test on the right side of the distribution.

$$p\text{-value} = P\left(z_{\bar{p}_1 - \bar{p}_2} > 1.95\right) = 0.50 - 0.4744 = 0.0256$$

The significance level $\alpha = 0.05$ exceeds the *p*-value 0.0256, so you reject the null hypothesis.

Note: Problems 11.54–11.56 refer to a pair of samples in which 85 of 210 adult men and 60 of 225 adult women were overweight.

11.56 Construct a 90% confidence interval for the difference between the population proportions.

A 90% confidence interval has a critical z-score of $z_c = 1.64$. According to Problem 11.54, $\hat{\sigma}_{\bar{p}_1 - \bar{p}_2} = 0.0452$; the difference between the sample proportions is $\bar{p}_1 - \bar{p}_2 = 0.405 - 0.267 = 0.138$.

The entire interval is greater than 0.05, so you're 95% confident that the true population proportions differ by more than 5%.

$$\text{lower limit} = \left(\bar{p}_1 - \bar{p}_2\right) - z_c \hat{\sigma}_{\bar{p}_1 - \bar{p}_2}$$
$$= 0.138 - (1.64)(0.0452)$$
$$= 0.0639$$

$$\text{upper limit} = \left(\bar{p}_1 - \bar{p}_2\right) + z_c \hat{\sigma}_{\bar{p}_1 - \bar{p}_2}$$
$$= 0.138 + (1.64)(0.0452)$$
$$= 0.2121$$

Based on these samples, you are 95% confident that the difference between the proportions is between 0.0639 and 0.2121.

Note: Problems 11.57–11.59 refer to a sample of 75 flights on Airline A, in which 17 arrived late, and a sample of 85 flights on Airline B, in which 30 arrived late.

11.57 Test Airline B's claim that, despite a higher proportion of late flights, the difference between the proportions is not statistically significant when $\alpha = 0.02$.

State the null and alternative hypotheses, using Airline A as population 1 and Airline B as population 2.

$$H_0 : p_1 - p_2 = 0$$
$$H_1 : p_1 - p_2 \neq 0$$

You're testing to see if the difference IS significant. If you reject H_0, then Airline B is wrong.

A two-tailed test with $\alpha = 0.02$ has critical z-scores of $z_c = \pm 2.33$. Calculate the sample and overall proportions.

$$\bar{p}_1 = \frac{x_1}{n_1} = \frac{17}{75} = 0.227$$

$$\bar{p}_2 = \frac{x_2}{n_2} = \frac{30}{85} = 0.353$$

$$\hat{p} = \frac{x_1 + x_2}{n_1 + n_2} = \frac{17 + 30}{75 + 85} = \frac{47}{160} = 0.294$$

Determine the estimated standard error of the difference between the proportions.

$$\hat{\sigma}_{\bar{p}_1 - \bar{p}_2} = \sqrt{\left(\hat{p}\right)\left(1 - \hat{p}\right)\left(\frac{1}{n_1} + \frac{1}{n_2}\right)} = \sqrt{(0.294)(1 - 0.294)\left(\frac{1}{75} + \frac{1}{85}\right)} = \sqrt{(0.20756)(0.02510)} = 0.0722$$

Calculate the z-score of the difference between sample proportions.

$$z_{\bar{p}_1 - \bar{p}_2} = \frac{\left(\bar{p}_1 - \bar{p}_2\right) - \left(p_1 - p_2\right)}{\hat{\sigma}_{\bar{p}_1 - \bar{p}_2}} = \frac{(0.227 - 0.353) - 0}{0.0722} = \frac{-0.126}{0.0722} = -1.75$$

Because $z_{\bar{p}_1 - \bar{p}_2} = 1.75$ is neither less than $z_c = -2.33$ nor greater than $z_c = 2.33$, you fail to reject H_0. Airline B is correct in its assertion that the difference between the proportions is statistically insignificant when $\alpha = 0.02$.

Note: Problems 11.57–11.59 refer to a sample of 75 flights on Airline A, in which 17 arrived late, and a sample of 85 flights on Airline B, in which 30 arrived late.

11.58 Verify your answer to Problem 11.57 by comparing the *p*-value to the level of significance $\alpha = 0.02$.

According to Problem 11.57, $z_{\bar{p}_1 - \bar{p}_2} = 1.75$. Calculate the *p*-value of the two-tailed test.

$$p\text{-value} = 2 \cdot P\left(z_{\bar{p}_1 - \bar{p}_2} > 1.75\right) = 2(0.50 - 0.4599) = 0.0802$$

Because the *p*-value 0.0802 is greater than $\alpha = 0.02$, you fail to reject the null hypothesis.

Note: Problems 11.57–11.59 refer to a sample of 75 flights on Airline A, in which 17 arrived late, and a sample of 85 flights on Airline B, in which 30 arrived late.

11.59 Construct a 95% confidence interval for the difference between the population proportions.

A 95% confidence interval has a corresponding critical *z*-score of $z_c - 1.96$. According to Problem 11.58, $\bar{p}_1 - \bar{p}_2 = -0.126$ and $\hat{\sigma}_{\bar{p}_1 - \bar{p}_2} = 0.0722$.

$$\text{lower limit} = \left(\bar{p}_1 - \bar{p}_2\right) - z_c \hat{\sigma}_{\bar{p}_1 - \bar{p}_2} \qquad \text{upper limit} = \left(\bar{p}_1 - \bar{p}_2\right) + z_c \hat{\sigma}_{\bar{p}_1 - \bar{p}_2}$$
$$= -0.126 - (1.96)(0.0722) \qquad\qquad = -0.126 + (1.96)(0.0722)$$
$$= -0.2675 \qquad\qquad\qquad\qquad = 0.0155$$

Based on these samples, you are 95% confident that the difference between the proportions is between −0.2675 and 0.0155.

Chapter 12
CHI-SQUARE AND VARIANCE TESTS

Testing categorical data for variation

Tests discussed in preceding chapters have sometimes required you to assume that a population has a specific probability distribution, such as the normal or binomial distribution. This chapter introduces the chi-square distribution, a technique to test these assumptions. This chapter also explores the chi-square distribution as a means of performing one- and two-population hypothesis testing.

The goodness-of-fit test determines whether a population follows a particular distribution. The chi-square test for independence will tell you if two categorical variables are related. This chapter also includes two types of variance tests: a single population test that uses the chi-square distribution and a hypothesis test that compares two variances using the F-distribution.

Chi-Square Goodness-of-Fit Test
Is the data distributed the way you thought it would be?

12.1 Explain how to perform the chi-square goodness-of-fit test.

The goodness-of-fit test is a hypothesis test that uses the chi-square distribution to test whether a frequency distribution fits a predicted distribution. The null hypothesis states that the sample of observed frequencies supports the claim about the expected frequencies. As usual, the alternative hypothesis states the opposite, that there is no support for the claim.

The chi-square test compares observed (O) and expected (E) frequencies to determine whether there is a statistically significant difference. Apply the following formula to calculate the χ^2 statistic.

$$\chi^2 = \sum \frac{(O-E)^2}{E}$$

The value of χ^2 is then compared to the critical chi-square score obtained from Reference Table 3 to determine whether to reject the null hypothesis.

Note: Problems 12.2–12.3 refer to the data set below, the number of students absent over a five-day period.

Day	1	2	3	4	5
Number of Students	15	17	12	10	6

These are the observed values. The expected values are the numbers that you think belong here.

12.2 Calculate the expected number of absences per day, assuming the population of absenteeism is uniformly distributed.

Assuming the absences are uniformly distributed means assuming that roughly the same number of students are absent each day. Calculate the average of the data values.

$$\bar{x} = \frac{15+17+12+10+6}{5} = \frac{60}{5} = 12$$

You expect 12 students to be absent each day, whereas the data provides the actual observed number of absences each day.

Day	1	2	3	4	5
Observed (O)	15	17	12	10	6
Expected (E)	12	12	12	12	12

Note: Problems 12.2–12.3 refer to the data set in Problem 12.2, showing the number of students absent over a five-day period.

12.3 Test the claim that the distribution of absences is uniformly distributed at the $\alpha = 0.05$ significance level.

State the null and alternative hypotheses.

H_0 : Daily absences are uniformly distributed

H_1 : Daily absences are not uniformly distributed

Use the table below to calculate χ^2, the chi-square statistic.

Day	O	E	$O - E$	$(O - E)^2$	$\dfrac{(O - E)^2}{E}$
1	15	12	3	9	0.75
2	17	12	5	25	2.08
3	12	12	0	0	0.00
4	10	12	−2	4	0.33
5	6	12	−6	36	3.00
Total					**6.16**

Divide O − E = 9 by E = 12 to get 0.75.

The sum of the values in the rightmost column is the chi-square statistic: $\chi^2 = 6.16$. Five categories (in this case, $k = 5$ different days) of observed data are provided by the problem, so there are df $= k - 1 = 5 - 1 = 4$ degrees of freedom.

Consider Reference Table 3 (an excerpt of which is shown below). To identify the critical chi-square value χ_c^2, identify the value at the place where the $\alpha = 0.05$ column and df = 4 row meet: $\chi_c^2 = 9.488$ (underlined in the excerpt from Reference Table 3 below).

Probabilities Under the Chi-Square Distribution										
df	0.995	0.99	0.975	0.95	0.90	0.10	0.05	0.025	0.01	0.005
1	—	—	0.001	0.004	0.016	2.706	3.841	5.024	6.635	7.879
2	0.010	0.020	0.051	0.103	0.211	4.605	5.991	7.378	9.210	10.597
3	0.072	0.115	0.216	0.352	0.584	6.251	7.815	9.348	11.345	12.838
4	0.207	0.297	0.484	0.711	1.064	7.779	9.488	11.143	13.277	14.860

The critical chi-square value χ_c^2 defines the lower boundary of the rejection region, as illustrated in the following diagram. Because $\chi^2 = 6.16$ is less than $\chi_c^2 = 9.488$, you fail to reject H_0; you conclude the distribution of absences is uniformly distributed for $\alpha = 0.05$.

The chi-square goodness-of-fit test is always a one-tailed test on the right side of the distribution.

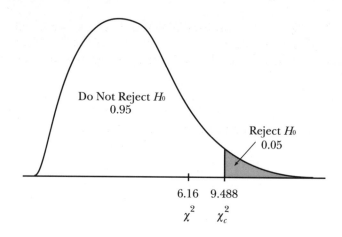

Do Not Reject H_0
0.95

Reject H_0
0.05

6.16   9.488
χ^2 χ_c^2

Note: Problems 12.4–12.5 refer to the data set below, the number of games in which a particular Major League Baseball player had 0, 1, 2, 3, or 4 hits per game over the last 100 games. Assume the player had four at-bats per game.

Number of Hits	Number of Games
0	26
1	34
2	30
3	7
4	3
Total	**100**

12.4 Calculate the expected number of games in which the player has each number of hits, assuming the hitting distribution is binomial and the player has a 0.300 batting average.

See Problems 6.1–6.16 to review the binomial distribution.

Calculate the binomial probabilities for the number of hits per game assuming $n = 4$ trials and a 0.300 (or, more simply, 0.3) probability of success.

$$P(r) = \binom{n}{r} p^r q^{n-r}$$

$$P(0) = \binom{4}{0}(0.3)^0 (0.7)^{4-0} = 0.2401$$

$$P(1) = \binom{4}{1}(0.3)^1 (0.7)^{4-1} = 0.4116$$

$$P(2) = \binom{4}{2}(0.3)^2 (0.7)^{4-2} = 0.2646$$

$$P(3) = \binom{4}{3}(0.3)^3 (0.7)^{4-3} = 0.0756$$

$$P(4) = \binom{4}{4}(0.3)^4 (0.7)^{4-4} = 0.0081$$

If there is a $P(0) = 0.2401$ probability that the baseball player will have 0 hits in a game, then you expect him to have 0 hits in $(0.2401)(100) = 24.01$ games. Calculate the remainder of the expected frequencies by multiplying the probabilities by 100.

Number of Hits	Binomial Probability	Number of Games	Expected Frequency (E)
0	0.2401	100	24.01
1	0.4116	100	41.16
2	0.2646	100	26.46
3	0.0756	100	7.56
4	0.0081	100	0.81
Total			**100**

Before you can calculate the chi-square statistic, you must first verify that all of the expected frequencies are greater than 5. Because the expected frequency in the final category is less than 5, you need to combine the last two rows of the table.

> The player is expected to have 4 hits in 0.81 games (which is fewer than 5 games).

Number of Hits	Binomial Probability	Number of Games	Expected Frequency (E)
0	0.2401	100	24.01
1	0.4116	100	41.16
2	0.2646	100	26.46
3–4	0.0837	100	8.37
Total			**100**

Note: Problems 12.4–12.5 refer to the data set in Problem 12.4, showing the number of games in which a particular Major League Baseball player had 0, 1, 2, 3, or 4 hits per game over the last 100 games. Assume the player had four at-bats per game.

12.5 Test the claim that the number of hits per game is binomially distributed at the $\alpha = 0.05$ significance level if $p = 0.3$.

State the null and alternative hypotheses.

H_0 : The number of hits per game is binomially distributed for $p = 0.3$

H_1 : The number of hits per game is not binomially distributed for $p = 0.3$

Use the following table to calculate the chi-square statistic.

Hits	O	E	$O - E$	$(O - E)^2$	$\dfrac{(O - E)^2}{E}$
0	26	24.01	1.99	3.96	0.16
1	34	41.16	–7.16	51.27	1.25
2	30	26.46	3.54	12.53	0.47
3–4	10	8.37	1.63	2.66	0.32
Total					**2.2**

Add the values in the right column: $\chi^2 = 2.2$. According to Reference Table 3, the critical chi-square value with df = 4 – 1 = 3 degrees of freedom at the $\alpha = 0.05$ significance level is 7.815. Because $\chi^2 = 2.2$ is less than $\chi^2_c = 7.815$, you fail to reject H_0; the player's hits per game are binomially distributed assuming $p = 0.3$.

Note: Problems 12.6–12.7 refer to the data set below, the number of customer visits per minute for an online store.

Number of Visits per Minute	0	1	2	3	4 or More	Total
Frequency	115	148	80	25	12	380

12.6 Calculate the expected number of visits per minute for each category assuming the data is Poisson-distributed and $\lambda = 1$.

> See Problems 6.17–6.31 to review the Poisson distribution.

Calculate the probabilities of 0, 1, 2, or 3 customer visits.

$$P(x) = \frac{\lambda^x e^{-\lambda}}{x!}$$

$$P(0) = \frac{1^0 \left(e^{-1}\right)}{0!} = \frac{1\left(e^{-1}\right)}{1} = 0.3679$$

$$P(1) = \frac{1^1 \left(e^{-1}\right)}{1!} = \frac{1\left(e^{-1}\right)}{1} = 0.3679$$

$$P(2) = \frac{1^2 \left(e^{-1}\right)}{2!} = \frac{1\left(e^{-1}\right)}{2} = 0.1839$$

$$P(3) = \frac{1^3 \left(e^{-1}\right)}{3!} = \frac{1\left(e^{-1}\right)}{6} = 0.0613$$

All of the Poisson probabilities must add up to 1. Thus, the probability that 4 or more customers visit is the complement of exactly 0, 1, 2, or 3 customers visiting.

$$P(4 \text{ or more}) = 1 - \left[P(0) + P(1) + P(2) + P(3)\right]$$
$$= 1 - (0.3679 + 0.3679 + 0.1839 + 0.0613)$$
$$= 1 - 0.981$$
$$= 0.0190$$

You are given the observed frequencies of 380 visits. Calculate the expected frequencies by multiplying each of the Poisson probabilities by 380.

Number of Visits	Poisson Probability	Total Visits	Expected Frequency (E)
0	0.3679	380	139.80
1	0.3679	380	139.80
2	0.1839	380	69.89
3	0.0613	380	23.29
4 or more	0.0190	380	7.22
Total	1.0000		380

Note: Problems 12.6–12.7 refer to the data set in Problem 12.6, the number of customer visits per minute for an online store.

12.7 Test the claim that the number of visits per minute is Poisson-distributed with $\lambda = 1$ at the $\alpha = 0.10$ level of significance.

State the null and alternative hypotheses.

H_0 : The number of visits per minute is Poisson-distributed when $\lambda = 1$

H_1 : The number of visits per minute is not Poisson-distributed when $\lambda = 1$

Calculate X^2 using the following table.

Visits	O	E	$O - E$	$(O - E)^2$	$\dfrac{(O - E)^2}{E}$
0	115	139.80	−24.80	615.04	4.40
1	148	139.80	8.20	67.24	0.48
2	80	69.89	10.11	102.21	1.46
3	25	23.29	1.71	2.92	0.13
4 or more	12	7.22	4.78	22.85	3.16
Total					9.63

The sum of the right column of the table is $X^2 = 9.63$. According to Reference Table 3, the critical chi-square value given df = 5 – 1 = 4 degrees of freedom and $\alpha = 0.10$ is 7.779. Because $X^2 = 9.63$ is greater than $\chi_c^2 = 7.779$, you reject H_0; the customer visits are not Poisson-distributed. ←

Remember, H_0 claims the Poisson distribution, so rejecting H_0 means the data ~~is~~ are not distributed that way.

Note: Problems 12.8–12.9 refer to the data set below, the speeds at which cars pass through a checkpoint during a one-hour period.

Speed (mph)	Frequency
Under 56	20
56–under 62	26
62–under 68	65
68–under 74	37
74 and over	27
Total	**175**

> See Problems 7.1–7.22 to review the normal distribution.

12.8 Calculate the expected number of cars for each category assuming the data is normally distributed with a mean of 65 miles per hour (mph) and a standard deviation of 6 mph.

Calculate the z-scores for the categories, assuming $\mu = 65$ and $\sigma = 6$.

> The subscript of each z-score is the maximum speed for that category. Remember that $z_x = \dfrac{x - \mu}{\sigma}$.

$$z_{56} = \frac{56 - 65}{6} = -1.5 \qquad P(z < -1.5) = 0.5 - 0.4332 = 0.0668$$

$$z_{62} = \frac{62 - 65}{6} = -0.5 \qquad P(-1.5 \le z < -0.5) = 0.4332 - 0.1915 = 0.2417$$

$$z_{68} = \frac{68 - 65}{6} = 0.5 \qquad P(-0.5 \le z < 0.5) = 0.1915 + 0.1915 = 0.3830$$

$$z_{74} = \frac{74 - 65}{6} = 1.5 \qquad P(0.5 \le z < 1.5) = 0.4332 - 0.1915 = 0.2417$$

$$P(z \ge 1.5) = 0.5 - 0.4332 = 0.0668$$

Calculate the expected frequencies for each category.

Speed of Cars	Normal Probability	Total Cars	Expected Frequency (E)
Under 56	0.0668	175	11.69
56–under 62	0.2417	175	42.30
62–under 68	0.3830	175	67.03
68–under 74	0.2417	175	42.30
74 and over	0.0668	175	11.69
Total			**175**

Note: Problems 12.8–12.9 refer to the data set in Problem 12.8, the speeds at which cars pass through a checkpoint during a one-hour period.

12.9 Test the claim that the speeds of the cars are normally distributed with a mean of 65 mph and a standard deviation of 6 mph at the $\alpha = 0.05$ significance level.

State the null and alternative hypotheses.

H_0 : The speeds of the cars are normally distributed.

H_1 : The speeds of the cars are not normally distributed.

Calculate X^2 using the following table.

Speed	O	E	$O - E$	$(O - E)^2$	$\dfrac{(O - E)^2}{E}$
Under 56	20	11.69	8.31	69.06	5.91
56–under 62	26	42.30	−16.30	265.69	6.28
62–under 68	65	67.03	−2.03	4.12	0.06
68–under 74	37	42.30	−5.30	28.09	0.66
74 and over	27	11.69	15.31	234.40	20.05
Total					32.96

According to Reference Table 3, the critical chi-square value given 4 degrees of freedom and $\alpha = 0.05$ is 9.488. Because $X^2 = 32.96$ (as calculated above) is greater than $\chi_c^2 = 9.488$, you reject H_0; the distribution is not normally distributed with a mean of 65 and a standard deviation of 6.

Note: Problems 12.10–12.11 refer to the data set below, the frequency distribution for the volume of cups of coffee (in ounces) dispensed by a vending machine.

Volume (ounces)	Frequency
7.2–under 7.4	15
7.4–under 7.6	11
7.6–under 7.8	18
7.8–under 8.0	22
8.0–under 8.2	12
8.2–under 8.4	6
Total	84

12.10 Calculate the expected frequency for each category assuming the data is uniformly distributed over the given ranges.

The expected frequency for each category of uniformly distributed data is the average frequency of the categories.

$$\bar{x} = \frac{15+11+18+22+12+6}{6} = \frac{84}{6} = 14$$

Each of the categories has an expected frequency of 14.

> *Note: Problems 12.10–12.11 refer to the data set in Problem 12.10, the frequency distribution for the volume of cups of coffee (in ounces) dispensed by a vending machine.*

12.11 Test the claim that the data is uniformly distributed over the given ranges at the $\alpha = 0.10$ significance level.

State the null and alternative hypotheses.

H_0 : The volume of coffee dispensed is uniformly distributed over the ranges

H_1 : The volume of coffee dispensed is not uniformly distributed over the ranges

Calculate the chi-square statistic using the table below.

Speed	O	E	$O - E$	$(O - E)^2$	$\dfrac{(O-E)^2}{E}$
7.2–under 7.4	15	14	1	1	0.07
7.4–under 7.6	11	14	−3	9	0.64
7.6–under 7.8	18	14	4	16	1.14
7.8–under 8.0	22	14	8	64	4.57
8.0–under 8.2	12	14	−2	4	0.29
8.2–under 8.4	6	14	−8	64	4.57
Total					**11.28**

According to Reference Table 3, the critical chi-square value given 5 degrees of freedom and $\alpha = 0.10$ is 9.236. Because $\chi^2 = 11.28$ is greater than $\chi_c^2 = 9.236$, you reject H_0; the volume of coffee dispensed per cup is not uniformly distributed.

Note: Problems 12.12–12.13 refer to the data set below, the number of grocery shoppers (from a random sample of 4) who use the self-checkout line.

Number of Self-Checkout Shoppers	Frequency
0	32
1	86
2	60
3	10
4	12
Total	**200**

So the researchers pick four random people and watch to see if those people use the self-checkout line. They do this a total of 200 times. All four of their randomly selected people used the self-checkout a total of 12 times.

12.12 Calculate the expected frequency for each category, assuming the distribution of shoppers who use the self-checkout line is binomially distributed with a 40% chance that each shopper will choose the self-checkout option.

A binomial distribution is based on two possible outcomes, a "success" and a "failure." In this case, a shopper choosing the self-checkout line is considered a success, so $p = 0.4$ and the complement is $q = 1 - p = 0.6$. Calculate the binomial probabilities for each category.

$$P(r) = \binom{n}{r} p^r q^{n-r}$$

$$P(0) = \binom{4}{0}(0.4)^0 (0.6)^{4-0} = 0.1296$$

$$P(1) = \binom{4}{1}(0.4)^1 (0.6)^{4-1} = 0.3456$$

$$P(2) = \binom{4}{2}(0.4)^2 (0.6)^{4-2} = 0.3456$$

$$P(3) = \binom{4}{3}(0.4)^3 (0.6)^{4-3} = 0.1536$$

$$P(4) = \binom{4}{4}(0.4)^4 (0.6)^{4-4} = 0.0256$$

Multiply each of the probabilities by the total number of observations, 200.

Number of Shoppers	Binomial Probability	Number of Days	Expected Frequency
0	0.1296	200	25.92
1	0.3456	200	69.12
2	0.3456	200	69.12
3	0.1536	200	30.72
4	0.0256	200	5.12
Total			200

Note: Problems 12.12–12.13 refer to the data set in Problem 12.12, the number of grocery shoppers (from a random sample of 4) who use the self-checkout line.

12.13 Test the claim that the observed data is binomially distributed with $p = 0.4$ at the $\alpha = 0.01$ significance level.

State the null and alternative hypotheses.

H_0 : The number of shoppers that use the self-checkout line is binomially distributed.

H_1 : The number of shoppers that use the self-checkout line is not binomially distributed.

Calculate χ^2 using the following table.

Shoppers	O	E	$O - E$	$(O - E)^2$	$\dfrac{(O-E)^2}{E}$
0	32	25.92	6.08	39.97	1.43
1	86	69.12	16.88	284.93	4.12
2	60	69.12	−9.12	83.17	1.20
3	10	30.72	−20.72	429.32	13.98
4	12	5.12	6.88	47.33	9.24
Total					29.97

The critical chi-square value, given 3 degrees of freedom and $\alpha = 0.01$, is 13.277. Because $\chi^2 = 29.92$ is greater than $\chi_c^2 = 13.277$, you reject H_0 and conclude that the data is not binomially distributed.

This might feel a little backward, because in past chapters H_1 represented what you were claiming. In this case, H_1 is the opposite of the claim that the data is binomially distributed.

Note: Problems 12.14–12.15 refer to a lumber yard that inspects 150 boards and records the number of knots in each. The resulting frequencies are recorded in the table below.

Number of Knots per Board	0	1	2	3	4	5+	Total
Frequency	33	44	42	21	8	2	150

12.14 Determine the expected frequencies for each category assuming the data is Poisson-distributed with $\lambda = 1.4$.

Calculate the Poisson probabilities for the first four categories given $\lambda = 1.4$.

$$P(x) = \frac{\lambda^x e^{-\lambda}}{x!}$$

$$P(0) = \frac{(1.4^0)e^{-1.4}}{0!} = 0.2466$$

$$P(1) = \frac{(1.4^1)e^{-1.4}}{1!} = 0.3452$$

$$P(2) = \frac{(1.4^2)e^{-1.4}}{2!} = 0.2417$$

$$P(3) = \frac{(1.4^3)e^{-1.4}}{3!} = 0.1128$$

$$P(4) = \frac{(1.4^4)e^{-1.4}}{4!} = 0.0395$$

Apply the complement rule to calculate $P(5 \text{ or more})$.

$$P(5 \text{ or more}) = 1 - [P(0) + P(1) + P(2) + P(3) + P(4)]$$
$$= 1 - (0.2466 + 0.3452 + 0.2417 + 0.1128 + 0.0395)$$
$$= 0.0142$$

Multiply each of the categories by 150 to calculate the expected frequencies.

Number of Knots	Poisson Probability	Total Boards	Expected Frequency (E)
0	0.2466	150	36.99
1	0.3452	150	51.78
2	0.2417	150	36.26
3	0.1128	150	16.92
4	0.0395	150	5.93
5 or more	0.0142	150	2.13 ←
Total			**≈150**

This expected frequency is less than 5, so you need to combine the last two categories when you calculate the chi-square statistic, if your textbook requires it.

Note: Problems 12.14–12.15 refer to a lumber yard that inspects 150 boards and records the number of knots in each.

12.15 Test the claim that the data is Poisson-distributed with $\lambda = 1.4$ at the $\alpha = 0.05$ significance level.

State the null and alternative hypotheses.

H_0 : The data is Poisson-distributed with $\lambda = 1.4$.

H_1 : The data is not Poisson-distributed with $\lambda = 1.4$.

> Modify the table in Problem 12.14 by adding the probabilities and expected frequencies. The final category (5+ knots) has an expected frequency that's too small by itself (2.13).

Use the table below to calculate χ^2, ensuring that each of the categories has an expected frequency greater than 5.

Knots	O	E	$O - E$	$(O - E)^2$	$\dfrac{(O - E)^2}{E}$
0	33	36.99	−3.99	15.92	0.43
1	44	51.78	−7.78	60.53	1.17
2	42	36.26	5.74	32.95	0.91
3	21	16.92	4.08	16.65	0.98
4 or more	10	8.06	1.94	3.77	0.47
Total	**150**				**3.96**

According to Reference Table 3, the critical chi-square value given 4 degrees of freedom and $\alpha = 0.05$ is 9.488. Because $\chi^2 = 3.96$ is less than $\chi_c^2 = 9.488$, you fail to reject H_0 and conclude that the data is Poisson-distributed assuming $\lambda = 1.4$ knots per board.

Note: Problems 12.16–12.17 refer to a professor who claims that his typical grade distribution is 20% As, 25% Bs, 40% Cs, 10% Ds, and 5% Fs.

12.16 This semester, the professor's class contains 85 students. Estimate the number of students who will earn each grade, based on the professor's claim.

Multiply each percentage by 85 to calculate the expected values for each grade category.

Grade	Expected Percent of Students	Total Students	Expected Number of Students
A	20%	85	17
B	25%	85	21.25
C	40%	85	34
D	10%	85	8.5
F	5%	85	4.25
Total	**100%**		**80**

Note: Problems 12.16–12.17 refer to a professor who claims that his typical grade distribution is 20% As, 25% Bs, 40% Cs, 10% Ds, and 5% Fs.

12.17 At the end of the semester, the 85 students in the course are assigned grades as follows: 22 As, 29 Bs, 20 Cs, 10 Ds, and 4 Fs. Test the professor's claim at the $\alpha = 0.05$ significance level.

State the null and alternative hypotheses.

H_0 : The faculty member followed his stated grade distribution.

H_1 : The faculty member did not follow his stated grade distribution.

Use the following table to calculate χ^2.

Grade	O	E	$O - E$	$(O-E)^2$	$\dfrac{(O-E)^2}{E}$
A	22	17	5	25	1.47
B	29	21.25	7.75	60.06	2.83
C	20	34	−14	196	5.76
D/F	14	12.75	1.25	1.56	0.12
Total					**10.18**

You have to combine the D and F categories, because F has an expected frequency of 4.25, which is less than 5.

According to Reference Table 3, the critical chi-square value given df = 3 and $\alpha = 0.05$ is 7.815. Because $\chi^2 = 10.18$ is greater than $\chi_c^2 = 7.815$, you reject H_0 and conclude that the professor did not follow his stated grade distribution.

Chi-Square Test for Independence

Are the variables related?

12.18 Explain how to perform the chi-square test for independence.

The chi-square test for independence is used to determine whether two categorical variables affect each other. Begin by stating a null hypothesis that the variables are independent; the alternative hypothesis states that the variables are not independent—they are related in some way.

A contingency table should be constructed with rows that are the categories of one variable and columns that are the categories of the other. The cells at the intersections of the rows and columns contain the observed frequencies. A contingency table with r rows and c columns contains $r \cdot c$ cells. Calculate the expected frequencies using the following formula.

$$E_{r,c} = \frac{(\text{total of row } r)(\text{total of column } c)}{\text{total number of observations}}$$

Use the chi-square distribution to compare the observed and expected frequencies. The chi-square statistic χ^2 is calculated using the formula first introduced in Problem 12.1.

$$\chi^2 = \sum \frac{(O - E)^2}{E}$$

Use Reference Table 3 to determine the critical chi-square value χ_c^2 given $df = (r-1)(c-1)$ degrees of freedom and the stated significance level α. In order to reject the null hypothesis (and therefore claim that the variables are independent), χ^2 must be greater than χ_c^2. Otherwise, you fail to reject the null hypothesis and must conclude that the variables are dependent.

Note: Problems 12.19–12.20 refer to the data set below, the number of head-to-head tennis matches won by Bob and Deb given warm-up times of 0–10 minutes, 11–20 minutes, and more than 20 minutes.

	0–10 min	11–20 min	More than 20 min	Total
Deb Wins	4	10	9	23
Bob Wins	14	9	4	27
Total	**18**	**19**	**13**	**50**

12.19 Calculate the expected frequency for each cell of the table, assuming that the warm-up time and the match winner are independent variables.

Calculate the expected frequency for each cell in the table using the following equation.

$$E_{r,c} = \frac{(\text{total of row } r)(\text{total of column } c)}{\text{total number of observations}}$$

Calculate this for every cell except the bold-faced cells, which represent totals.

Calculate $E_{1,1}$.

$$E_{1,1} = \frac{(23)(18)}{50} = 8.28$$

You expect Deb to win $E_{1,1} = 8.28$ matches given 0–10 minutes of warm-up time. Calculate the expected frequencies for the other 5 cells using the same technique.

$$E_{1,1} = \frac{(23)(18)}{50} = 8.28 \quad E_{1,2} = \frac{(23)(19)}{50} = 8.74 \quad E_{1,3} = \frac{(23)(13)}{50} = 5.98$$

$$E_{2,1} = \frac{(27)(18)}{50} = 9.72 \quad E_{2,2} = \frac{(27)(19)}{50} = 10.26 \quad E_{2,3} = \frac{(27)(13)}{50} = 7.02$$

Note: Problems 12.19–12.20 refer to the data set in Problem 12.19, the number of head-to-head tennis matches won by Bob and Deb given warm-up times of 0–10 minutes, 11–20 minutes, and more than 20 minutes.

12.20 Determine whether the length of time the players warm up affects the winner of the match at the $\alpha = 0.10$ significance level.

State the null and alternative hypotheses.

H_0: Warm-up time is independent of the eventual winner of the match.

H_1: Warm-up time is not independent of the eventual winner of the match.

> *This part is no different from Problems 12.2–12.17.*

Calculate χ^2 using the following table. ⟵

Row	Column	O	E	$O - E$	$(O - E)^2$	$\dfrac{(O - E)^2}{E}$
1	1	4	8.28	−4.28	18.32	2.21
1	2	10	8.74	1.26	1.59	0.18
1	3	9	5.98	3.02	9.12	1.53
2	1	14	9.72	4.28	18.32	1.88
2	2	9	10.26	−1.26	1.59	0.15
2	3	4	7.02	−3.02	9.12	1.30
Total						**7.25**

The table in Problem 12.19 (excluding the boldfaced totals) contains $r = 2$ rows (representing the players) and $c = 3$ columns (representing the different warm-up periods). Thus, there are df $= (2 - 1)(3 - 1) = 2$ degrees of freedom.

> *This is the sum of the numbers in the rightmost column of the last table.*

According to Reference Table 3, the critical chi-square value given df $= 2$ and $\alpha = 0.10$ is 4.605. Because $\chi^2 = 7.25$ is more than $\chi_c^2 = 4.605$, you reject H_0; it appears there is some sort of relationship between the warm-up time and the eventual winner of the match. The variables are not independent at the $\alpha = 0.10$ significance level.

Note: Problems 12.21–12.22 refer to the data set below, the number of men and women who decided to purchase or not to purchase an extended warranty for a digital camera at an electronics store.

	Warranty	No Warranty	Total
Men	7	50	**57**
Women	9	19	**28**
Total	**16**	**69**	**85**

12.21 Calculate the expected frequencies for each cell, assuming that the warranty decision and the gender of customer are independent variables.

Calculate the expected frequencies of each cell by multiplying its row total by its column total and dividing by the overall total.

If gender and warranty decision are independent, you would expect men to purchase the warranty 10.73 times.

$$E_{1,1} = \frac{(57)(16)}{85} = 10.73 \quad E_{1,2} = \frac{(57)(69)}{85} = 46.27$$

$$E_{2,1} = \frac{(28)(16)}{85} = 5.27 \quad E_{2,2} = \frac{(28)(69)}{85} = 22.73$$

Note: Problems 12.21–12.22 refer to the data set in Problem 12.21, the number of men and women who decided to purchase or not to purchase an extended warranty for a digital camera at an electronics store.

12.22 Determine whether the warranty decision and the gender of the customer are independent variables at the $\alpha = 0.05$ significance level.

State the null and alternative hypotheses.

H_0 : Warranty choice is independent of customer gender.

H_1 : Warranty choice is not independent of customer gender.

Use the following table to calculate χ^2.

Wrong: Taken from problem 12.20 →

Row	Column	O	E	$O - E$	$(O - E)^2$	$\dfrac{(O - E)^2}{E}$
~~1~~	~~1~~	~~4~~	~~8.28~~	~~-4.28~~	~~18.32~~	~~2.21~~
1	1	7	10.73	−3.73	13.91	1.30
1	2	50	46.27	3.73	13.91	0.30
2	1	9	5.27	3.73	13.91	2.64
2	2	19	22.73	−3.73	13.91	0.61
Total						~~7.06~~ 4.85

According to Reference Table 3, the critical chi-square value given df = (2 − 1)(2 − 1) = 1 degree of freedom and $\alpha = 0.05$ is 3.841. Because $\chi^2 = $ ~~7.06~~ *4.85* is greater than $\chi_c^2 = 3.841$, you reject H_0; there is a relationship between warranty decision and gender.

Note: Problems 12.23–12.24 refer to the data set below, the final exam grade distribution for 215 graduate students and the number of hours the students spent studying for the exam.

	A	B	C	Total
Less than 3 hr	18	48	16	82
3–5 hr	30	28	12	70
More than 5 hr	33	25	5	63
Total	81	101	33	215

12.23 Calculate the expected frequencies for each cell, assuming that the final exam grade and the time spent studying are independent variables.

Excluding the total column and the total row, there are $r = 3$ rows and $c = 3$ columns. Calculate the expected frequency for each cell by multiplying its row total by its column total and dividing by the overall total.

$$E_{1,1} = \frac{(82)(81)}{215} = 30.89 \quad E_{1,2} = \frac{(82)(101)}{215} = 38.52 \quad E_{1,3} = \frac{(82)(33)}{215} = 12.59$$

$$E_{2,1} = \frac{(70)(81)}{215} = 26.37 \quad E_{2,2} = \frac{(70)(101)}{215} = 32.88 \quad E_{2,3} = \frac{(70)(33)}{215} = 10.74$$

$$E_{3,1} = \frac{(63)(81)}{215} = 23.73 \quad E_{3,2} = \frac{(63)(101)}{215} = 29.60 \quad E_{3,3} = \frac{(63)(33)}{215} = 9.67$$

If the grade you get on the exam doesn't depend on the length of time you study, then 30.89 students who study less than 3 hours should get an A.

Note: Problems 12.23–12.24 refer to the data set in Problem 12.23, the final exam grade distribution for 215 graduate students and the number of hours the students spent studying for the exam.

12.24 Determine whether the exam grade and the time spent studying for the exam are independent variables at the $\alpha = 0.01$ significance level.

State the null and alternative hypotheses.

H_0 : The final exam grade is independent of the hours spent studying for it.

H_1 : The final exam grade is not independent of the hours spent studying for it.

Calculate χ^2 by adding the values in the right column of the table below.

Row	Column	O	E	O – E	(O – E)²	$\frac{(O-E)^2}{E}$
1	1	18	30.89	–12.89	166.15	5.38
1	2	48	38.52	9.48	89.87	2.33
1	3	16	12.59	3.41	11.63	0.92
2	1	30	26.37	3.63	13.18	0.50
2	2	28	32.88	–4.88	23.81	0.72
2	3	12	10.74	1.26	1.59	0.15
3	1	33	23.73	9.27	85.93	3.62
3	2	25	29.60	–4.60	21.16	0.71
3	3	5	9.67	–4.67	21.81	2.26
Total						**16.59**

According to Reference Table 3, the critical chi-square value given df = $(3 – 1)(3 – 1) = 4$ degrees of freedom and $\alpha = 0.01$ is 13.277. Because $\chi^2 = 16.59$ is greater than $\chi_c^2 = 13.277$, you reject H_0; it appears that the length of time spent studying for the final exam has an effect on the final exam grade.

Note: Problems 12.25–12.26 refer to the data set below, the number of voters who are satisfied and unsatisfied with the current economy and their party affiliations.

Party	Satisfied	Unsatisfied	Total
Democrat	140	172	**312**
Republican	135	163	**298**
Independent	30	22	**52**
Total	**305**	**357**	**662**

12.25 Calculate the expected frequencies for each cell, assuming that economic satisfaction and party affiliation are independent variables.

Calculate the expected frequency of each cell by multiplying its row total by its column total and dividing by the overall total.

$$E_{1,1} = \frac{(312)(305)}{662} = 143.75 \quad E_{1,2} = \frac{(312)(357)}{662} = 168.25$$

$$E_{2,1} = \frac{(298)(305)}{662} = 137.30 \quad E_{2,2} = \frac{(298)(357)}{662} = 160.70$$

$$E_{3,1} = \frac{(52)(305)}{662} = 23.96 \quad E_{3,2} = \frac{(52)(357)}{662} = 28.04$$

Note: Problems 12.25–12.26 refer to the data set in Problem 12.25, the number of voters who are satisfied and unsatisfied with the current economy and their party affiliations.

12.26 Determine whether satisfaction with the economy and the party affiliation of the voter are independent variables at the $\alpha = 0.05$ significance level.

State the null and alternative hypotheses.

H_0 : Satisfaction with the economy is independent of party affiliation.

H_1 : Satisfaction with the economy is not independent of party affiliation.

Calculate χ^2 using the following table.

Row	Column	O	E	$O - E$	$(O - E)^2$	$\dfrac{(O - E)^2}{E}$
1	1	140	143.75	−3.75	14.06	0.10
1	2	172	168.25	3.75	14.06	0.08
2	1	135	137.30	−2.30	5.29	0.04
2	2	163	160.70	2.30	5.29	0.03
3	1	30	23.96	6.04	36.48	1.52
3	2	22	28.04	−6.04	36.48	1.30
Total						**3.07**

There are $df = (3 - 1)(2 - 1) = 2$ degrees of freedom.

Because $\chi^2 = 3.07$ is less than $\chi_c^2 = 5.991$, you fail to reject H_0; there is no relationship between satisfaction and party affiliation.

Note: Problems 12.27–12.28 refer to the data set below, the arrival status of 300 flights that originated from New York, Chicago, or Los Angeles airports.

Status	NY	Chi	LA	Total
Early	18	24	22	64
On time	62	45	50	157
Late	25	40	14	79
Total	**105**	**109**	**86**	**300**

Should be bold.

12.27 Calculate the expected frequency for each cell, assuming that arrival status and flight origin are independent variables.

Calculate the expected frequency of each cell by multiplying its row total by its column total and dividing by the overall total.

$$E_{1,1} = \frac{(64)(105)}{300} = 22.40 \qquad E_{1,2} = \frac{(64)(109)}{300} = 23.25 \qquad E_{1,3} = \frac{(64)(86)}{300} = 18.35$$

$$E_{2,1} = \frac{(157)(105)}{300} = 54.95 \qquad E_{2,2} = \frac{(157)(109)}{300} = 57.04 \qquad E_{2,3} = \frac{(157)(86)}{300} = 45.01$$

$$E_{3,1} = \frac{(79)(105)}{300} = 27.65 \qquad E_{3,2} = \frac{(79)(109)}{300} = 28.70 \qquad E_{3,3} = \frac{(79)(86)}{300} = 22.65$$

Note: Problems 12.27–12.28 refer to the data set in Problem 12.27, the arrival status of 300 flights that originated from New York, Chicago, or Los Angeles airports.

12.28 Determine whether arrival status and flight origin are independent at the $\chi = 0.10$ significance level.

State the null and alternative hypotheses.

H_0 : Arrival status is independent of flight origin.

H_1 : Arrival status is not independent of flight origin.

Use the following table to calculate χ^2.

Row	Column	O	E	$O - E$	$(O - E)^2$	$\dfrac{(O - E)^2}{E}$
1	1	18	22.40	−4.40	19.36	0.86
1	2	24	23.35	0.65	0.42	0.02
1	3	22	18.35	3.65	13.32	0.73
2	1	62	54.95	7.05	49.70	0.90
2	2	45	57.04	−12.04	144.96	2.54
2	3	50	45.01	4.99	24.90	0.55
3	1	25	27.65	−2.65	7.02	0.25
3	2	40	28.70	11.30	127.69	4.45
3	3	14	22.65	−8.65	74.82	3.30
Total						**13.60**

According to Reference Table 3, the critical chi-square value given df = $(3 - 1)(3 - 1) = 4$ degrees of freedom and $\alpha = 0.10$ is 7.779. Because $\chi^2 = 13.60$ is greater than $\chi_c^2 = 7.779$, you reject H_0; there appears to be a relationship between arrival status and flight origin.

Hypothesis Test for a Single Population Variance
Testing variation instead of the mean

12.29 Describe the procedure for hypothesis testing a single population variance, including the formula used to calculate χ^2.

The hypothesis test for population variance is similar to the test for population mean in that the null and alternative hypotheses are subject to one- and two-tailed tests. The chi-square distribution determines the outcome of the test with $n - 1$ degrees of freedom.

As opposed to all of the chi-square problems in this chapter so far, which were all one-tailed tests on the right side of the distribution.

If the variance of the sample is s^2 and σ^2 is the population variance (stated in the null hypothesis), then χ^2—the test statistic—is calculated according to the following formula.

$$\chi^2 = \frac{(n-1)s^2}{\sigma^2}$$

This hypothesis test applies only when the population is normally distributed; thus, each problem in this section makes that assumption.

Note: Problems 12.30–12.31 refer to the data set below, the number of minutes 7 randomly chosen customers waited on hold for phone support for a particular company.

Number of Minutes						
5	14	4	6	10	6	3

12.30 The company claims that the standard deviation of the wait time is less than 5 minutes. Test this claim at the $\alpha = 0.05$ significance level.

No hypothesis test exists for the standard deviation; you need to test for the population variance. To construct the null hypothesis, convert the standard deviation to variance by squaring it.

> Remember, the square root of the variance is the standard deviation.

$$H_0 : \sigma^2 \geq 25$$
$$H_1 : \sigma^2 < 25 \longleftarrow$$

> You want to prove $\sigma < 5$, which is the same as proving $\sigma^2 < 25$.

Calculate the sample variance, as explained in Problem 3.45.

	x	x^2
	5	25
	14	196
	4	16
	6	36
	10	100
	6	36
	3	9
Total	**48**	**418**

$$s^2 = \frac{\sum x^2 - \frac{\left(\sum x\right)^2}{n}}{n-1} = \frac{418 - \frac{(48)^2}{7}}{7-1} = \frac{418 - 329.143}{6} = 14.81$$

Substitute the proposed population deviance boundary ($\sigma = 5$), the sample size ($n = 7$), and the sample variance ($s^2 = 14.81$) into the chi-square formula presented in Problem 12.29.

$$\chi^2 = \frac{(n-1)s^2}{\sigma^2} = \frac{(7-1)(14.81)}{(5)^2} = \frac{88.86}{25} = 3.55$$

When H_1 contains <, you look up $1 - \alpha$ in Reference Table 3. When H_1 contains >, you look up α.

The chi-square distribution is not symmetrical, so each tail has its own critical score. The column headings in Reference Table 3 indicate the area in the right tail of the distribution. However, in this problem you are performing a left-tailed test. The significance level is $\alpha = 0.05$, so the area to the right of the rejection region is $1 - 0.05 = 0.95$.

Use $\alpha = 0.95$ and $df = n - 1 = 7 - 1 = 6$ to identify the critical chi-square value in Reference Table 3: $\chi_c^2 = 1.635$. In order to reject the null hypothesis, χ^2 must be less than 1.635 (as illustrated below).

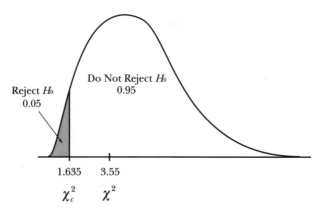

Reject H_0
0.05

Do Not Reject H_0
0.95

1.635   3.55
χ_c^2 χ^2

However, $\chi^2 = 3.55$ is greater than $\chi_c^2 = \chi_{0.95}^2 = 1.635$, so you fail to reject H_0. The population variance is not less than 25, so the standard deviation is not less than 5 minutes.

Note: Problems 12.30–12.31 refer to the data set in Problem 12.30, the number of minutes 7 randomly chosen customers waited on hold for phone support for a particular company.

12.31 Construct a 95% confidence interval around this sample variance to estimate the population standard deviation.

Apply the following equation to calculate the boundaries of the confidence interval for population variance, where χ_U^2 represents the upper chi-square critical score and χ_L^2 represents the lower chi-square critical score.

$$\frac{(n-1)s^2}{\chi_U^2} \le \sigma^2 \le \frac{(n-1)s^2}{\chi_L^2}$$

A 95% confidence interval produces two tails of equal area that contain the remaining 5%. Divide 0.05 by 2 to get a 0.025 area for each of the tails. Use $\alpha = 0.025$ for the upper chi-square critical score and $1 - 0.025 = 0.975$ for the lower chi-square critical score.

According to Reference Table 3, given $df = 7 - 1 = 6$ degrees of freedom, the critical scores are $\chi_U^2 = \chi_{0.025}^2 = 14.449$ and $\chi_L^2 = \chi_{0.975}^2 = 1.237$. Recall that $s^2 = 14.81$, according to Problem 12.30. Substitute these values into the confidence interval boundary formula.

$$\frac{(n-1)s^2}{\chi_U^2} \le \sigma^2 \le \frac{(n-1)s^2}{\chi_L^2}$$

$$\frac{(7-1)(14.81)}{14.449} \le \sigma^2 \le \frac{(7-1)(14.81)}{1.237}$$

$$\frac{88.86}{14.449} \le \sigma^2 \le \frac{88.86}{1.237}$$

$$6.15 \le \sigma^2 \le 71.84$$

Take the square root of each value to identify the 95% confidence interval for the population standard deviation.

$$\sqrt{6.15} \le \sigma \le \sqrt{71.84}$$

$$2.48 \le \sigma \le 8.48$$

You are 95% confident that the true population standard deviation is between 2.48 and 8.48 minutes.

Note: In Problems 12.32–12.33, the manager of a repair shop that services office copy machines is concerned that the standard deviation for the arrival time of a repairman has exceeded 30 minutes. A random sample of 24 service calls has a standard deviation of 33.4 minutes.

12.32 Investigate the manager's concern using a hypothesis test with an $\alpha = 0.05$ significance level.

State the null and alternative hypotheses in terms of the population variance.

$$H_0 : \sigma^2 \le 900$$

$$H_1 : \sigma^2 > 900$$

Calculate χ^2.

$$\chi^2 = \frac{(n-1)s^2}{\sigma^2} = \frac{(24-1)(33.4)^2}{(30)^2} = \frac{(23)(1{,}115.56)}{900} = 28.51$$

The manager wants to know if the standard deviation is greater than 30. That would mean the variance is greater than $(30)^2 = 900$.

The sample size is $n = 24$, so there are df $= 24 - 1 = 23$ degrees of freedom. The alternate hypothesis contains "greater than," so you perform a right-tailed test: $\alpha = 0.05$. According to Reference Table 3, the critical chi-square value is $\chi_c^2 = \chi_{0.05}^2 = 35.172$.

Because $\chi^2 = 28.51$ is less than $\chi_{0.05}^2 = 35.172$, you fail to reject H_0; the population variance is not more than 900, so the standard deviation is not more than 30 minutes.

If you're 90% confident, then you're 10% not confident. Divide that 0.10 by 2 and you get 0.05 for each end of the distribution. One end is 0.05 and the other is 1 – 0.05 = 0.95.

Note: In Problems 12.32–12.33, the manager of a repair shop that services office copy machines is concerned that the standard deviation for the arrival time of a repairman has exceeded 30 minutes. A random sample of 24 service calls has a standard deviation of 33.4 minutes.

12.33 Construct a 90% confidence interval around the sample variance to estimate the population standard deviation.

A 90% confidence interval produces two tails of area 0.05 at each end of the chi-square distribution. Given 23 degrees of freedom, the critical chi-square scores are $\chi_U^2 = \chi_{0.05}^2 = 35.172$ and $\chi_L^2 = \chi_{0.95}^2 = 13.091$. Calculate the boundaries of the confidence interval for the population variance.

$$\frac{(n-1)s^2}{\chi_U^2} \le \sigma^2 \le \frac{(n-1)s^2}{\chi_L^2}$$

$$\frac{(24-1)(33.4)^2}{35.172} \le \sigma^2 \le \frac{(24-1)(33.4)^2}{13.091}$$

$$\frac{(23)(1,115.56)}{35.172} \le \sigma^2 \le \frac{(23)(1,115.56)}{13.091}$$

$$729.50 \le \sigma^2 \le 1,959.96$$

Take the square root of all three expressions to identify the confidence interval for the population standard deviation.

$$\sqrt{729.50} \le \sigma \le \sqrt{1,959.96}$$

$$27.01 \le \sigma \le 44.27$$

You are 90% confident that the true population standard deviation is between 27.01 and 44.27 minutes.

Note: In Problems 12.34–12.35, a professor tries to design a 100-point test so that the scores will have a standard deviation of 10 points. A recent sample of 20 exams has a sample standard deviation of 13.9.

12.34 Determine whether the professor met his goal at the $\alpha = 0.05$ significance level.

When H_1 contains "not equal to," it means a two-tailed test is on the way. (H_1 can never contain "equal to," only "greater than," "less than," or "not equal to."

State the null and alternative hypotheses in terms of the population variance.

$$H_0 : \sigma^2 = 100$$
$$H_1 : \sigma^2 \ne 100$$

Calculate χ^2.

$$\chi^2 = \frac{(n-1)s^2}{\sigma^2} = \frac{(20-1)(13.9)^2}{(10)^2} = \frac{(19)(193.21)}{100} = 36.71$$

The sample size is $n = 20$, so df $= 20 - 1 = 19$. You are performing a two-tailed test, so split $\alpha = 0.05$ into two equal halves at the right and left ends of the distribution. Each tail contains an area of 0.025 with its own critical chi-square score.

> Just like you'd split up the confidence level to calculate the confidence interval boundaries.

According to Reference Table 3, the critical chi-square values are $\chi^2_{0.025} = 32.852$ and $\chi^2_{0.975} = 8.907$. Thus, you reject the hypothesis only if χ^2 is less than 8.907 or greater than 32.852, as illustrated below. Recall that $\chi^2 = 36.71$, so you reject H_0; the standard deviation is not equal to 10.

> $1 - 0.025 = 0.975$

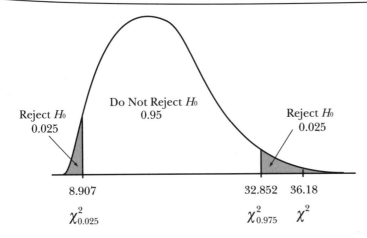

Reject H_0
0.025

Do Not Reject H_0
0.95

Reject H_0
0.025

8.907

$\chi^2_{0.025}$

32.852   36.18

$\chi^2_{0.975}$ χ^2

Note: In Problems 12.34–12.35, a professor tries to design a 100-point test so that the scores will have a standard deviation of 10 points. A recent sample of 20 exams has a sample standard deviation of 13.9.

12.35 Construct a 98% confidence interval around the sample variance to estimate the population standard deviation.

A 98% confidence interval produces two tails with area 0.01 at each end of the chi-square distribution. Given 19 degrees of freedom and the α values 0.01 and $1 - 0.01 = 0.99$, the corresponding critical chi-square scores are $\chi^2_U = \chi^2_{0.01} = 36.191$ and $\chi^2_L = \chi^2_{0.99} = 7.633$.

Identify the boundaries of the population variance confidence interval and the resulting confidence interval for the population standard deviation.

$$\frac{(n-1)s^2}{\chi^2_U} \leq \sigma^2 \leq \frac{(n-1)s^2}{\chi^2_L}$$

$$\frac{(20-1)(13.9)^2}{36.191} \leq \sigma^2 \leq \frac{(20-1)(13.9)^2}{7.633}$$

$$101.43 \leq \sigma^2 \leq 480.94$$

$$\sqrt{101.43} \leq \sigma \leq \sqrt{480.94}$$

$$10.07 \leq \sigma \leq 21.93$$

> The glass should have a standard deviation of 0.5 mm or less. You're trying to show that the standard deviation is actually larger, so make that the alternative hypothesis, H_1.

Note: In Problems 12.36–12.37, a glass manufacturer is concerned that the standard deviation of the glass thickness is exceeding the company standard of 0.5 mm. A recent sample of 27 panes of glass has a sample standard deviation of 0.67 mm.

12.36 Investigate the manufacturer's concern at the $\alpha = 0.10$ significance level.

State the null and alternative hypotheses in terms of the population variance.

$$H_0 : \sigma^2 \le 0.25$$
$$H_1 : \sigma^2 > 0.25$$

Calculate χ^2.

$$\chi^2 = \frac{(n-1)s^2}{\sigma^2} = \frac{(27-1)(0.67)^2}{(0.50)^2} = \frac{(26)(0.4489)}{0.25} = 46.69$$

You are performing a right-tailed test with 26 degrees of freedom, so the critical chi-square value is 35.563. Because $\chi^2 = 46.69$ is greater than $\chi^2_{0.90} = 35.563$, you reject H_0; the standard deviation for glass thickness exceeds 0.5 mm.

Note: In Problems 12.36–12.37, a glass manufacturer is concerned that the standard deviation of the glass thickness is exceeding the company standard of 0.5 mm. A recent sample of 27 panes of glass has a sample standard deviation of 0.67 mm.

12.37 Construct a 95% confidence interval to estimate the true process standard deviation for glass thickness.

A 95% confidence interval has a corresponding area of 0.025 in each tail of the chi-square distribution. Given 26 degrees of freedom, the critical chi-square values are $\chi^2_U = \chi^2_{0.025} = 41.923$ and $\chi^2_L = \chi^2_{0.975} = 13.844$. Construct the boundaries for the population standard deviation confidence intervals.

$$\frac{(n-1)s^2}{\chi^2_U} \le \sigma^2 \le \frac{(n-1)s^2}{\chi^2_L}$$
$$\frac{(27-1)(0.67)^2}{41.923} \le \sigma^2 \le \frac{(27-1)(0.67)^2}{13.844}$$
$$0.28 \le \sigma^2 \le 0.84$$
$$\sqrt{0.28} \le \sigma \le \sqrt{0.84}$$
$$0.53 \le \sigma \le 0.92$$

> The entire interval is larger than 0.5 mm, which verifies your answer to Problem 12.36.

Note: Problems 12.38–12.39 refer to a process that fills boxes with 18 ounces of cereal. Under normal operating conditions, the standard deviation of the weights of the boxes is 1.1 ounces. A random sample of 30 boxes has a sample standard deviation of 0.81 ounces.

12.38 Determine whether the standard deviation of the filling process is operating under normal conditions at the $\alpha = 0.10$ significance level.

State the null and alternative hypotheses in terms of the population variance.

$$H_0 : \sigma^2 = 1.21$$
$$H_1 : \sigma^2 \neq 1.21$$

Calculate χ^2.

If the process is not operating under normal conditions, then the standard deviation isn't 1.1 and the variance isn't 1.21.

$$\chi^2 = \frac{(n-1)s^2}{\sigma^2} = \frac{(30-1)(0.81)^2}{(1.1)^2} = \frac{(29)(0.6561)}{1.21} = 15.72$$

You are performing a two-tailed test with 29 degrees of freedom. The corresponding critical chi-square values are $\chi^2_{0.05} = 42.557$ and $\chi^2_{0.95} = 17.708$. Notice that $\chi^2 = 15.72$ is less than 17.708 and therefore lies in the left rejection region of the distribution. The standard deviation of the filling process is not 1.1 ounces, so the process is not operating under normal conditions.

Note: Problems 12.38–12.39 refer to a process that fills boxes with 18 ounces of cereal. Under normal operating conditions, the standard deviation of the weights of the boxes is 1.1 ounces. A random sample of 30 boxes has a sample standard deviation of 0.81 ounces.

12.39 Construct a 95% confidence interval to estimate the true standard deviation for the cereal box filling process.

A 95% confidence interval produces two tails of area 0.025 at each end of the chi-square distribution. Given 29 degrees of freedom, the critical chi-square values are $\chi^2_U = \chi^2_{0.025} = 45.772$ and $\chi^2_L = \chi^2_{0.975} = 16.047$. Identify the boundaries of the process standard deviation confidence interval.

$$\frac{(n-1)s^2}{\chi^2_U} \leq \sigma^2 \leq \frac{(n-1)s^2}{\chi^2_L}$$
$$\frac{(30-1)(0.81)^2}{45.772} \leq \sigma^2 \leq \frac{(30-1)(0.81)^2}{16.047}$$
$$\frac{(29)(0.6561)}{45.772} \leq \sigma^2 \leq \frac{(29)(0.6561)}{16.047}$$
$$0.42 \leq \sigma^2 \leq 1.19$$
$$\sqrt{0.42} \leq \sigma \leq \sqrt{1.19}$$
$$0.65 \leq \sigma \leq 1.09$$

Hypothesis Test for Two Population Variances
Introducing the F-distribution

12.40 Describe the hypothesis testing procedure used to compare two population variances, including the formula for the calculated F-score.

> s_1 and s_2 are the sample standard deviations.

A two-population variance hypothesis test may be one- or two-tailed, much like hypothesis tests for a single population. However, rather than the chi-square distribution, the F-distribution is used to determine whether to reject the null hypothesis. The calculated F-score is $F = \dfrac{s_1^2}{s_2^2}$, where s_1^2 and s_2^2 are the sample variances.

Note that this procedure assumes that the populations are normally distributed and that the two samples are independent. These assumptions are thus implicit in the remaining problems in this chapter.

12.41 Airport management is investigating procedures at different terminals to reduce the variability in the length of time required to pass through airport security. The following table summarizes sample data from two different terminals employing different procedures.

	Terminal A	Terminal B
Sample standard deviation	7.4 minutes	8.9 minutes
Sample size	10	9

Determine whether the procedures at Terminal A are more effective in reducing variability than those used at Terminal B, at the $\alpha = 0.05$ significance level.

State the null and alternative hypotheses using Terminal B as population 1 and Terminal A as population 2.

> Whichever population has the larger sample variance (in this case, Terminal B) needs to be population 1.

$$H_0 : \sigma_1^2 \leq \sigma_2^2$$
$$H_1 : \sigma_1^2 > \sigma_2^2$$

Calculate the F-score.

> A one-tailed test for two variances should always have this alternative hypothesis, not $H_1 : \sigma_1^2 < \sigma_2^2$.

$$F = \frac{s_1^2}{s_2^2} = \frac{(8.9)^2}{(7.4)^2} = \frac{79.21}{54.76} = 1.446$$

The F-distribution has two separate degrees of freedom, one for each population and both equal to one fewer than the sample size.

$$D_1 = n_1 - 1 = 9 - 1 = 8$$
$$D_2 = n_2 - 1 = 10 - 1 = 9$$

Reference Table 4 contains F-distribution tables, each representing a value for the area in the right tail of distribution. Locate the section of the table for $\alpha = 0.05$. Identify the critical F-score under the $D_1 = 8$ column and along the $D_2 = 9$ row. The value is underlined in the Reference Table 4 excerpt below.

Area in the Right Tail of Distribution = 0.05

					D_1				
D_2	1	2	3	4	5	6	7	8	9
1	161.448	199.500	215.707	224.583	230.162	233.986	236.768	238.883	240.543
2	18.513	19.000	19.164	19.247	19.296	19.330	19.353	19.371	19.385
3	10.128	9.552	9.277	9.117	9.013	8.941	8.887	8.845	8.812
4	7.709	6.944	6.591	6.388	6.256	6.163	6.094	6.041	5.999
5	6.608	5.786	5.409	5.192	5.050	4.950	4.876	4.818	4.772
6	5.987	5.143	4.757	4.534	4.387	4.284	4.207	4.147	4.099
7	5.591	4.737	4.347	4.120	3.972	3.866	3.787	3.726	3.677
8	5.318	4.459	4.066	3.838	3.687	3.581	3.500	3.438	3.388
9	5.117	4.256	3.863	3.633	3.482	3.374	3.293	3.230	3.179
10	4.965	4.103	3.708	3.478	3.326	3.217	3.135	3.072	3.020

The critical F-score $F_c = 3.230$ defines the lower bound for the rejection region on the right side of the distribution. Because $F = 1.446$ is less than $F_c = 3.230$, you fail to reject H_0 and conclude that the procedures at Terminal A are not more effective in reducing variability than Terminal B.

12.42 The table below contains temperature variation data for two brands of refrigerators. Perform a hypothesis test to determine whether the population variances are different at the $\alpha = 0.05$ significance level.

	Brand A	Brand B
Sample standard deviation	4.8 degrees	4.1 degrees
Sample size	20	17

State the null and alternative hypotheses, using Brand A as population 1 and Brand B as population 2.

$$H_0: \sigma_1^2 = \sigma_2^2$$
$$H_1: \sigma_1^2 \neq \sigma_2^2$$

Calculate the F-score.

$$F = \frac{s_1^2}{s_2^2} = \frac{(4.8)^2}{(4.1)^2} = \frac{23.04}{16.81} = 1.371$$

Population 1 should always have the larger variance. That means the numerator of F should always be bigger than the denominator.

Calculate the degrees of freedom for the F-distribution.

$$D_1 = n_1 - 1 = 20 - 1 = 19$$
$$D_2 = n_2 - 1 = 17 - 1 = 16$$

You are performing a two-tailed test, so divide $\alpha = 0.05$ into two equal halves of area 0.025 at both ends of the distribution. Use the section of Reference Table 4 labeled "Area in the Right Tail of Distribution = 0.025." Given degrees of freedom $D_1 = 19$ and $D_2 = 16$, the critical F-score is $F_c = 2.698$.

Because $F = 1.371$ is less than $F_c = 2.698$, you fail to reject H_0; the temperature variations of the refrigerator brands are not different.

> Even though this is a two-tailed test, you only have to worry about the right side. That's because you were careful to put the higher variance in the numerator of the F-score formula.

12.43 A tire company has developed a new brand that should exhibit more consistent tread life. The table below contains sample data for the existing and new brands of tire.

	Existing	New
Sample standard deviation	5,325 miles	3,560 miles
Sample size	15	18

Determine whether the tread life variance in the new brand is less than the variance in the existing brand at the $\alpha = 0.10$ significance level.

State the null and alternative hypotheses, using the existing brand as population 1 and the new brand as population 2.

$$H_0 : \sigma_1^2 \le \sigma_2^2$$
$$H_1 : \sigma_1^2 > \sigma_2^2$$

> Even though the problem uses the phrase "less than," the alternative hypothesis always contains "greater than" when testing variance. The sample variance of population 1 is bigger, so you want to try to prove that the population variance is as well.

Calculate the F-score and the degrees of freedom.

$$F = \frac{s_1^2}{s_2^2} = \frac{(5,325)^2}{(3,560)^2} = 2.237 \qquad \begin{aligned} D_1 &= n_1 - 1 = 15 - 1 = 14 \\ D_2 &= n_2 - 1 = 18 - 1 = 17 \end{aligned}$$

You are applying a one-tailed test with $\alpha = 0.10$, so refer to the section of Reference Table 4 labeled "Area in the Right Tail of Distribution = 0.10." The critical F-score is in column 14 and row 17: $F_c = 1.925$.

Because $F = 2.237$ is greater than $F_c = 1.925$, you reject H_0. The new brand of tire has a more consistent tread life.

12.44 A university examines the variability in the math SAT scores of students accepted to the business and engineering schools. The table below presents sample data from the most recent incoming class.

	Engineering	Business
Sample standard deviation	122.4	106.5
Sample size	16	19

Perform a hypothesis test to determine whether the variation in math SAT scores is different if $\alpha = 0.05$.

State the null and alternative hypotheses using the engineering students as population 1 and the business students as population 2.

$$H_0 : \sigma_1^2 = \sigma_2^2$$
$$H_1 : \sigma_1^2 \neq \sigma_2^2$$

Calculate the F-score and the degrees of freedom.

$$F = \frac{s_1^2}{s_2^2} = \frac{(122.4)^2}{(106.5)^2} = \frac{14,981.76}{11,342.25} = 1.321 \qquad \begin{aligned} D_1 &= n_1 - 1 = 16 - 1 = 15 \\ D_2 &= n_2 - 1 = 19 - 1 = 18 \end{aligned}$$

Because $F = 1.3221$ is less than $F_c = 2.667$, you fail to reject H_0. The variability in math SAT scores between the schools is not significantly different.

> Divide 0.05 by 2 to get 0.025. Use the section of Reference Table 4 labeled "Area in the Right Tail of Distribution = 0.025." Look in column 15 and row 18—you'll find 2.667.

12.45 The table below contains random samples of high school teacher salaries from Ohio and New York. Perform a hypothesis test to determine whether New York salaries vary more than Ohio salaries if $\alpha = 0.01$.

	Ohio	New York
Sample standard deviation	$6,180	$7,760
Sample size	14	12

State the null and alternative hypotheses, using New York as population 1 and Ohio as population 2.

$$H_0 : \sigma_1^2 \leq \sigma_2^2$$
$$H_1 : \sigma_1^2 > \sigma_2^2$$

> The bigger sample standard deviation (or sample variance) needs to be population 1.

Calculate the F-score and the degrees of freedom.

$$F = \frac{s_1^2}{s_2^2} = \frac{(7,760)^2}{(6,180)^2} = 1.577 \qquad \begin{aligned} D_1 &= n_1 - 1 = 12 - 1 = 11 \\ D_2 &= n_2 - 1 = 14 - 1 = 13 \end{aligned}$$

Because $F = 1.577$ is less than $F_c = 4.025$, you fail to reject H_0. The teacher salaries in New York do not vary more than the teacher salaries in Ohio.

Chapter 13
ANALYSIS OF VARIANCE

Comparing multiple means with the F-distribution

Chapter 11 outlined a procedure for comparing two population means to determine if the difference between them was statistically significant. This chapter introduces analysis of variance (ANOVA), which allows you to compare three or more population means. Once you determine that two or more of the population means differ, you apply pairwise comparison tests to identify those populations.

At the end of Chapter 12, you compared two population variances using the F-distribution. The analysis of variance test in this chapter uses the F-distribution to compare the variance that occurs within each sample to the variance between the samples. Once you figure out that a group of populations has at least two means that are different, Scheffé's pairwise comparison test and Tukey's method are used to figure out which of the populations contain those different means.

One-Way ANOVA: Completely Randomized Design

The most basic ANOVA procedure

13.1 Describe the purpose of one-way analysis of variance and explain the difference between randomized design and randomized block design.

One-way analysis of variance (ANOVA) performs a hypothesis test that compares three or more population means based on sample data. The null hypothesis always states that all the population means are equal, while the alternative hypothesis states that at least two population means are different.

The most common type of one-way ANOVA is the completely randomized design. Consider the data in the following table, randomly selected golf scores of three individuals.

Bob	Brian	John
93	85	80
98	87	88
89	82	84
90	80	82

One-way AVOVA can be applied to determine whether there is a statistically significant difference between Bob's, Brian's, and John's average golf scores. If the golf scores are assigned randomly within each of the three samples, this is considered a completely randomized design.

The second type of one-way AVOVA is the randomized block design, which is demonstrated in the following table.

Course	Bob	Brian	John
1	93	85	80
2	98	87	88
3	89	82	84
4	90	80	82

Now each row of golf scores is associated with a particular golf course. You've added another variable to the analysis, called a *blocking variable*, which provides additional context and changes the ANOVA procedure.

You don't know anything about the scores, such as how current they are or what course they were scored on. These are just four random scores from each golfer.

13.2 Identify the three conditions that must be met to perform analysis of variance tests.

In order to conduct ANOVA tests, the data sets must be normally distributed, must be independent, and must have equal population variances.

Note: Problems 13.3–13.7 refer to the data set below, the satisfaction ratings recorded by 15 customers for three different fast-food chains on a scale of 1 to 10.

Chain 1	Chain 2	Chain 3
7	8	9
7	9	7
6	7	8
5	6	10
3	9	8

13.3 State the one-way analysis of variance hypothesis and calculate the total sum of squares.

This problem considers three populations, so the null hypothesis is that all three population means are equal. If they are not equal, then at least two of them are different; this is the alternative hypothesis.

$$H_0 : \mu_1 = \mu_2 = \mu_3$$
$$H_1 : \text{At least two population means are different.}$$

> Each store chain represents a population of satisfaction ratings, so there are three populations.

Let x_i represent the ith data observation and n_T represent the total number of data observations. The total sum of squares (SST) is the total variation in the data set, and it is calculated using the formula below.

$$SST = \sum_{i=1}^{n_T} x_i^2 - \frac{\left(\sum_{i=1}^{n_T} x_i\right)^2}{n_T}$$

> Different books have different-looking SST equations because they alter them algebraically. They all give the same SST values, though.

The following table lists the ratings (x_i) and their squares (x_i^2).

x_i	x_i^2	x_i	x_i^2	x_i	x_i^2
7	49	8	64	9	81
7	49	9	81	7	49
6	36	7	49	8	64
5	25	6	36	10	100
3	9	9	81	8	64

Calculate the sum of the $n_T = 15$ data values and the sum of the squared data: $\sum x_i = 109$ and $\sum_{i=1}^{n_T} x_i^2 = 837$. Substitute these into the total sum of squares formula.

$$SST = \sum_{i=1}^{n_T} x_i^2 - \frac{\left(\sum_{i=1}^{n_T} x_i\right)^2}{n_T} = 837 - \frac{(109)^2}{15} = 837 - 792.07 = 44.93$$

Note: Problems 13.3–13.7 refer to the data set in Problem 13.3, the satisfaction ratings recorded by 15 customers for three different fast-food chains on a scale of 1 to 10.

13.4 Partition the total sum of squares SST, calculated in Problem 13.3, into the sum of squares within (SSW) and the sum of squares between (SSB).

The total sum of squares can be separated, or partitioned, into the sum of squares within (SSW) and the sum of squares between (SSB): SST = SSW + SSB. The sum of squares between measures the variation of the sample means with respect to the overall (or grand) mean.

If all the samples from the population had the same mean, the SSB would equal zero.

Define the following variables: n_i is the sample size of the ith sample, $\bar{x}_i$ is the mean of the ith sample, and $\bar{\bar{x}}$ is the grand mean, the average of all $n_T = 15$ data values. Apply the formula below to calculate the sum of squares between.

$$SSB = \sum_{i=1}^{k} n_i \left(\bar{x}_i - \bar{\bar{x}}\right)^2$$

Some books call the SSB the treatment sum of squares (SSTR).

To calculate the SSB, first calculate the means of all three samples.

$$\bar{x}_1 = \frac{7+7+6+5+3}{5} = \frac{28}{5} = 5.6$$
$$\bar{x}_2 = \frac{8+9+7+6+9}{5} = \frac{39}{5} = 7.8$$
$$\bar{x}_3 = \frac{9+7+8+10+8}{5} = \frac{42}{5} = 8.4$$

If the sample sizes aren't equal, add up all of the data and divide by n_T.

Next, identify the grand mean. Because the sample sizes are equal, you can calculate the mean of the sample means.

$$\bar{\bar{x}} = \frac{\bar{x}_1 + \bar{x}_2 + \bar{x}_3}{3} = \frac{5.6+7.8+8.4}{3} = \frac{21.8}{3} = 7.27$$

Substitute the means into the SSB formula, multiplying each sample size by the difference between the sample mean and the grand mean and then adding those products together. This example includes three different samples, so three products are summed.

$$SSB = \sum_{i=1}^{k} n_i \left(\bar{x}_i - \bar{\bar{x}} \right)^2$$

$$= (5)(5.6 - 7.27)^2 + (5)(7.8 - 7.27)^2 + (5)(8.4 - 7.27)^2$$

$$= (5)(-1.67)^2 + (5)(0.53)^2 + (5)(1.13)^2$$

$$= 13.95 + 1.40 + 6.38$$

$$= 21.73$$

The sum of squares within (SSW) measures the variation of each data point with respect to its sample mean. In the formula below, n_i is the size and s_i^2 is the variance of the nth sample.

$$SSW = \sum_{i=1}^{k} (n_i - 1) s_i^2$$

Rather than calculate the standard deviation of each sample, recall that the SST is equal to the sum of SSW and SSB. According to Problem 13.3, SST = 44.93.

$$SST = SSW + SSB$$

$$44.93 = SSW + 21.73$$

$$44.93 - 21.73 = SSW$$

$$23.2 = SSW$$

Note: Problems 13.3–13.7 refer to the data set in Problem 13.3, the satisfaction ratings recorded by 15 customers for three different fast-food chains on a scale of 1 to 10.

13.5 Perform a hypothesis test to determine whether there is a difference in the customer satisfaction ratings of the stores at the $\alpha = 0.05$ significance level.

Consider the hypotheses stated in Problem 13.3 and the values SSB = 21.73 and SSW = 23.2 calculated in Problem 13.4 for the $k = 3$ populations with a total of $n_T = 15$ data values.

> The variable k represents the number of populations you're comparing. Some books call it the number of levels or number of treatments.

To test the hypothesis for ANOVA, you will apply the F-distribution. The F-score for the data is the quotient of the mean square between (MSB) and the mean square within (MSW), as defined below.

$$MSB = \frac{SSB}{k-1} = \frac{21.73}{3-1} = 10.87$$

$$MSW = \frac{SSW}{n_T - k} = \frac{23.2}{15-3} = 1.93$$

Calculate the F-score of the data.

$$F = \frac{MSB}{MSW} = \frac{10.87}{1.93} = 5.63$$

As was the case in Chapter 12, the critical F-score F_c requires D_1 and D_2, two different degrees of freedom, as defined below.

$$D_1 = k - 1 \quad = 3 - 1 = 2$$
$$D_2 = n_T - k = 15 - 3 = 12$$

Locate the 0.05 section of Reference Table 4. The critical F-score is in the $D_1 = 2$ column and the $D_2 = 12$ row: $F_c = 3.885$. Because $F = 5.63$ is greater than $F_c = 3.885$, you reject H_0. The means of the populations do not appear to be equal; there is a difference in customer satisfaction between the chains.

> The F-score has to be greater than F_c to reject an ANOVA hypothesis.

Note: Problems 13.3–13.7 refer to the data set in Problem 13.3, the satisfaction ratings recorded by 15 customers for three different fast-food chains on a scale of 1 to 10.

13.6 Construct a one-way ANOVA table for completely randomized design summarizing the findings in Problems 13.3–13.5.

A one-way ANOVA table is often generated by statistical computer software and takes the following form.

Source of Variation	SS	df	MS	F
Between Samples	SSB	$k - 1$	MSB	F
Within Samples	SSW	$n_T - k$	MSW	
Total	**SST**	$n_T - 1$		

Substitute the values calculated in Problems 13.3–13.5 into the ANOVA table.

Source of Variation	SS	df	MS	F
Between Samples	21.73	2	10.87	5.63
Within Samples	23.2	12	1.93	
Total	**44.93**	**14**		

Note: Problems 13.3–13.7 refer to the data set in Problem 13.3, the satisfaction ratings recorded by 15 customers for three different fast-food chains on a scale of 1 to 10.

13.7 Perform Scheffé's pairwise comparison test to identify the population means that are not equal at the $\alpha = 0.05$ significance level.

> You should use the same alpha value as in the original ANOVA test.

Problem 13.5 only states that at least two population means are different—it does not identify those populations. Scheffé's test uses F_s values to compare two sample means, a and b, using the following formula.

$$F_s = \frac{\left(\bar{x}_a - \bar{x}_b \right)^2}{MSW \left(\dfrac{1}{n_a} + \dfrac{1}{n_b} \right)}$$

Compare the sample means for Chain 1 and Chain 2. ⟵

$$F_S = \frac{\left(\overline{x}_1 - \overline{x}_2\right)^2}{\text{MSW}\left(\dfrac{1}{n_1} + \dfrac{1}{n_2}\right)} = \frac{(5.6 - 7.8)^2}{(1.93)\left(\dfrac{1}{5} + \dfrac{1}{5}\right)} = 6.27$$

There are three fast-food chains and you are comparing them two at a time. You'll have to do this a total of $_3C_2 = 3$ times.

Compare the sample means for Chain 1 and Chain 3.

$$F_S = \frac{\left(\overline{x}_1 - \overline{x}_3\right)^2}{\text{MSW}\left(\dfrac{1}{n_1} + \dfrac{1}{n_3}\right)} = \frac{(5.6 - 8.4)^2}{(1.93)\left(\dfrac{1}{5} + \dfrac{1}{5}\right)} = 10.16$$

Compare the sample means for Chain 2 and Chain 3.

$$F_S = \frac{\left(\overline{x}_2 - \overline{x}_3\right)^2}{\text{MSW}\left(\dfrac{1}{n_2} + \dfrac{1}{n_3}\right)} = \frac{(7.8 - 8.4)^2}{(1.93)\left(\dfrac{1}{5} + \dfrac{1}{5}\right)} = 0.47$$

The critical value F_{SC} for Scheffé's test is the product of F_c (the critical F-score from the ANOVA test) and $(k - 1)$.

$$F_{SC} = (k - 1)F_c = (3 - 1)(3.885) = 7.770$$

If $F_S \leq F_{SC}$, you conclude there is no difference between sample means; otherwise, there is a difference. The following table summarizes the findings of Scheffé's pairwise comparison test for this problem.

Sample Pair	F_S	F_{SC}	Conclusion
1 and 2	6.27	7.770	No difference
1 and 3	10.16	7.770	Difference
2 and 3	0.47	7.770	No difference

According to Scheffé's pairwise comparison test, Chains 1 and 3 have statistically significant differences in mean customer satisfaction scores.

Note: Problems 13.8–13.12 refer to the data set below, the gas mileage of three different cars based on random samples.

Model 1	Model 2	Model 3
24.5	23.7	17.2
20.8	19.8	18.0
22.6	24.0	21.1
23.6	23.1	19.8
21.0	24.9	

The sample sizes don't have to be equal for a completely randomized ANOVA. When they're different, it's called an unbalanced design.

13.8 Calculate the total sum of squares.

The following table lists the squares of the data values.

x_i	x_i^2	x_i	x_i^2	x_i	x_i^2
24.5	600.25	23.7	561.69	17.2	295.84
20.8	432.64	19.8	392.04	18.0	324.00
22.6	510.76	24.0	576.00	21.1	445.21
23.6	556.96	23.1	533.61	19.8	392.04
21.0	441.00	24.9	620.01		

The sum of the $n_T = 14$ data values is $\sum x_i = 304.1$ and the sum of the squared data is $\sum_{i=1}^{n_T} x_i^2 = 6,682.05$. Calculate SST.

$$\text{SST} = \sum_{i=1}^{n_T} x_i^2 - \frac{\left(\sum_{i=1}^{n_T} x_i\right)^2}{n_T} = 6,682.05 - \frac{(304.1)^2}{14} = 6,682.05 - 6,605.49 = 76.56$$

Note: Problems 13.8–13.12 refer to the data set in Problem 13.8, the gas mileage of three different cars based on random samples.

13.9 Partition the total sum of squares, calculated in Problem 13.8, into the sum of squares within (SSW) and the sum of squares between (SSB).

Calculate the mean of each sample.

$$\bar{x}_1 = \frac{24.5+20.8+22.6+23.6+21.0}{5} = \frac{112.5}{5} = 22.5$$

$$\bar{x}_2 = \frac{23.7+19.8+24.0+23.1+24.9}{5} = \frac{115.5}{5} = 23.1$$

$$\bar{x}_3 = \frac{17.2+18.0+21.1+19.8}{4} = \frac{76.1}{4} = 19.025$$

The sample sizes are different, so apply a weighted average to calculate the grand mean.

$$\overline{\overline{x}} = \frac{n_1 \overline{x}_1 + n_2 \overline{x}_2 + n_3 \overline{x}_3}{n_T} = \frac{(5)(22.5) + (5)(23.1) + (4)(19.025)}{14} = \frac{304.1}{14} = 21.72$$

Substitute these values into the sum of squares between formula.

$$
\begin{aligned}
SSB &= \sum_{i=1}^{k} n_i \left(\overline{x}_i - \overline{\overline{x}} \right)^2 \\
&= (5)(22.5 - 21.72)^2 + (5)(23.1 - 21.72)^2 + (4)(19.025 - 21.72)^2 \\
&= (5)(0.78)^2 + (5)(1.38)^2 + (4)(-2.695)^2 \\
&= 3.04 + 9.52 + 29.05 \\
&= 41.61
\end{aligned}
$$

Calculate the sum of squares within by subtracting SSB from SST.

$$SSW = SST - SSB = 76.56 - 41.61 = 34.95$$

Note: Problems 13.8–13.12 refer to the data set Problem 13.8, the gas mileage of three different cars based on random samples.

13.10 Perform a hypothesis test to determine whether there is a difference in gas mileage between the three car models at the $\alpha = 0.05$ significance level.

State the null and alternative hypotheses.

$$H_0 : \mu_1 = \mu_2 = \mu_3$$

H_1 : At least two population means are different.

According to Problems 13.8 and 13.9, SSW = 34.95 and SSB = 41.61. There are $k = 3$ populations and $n_T = 14$ total observations. Calculate the mean square between (MSB) and mean square within (MSW); use these values to identify the F-score of the data.

$$MSB = \frac{SSB}{k-1} = \frac{41.61}{3-1} = 20.81$$

$$MSW = \frac{SSW}{n_T - k} = \frac{34.95}{14-3} = 3.18$$

$$F = \frac{MSB}{MSW} = \frac{20.81}{3.18} = 6.54$$

There are $D_1 = k - 1 = 2$ and $D_2 = n_T - k = 11$ degrees of freedom and $\alpha = 0.05$. According to Reference Table 4, $F_c = 3.982$. Because $F = 6.54$ is greater than $F_c = 3.982$, you reject H_0 and conclude that at least two of the three sample means are different.

Note: Problems 13.8–13.12 refer to the data set in Problem 13.8, the gas mileage of three different cars based on random samples.

13.11 Construct a one-way ANOVA table summarizing the findings in Problems 13.8–13.10.

Source of Variation	SS	df	MS	F
Between Samples	41.61	2	20.81	6.54
Within Samples	34.95	11	3.18	
Total	**76.56**	**13**		

Note: Problems 13.8–13.12 refer to the data set in Problem 13.8, the gas mileage of three different cars based on random samples.

13.12 Perform Scheffé's pairwise comparison test to identify the unequal means using $\alpha = 0.05$.

Compare two sample means at a time until all possible combinations are complete.

> The first formula compares Models 1 and 2, the second compares Models 1 and 3, and the third compares Models 2 and 3.

$$F_S = \frac{\left(\bar{x}_1 - \bar{x}_2\right)^2}{\text{MSW}\left(\dfrac{1}{n_1} + \dfrac{1}{n_2}\right)} = \frac{(22.5 - 23.1)^2}{(3.18)\left(\dfrac{1}{5} + \dfrac{1}{5}\right)} = 0.28$$

$$F_S = \frac{\left(\bar{x}_1 - \bar{x}_3\right)^2}{\text{MSW}\left(\dfrac{1}{n_1} + \dfrac{1}{n_3}\right)} = \frac{(22.5 - 19.025)^2}{(3.18)\left(\dfrac{1}{5} + \dfrac{1}{4}\right)} = 8.44$$

$$F_S = \frac{\left(\bar{x}_2 - \bar{x}_3\right)^2}{\text{MSW}\left(\dfrac{1}{n_2} + \dfrac{1}{n_3}\right)} = \frac{(23.1 - 19.025)^2}{(3.18)\left(\dfrac{1}{5} + \dfrac{1}{4}\right)} = 11.61$$

Calculate the critical value for Scheffé's test, F_{SC}.

$$F_{SC} = (k-1)F_c = (3-1)(3.982) = 7.964$$

The following table summarizes the findings of Scheffé's pairwise comparison test.

Sample Pair	F_S	F_{SC}	Conclusion
1 and 2	0.28	7.964	No difference
1 and 3	8.44	7.964	Difference
2 and 3	11.61	7.964	Difference

> Because Models 1 and 2 have higher sample means than Model 3—they can travel farther on a gallon of gas.

According to Scheffé's pairwise comparison test, the gas mileage for Model 3 is significantly different than Model 1 and Model 2. It appears that Model 1 and 2 both provide better gas mileage than Model 3.

Note: Problems 13.13–13.16 refer to the table below, a partially completed ANOVA hypothesis test using a completely randomized design.

Source of Variation	SS	df	MS	F
Between Samples		4	4	
Within Samples	56.68			
Total	72.11	24		

13.13 Determine the total number of observations in this ANOVA test.

The ANOVA test has df = $n_T - 1$ total degrees of freedom. According to the table, df = 24.

$$df = n_T - 1$$
$$24 = n_T - 1$$
$$24 + 1 = n_T$$
$$25 = n_T$$

There are a total of $n_T = 25$ observations.

Note: Problems 13.13–13.16 refer to the table in Problem 13.13, a partially completed ANOVA hypothesis test using a completely randomized design.

13.14 How many populations are compared in the ANOVA test?

The table states that there are $D_1 = 4$ degrees of freedom between samples. Recall that $D_1 = k - 1$.

$$D_1 = k - 1$$
$$4 = k - 1$$
$$4 + 1 = k$$
$$5 = k$$

$D_2 = n_T - k$ represents the degrees of freedom within samples.

A total of $k = 5$ populations are compared.

Note: Problems 13.13–13.16 refer to the table in Problem 13.13, a partially completed ANOVA hypothesis test using a completely randomized design.

13.15 Complete the ANOVA table.

The total sum of squares is the sum of the squares between and the squares within.

$$SST = SSB + SSW$$
$$72.11 = SSB + 56.68$$
$$72.11 - 56.68 = SSB$$
$$15.43 = SSB$$

Once you have partitioned SST into SSB and SSW, you can calculate the mean square between, the mean square within, and the F-score.

$$MSB = \frac{SSB}{k-1} = \frac{15.43}{5-1} = 3.86$$

$$MSW = \frac{SSW}{n_T - k} = \frac{56.68}{25-5} = 2.83$$

$$F = \frac{MSB}{MSW} = \frac{3.86}{2.83} = 1.36$$

Complete the table using the values calculated above.

Source of Variation	SS	df	MS	F
Between Samples	15.43	4	3.86	1.36
Within Samples	56.68	20	2.83	
Total	72.11	24		

Note: Problems 13.13–13.16 refer to the table in Problem 13.13, a partially completed ANOVA hypothesis test using a completely randomized design.

13.16 Use the completed ANOVA table from Problem 13.15 to draw conclusions about the hypothesis test, given $\alpha = 0.10$.

State the null and alternative hypotheses.

$$H_0 : \mu_1 = \mu_2 = \mu_3 = \mu_4 = \mu_5$$
$$H_1 : \text{At least two population means are different.}$$

Problem 3.14 said there were five populations, so the null hypothesis is that all five population means are equal.

Calculate the degrees of freedom for the critical F-score. Note that $D_1 = 4$ and $D_2 = 20$. According to Reference Table 4, the critical F-score is $F_c = 2.249$. Because $F = 1.36$ is less than F_c, you fail to reject H_0; the population means are not different.

Note: Problems 13.17–13.21 refer to the data set below, the amount of bananas sold per week (in pounds) at a grocery store when the banana display was located in the produce, milk, and cereal sections of the store.

Produce	Milk	Cereal
61	39	26
40	18	55
65	32	53
50	55	50
	39	

13.17 Calculate the total sum of squares.

Square each of the data values.

x_i	x_i^2	x_i	x_i^2	x_i	x_i^2
61	3,721	39	1,521	26	676
40	1,600	18	324	55	3,025
65	4,225	32	1,024	53	2,809
50	2,500	55	3,025	50	2,500
		39	1,521		

The sum of the $n_T = 13$ data values is $\sum x_i = 583$; the sum of the squared data is $\sum_{i=1}^{n_T} x_i^2 = 28,471$. Calculate SST.

$$\text{SST} = \sum_{i=1}^{n_T} x_i^2 - \frac{\left(\sum_{i=1}^{n_T} x_i\right)^2}{n_T} = 28,471 - \frac{(583)^2}{13} = 28,471 - 26,145.31 = 2,325.69$$

> SST is always a positive number, because squared numbers are positive and SST is a sum of squared numbers. Okay, technically 0^2 isn't positive, but you know what I mean.

Note: Problems 13.17–13.21 refer to the data set in Problem 13.17, the amount of bananas sold per week (in pounds) at a grocery store when the banana display was located in the produce, milk, and cereal sections of the store.

13.18 Partition the total sum of squares into the sum of squares within (SSW) and the sum of squares between (SSB).

Make sample 1 the produce section, sample 2 the milk section, and sample 3 the cereal section. Calculate the sample means.

$$\bar{x}_1 = \frac{61+40+65+50}{4} = \frac{216}{4} = 54$$

$$\bar{x}_2 = \frac{39+18+32+55+39}{5} = \frac{183}{5} = 36.6$$

$$\bar{x}_3 = \frac{26+55+53+50}{4} = \frac{184}{4} = 46$$

The grand mean is the weighted mean of $\bar{x}_1$, $\bar{x}_2$, and $\bar{x}_3$.

$$\bar{\bar{x}} = \frac{n_1\bar{x}_1 + n_2\bar{x}_2 + n_3\bar{x}_3}{n_T} = \frac{(4)(54)+(5)(36.6)+(4)(46)}{13} = \frac{583}{13} = 44.85$$

Calculate the sum of squares between.

$$SSB = \sum_{i=1}^{k} n_i \left(\bar{x}_i - \bar{\bar{x}} \right)^2$$

$$= (4)(54.0 - 44.85)^2 + (5)(36.6 - 44.85)^2 + (4)(46.0 - 44.85)^2$$

$$= (4)(9.15)^2 + (5)(-8.25)^2 + (4)(1.15)^2$$

$$= 334.89 + 340.31 + 5.29$$

$$= 680.49$$

Subtract SSB from SST to calculate the sum of squares within:
SSW = 2,325.69 − 680.49 = 1,645.2.

Note: Problems 13.17–13.21 refer to the data set in Problem 13.17, the amount of bananas sold per week (in pounds) at a grocery store when the banana display was located in the produce, milk, and cereal sections of the store.

13.19 Perform a hypothesis test to determine whether mean banana sales differ based upon their location, assuming $\alpha = 0.01$.

State the null and alternative hypotheses.

$$H_0 : \mu_1 = \mu_2 = \mu_3$$

H_1 : At least two population means are different.

Calculate the mean square between, the mean square within, and the *F*-score of the data.

$$MSB = \frac{SSB}{k-1} = \frac{680.49}{3-1} = 340.25$$

$$MSW = \frac{SSW}{n_T - k} = \frac{1,645.2}{13-3} = 164.52$$

$$F = \frac{MSB}{MSW} = \frac{340.25}{164.52} = 2.07$$

The formulas are $D_1 = k - 1$ and $D_2 = n_T - k$.

The critical *F*-score has $D_1 = 3 - 1 = 2$ and $D_2 = 13 - 3 = 10$ degrees of freedom. Because $F = 2.07$ is less than $F_c = 7.559$ (the critical *F*-score from Reference Table 4), you fail to reject H_0. Banana sales seem to be the same regardless of where in the store they are displayed.

Note: Problems 13.17–13.21 refer to the data set in Problem 13.17, the amount of bananas sold per week (in pounds) at a grocery store when the banana display was located in the produce, milk, and cereal sections of the store.

13.20 Construct a one-way ANOVA table summarizing the data.

Source of Variation	SS	df	MS	F
Between Samples	680.49	2	340.25	2.07
Within Samples	1,645.2	10	164.52	
Total	**2,325.69**	**12**		

In this problem, the samples themselves contain too much variation (MSW) compared to the variation between the samples (MSB). Thus, you are not able to reject the null hypothesis and cannot conclude that the population means are different.

> MSW is the denominator of the F-score. The bigger MSW is, the smaller the F-score will be, and the less likely it becomes that F will be larger than F_c.

Note: Problems 13.17–13.21 refer to the data set in Problem 13.17, the amount of bananas sold per week (in pounds) at a grocery store when the banana display was located in the produce, milk, and cereal sections of the store.

13.21 Perform Scheffé's pairwise comparison test to verify that none of the population means are different.

Compare the sample means, two at a time. Recall that sample 1 is the produce section, sample 2 is the milk section, and sample 3 is the cereal section.

$$F_S = \frac{\left(\bar{x}_1 - \bar{x}_2\right)^2}{\text{MSW}\left(\dfrac{1}{n_1} + \dfrac{1}{n_2}\right)} = \frac{(54 - 36.6)^2}{(164.52)\left(\dfrac{1}{4} + \dfrac{1}{5}\right)} = 4.09$$

$$F_S = \frac{\left(\bar{x}_1 - \bar{x}_3\right)^2}{\text{MSW}\left(\dfrac{1}{n_1} + \dfrac{1}{n_3}\right)} = \frac{(54 - 46)^2}{(164.52)\left(\dfrac{1}{4} + \dfrac{1}{4}\right)} = 0.78$$

$$F_S = \frac{\left(\bar{x}_2 - \bar{x}_3\right)^2}{\text{MSW}\left(\dfrac{1}{n_2} + \dfrac{1}{n_3}\right)} = \frac{(36.6 - 46)^2}{(164.52)\left(\dfrac{1}{5} + \dfrac{1}{4}\right)} = 1.19$$

Calculate the critical value of Scheffé's test.

$$F_{SC} = (k-1)F_c = (3-1)(7.559) = 15.118$$

According to Scheffé's pairwise comparison test, summarized in the following table, none of the population means are significantly different. Each value of F_S is less than $F_{SC} = 15.118$.

Sample Pair	F_s	F_{sc}	Conclusion
1 and 2	4.09	15.118	No difference
1 and 3	0.78	15.118	No difference
2 and 3	1.19	15.118	No difference

Note: Problems 13.22–13.25 refer to the table below, a partially completed ANOVA hypothesis test using a completely randomized design.

Source of Variation	SS	df	MS	F
Between Samples	419.25			
Within Samples		19		
Total	**887.06**	**22**		

13.22 Determine the total number of observations in this ANOVA test.

The ANOVA test has df = 22 total degrees of freedom. Recall that the number of total observations is exactly one more than the degrees of freedom (df = $n_T - 1$), so there are a total of $n_T = 22 + 1 = 23$ observations.

Note: Problems 13.22–13.25 refer to the table in Problem 13.22, a partially completed ANOVA hypothesis test using a completely randomized design.

13.23 Determine the total number of populations compared in this ANOVA test.

The total degrees of freedom is equal to the sum of the between (D_1) and within (D_2) degrees of freedom. The table states that $D_2 = 19$.

$$df = D_1 + D_2$$
$$22 = D_1 + 19$$
$$22 - 19 = D_1$$
$$3 = D_1$$

Recall that $D_1 = k - 1$. Substitute $D_1 = 3$ into the equation and solve for k, the total number of populations.

$$D_1 = k - 1$$
$$3 = k - 1$$
$$3 + 1 = k$$
$$4 = k$$

Note: Problems 13.22–13.25 refer to the table in Problem 13.22, a partially completed ANOVA hypothesis test using a completely randomized design.

13.24 Complete the ANOVA table.

The total sum of squares is the sum of the squares between and the squares within.

$$SST = SSB + SSW$$
$$887.06 = 419.25 + SSW$$
$$887.06 - 419.25 = SSB$$
$$467.81 = SSB$$

Once you have partitioned SST into SSB and SSW, you can calculate the mean square between, the mean square within, and the *F*-score.

$$MSB = \frac{SSB}{k-1} = \frac{419.25}{4-1} = 139.75$$

$$MSW = \frac{SSW}{n_T - k} = \frac{467.81}{23-4} = 24.62$$

$$F = \frac{MSB}{MSW} = \frac{139.75}{24.62} = 5.68$$

Source of Variation	SS	df	MS	F
Between Samples	419.25	3	139.75	5.68
Within Samples	467.81	19	24.62	
Total	**887.06**	**22**		

Note: Problems 13.22–13.25 refer to the table in Problem 13.22, a partially completed ANOVA hypothesis test using a completely randomized design.

13.25 Based on the completed ANOVA table in Problem 13.24, state the conclusions of the hypothesis test if $\alpha = 0.01$.

State the null and alternative hypotheses.

$$H_0 : \mu_1 = \mu_2 = \mu_3 = \mu_4$$
$$H_1 : \text{At least two population means are different.}$$

The critical *F*-score given $D_1 = 3$ and $D_2 = 19$ degrees of freedom and a significance level of $\alpha = 0.10$ is $F_c = 5.010$. Because $F = 5.67$ is greater than $F_c = 5.010$, you reject H_0 and conclude that at least one pair of sample means is significantly different. ←

Without the actual samples and their sample means, you won't be able to compare them and figure out which of the four populations have different means.

Note: Problems 13.26–13.30 refer to the data set below, customer satisfaction ratings at a retail store on a scale of 1 to 20 when three different environmental scents were used.

Lavender	Citrus	Vanilla
13	18	12
16	20	16
18	15	10
16	15	15
15	19	14
	18	12

13.26 Calculate the total sum of squares.

Square each of the data values.

x_i	x_i^2	x_i	x_i^2	x_i	x_i^2
13	169	18	324	12	144
16	256	20	400	16	256
18	324	15	225	10	100
16	256	15	225	15	225
15	225	19	361	14	196
		18	324	12	144

The sum of the $n_T = 17$ data values is $\sum x_i = 262$ and the sum of the squares is $\sum_{i=1}^{n_T} x_i^2 = 4{,}154$; calculate SST.

$$\text{SST} = \sum_{i=1}^{n_T} x_i^2 - \frac{\left(\sum_{i=1}^{n_T} x_i\right)^2}{n_T} = 4{,}154 - \frac{(262)^2}{17} = 4{,}154 - 4{,}037.88 = 116.12$$

Note: Problems 13.26–13.30 refer to the data in Problem 13.26, customer satisfaction ratings at a retail store on a scale of 1 to 20 when three different environmental scents were used.

13.27 Partition the total sum of squares calculated in Problem 13.26 into the sum of squares within (SSW) and the sum of squares between (SSB).

Assume lavender is sample 1, citrus is sample 2, and vanilla is sample 3. Calculate the sample means.

$$\bar{x}_1 = \frac{13+16+18+16+15}{5} = \frac{78}{5} = 15.6$$

$$\bar{x}_2 = \frac{18+20+15+15+19+18}{6} = \frac{105}{6} = 17.5$$

$$\bar{x}_3 = \frac{12+16+10+15+14+12}{6} = \frac{79}{6} = 13.17$$

Now calculate the grand mean.

$$\bar{\bar{x}} = \frac{n_1\bar{x}_1 + n_2\bar{x}_2 + n_3\bar{x}_3}{n_T} = \frac{(5)(15.6)+(6)(17.5)+(6)(13.17)}{17} = \frac{262.02}{17} = 15.41$$

Square the difference between each sample mean and the grand mean and multiply it by the sample size. The sum of these values is the sum of squares between.

$$SSB = \sum_{i=1}^{k} n_i\left(\bar{x}_i - \bar{\bar{x}}\right)^2$$

$$= (5)(15.6-15.41)^2 + (6)(17.5-15.41)^2 + (6)(13.17-15.41)^2$$

$$= (5)(0.19)^2 + (6)(2.09)^2 + (6)(-2.24)^2$$

$$= 0.18 + 26.21 + 30.11$$

$$= 56.5$$

According to Problem 13.26, SST = 116.12.

$$SSW = SST - SSB = 116.12 - 56.5 = 59.62$$

Note: Problems 13.26–13.30 refer to the data in Problem 13.26, customer satisfaction ratings at a retail store on a scale of 1 to 20 when three different environmental scents were used.

13.28 Perform a hypothesis test at the $\alpha = 0.05$ significance level to determine whether customer satisfaction rating means change according to the environmental scent used.

State the null and alternative hypotheses.

$$H_0 : \mu_1 = \mu_2 = \mu_3$$

H_1 : At least two population means are different.

According to Problems 13.26–13.27, SSB = 56.5 and SSW = 59.62 for the $n_T = 17$ data points in $k = 3$ different populations. Calculate the mean square between, the mean square within, and the corresponding F-score.

$$MSB = \frac{SSB}{k-1} = \frac{56.5}{3-1} = 28.25$$

$$MSW = \frac{SSW}{n_T - k} = \frac{59.62}{17-3} = 4.26$$

$$F = \frac{MSB}{MSW} = \frac{28.25}{4.26} = 6.63$$

Given $D_1 = k - 1 = 2$ and $D_2 = n_T - k = 14$ degrees of freedom and $\alpha = 0.05$, the critical F-score is $F_c = 3.739$. Because $F > F_c$, you reject H_0 and conclude that the population means are different.

Note: Problems 13.26–13.30 refer to the data in Problem 13.26, customer satisfaction ratings at a retail store on a scale of 1 to 20 when three different environmental scents were used.

13.29 Construct a one-way ANOVA table summarizing the findings in Problems 13.26–13.28.

Source of Variation	SS	df	MS	F
Between Samples	56.5	2	28.25	6.63
Within Samples	59.62	14	4.26	
Total	116.12	16		

Note: Problems 13.26–13.30 refer to the data in Problem 13.26, customer satisfaction ratings at a retail store on a scale of 1 to 20 when three different environmental scents were used.

13.30 Perform Scheffé's pairwise comparison test to identify the unequal population means when $\alpha = 0.05$.

Recall that lavender is sample 1, citrus is sample 2, and vanilla is sample 3.

$$F_S = \frac{(\bar{x}_1 - \bar{x}_2)^2}{\text{MSW}\left(\frac{1}{n_1} + \frac{1}{n_2}\right)} = \frac{(15.6 - 17.5)^2}{(4.26)\left(\frac{1}{5} + \frac{1}{6}\right)} = 2.31$$

$$F_S = \frac{(\bar{x}_1 - \bar{x}_3)^2}{\text{MSW}\left(\frac{1}{n_1} + \frac{1}{n_3}\right)} = \frac{(15.6 - 13.17)^2}{(4.26)\left(\frac{1}{5} + \frac{1}{6}\right)} = 3.78$$

$$F_S = \frac{(\bar{x}_2 - \bar{x}_3)^2}{\text{MSW}\left(\frac{1}{n_2} + \frac{1}{n_3}\right)} = \frac{(17.5 - 13.17)^2}{(4.26)\left(\frac{1}{6} + \frac{1}{6}\right)} = 13.20$$

Calculate the critical value F_{SC}.

$$F_{SC} = (k - 1)F_c = (3 - 1)(3.739) = 7.478$$

According to Scheffé's pairwise comparison test, the only significantly different means are sample 2 (citrus) and sample 3 (vanilla).

Sample Pair	F_s	F_{sc}	Conclusion
1 and 2	2.31	7.478	No difference
1 and 3	3.78	7.478	No difference
2 and 3	13.20	7.478	Difference

One-Way ANOVA: Randomized Block Design

Adding a blocking variable to the test

Note: Problems 13.31–13.37 refer to the data set below, golf scores for 3 people at 4 different golf courses.

Course	Bob	Brian	John
1	93	85	80
2	98	87	88
3	89	82	84
4	90	84	82

> There's an extra column in this table (compared to the tables in Problems 13.3–13.30). You've got three columns of golf score data and one column (the blocking variable) that provides context—the course on which each score was recorded.

13.31 Calculate the total sum of squares (SST).

Randomized block design and completely random design SST values are calculated in exactly the same way (a technique first demonstrated in Problem 13.3). Square each of the data values, calculate the sum of the data and the sum of the squares, and then substitute those sums into the same SST formula that was applied in the preceding section.

x_i	x_i^2	x_i	x_i^2	x_i	x_i^2
93	8,649	85	7,225	80	6,400
98	9,604	87	7,569	88	7,744
89	7,921	82	6,724	84	7,056
90	8,100	84	7,056	82	6,724

The sum of the $n_T = 12$ data values is $\sum x_i = 1,042$; the sum of the squares is $\sum_{i=1}^{n_T} x_i^2 = 90,772$. Calculate the total sum of squares.

$$\text{SST} = \sum_{i=1}^{n_T} x_i^2 - \frac{\left(\sum_{i=1}^{n_T} x_i\right)^2}{n_T} = 90,772 - \frac{(1,042)^2}{12} = 90,772 - 90,480.33 = 291.67$$

You'll calculate SSB just like you did way back in Problem 13.4. There's no change, even though you're using the randomized block design now. SSW, however, is calculated slightly differently, as you'll see in Problem 13.33.

Note: Problems 13.31–13.37 refer to the data set in Problem 13.31, golf scores for 3 people at 4 different golf courses.

13.32 Calculate the sum of squares between (SSB).

Assume Bob's scores are sample 1, Brian's are sample 2, and John's are sample 3. Calculate the sample means.

$$\bar{x}_1 = \frac{93+98+89+90}{4} = \frac{370}{4} = 92.5$$

$$\bar{x}_2 = \frac{85+87+82+84}{4} = \frac{338}{4} = 84.5$$

$$\bar{x}_3 = \frac{80+88+84+82}{4} = \frac{334}{4} = 83.5$$

The sample sizes are equal, so the grand mean is the average of the sample means.

$$\bar{\bar{x}} = \frac{92.5+84.5+83.5}{3} = \frac{260.5}{3} = 86.83$$

Substitute the above values into the SSB formula.

$$SSB = \sum_{i=1}^{k} n_i \left(\bar{x}_i - \bar{\bar{x}} \right)^2$$

$$= (4)(92.5-86.83)^2 + (4)(84.5-86.83)^2 + (4)(83.5-86.83)^2$$

$$= (4)(5.67)^2 + (4)(-2.33)^2 + (4)(-3.33)^2$$

$$= 128.60 + 21.72 + 44.36$$

$$= 194.68$$

Note: Problems 13.31–13.37 refer to the data set in Problem 13.31, golf scores for 3 people at 4 different golf courses.

13.33 Calculate the sum of squares for the blocking variable (SSBL) and the sum of squares within (SSW).

The blocking variable in this data set is the golf course on which each score was recorded. Randomized block design is used to determine whether the variation in course averages has an impact on the sample means.

Assume $\bar{x}_j$ is the average of the *j*th block, *b* is the number of blocks, *k* is the number of populations, and $\bar{\bar{x}}$ is the grand mean. The sum of squares for the blocking variable is calculated using the formula below.

$$SSBL = \sum_{j=1}^{b} k \left(\bar{x}_j - \bar{\bar{x}} \right)^2$$

This data set contains information about $b = 4$ different courses and $k = 3$ different players.

The samples each represent a population. There are 3 samples (golf scores from 3 golfers) listed in 4 different blocks (each row represents a different block, a different golf course).

Calculate the mean of each block, in this case the average score at each course.

$$\bar{x}_1 = \frac{93+85+80}{3} = \frac{258}{3} = 86 \qquad \bar{x}_2 = \frac{98+87+88}{3} = \frac{273}{3} = 91$$

$$\bar{x}_3 = \frac{89+82+84}{3} = \frac{255}{3} = 85 \qquad \bar{x}_4 = \frac{90+84+82}{3} = \frac{256}{3} = 85.33$$

Square the difference between each mean and the grand mean $\bar{\bar{x}} = 86.83$ and multiply it by k, the number of populations. The sum of those products is the sum of squares for the blocking variable, SSBL.

> The grand mean is not the average of the four means you just calculated. It's still the value from Problem 13.32.

$$SSBL = \sum_{j=1}^{b} k\left(\bar{x}_j - \bar{\bar{x}}\right)^2$$

$$= (3)(86-86.83)^2 + (3)(91-86.83)^2 + (3)(85-86.83)^2 + (3)(85.33-86.83)^2$$

$$= (3)(-0.83)^2 + (3)(4.17)^2 + (3)(-1.83)^2 + (3)(-1.5)^2$$

$$= 2.07 + 52.17 + 10.05 + 6.75$$

$$= 71.04$$

The total sum of squares for a randomized block design is the sum of squares within, the sum of squares between, and the sum of squares of the blocking variable.

$$SST = SSW + SSB + SSBL$$

Calculate SSW.

$$291.67 = SSW + 194.68 + 71.04$$
$$291.67 = SSW + 265.72$$
$$291.67 - 265.72 = SSW$$
$$25.95 = SSW$$

Note: Problems 13.31–13.37 refer to the data set in Problem 13.31, golf scores for 3 people at 4 different golf courses.

13.34 Perform a hypothesis test to determine whether the blocking variable was effective in the ANOVA at the $\alpha = 0.05$ level of significance.

Randomized block design consists of two hypothesis tests. The first is the primary hypothesis test for a difference in population means that was explored in Problems 13.3–13.30. This familiar test will be revisited in Problem 13.35 for this set of data.

The second hypothesis test for randomized block design investigates the effectiveness of the blocking variable. Like the primary test, its hypotheses are stated in a standard way: the null hypothesis claims the blocking means are equal and the alternative hypothesis makes the opposite claim.

> Use a prime symbol (it looks like an apostrophe) on the null and alternative hypotheses for the second test, so you can tell the two sets of hypotheses apart. Some textbooks skip this secondary hypothesis step altogether.

H_0': The blocking means are all equal.

H_1': Not all of the blocking means are equal.

If you reject this secondary null hypothesis, you conclude that the blocking variable is effective and should be used for the primary hypothesis test. If, however, you fail to reject this null hypothesis, the blocking variable should be removed and the completely randomized design should be used to test the primary hypothesis.

Calculate the mean square within (MSW), the mean square blocking (MSBL), and the corresponding F'-score using the formulas below.

$$\text{MSBL} = \frac{\text{SSBL}}{b-1} = \frac{71.04}{4-1} = 23.68$$

$$\text{MSW} = \frac{\text{SSW}}{(k-1)(b-1)} = \frac{25.95}{(3-1)(4-1)} = 4.33$$

$$F' = \frac{\text{MSBL}}{\text{MSW}} = \frac{23.68}{4.33} = 5.47$$

In order to identify the critical F'-score in Reference Table 4, you must first compute the degrees of freedom.

$$D_1' = b - 1 = 4 - 1 = 3$$

$$D_2' = (k-1)(b-1) = (3-1)(4-1) = 6$$

The critical F' score is 4.757. Because $F' = 5.47$ is greater than $F_c' = 4.757$, you reject H_0' and conclude that the blocking means are not equal. The courses themselves have an effect on the averages. Course 2, for instance, was the most difficult for all four golfers.

Note: Problems 13.31–13.37 refer to the data set in Problem 13.31, golf scores for 3 people at 4 different golf courses.

13.35 Perform a hypothesis test to determine whether the players' mean golf scores are different at a $\alpha = 0.05$ significance level.

Having completed Problem 13.34, you can now proceed to the primary hypothesis test, determining whether the golf scores have different population means.

$$H_0 : \mu_1 = \mu_2 = \mu_3$$

H_1 : At least two population means are different.

According to Problems 13.33–13.34, SSB = 194.68, MSW = 4.33, and $k = 3$. Calculate the mean square between (MSB) using the same formula applied in completely randomized design.

$$\text{MSB} = \frac{\text{SSB}}{k-1} = \frac{194.68}{3-1} = 97.34$$

Calculate the corresponding *F*-score.

$$F = \frac{MSB}{MSW} = \frac{97.34}{4.33} = 22.48$$

Calculate the degrees of freedom for the critical *F*-score.

$$D_1 = k - 1 = 3 - 1 = 2$$
$$D_2 = (k-1)(b-1) = (3-1)(4-1) = 6$$

According to Reference Table 4, $F_c = 5.143$. Because $F = 22.48$ is greater than $F_c = 5.143$, you reject H_0; there is a difference in average golf score.

Note: Problems 13.31–13.37 refer to the data set in Problem 13.31, golf scores for 3 people at 4 different golf courses.

13.36 Construct a one-way ANOVA table for a randomized block design summarizing the findings of Problems 13.31–13.35.

In the table below, rows represent variation in the blocking variable, columns represent variation between samples, and errors represent variation within samples.

Source of Variation	SS	df	MS	F
Rows	SSBL	$b - 1$	MSBL	$\dfrac{MSBL}{MSW}$
Columns	SSB	$k - 1$	MSB	$\dfrac{MSB}{MSW}$
Errors	SSW	$(b-1)(k-1)$	MSW	
Total	**SST**	$n_T - 1$		

Complete the table by substituting the values calculated in Problems 13.31–13.35.

Source of Variation	SS	df	MS	F
Rows	71.04	3	23.68	5.47
Columns	194.68	2	97.34	22.48
Errors	25.95	6	4.33	
Total	**291.67**	**11**		

MSB is much higher than MSW, which results in a large F-score (because F = MSB ÷ MSW). There's much more variation between the golfers than within the individual golfers' scores, which means the population means aren't equal.

Note: Problems 13.31–13.37 refer to the data set in Problem 13.31, golf scores for 3 people at 4 different golf courses.

13.37 Use Tukey's method to identify the unequal population means, assuming $\alpha = 0.05$.

Tukey's method is a good pairwise comparison test for the means in a randomized block design. Begin by calculating the following degrees of freedom.

$$D_1 = k = 3$$
$$D_2 = (k-1)(b-1) = (3-1)(4-1) = (2)(3) = 6$$

Consult Reference Table 5, using the first section of the table ($\alpha = 0.05$) to identify the cell at which $D_1 = 3$ and $D_2 = 6$ intersect. That critical value is $q_\alpha = 4.339$. Recall that MSW = 4.33 and $b = 4$ for this data set and apply the following formula to calculate the critical range for Tukey's method.

$$\text{critical range} = q_\alpha \sqrt{\frac{\text{MSW}}{b}} = 4.339 \sqrt{\frac{4.33}{4}} = 4.51$$

To determine which pairs of means are significantly different, compare the critical range to the absolute value of the difference in sample means. According to Problem 13.32, $\bar{x}_1 = 92.5$, $\bar{x}_2 = 84.5$, and $\bar{x}_3 = 83.5$. If the difference in sample means exceeds the critical range, the means are significantly different.

Absolute Difference	Critical Range	Conclusion
$\left\|\bar{x}_1 - \bar{x}_2\right\| = \left\|92.5 - 84.5\right\| = \left\|8\right\| = 8$	4.51	Difference
$\left\|\bar{x}_1 - \bar{x}_3\right\| = \left\|92.5 - 83.5\right\| = \left\|9\right\| = 9$	4.51	Difference
$\left\|\bar{x}_2 - \bar{x}_3\right\| = \left\|84.5 - 83.5\right\| = \left\|1\right\| = 1$	4.51	No difference

You can conclude that Bob and Brian (samples 1 and 2, respectively) have different average golf scores, as do Bob and John (samples 1 and 3, respectively). There is no significant difference between the mean golf scores of Brian and John.

Note: Problems 13.38–13.41 refer to the table below, a partially completed ANOVA hypothesis test using a randomized block design.

Source of Variation	SS	df	MS	F
Rows		6		
Columns	23	3		
Error	105			
Total	269	27		

13.38 Calculate the total number of blocking levels in this ANOVA test.

The rows have 6 degrees of freedom. According to Problem 13.36, the formula for this cell is $b - 1$.

$$b - 1 = 6$$
$$b = 6 + 1$$
$$b = 7$$

There are a total of $b = 7$ blocking levels in this ANOVA test.

> The blocking levels are the rows of original data, like the 4 rows of golf course data in Problems 13.31-13.37.

Note: Problems 13.38–13.41 refer to the table in Problem 13.38, a partially completed ANOVA hypothesis test using a randomized block design.

13.39 Complete the ANOVA table.

The partially completed table contains the following information: SSB = 23, SSW = 105, SST = 269, $b - 1 = 6$, $k - 1 = 3$, and $n_T - 1 = 27$. Given SST, SSW, and SSB, you can calculate SSBL.

$$SST = SSW + SSB + SSBL$$
$$269 = 105 + 23 + SSBL$$
$$269 - 105 - 23 = SSBL$$
$$141 = SSBL$$

> Refer back to Problem 13.36 and compare each of the values present to the ANOVA table template. For instance, SS intersects Columns at cell SSB.

Similarly, there are $27 - 6 - 3 = 18$ degrees of freedom in the Error row. Calculate the mean square blocking, mean square between, and mean square within.

$$MSBL = \frac{SSBL}{b-1} = \frac{141}{7-1} = 23.5$$

$$MSB = \frac{SSB}{k-1} = \frac{23}{4-1} = 7.67$$

$$MSW = \frac{SSW}{(b-1)(k-1)} = \frac{105}{(7-1)(4-1)} = 5.83$$

Complete the table by calculating the *F*-scores.

$$F' = \frac{MSBL}{MSW} = \frac{23.5}{5.83} = 4.03$$

$$F = \frac{MSB}{MSW} = \frac{7.67}{5.83} = 1.32$$

Source of Variation	SS	df	MS	F
Rows	141	6	23.5	4.03
Columns	23	3	7.67	1.32
Error	105	18	5.83	
Total	**269**	**27**		

Note: Problems 13.38–13.41 refer to the table in Problem 13.38, a partially completed ANOVA hypothesis test using a randomized block design.

13.40 Determine whether the blocking variable was effective given $\alpha = 0.05$.

> Use primes on the hypotheses that test the effectiveness of a blocking variable.

State the null and alternative hypotheses.

$$H_0' : \text{The blocking means are all equal.}$$
$$H_1' : \text{Not all of the blocking means are equal.}$$

Consult the table completed in Problem 13.39 to identify the F'-score: $F' = 4.03$. There are $D_1' = 6$ and $D_2' = 18$ degrees of freedom. According to Reference Table 4, $F_c' = 2.661$. Because $F' > F_c'$, you reject H_0'; the blocking means are different and the blocking variable is effective.

Note: Problems 13.38–13.41 refer to the table in Problem 13.38, a partially completed ANOVA hypothesis test using a randomized block design.

13.41 State the conclusions of the primary hypothesis test given $\alpha = 0.05$.

State the null and alternative hypotheses.

$$H_0 : \mu_1 = \mu_2 = \mu_3 = \mu_4$$
$$H_1 : \text{At least two population means are different.}$$

> Notice that $D_2 = 18$ in both Problems 13.40 and 13.41, prime or no prime.

The F-score for the primary hypothesis, according to the completed table in Problem 13.39, is $F = 1.32$. There are $D_1 = 3$ and $D_2 = 18$ degrees of freedom. According to Reference Table 4, $F_c = 3.160$. Because $F < F_c$, you fail to reject H_0 and conclude that there is no difference between the population means. Even though the blocking variable was effective (according to Problem 13.40), the primary hypothesis can still be rejected.

> Secret shoppers are hired by retail chains to act like regular customers but secretly rate their experiences in the stores.

Note: Problems 13.42–13.48 refer to the data set below, secret shopper ratings for the cleanliness of three retail stores on a scale of 1 to 100. Each secret shopper rated all three stores.

Shopper	Store 1	Store 2	Store 3
1	75	81	75
2	82	85	88
3	72	70	74
4	90	89	88
5	64	90	77

13.42 Calculate the total sum of squares.

Square each of the data values.

x_i	x_i^2	x_i	x_i^2	x_i	x_i^2
75	5,625	81	6,561	75	5,625
82	6,724	85	7,225	88	7,744
72	5,184	70	4,900	74	5,476
90	8,100	89	7,921	88	7,744
64	4,096	90	8,100	77	5,929

The sum of the $n_T = 15$ data values is $\sum x_i = 1,200$; the sum of the squares is $\sum x_i^2 = 96,954$. Calculate SST.

$$SST = \sum_{i=1}^{n_T} x_i^2 - \frac{\left(\sum_{i=1}^{n_T} x_i\right)^2}{n_T} = 96,954 - \frac{(1,200)^2}{15} = 96,954 - 96,000 = 954$$

> The sample sizes are equal (each secret shopper rated each store), so you don't need to calculate a weighted mean to get the grand mean.

Note: Problems 13.42–13.48 refer to the data set in Problem 13.42, secret shopper ratings for the cleanliness of three retail stores on a scale of 1 to 100. Each secret shopper rated all three stores.

13.43 Calculate the sum of squares between.

Calculate the sample means; their average is the grand mean.

$$\bar{x}_1 = \frac{75 + 82 + 72 + 90 + 64}{5} = \frac{383}{5} = 76.6$$

$$\bar{x}_2 = \frac{81 + 85 + 70 + 89 + 90}{5} = \frac{415}{5} = 83$$

$$\bar{x}_3 = \frac{75 + 88 + 74 + 88 + 77}{5} = \frac{402}{5} = 80.4$$

$$\bar{\bar{x}} = \frac{\bar{x}_1 + \bar{x}_2 + \bar{x}_3}{3} = \frac{76.6 + 83 + 80.4}{3} = \frac{240}{3} = 80$$

Calculate SSB.

$$SSB = \sum_{i=1}^{k} n_i \left(\bar{x}_i - \bar{\bar{x}}\right)^2$$
$$= (5)(76.6 - 80)^2 + (5)(83 - 80)^2 + (5)(80.4 - 80)^2$$
$$= (5)(-3.4)^2 + (5)(3)^2 + (5)(0.4)^2$$
$$= 57.8 + 45 + 0.8$$
$$= 103.6$$

Note: Problems 13.42–13.48 refer to the data set in Problem 13.42, secret shopper ratings for the cleanliness of three retail stores on a scale of 1 to 100. Each secret shopper rated all three stores.

13.44 Calculate the sum of squares for the blocking variable (SSBL) and the sum of squares within (SSW).

Each of the $b = 5$ shoppers rated $k = 3$ stores. Calculate the mean of each block.

$$\bar{x}_1 = \frac{75+81+75}{3} = \frac{231}{3} = 77 \qquad \bar{x}_2 = \frac{82+85+88}{3} = \frac{255}{3} = 85$$

$$\bar{x}_3 = \frac{72+70+74}{3} = \frac{216}{3} = 72 \qquad \bar{x}_4 = \frac{90+89+88}{3} = \frac{267}{3} = 89$$

$$\bar{x}_5 = \frac{64+90+77}{3} = \frac{231}{3} = 77$$

Calculate SSBL.

$$\text{SSBL} = \sum_{j=1}^{b} k\left(\bar{x}_j - \bar{\bar{x}}\right)^2$$

$$= (3)(77-80)^2 + (3)(85-80)^2 + (3)(72-80)^2 + (3)(89-80)^2 + (3)(77-80)^2$$

$$= (3)(-3)^2 + (3)(5)^2 + (3)(-8)^2 + (3)(9)^2 + (3)(-3)^2$$

$$= 27 + 75 + 192 + 243 + 27$$

$$= 564$$

Given the total sum of squares and two of its three partitions, you can calculate the remaining partition, SSW.

$$\text{SSW} = \text{SST} - \text{SSB} - \text{SSBL} = 954 - 103.6 - 564 = 286.4$$

Note: Problems 13.42–13.48 refer to the data set in Problem 13.42, secret shopper ratings for the cleanliness of three retail stores on a scale of 1 to 100. Each secret shopper rated all three stores.

13.45 Perform a hypothesis test to determine whether the blocking variable was effective in the ANOVA model, assuming $\alpha = 0.05$.

State the secondary hypotheses for the effectiveness of the blocking variable.

H_0' : The blocking means are all equal.

H_1' : Not all of the blocking means are equal.

SSBL = 564 and SSW = 286.4, according to Problem 13.44.

Calculate the mean square within, the mean square blocking, and the corresponding F'-score.

$$\text{MSBL} = \frac{\text{SSBL}}{b-1} = \frac{564}{5-1} = 141$$

$$\text{MSW} = \frac{\text{SSW}}{(k-1)(b-1)} = \frac{286.4}{(3-1)(5-1)} = 35.8$$

$$F' = \frac{\text{MSBL}}{\text{MSW}} = \frac{141}{35.8} = 3.94$$

Given $D_1' = 5 - 1 = 4$ and $D_2' = (3-1)(5-1) = 8$ degrees of freedom (and $\alpha = 0.05$), the critical F-score is 3.838. Because $F' = 3.94$ is greater than $F_c' = 3.838$, you reject H_0' and conclude that the blocking means are different. The blocking variable (secret shopper) is effective in this model.

Note: Problems 13.42–13.48 refer to the data set in Problem 13.42, secret shopper ratings for the cleanliness of three retail stores on a scale of 1 to 100. Each secret shopper rated all three stores.

13.46 Perform a hypothesis test to determine whether the three stores have different average ratings, assuming $\alpha = 0.05$.

State the primary hypotheses.

$H_0 : \mu_1 = \mu_2 = \mu_3$

H_1 : At least two population means are different.

Calculate the mean square between and the corresponding F-score. Recall that SSB = 103.6 and MSW = 35.8.

$$\text{MSB} = \frac{\text{SSB}}{k-1} = \frac{103.6}{3-1} = 51.8$$

$$F = \frac{\text{MSB}}{\text{MSW}} = \frac{51.8}{35.8} = 1.45$$

Given $D_1 = 3 - 1 = 2$ and $D_2 = (3-1)(5-1) = 8$ degrees of freedom, the critical F-score is 4.459. Because $F = 1.45$ is less than $F_c = 4.459$, you fail to reject H_0 and conclude that the stores' average ratings are not different.

Note: Problems 13.42–13.48 refer to the data set in Problem 13.42, secret shopper ratings for the cleanliness of three retail stores on a scale of 1 to 100. Each secret shopper rated all three stores.

13.47 Construct a one-way ANOVA table summarizing the findings of Problems 13.42–13.46.

Source of Variation	SS	df	MS	F
Rows	564	4	141	3.94
Columns	103.6	2	51.8	1.45
Errors	286.4	8	35.8	
Total	954	14		

Note: Problems 13.42–13.48 refer to the data set in Problem 13.42, secret shopper ratings for the cleanliness of three retail stores on a scale of 1 to 100. Each secret shopper rated all three stores.

13.48 Confirm that no pairs of store means are different using Tukey's method and $\alpha = 0.05$.

Tukey's method uses $D_1 = k = 3$ and $D_2 = (k-1)(b-1) = (3-1)(5-1) = 8$ degrees of freedom. According to Reference Table 5, $q_\alpha = 4.041$. Calculate the critical range.

$$\text{critical range} = q_\alpha \sqrt{\frac{\text{MSW}}{b}} = 4.041 \sqrt{\frac{35.8}{5}} = 10.81$$

The following sample means were calculated in Problem 13.43.

$$\bar{x}_1 = 76.6 \qquad \bar{x}_2 = 83.0 \qquad \bar{x}_3 = 80.4$$

Calculate the absolute difference between each pair of sample means and compare those differences to the critical range.

Absolute Difference	Critical Range	Conclusion
$\lvert \bar{x}_1 - \bar{x}_2 \rvert = \lvert 76.6 - 83 \rvert = \lvert -6.4 \rvert = 6.4$	10.81	No difference
$\lvert \bar{x}_1 - \bar{x}_3 \rvert = \lvert 76.6 - 80.4 \rvert = \lvert -3.8 \rvert = 3.8$	10.81	No difference
$\lvert \bar{x}_2 - \bar{x}_3 \rvert = \lvert 83 - 80.4 \rvert = \lvert 2.6 \rvert = 2.6$	10.81	No difference

Each absolute difference is less than the critical range of 10.81, so no pair of stores has significantly different ratings. This verifies the conclusion reached in Problem 13.46.

Note: Problems 13.49–13.50 refer to the table below, an ANOVA hypothesis test using a randomized block design.

Source of Variation	SS	df	MS	F
Rows	60	4	15	2.5
Columns	150	5	30	5.45
Error	110	20	5.5	
Total	**320**	**29**		

13.49 Determine whether the blocking variable was effective given $\alpha = 0.10$.

State the hypotheses.

$$H_0' : \text{The blocking means are all equal.}$$
$$H_1' : \text{Not all of the blocking means are equal.}$$

According to the table, $F' = 2.5$. There are $D_1' = 4$ and $D_2' = 20$ degrees of freedom, so $F_c' = 4.432$. Because $F' < F_c'$, you fail to reject H_0' and conclude that the blocking variable is not effective.

Note: Problems 13.49–13.50 refer to the table in Problem 13.49, an ANOVA hypothesis test using a randomized block design.

13.50 State the conclusions of the primary hypothesis test given $\alpha = 0.10$.

State the primary null and alternative hypotheses.

$$H_0 : \mu_1 = \mu_2 = \mu_3 = \mu_4 = \mu_5 = \mu_6$$
$$H_1 : \text{At least two population means are different.}$$

According to the table, $F = 5.45$. There are $D_1 = 5$ and $D_2 = 20$ degrees of freedom, so the critical F-score is 4.103. Because $F > F_c$, you reject H_0 and conclude that the population means are different. ←

You can still reject the primary null hypothesis even though you couldn't reject the null hypothesis in Problem 13.51.

Note: Problems 13.51–13.58 refer to the data set below, the volume of grass clippings cut from identically sized areas of lawn with different types of fertilizer applied.

Lawn	Fertilizer 1	Fertilizer 2	Fertilizer 3	Fertilizer 4
1	10	12	8	10
2	9	13	9	8
3	8	9	13	9
4	11	10	9	9
5	8	13	11	11
6	8	12	10	10

13.51 Calculate the total sum of squares.

The sum of the $n_T = 24$ data values is $\sum x_i = 240$; the sum of the squares is $\sum x_i^2 = 2{,}464$. Calculate SST.

$$\text{SST} = \sum_{i=1}^{n_T} x_i^2 - \frac{\left(\sum_{i=1}^{n_T} x_i\right)^2}{n_T} = 2{,}464 - \frac{(240)^2}{24} = 2{,}464 - 2{,}400 = 64$$

Note: Problems 13.51–13.58 refer to the data set in Problem 13.51, the volume of grass clippings cut from identically sized areas of lawn with different types of fertilizer applied.

13.52 Calculate the sum of squares between.

Calculate the sample means and the grand mean.

$$\bar{x}_1 = \frac{10+9+8+11+8+8}{6} = \frac{54}{6} = 9 \qquad \bar{x}_2 = \frac{12+13+9+10+13+12}{6} = \frac{69}{6} = 11.5$$

$$\bar{x}_3 = \frac{8+9+13+9+11+10}{6} = \frac{60}{6} = 10 \qquad \bar{x}_4 = \frac{10+8+9+9+11+10}{6} = \frac{57}{6} = 9.5$$

The grand mean of the data is the average of the sample means.

$$\bar{\bar{x}} = \frac{\bar{x}_1 + \bar{x}_2 + \bar{x}_3 + \bar{x}_4}{4} = \frac{9.0+11.5+10.0+9.5}{4} = 10$$

Calculate SSB.

$$SSB = \sum_{i=1}^{k} n_i \left(\bar{x}_i - \bar{\bar{x}}\right)^2$$

$$= (6)(9-10)^2 + (6)(11.5-10)^2 + (6)(10-10)^2 + (6)(9.5-10.0)^2$$

$$= (6)(-1)^2 + (6)(1.5)^2 + (6)(0)^2 + (6)(-0.5)^2$$

$$= 6+13.5+0+1.5$$

$$= 21$$

Note: Problems 13.51–13.58 refer to the data set in Problem 13.51, the volume of grass clippings cut from identically sized areas of lawn with different types of fertilizer applied.

13.53 Calculate the sum of squares for the blocking variable and the sum of squares within.

There are $b = 6$ lawns, each of which is treated with $k = 4$ fertilizers. Calculate the mean of each block.

$$\bar{x}_1 = \frac{10+12+8+10}{4} = \frac{40}{4} = 10 \qquad \bar{x}_2 = \frac{9+13+9+8}{4} = \frac{39}{4} = 9.75$$

$$\bar{x}_3 = \frac{8+9+13+9}{4} = \frac{39}{4} = 9.75 \qquad \bar{x}_4 = \frac{11+10+9+9}{4} = \frac{39}{4} = 9.75$$

$$\bar{x}_5 = \frac{8+13+11+11}{4} = \frac{43}{4} = 10.75 \qquad \bar{x}_6 = \frac{8+12+10+10}{4} = \frac{40}{4} = 10$$

Calculate SSBL.

$$SSBL = \sum_{j=1}^{b} k \left(\bar{x}_j - \bar{\bar{x}} \right)^2$$

$$= (4)(10-10)^2 + (4)(9.75-10)^2 + (4)(9.75-10)^2 +$$

$$(4)(9.75-10)^2 + (4)(10.75-10)^2 + (4)(10-10)^2$$

$$= (4)(0)^2 + (4)(-0.25)^2 + (4)(-0.25)^2 + (4)(-0.25)^2 + (4)(0.75)^2 + (4)(0)^2$$

$$= 0 + 0.25 + 0.25 + 0.25 + 2.25 + 0$$

$$= 3$$

Subtract SSB and SSBL from SST to calculate SSW.

$$SSW = SST - SSB - SSBL = 64 - 21 - 3 = 40$$

Note: Problems 13.51–13.58 refer to the data set in Problem 13.51, the volume of grass clippings cut from identically sized areas of lawn with different types of fertilizer applied.

13.54 Perform a hypothesis test to determine whether the blocking variable was effective in the ANOVA model when $\alpha = 0.05$.

State the secondary hypotheses.

H_0' : The blocking means are all equal.

H_1' : Not all of the blocking means are equal.

Calculate MSW, MSBL, and the corresponding F'-score.

$$MSBL = \frac{SSBL}{b-1} = \frac{3}{6-1} = 0.6$$

$$MSW = \frac{SSW}{(k-1)(b-1)} = \frac{40}{(4-1)(6-1)} = 2.67$$

$$F' = \frac{MSBL}{MSW} = \frac{0.6}{2.67} = 0.22$$

Given $D_1' = 6 - 1 = 5$ and $D_2' = (4-1)(6-1) = 15$ degrees of freedom, the critical F-score is $F_c' = 2.901$. Because $F' < F_c'$, you fail to reject H_0' and conclude that the blocking variable is not effective in this model.

Note: Problems 13.51–13.58 refer to the data set in Problem 13.51, the volume of grass clippings cut from identically sized areas of lawn with different types of fertilizer applied.

13.55 Perform a hypothesis test to determine whether there is a difference in average volume of grass clippings when $\alpha = 0.05$.

State the primary hypotheses.

$H_0 : \mu_1 = \mu_2 = \mu_3 = \mu_4$

H_1 : At least two population means are different.

Calculate MSB and the corresponding *F*-score.

$$MSB = \frac{SSB}{k-1} = \frac{21}{4-1} = 7$$

$$F = \frac{MSB}{MSW} = \frac{7}{2.67} = 2.62$$

Given $D_1 = 4 - 1 = 3$ and $D_2 = (4-1)(6-1) = 15$ degrees of freedom, the critical *F*-score is $F_c = 3.287$. Because $F < F_c$, you fail to reject H_0; the different fertilizers do not produce different volumes of grass.

When the primary null hypothesis is not rejected using a randomized block design, the hypothesis should be retested using a completely randomized design. Sometimes the blocking variable will mask a true difference in population means.

See the next problem for more information.

Note: Problems 13.51–13.58 refer to the data set in Problem 13.51, the volume of grass clippings cut from identically sized areas of lawn with different types of fertilizer applied.

13.56 Perform a hypothesis test using completely randomized design to determine whether there is a difference in average volume of grass clippings when $\chi = 0.05$.

State the primary null and alternative hypotheses.

$$H_0 : \mu_1 = \mu_2 = \mu_3 = \mu_4$$

$$H_1 : \text{At least two population means are different.}$$

There is no SSBL when you're ignoring the blocking variable.

Recall that the total sum of squares in a completely randomized design is equal to the sum of squares within plus the sum of squares between.

$$SST = SSB + SSW$$

$$64 = 21 + SSW$$

$$43 = SSW$$

Calculate MSW and the corresponding *F*-score.

$$MSW = \frac{SSW}{n_T - k} = \frac{43}{24-4} = 2.15$$

$$F = \frac{MSB}{MSW} = \frac{7}{2.15} = 3.26$$

Given $D_1 = k - 1 = 3$ and $D_2 = n_T - k = 20$ degrees of freedom, the critical *F*-score is $F_c = 3.098$. Because $F = 3.26$ is greater than $F_c = 3.098$, you reject H_0 and conclude that different fertilizers produce different volumes of grass. In this case, the blocking variable masked the difference in population means.

Note: Problems 13.51–13.58 refer to the data set in Problem 13.51, the volume of grass clippings cut from identically sized areas of lawn with different types of fertilizer applied.

13.57 Construct a one-way ANOVA table for a randomized block design summarizing Problems 13.51–13.55 and compare it to an ANOVA table summarizing Problem 13.56.

An ANOVA table for a randomized block design includes rows, columns, and errors as sources of variation.

Source of Variation	SS	df	MS	F
Rows	3	5	0.6	0.22
Columns	21	3	7	2.62
Errors	40	15	2.67	
Total	**64**	**23**		

The completely randomized design table combines the Rows and Errors rows into a row labeled "Within Samples."

Source of Variation	SS	df	MS	F
Between Samples	21	3	7	3.26
Within Samples	43	20	2.15	
Total	**64**	**23**		

Note: Problems 13.51–13.58 refer to the data set in Problem 13.51, the volume of grass clippings cut from identically sized areas of lawn with different types of fertilizer applied.

13.58 Apply Tukey's method to the completely randomized design to identify the significantly different means when $\alpha = 0.05$.

Tukey's method has been applied in preceding problems in order to analyze randomized block design, using the formula below.

$$\text{critical range} = q_\alpha \sqrt{\frac{\text{MSW}}{b}}$$

In those examples, b represented the number of blocking levels in the design. Notice that b describes the sample size of all samples if the design is balanced. Let n represent the shared sample size to modify Tukey's method for completely randomized design.

> In a balanced design, all of the sample sizes are equal.

$$\text{critical range} = q_\alpha \sqrt{\frac{\text{MSW}}{n}}$$

According to Problem 13.57, the $k = 4$ populations of sample size $n = 6$ result in $n_T = 24$ total data values. Given $D_1 = k = 4$ and $D_2 = n_T - 1 = 24 - 1 = 23$ degrees of freedom, Reference Table 5 states that $q_\alpha = q_{0.05} = 3.902$.

> These are not the same degrees of freedom you calculated in Problem 13.56, because in this case you're using Reference Table 5.

$$\text{critical range} = q_\alpha \sqrt{\frac{\text{MSW}}{b}} = 3.902 \sqrt{\frac{2.15}{6}} = 2.34$$

The following sample means were calculated in Problem 13.52.

$$\bar{x}_1 = 9 \qquad \bar{x}_2 = 11.5 \qquad \bar{x}_3 = 10 \qquad \bar{x}_4 = 9.5$$

Compare the absolute difference between pairs of sample means to the critical range in order to identify the unequal means, as demonstrated in the table below.

Absolute Difference	Critical Range	Conclusion
$\left\lvert \bar{x}_1 - \bar{x}_2 \right\rvert = \lvert 9 - 11.5 \rvert = \lvert -2.5 \rvert = 2.5$	2.34	Difference
$\left\lvert \bar{x}_1 - \bar{x}_3 \right\rvert = \lvert 9 - 10 \rvert = \lvert -1 \rvert = 1$	2.34	No Difference
$\left\lvert \bar{x}_1 - \bar{x}_4 \right\rvert = \lvert 9 - 9.5 \rvert = \lvert -0.5 \rvert = 0.5$	2.34	No Difference
$\left\lvert \bar{x}_2 - \bar{x}_3 \right\rvert = \lvert 11.5 - 10 \rvert = \lvert 1.5 \rvert = 1.5$	2.34	No Difference
$\left\lvert \bar{x}_2 - \bar{x}_4 \right\rvert = \lvert 11.5 - 9.5 \rvert = \lvert 2 \rvert = 2$	2.34	No Difference
$\left\lvert \bar{x}_3 - \bar{x}_4 \right\rvert = \lvert 10 - 9.5 \rvert = \lvert 0.5 \rvert = 0.5$	2.34	No Difference

Only the absolute difference between samples 1 and 2 exceeds the critical range, so Fertilizers 1 and 2 produce significantly different average volumes of grass.

Chapter 14
CORRELATION AND SIMPLE REGRESSION ANALYSIS

Finding relationships between two variables

This chapter explores the influence of one random variable on the value of a second variable. Correlation measures the strength and direction of the relationship between variables. Once the correlation between variables is identified, simple regression analysis can be used to construct a linear equation that models the relationship. Given a value of one variable, you can approximate the value of the correlated variable using the regression line, assuming that the correlation is linear.

This chapter is a hodgepodge of concepts you're already familiar with (if you've studied the previous chapters), including hypothesis tests, confidence intervals, Student's t-distribution, and the F-distribution. Your goal will be to find relationships between variables. Is an adult's height related to his shoe size? Does the amount of homework a teacher assigns relate to how well the students understand the concepts in the assignment? These are correlation questions.

Just remember, correlation does not imply causation. Just because two variables are related, it doesn't mean that some other unidentified factor isn't actually at work and influencing both of them.

Correlation
Describing the strength and direction of a relationship

14.1 Describe the difference between an independent variable and a dependent variable.

An independent variable x causes variation in the dependent variable y. This causal relationship exists only in one direction.

$$\text{independent variable } (x) \rightarrow \text{dependent variable } (y)$$

For example, the price of a used car is heavily influenced by the car's mileage. Thus, the mileage is the independent variable and the price of the car is the dependent variable. As the car's mileage increases, you would expect the price of the car to decrease.

> These are also called the explanatory (x) and response (y) variables. This may be more useful as it isn't always clear which variable depends on which.

14.2 Define *correlation* between two variables and explain how to interpret different values of r.

Correlation measures both the strength and direction of the relationship between two variables. The correlation coefficient r for an independent variable and a dependent variable is calculated using the following formula.

$$r = \frac{n\sum xy - \left(\sum x\right)\left(\sum y\right)}{\sqrt{\left[n\sum x^2 - \left(\sum x\right)^2\right]\left[n\sum y^2 - \left(\sum y\right)^2\right]}}$$

Values for r range between -1.0 and 1.0. A positive r indicates a positive relationship between the two variables; as x increases, y also increases. For example, length of employment and salary usually exhibit a positive correlation. The longer you work for a company, the more you are likely to earn.

If the correlation coefficient r is negative, a negative relationship exists between the two variables; as x increases, y decreases. The age of a used car and its retail value are negatively correlated, because as a car's age increases, its retail value decreases.

The closer the correlation coefficient is to the boundaries of 1 and -1, the more strongly the variables are correlated. Consider the scatter plots of an independent variable x and a dependent variable y below.

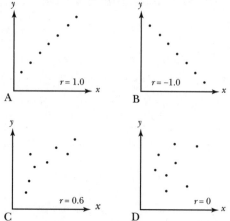

Graph A illustrates a perfect positive correlation between x and y with $r = 1.0$.
Graph B illustrates a perfect negative correlation between x and y with $r = -1.0$.
Graph C exhibits positive correlation that is weaker than Graph A
(as $r = 0.6 < 1$). Graph D exhibits no correlation between x and y, as $r = 0$.

> As x gets bigger (as you go right from the origin), y gets bigger (the points go up). Not only do they go up, but they appear to do so along a straight line in a perfectly predictable way.

14.3 Explain how to test the significance of the correlation coefficient.

The significance test for the correlation coefficient is used to determine whether the population correlation coefficient ρ is significantly different from zero based on the sample correlation coefficient r. The test uses the following hypotheses.

$$H_0 : \rho = 0$$
$$H_1 : \rho \neq 0$$

> A correlation coefficient of zero means that there is no relationship between x and y.

If you reject the null hypothesis, you conclude that a relationship exists between the two variables of interest. The test statistic for this hypothesis test uses the Student's t-distribution. Given a sample with size n and a correlation coefficient r, use the formula below to calculate the t-score.

$$t = \frac{r}{\sqrt{\dfrac{1 - r^2}{n - 2}}}$$

The calculated t-score is compared to the critical t-score t_c, which is based on $n - 2$ degrees of freedom and Reference Table 2.

> See Problem 9.25 to review the t-distribution.

Note: Problems 14.4–14.5 refer to the data set below, the number of hours six students studied for a final exam and their final exam scores.

Hours of Study	Exam Score
3	86
5	95
4	92
4	83
2	78
3	82

14.4 Calculate the correlation coefficient between hours studied and exam score.

Assign hours of study as the independent variable (x) and exam score as the dependent variable (y). Calculating the correlation coefficient requires a number of prerequisite calculations, completed in the table below.

> The longer you study, the higher your exam score should be. You have no direct control over the exam score because it depends on the amount of time you study (if a correlation exists, that is).

	Hours of Study	Exam Score			
	x	y	xy	x^2	y^2
	3	86	258	9	7,396
	5	95	475	25	9,025
	4	92	368	16	8,464
	4	83	332	16	6,889
	2	78	156	4	6,084
	3	82	246	9	6,724
Total	**21**	**516**	**1,835**	**79**	**44,582**

There are $n = 6$ pairs of data points. Calculate the correlation coefficient by substituting the sums calculated above into the formula presented in Problem 14.2.

$$r = \frac{n\sum xy - \left(\sum x\right)\left(\sum y\right)}{\sqrt{\left[n\sum x^2 - \left(\sum x\right)^2\right]\left[n\sum y^2 - \left(\sum y\right)^2\right]}}$$

$$= \frac{(6)(1,835) - (21)(516)}{\sqrt{\left[(6)(79) - (21)^2\right]\left[(6)(44,582) - (516)^2\right]}}$$

$$= \frac{11,010 - 10,836}{\sqrt{(474 - 441)(267,492 - 266,256)}}$$

$$= \frac{174}{\sqrt{(33)(1,236)}}$$

$$= 0.862$$

Note: Problems 14.4–14.5 refer to the data set in Problem 14.4, the number of hours six students studied for a final exam and their final exam scores.

14.5 Test the significance of the correlation coefficient between hours studied and exam score using $\alpha = 0.05$.

State the null and alternative hypotheses.

$$H_0 : \rho = 0$$
$$H_1 : \rho \neq 0$$

According to Problem 14.4, $r = 0.862$. Calculate the t–score for the correlation coefficient.

$$t = \frac{r}{\sqrt{\dfrac{1 - r^2}{n - 2}}} = \frac{0.862}{\sqrt{\dfrac{1 - (0.862)^2}{6 - 2}}} = \frac{0.862}{\sqrt{\dfrac{0.25696}{4}}} = \frac{0.862}{0.25346} = 3.40$$

There are df = $n - 2 = 6 - 2 = 4$ degrees of freedom and $\alpha = 0.05$. According to Reference Table 2, the critical *t*-scores are $t_c = \pm 2.776$. Because $t = 3.40$ is greater than $t_c = 2.776$, you reject the null hypothesis and conclude that the population correlation coefficient is not equal to zero. There appears to be a relationship between the number of hours studied and the resulting exam score.

> The null hypothesis contains "not equal," so this is a two-tailed test. There are two rejection regions—one left of $t = -2.776$ and one right of $t = 2.776$.

Note: Problems 14.6–14.7 refer to the data set below, the monthly demand for a specific computer printer at various price levels.

Demand	Price	Demand	Price
36	$70	14	$110
23	$80	10	$120
12	$90	5	$130
16	$100	2	$140

14.6 Calculate the correlation coefficient between demand and price.

While the retail store can set the price of a product, it cannot directly manipulate the demand. Thus, price is the independent variable and demand is the dependent variable.

> In this problem, you adjust something and stand back to see how it affects something else. The thing you adjust is the independent variable, and you look for results in the dependent variable.

Price (x)	Demand (y)			
x	y	xy	x^2	y^2
70	36	2,520	4,900	1,296
80	23	1,840	6,400	529
90	12	1,080	8,100	144
100	16	1,600	10,000	256
110	14	1,540	12,100	196
120	10	1,200	14,400	100
130	5	650	16,900	25
140	2	280	19,600	4
Total	**840**	**118**	**10,710**	**92,400**	**2,550**

Calculate the correlation coefficient.

$$r = \frac{n\sum xy - (\sum x)(\sum y)}{\sqrt{\left[n\sum x^2 - (\sum x)^2\right]\left[n\sum y^2 - (\sum y)^2\right]}}$$

$$= \frac{(8)(10,710) - (840)(118)}{\sqrt{\left[(8)(92,400) - (840)^2\right]\left[(8)(2,550) - (118)^2\right]}}$$

$$= \frac{85,680 - 99,120}{\sqrt{(739,200 - 705,600)(20,400 - 13,924)}}$$

$$= \frac{-13,440}{\sqrt{(33,600)(6,476)}}$$

$$= -0.911$$

The closer r is to 1 or −1, the more strongly the variables are correlated. Here, $r = -0.911$, so there is a very, very strong correlation in the negative direction.

The correlation coefficient is negative because as price of the printer increases, the demand decreases.

Note: Problems 14.6–14.7 refer to the data set in Problem 14.6, the monthly demand for a specific computer printer at various price levels.

14.7 Test the significance of the correlation coefficient between price and demand using $\alpha = 0.01$.

State the null and alternative hypotheses.

$$H_0 : \rho = 0$$
$$H_1 : \rho \neq 0$$

According to Problem 14.6, $r = -0.911$. Calculate the *t*-score for the correlation coefficient.

$$t = \frac{r}{\sqrt{\dfrac{1 - r^2}{n - 2}}} = \frac{-0.911}{\sqrt{\dfrac{1 - (0.911)^2}{8 - 2}}} = \frac{-0.911}{\sqrt{\dfrac{0.17008}{6}}} = -5.41$$

There are df $= n - 2 = 6$ degrees of freedom and $\alpha = 0.05$. According to Reference Table 2, the critical *t*-scores are $t_c = \pm 3.707$. Because $t = -5.41$ is less than $t_c = -3.707$, it lies in the left rejection region. You reject H_0 and conclude that there is a relationship between price and demand.

Note: Problems 14.8–14.9 refer to the data set below, the GMAT scores for five MBA students and the students' grade point averages (GPAs) upon graduation.

GMAT	GPA
660	3.7
580	3.0
480	3.2
710	4.0
600	3.5

14.8 Calculate the correlation coefficient between GMAT score and GPA.

Many schools admit students based on standardized test scores because they believe a correlation exists between the exam score and how well the student is likely to perform at the school. Thus, they believe that GMAT score is an independent variable that affects GPA, the dependent variable.

Calculate the sums of the variables, the squares, and the products of the paired data.

GMAT	GPA			
x	y	xy	x^2	y^2
660	3.7	2,442	435,600	13.69
580	3.0	1,740	336,400	9.0
480	3.2	1,536	230,400	10.24
710	4.0	2,840	504,100	16.00
600	3.5	2,100	360,000	12.25
Total 3,030	17.4	10,658	1,866,500	61.18

Calculate the correlation coefficient.

$$r = \frac{n\sum xy - \left(\sum x\right)\left(\sum y\right)}{\sqrt{\left[n\sum x^2 - \left(\sum x\right)^2\right]\left[n\sum y^2 - \left(\sum y\right)^2\right]}}$$

$$= \frac{(5)(10,658) - (3,030)(17.4)}{\sqrt{\left[(5)(1,866,500) - (3,030)^2\right]\left[(5)(61.18) - (17.4)^2\right]}}$$

$$= \frac{53,290 - 52,722}{\sqrt{(9,332,500 - 9,180,900)(305.90 - 302.76)}}$$

$$= \frac{568}{\sqrt{(151,600)(3.14)}}$$

$$= 0.823$$

Note: Problems 14.8–14.9 refer to the data set below, the GMAT scores for five MBA students and the students' grade point averages (GPAs) upon graduation.

14.9 Test the significance of the correlation coefficient between GMAT score and GPA using $\alpha = 0.10$.

State the null and alternative hypotheses.

$$H_0 : \rho = 0$$
$$H_1 : \rho \neq 0$$

According to Problem 14.8, $r = 0.823$. Calculate the *t*-score for the correlation coefficient.

> **A high correlation does not necessarily mean that one variable causes the effect on another. It just means the values move together.**

$$t = \frac{r}{\sqrt{\frac{1-r^2}{n-2}}} = \frac{0.823}{\sqrt{\frac{1-(0.823)^2}{5-2}}} = \frac{0.823}{\sqrt{\frac{0.322671}{3}}} = 2.51$$

Given df $= 5 - 2 = 3$ degrees of freedom and $\alpha = 0.10$, the critical *t*-scores are $t_c = \pm 2.353$. Because $t = 2.51$ is greater than $t_c = 2.353$, you reject the null hypothesis and conclude that GMAT score and graduating GPA are related.

Simple Regression Analysis

Line of best fit

14.10 Describe the procedure used to identify the line of best fit $\hat{y} = a + bx$, including the formulas for *a* and *b*.

> **A scatter plot is a collection of points (x, y), where x is the independent variable and y is the corresponding dependent variable.**

A simple linear regression is a straight line that best describes a scatter plot. Let *x* represent an independent variable value and $\hat{y}$ be the predicted dependent variable that corresponds to *x*. The regression equation for the data is $\hat{y} = a + bx$, where *a* is the *y*-intercept of the line and *b* is the slope.

Simple regression is also known as the least squares method. It requires separate formulas to calculate *a* and *b*, which are then substituted into the regression equation $\hat{y} = a + bx$. The *b* formula is listed first because it should be calculated first—once evaluated, it is substituted into the formula for *a*.

$$b = \frac{n\sum xy - \left(\sum x\right)\left(\sum y\right)}{n\sum x^2 - \left(\sum x\right)^2}$$

$$a = \frac{\sum y}{n} - b\left(\frac{\sum x}{n}\right)$$

The figure below illustrates a line of best fit for a scatter plot representing a set of ordered pairs.

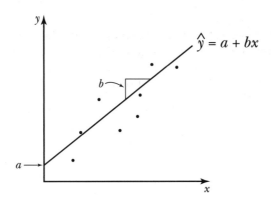

Note: Problems 14.11–14.21 refer to the data set below, the payroll (in millions of dollars) for eight Major League Baseball teams for a particular season and the number of times the teams won that season.

Payroll	Wins	Payroll	Wins
$209	89	$67	68
$139	74	$49	67
$101	86	$119	97
$74	74	$98	92

14.11 Identify the linear equation that best fits the data and interpret the results.

Because you expect an increase in team payroll to correlate with an increased number of wins, let payroll be the independent variable x and wins be the dependent variable y. The table below contains the sums necessary to complete the least squares calculations.

	Payroll	Wins			
	x	y	xy	x^2	y^2
	209	89	18,601	43,681	7,921
	139	74	10,286	19,321	5,476
	101	86	8,686	10,201	7,396
	74	74	5,476	5,476	5,476
	67	68	4,556	4,489	4,624
	49	67	3,283	2,401	4,489
	119	97	11,543	14,161	9,409
	98	92	9,016	9,604	8,464
Total	**856**	**647**	**71,447**	**109,334**	**53,255**

You'll need the y^2 column in Problem 4.13.

Calculate the slope b of the regression equation.

$$b = \frac{n\sum xy - \left(\sum x\right)\left(\sum y\right)}{n\sum x^2 - \left(\sum x\right)^2} = \frac{(8)(71,447) - (856)(647)}{(8)(109,334) - (856)^2} = \frac{571,576 - 553,832}{874,672 - 732,736} = \frac{17,744}{141,936} = 0.125$$

The value of b represents the expected increase in y when x increases by one unit. Thus, every \$1 million invested in payroll produces an average of 0.125 wins during the season. Calculate the y-intercept a for the regression equation.

The value of a represents the expected value for y if $x = 0$. In this problem, the y-intercept does not have any particular meaning, because a baseball team can't have a payroll of zero dollars.

$$a = \frac{\sum y}{n} - b\left(\frac{\sum x}{n}\right) = \frac{647}{8} - (0.125)\left(\frac{856}{8}\right) = 80.875 - 13.375 = 67.5$$

The line of best fit is $\hat{y} = 67.5 + 0.125x$.

Note: Problems 14.11–14.21 refer to the data set in Problem 14.11, the payroll (in millions of dollars) for eight Major League Baseball teams for a particular season and the number of times the teams won that season.

14.12 Predict the number of wins for a team that has invested \$90 million in payroll.

Salary is the independent variable x, so substitute $x = 90$ into the regression equation generated in Problem 14.13.

$$\hat{y} = 67.5 + 0.125x = 67.5 + 0.125(90) = 67.5 + 11.25 = 78.75 \approx 79 \text{ wins}$$

The model predicts that a team with a \$90 million payroll will have approximately 79 wins.

Note: Problems 14.11–14.21 refer to the data set in Problem 14.11, the payroll (in millions of dollars) for eight Major League Baseball teams for a particular season and the number of times the teams won that season.

14.13 Calculate the total sum of squares for the model.

The total sum of squares (SST) of the $n = 8$ pairs of data measures the total variation in the dependent variable y according to the following formula.

$$SST = \sum y^2 - \frac{\left(\sum y\right)^2}{n}$$

According to Problem 14.11, $\sum y = 647$ and $\sum y^2 = 53,255$. Calculate SST.

$$SST = \sum y^2 - \frac{\left(\sum y\right)^2}{n} = 53,255 - \frac{(647)^2}{8} = 53,255 - 52,326.125 = 928.88$$

Note: Problems 14.11–14.21 refer to the data set in Problem 14.11, the payroll (in millions of dollars) for eight Major League Baseball teams for a particular season and the number of times the teams won that season.

14.14 Partition the total sum of squares into the sum of squares regression (SSR) and the sum of squares error (SSE).

The total sum of squares can be partitioned into the sum of squares regression and the sum of squares error: SST = SSR + SSE. The sum of squares regression measures the variation in the dependent variable that is explained by the independent variable. The sum of squares error measures the variation in the dependent variable due to other unidentified variables and is calculated using the following formula. ←

$$SSE = \sum y^2 - a \sum y - b \sum xy$$

Calculate the sum of squares error. ←

$$
\begin{aligned}
SSE &= \sum y^2 - a \sum y - b \sum xy \\
&= 53,255 - (67.5)(647) - (0.125)(71,447) \\
&= 53,255 - 43,672.5 - 8,930.875 \\
&= 651.63
\end{aligned}
$$

> *Once you calculate SSE, you can subtract it from SST to get SSR. That's why the book only gives you a formula for SSE.*

Calculate the sum of squares regression.

$$SSR = SST - SSE = 928.88 - 651.63 = 277.25$$

> *The sums, as well as a and b, are calculated in Problems 14.11 and 14.13.*

Note: Problems 14.11–14.21 refer to the data set in Problem 14.11, the payroll (in millions of dollars) for eight Major League Baseball teams for a particular season and the number of times the teams won that season.

14.15 Calculate the coefficient of determination for the model.

The coefficient of determination R^2 is a value between 0 and 1 that measures how well a regression model predicts the data. If $R^2 = 1$, the model predicts the data perfectly. To calculate R^2, divide the sum of squares regression by the total sum of squares.

$$R^2 = \frac{SSR}{SST} = \frac{277.25}{928.88} = 0.298$$

In the baseball regression model, 29.8% of the variation in wins is explained by team payroll. ←

> *R^2 also tells you how much of the variation in the dependent variable is explained by the independent variable.*

Note: Problems 14.11–14.21 refer to the data set in Problem 14.11, the payroll (in millions of dollars) for eight Major League Baseball teams for a particular season and the number of times the teams won that season.

14.16 Test the significance of the coefficient of determination calculated in Problem 14.17 using $\alpha = 0.05$.

The coefficient of determination significance test is used to verify that the population coefficient of determination ρ^2 is significantly different from zero, based on the sample coefficient of determination R^2.

$$H_0 : \rho^2 = 0$$
$$H_1 : \rho^2 \neq 0$$

If the population coefficient of determination equals zero, there is no relationship between payroll and wins because none of the variation in wins is explained by payroll.

The test statistic for this hypothesis test is the F-score, as calculated below.

$$F = \frac{\text{SSR}}{\left(\dfrac{\text{SSE}}{n-2}\right)} = \frac{277.25}{\left(\dfrac{651.63}{8-2}\right)} = \frac{277.25}{108.605} = 2.55$$

$D_1 = 1$ when you test the significance of the coefficient of determination. There's no formula—it's just one.

There are $D_1 = 1$ and $D_2 = n - 2 = 6$ degrees of freedom for the critical F-score F_c. Given $\alpha = 0.05$, $F_c = \pm 5.987$, according to Reference Table 4.

Because $F = 2.55$ is neither less than $F_c = -5.987$ nor greater than $F_c = 5.987$, you fail to reject H_0; the coefficient of determination is not different from zero. There appears to be no support for a relationship between payroll and wins in Major League Baseball.

It's a two-tailed test, so there are two rejection regions.

Note: Problems 14.11–14.21 refer to the data set in Problem 14.11, the payroll (in millions of dollars) for eight Major League Baseball teams for a particular season and the number of times the teams won that season.

14.17 Calculate the standard error of the estimate s_e for the regression model.

The standard error of the estimate s_e measures the dispersion of the observed data around the regression line. If the data points are very close to the regression line, the standard error of the estimate is relatively low, and vice versa, as illustrated below.

Small s_e

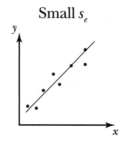

Large s_e

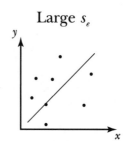

Calculate s_e using the formula below.

$$s_e = \sqrt{\frac{\text{SSE}}{n-2}} = \sqrt{\frac{651.63}{8-2}} = \sqrt{108.605} = 10.42$$

Note: Problems 14.11–14.21 refer to the data set in Problem 14.11, the payroll (in millions of dollars) for eight Major League Baseball teams for a particular season and the number of times the teams won that season.

14.18 Construct a 95% confidence interval for the average number of wins for a Major League Baseball team that has a payroll of $100 million.

Begin by calculating the predicted number of wins for a $100 million payroll using the regression equation generated in Problem 14.11.

$$y = 67.5 + 0.125 \, (100) = 80$$

Now calculate the sample mean of the independent variable—the average payroll of the eight teams.

$$\bar{x} = \frac{\sum x}{n} = \frac{856}{8} = 107$$

The confidence interval (CI) is computed according to the formula below. Note that the critical *t*-score t_c for this 95% confidence interval has $n - 2 = 8 - 2 = 6$ degrees of freedom. According to Reference Table 2, $t_c = 2.447$.

$$\text{CI} = \hat{y} \pm t_c s_e \sqrt{\frac{1}{n} + \frac{\left(x - \bar{x}\right)^2}{\left(\sum x^2\right) - \frac{\left(\sum x\right)^2}{n}}}$$

$$= 80 \pm (2.447)(10.42) \sqrt{\frac{1}{8} + \frac{(100 - 107)^2}{109,334 - \frac{(856)^2}{8}}}$$

$$= 80 \pm \sqrt{0.125 + \frac{49}{109,334 - 91,592}}$$

$$= 80 \pm 25.498\sqrt{0.12776}$$

$$= 80 \pm 9.11$$

$$= 70.9 \text{ and } 89.1 \text{ wins}$$

You are 95% confident that a team with a $100 million payroll will average between 70.9 and 89.1 wins.

Note: Problems 14.11–14.21 refer to the data set in Problem 14.11, the payroll (in millions of dollars) for eight Major League Baseball teams for a particular season and the number of times the teams won that season.

14.19 Calculate the standard error of the slope s_b for the regression model.

The standard error of the slope measures how consistent the slope of the regression equation b is when several sets of samples from the population are selected and the regression equation is constructed for each. A large error indicates that the slopes vary based on the subset of the data you choose.

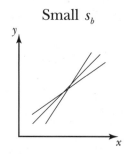
Small s_b

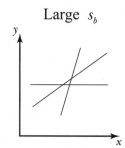
Large s_b

Apply the formula below to calculate s_b.

$$s_b = \frac{s_e}{\sqrt{\sum x^2 - n\left(\overline{x}\right)^2}} = \frac{10.42}{\sqrt{109{,}334 - (8)(107)^2}} = \frac{10.42}{\sqrt{109{,}334 - 91{,}592}} = 0.078$$

Note: Problems 14.11–14.21 refer to the data set in Problem 14.11, the payroll (in millions of dollars) for eight Major League Baseball teams for a particular season and the number of times the teams won that season.

14.20 Test the significance of slope b using $\chi = 0.05$.

The significance test for the slope of the regression equation determines whether the population slope β is significantly different from zero, based on the sample coefficient of determination R^2. Begin by stating the hypotheses.

$$H_0 : \beta = 0$$
$$H_1 : \beta \neq 0$$

If the population slope equals zero, there is no relationship between payroll and wins. A zero slope indicates that a change in payroll will have no impact on the number of wins.

Rejecting the null hypothesis indicates that there is a significant relationship between the independent and dependent variables. The test statistic for this hypothesis test is the *t*-distribution, as calculated below.

$$t = \frac{b - \beta}{s_b} = \frac{0.125 - 0}{0.078} = 1.60$$

This is zero because the null hypothesis assumes it is zero.

Given df = $n - 2 = 6$ degrees of freedom, the critical *t*-scores are $t_c = \pm 2.447$. Because $t = 1.60$ is neither less than -2.447 nor greater than 2.447, you fail to reject H_0 and conclude that the slope of the regression equation is not different from zero. There appears to be no support for a linear relationship between payroll and wins in Major League Baseball. This verifies the solution to Problem 14.16.

Note: Problems 14.11–14.21 refer to the data set in Problem 14.11, the payroll (in millions of dollars) for eight Major League Baseball teams for a particular season and the number of times the teams won that season.

14.21 Construct a 95% confidence interval for the slope of the regression equation.

The critical *t*-score for the 95% confidence interval has $n - 2 = 8 - 2 = 6$ degrees of freedom, so $t_c = 2.447$ according to Reference Table 2. Apply the formula below to calculate the boundaries of the confidence interval.

$$\text{CI} = b \pm t_c s_b = 0.125 \pm (2.447)(0.078) = 0.125 \pm 0.191$$

Thus, the lower boundary is $0.125 - 0.191 = -0.066$ and the upper boundary is $0.125 + 0.191 = 0.316$. Because this confidence interval includes zero, you can conclude that there is no linear relationship between payroll and wins in Major League Baseball.

Note: Problems 14.22–14.32 refer to the data set below, the mileage and selling prices of eight used cars of the same model.

Mileage	Price	Mileage	Price
21,000	$16,000	65,000	$10,000
34,000	$11,000	72,000	$12,000
41,000	$13,000	76,000	$7,000
43,000	$14,000	84,000	$7,000

14.22 Construct the linear equation of best fit and interpret the results.

Higher mileage should correlate negatively with selling price, so assign mileage to be the independent variable *x* and price to be the dependent variable *y*. In the table below, mileage and price are recorded in thousands.

	Mileage	Price			
	x	y	xy	x²	y²
	21	16	336	441	256
	34	11	374	1,156	121
	41	13	533	1,681	169
	43	14	602	1,849	196
	65	10	650	4,225	100
	72	12	864	5,184	144
	76	7	532	5,776	49
	84	7	588	7,056	49
Total	436	90	4,479	27,368	1,084

Compute the slope of the regression equation.

An increase in mileage means a decrease in price, so the line has a negative slope. The larger x gets, the smaller y gets, so the line goes down from left to right.

$$b = \frac{n\sum xy - \left(\sum x\right)\left(\sum y\right)}{n\sum x^2 - \left(\sum x\right)^2} = \frac{(8)(4,479) - (436)(90)}{(8)(27,368) - (436)^2} = \frac{-3,408}{28,848} = -0.118$$

Calculate the y-intercept.

$$a = \frac{\sum y}{n} - b\left(\frac{\sum x}{n}\right) = \frac{90}{8} - (-0.118)\left(\frac{436}{8}\right) = 11.25 - (-6.431) = 17.681$$

Extending the regression model beyond the data it comes from can lead to questionable results.

The value a represents the expected value for y when $x = 0$. In this case, it would represent the price of a used car with zero mileage. However, used cars have been driven, so it would not be appropriate to claim a new car's price would be $17,680.

The linear regression model for the car price (in thousands of dollars) based on x (in thousands of miles) is $\hat{y} = 17.681 - 0.118x$. Every mile on a used car subtracts an average of $0.118 from its value.

Note: Problems 14.22–14.32 refer to the data set in Problem 14.22, the mileage and selling prices of eight used cars of the same model.

14.23 Predict the selling price for a used car of this model with an odometer that reads 50,000 miles.

Substitute $x = 50$ into the regression equation constructed in Problem 14.22.

$$\hat{y} = 17.681 - 0.118x = 17.681 - 0.118(50) = 17.681 - 5.9 = 11.781$$

The variables are in thousands: 11.781(1,000) = 11,781.

The car has an expected value of $11,781.

Note: Problems 14.22–14.32 refer to the data set in Problem 14.22, the mileage and selling prices of eight used cars of the same model.

14.24 Calculate the total sum of squares for the model.

The total sum of squares (SST) of the $n = 8$ pairs of data measures the total variation in the dependent variable using the formula presented in Problem 14.13.

$$\text{SST} = \sum y^2 - \frac{\left(\sum y\right)^2}{n} = 1{,}084 - \frac{(90)^2}{8} = 1{,}084 - 1{,}012.5 = 71.5$$

Note: Problems 14.22–14.32 refer to the data set in Problem 14.22, the mileage and selling prices of eight used cars of the same model.

14.25 Partition the total sum of squares calculated in Problem 14.24 into the sum of squares regression and the sum of squares error.

Apply the formula presented in Problem 14.14 to calculate the sum of squares error.

$$\begin{aligned}
\text{SSE} &= \sum y^2 - a \sum y - b \sum xy \\
&= 1{,}084 - (17.681)(90) - (-0.118)(4{,}479) \\
&= 1{,}084 - (1{,}591.29) - (-528.522) \\
&= 21.23
\end{aligned}$$

Calculate the sum of squares regression by subtracting SSE from SST.

$$\text{SSR} = \text{SST} - \text{SSE} = 71.5 - 21.23 = 50.27$$

Note: Problems 14.22–14.32 refer to the data set in Problem 14.22, the mileage and selling prices of eight used cars of the same model.

14.26 Calculate the coefficient of determination for the car pricing model.

The coefficient of determination is the quotient of the sum of squares regression and the total sum of squares.

$$R^2 = \frac{\text{SSR}}{\text{SST}} = \frac{50.27}{71.5} = 0.703$$

70.3% of the variation in price is explained by the car's mileage.

A larger SSR means a smaller SSE, and the larger the SSR, the better the regression model is.

Note: Problems 14.22–14.32 refer to the data set in Problem 14.22, the mileage and selling prices of eight used cars of the same model.

14.27 Test the significance of the coefficient of determination calculated in Problem 14.28 using $\chi = 0.10$.

State the hypotheses for the two-tailed test.

$$H_0 : \rho^2 = 0$$
$$H_1 : \rho^2 \neq 0$$

Calculate the *F*-score for the data.

$$F = \frac{\text{SSR}}{\left(\dfrac{\text{SSE}}{n-2}\right)} = \frac{50.27}{\left(\dfrac{21.23}{8-2}\right)} = \frac{50.27}{3.538} = 14.21$$

Given $D_1 = 1$ and $D_2 = n - 2 = 6$ degrees of freedom and $\alpha = 0.10$, $F_c = \pm 3.776$. Because $F = 14.21$ is greater than $F_c = 3.776$, you reject H_0 and conclude that the coefficient of determination is different than zero. The data supports a relationship between mileage and selling price.

Note: Problems 14.22–14.32 refer to the data set in Problem 14.22, the mileage and selling prices of eight used cars of the same model.

14.28 Calculate the standard error of the estimate ~~sé~~ s_e for the regression model.

$$s_e = \sqrt{\frac{\text{SSE}}{n-2}} = \sqrt{\frac{21.23}{8-2}} = \sqrt{3.53833} = 1.88$$

Note: Problems 14.22–14.32 refer to the data set in Problem 14.22, the mileage and selling prices of eight used cars of the same model.

14.29 Construct a 90% confidence interval for the average price of a car of this model that has been driven 62,500 miles.

Calculate the expected price of a used car with 62,500 miles using the regression equation from Problem 14.22.

$$\hat{y} = 17.681 - 0.118x = 17.681 - (0.118)(62.5) = 17.681 - 7.375 = 10.306$$

Calculate the average mileage of the eight cars in the data set.

$$\bar{x} = \frac{\sum x}{n} = \frac{436}{8} = 54.5$$

Given df $= n - 2 = 6$ degrees of freedom, $t_c = 1.943$ for the 90% confidence interval, according to Reference Table 2. Calculate boundaries of the confidence interval.

$$CI = \hat{y} \pm t_c s_e \sqrt{\frac{1}{n} + \frac{\left(x - \bar{x}\right)^2}{\left(\sum x^2\right) - \frac{\left(\sum x\right)^2}{n}}}$$

$$= 10.306 \pm (1.943)(1.88) \sqrt{\frac{1}{8} + \frac{(62.5 - 54.5)^2}{27,368 - \frac{(436)^2}{8}}}$$

$$= 10.306 \pm 3.653 \sqrt{0.125 + \frac{64}{27,368 - 23,762}}$$

$$= 10.306 \pm 3.653 \sqrt{0.125 + 0.0177}$$

$$= 10.306 \pm 1.380$$

$$= 8.926 \text{ and } 11.686$$

You are 90% confident that the average price for a used car of this model with 62,500 miles is between $8,926 and $11,686.

Note: Problems 14.22–14.32 refer to the data set in Problem 14.22, the mileage and selling prices of eight used cars of the same model.

14.30 Calculate the standard error of the slope s_b.

$$s_b = \frac{s_e}{\sqrt{\sum x^2 - n\left(\bar{x}\right)^2}} = \frac{1.88}{\sqrt{27,368 - (8)(54.5)^2}} = \frac{1.88}{\sqrt{27,368 - 23,762}} = \frac{1.88}{\sqrt{3,606}} = 0.0313$$

Note: Problems 14.22–14.32 refer to the data set in Problem 14.22, the mileage and selling prices of eight used cars of the same model.

14.31 Test the significance of b using $\chi = 0.10$.

State the hypotheses.

$$H_0 : \beta = 0$$
$$H_1 : \beta \neq 0$$

Calculate the t-score for the test.

$$t = \frac{b - \beta}{s_b} = \frac{-0.118 - 0}{0.0313} = -3.77$$

Given df $= n - 2 = 6$ degrees of freedom and $\alpha = 0.10$, the critical t-scores are $t_c = \pm 1.943$. Because $t = -3.77$ is less than $t_c = -1.943$, you reject H_0. The data supports a linear relationship between mileage and price.

Note: Problems 14.22–14.32 refer to the data set in Problem 14.22, the mileage and selling prices of eight used cars of the same model.

14.32 Construct a 90% confidence interval for the slope of the regression equation.

The critical *t*-score for the 90% confidence interval given df = $n - 2 = 6$ degrees of freedom is $t_c = 1.943$.

$$CI = b \pm t_c s_b = -0.118 \pm (1.943)(0.0313) = -0.118 \pm 0.0608$$

You are 90% confident that the true population slope for the model is between $-0.118 - 0.0608 = -0.179$ and $-0.118 + 0.0608 = -0.0572$.

> The interval is restricted to negative numbers and does not contain zero, so the population regression equation has a negative slope.

Note: Problems 14.33–14.43 refer to the data set below, the number of wins for seven NFL teams and the average number of points they allow per game.

Points	Wins
21	12
21	10
27	9
28	8
21	7
22	4
32	0

14.33 Construct a regression equation that models the number of wins during the season based on the average number of points allowed per game.

> Basically, you're trying to prove that allowing your opponent to score a lot of points means you're going to win fewer games.

An increase in the average number of points allowed per game should correlate negatively with the number of wins during the season. Thus, the number of points allowed is the independent variable and wins is the dependent variable.

	Points	Wins			
	x	*y*	*xy*	*x²*	*y²*
	21	12	252	441	144
	21	10	210	441	100
	27	9	243	729	81
	28	8	224	784	64
	21	7	147	441	49
	22	4	88	484	16
	32	0	0	1,024	0
Total	172	50	1,164	4,344	454

Calculate the slope b and the y-intercept a of the regression equation.

$$b = \frac{n \sum xy - \left(\sum x\right)\left(\sum y\right)}{n \sum x^2 - \left(\sum x\right)^2} = \frac{(7)(1,164) - (172)(50)}{(7)(4,344) - (172)^2} = \frac{8,148 - 8,600}{30,408 - 29,584} = \frac{-452}{824} = -0.549$$

$$a = \frac{\sum y}{n} - b\left(\frac{\sum x}{n}\right) = \frac{50}{7} - (-0.549)\left(\frac{172}{7}\right) = 20.63$$

The linear regression model is $\hat{y} = 20.63 - 0.549x$.

Note: Problems 14.33–14.43 refer to the data set in Problem 14.33, the number of wins for seven NFL teams and the average number of points they allow per game.

14.34 Predict the expected number of wins for a team that allows an average of 25 points per game.

Substitute $x = 25$ into the regression model constructed in Problem 14.33.

$$\hat{y} = 20.63 - 0.549(25) = 20.63 - 13.725 = 6.905 \approx 7 \text{ wins}$$

Note: Problems 14.33–14.43 refer to the data set in Problem 14.33, the number of wins for seven NFL teams and the average number of points they allow per game.

14.35 Calculate the total sum of squares.

$$\text{SST} = \sum y^2 - \frac{\left(\sum y\right)^2}{n} = 454 - \frac{(50)^2}{7} = 454 - 357.14 = 96.86$$

Note: Problems 14.33–14.43 refer to the data set in Problem 14.33, the number of wins for seven NFL teams and the average number of points they allow per game.

14.36 Partition the total sum of squares into the sum of squares regression and the sum of squares error.

Calculate SSE.

$$\begin{aligned}
\text{SSE} &= \sum y^2 - a\sum y - b\sum xy \\
&= 454 - (20.63)(50) - (-0.549)(1,164) \\
&= 454 - 1,031.5 + 639.04 \\
&= 61.54
\end{aligned}$$

Subtract SSE from SST to calculate SSR.

$$\text{SSR} = \text{SST} - \text{SSE} = 96.86 - 61.54 = 35.32$$

Note: Problems 14.33–14.43 refer to the data set in Problem 14.33, the number of wins for seven NFL teams and the average number of points they allow per game.

14.37 Calculate the coefficient of determination to determine the percentage of variation in wins that is explained by points allowed.

$$R^2 = \frac{\text{SSR}}{\text{SST}} = \frac{35.32}{96.86} = 0.365$$

The average points allowed per game explains 36.5% of the variation in wins.

Note: Problems 14.33–14.43 refer to the data set in Problem 14.33, the number of wins for seven NFL teams and the average number of points they allow per game.

14.38 Test the significance of the coefficient of determination using $\alpha = 0.05$.

State the hypotheses.

$$H_0 : \rho^2 = 0$$
$$H_1 : \rho^2 \neq 0$$

Calculate the *F*-score for the test.

$$F = \frac{\text{SSR}}{\left(\dfrac{\text{SSE}}{n-2}\right)} = \frac{35.32}{\left(\dfrac{61.54}{7-2}\right)} = \frac{35.32}{12.308} = 2.87$$

The team might have a lot of points scored against it, but it might score a lot of points as well. Maybe the best defense is a good offense!

Given $D_1 = 1$ and $D_2 = n - 2 = 5$ degrees of freedom and $\alpha = 0.05$, the critical *F*-scores are $F_c = \pm 6.608$. Because $F = 2.87$ is neither greater than 6.608 nor less than -6.608, you fail to reject H_0. The data does not support a relationship between wins per season and average points allowed per game.

Note: Problems 14.33–14.43 refer to the data set in Problem 14.33, the number of wins for seven NFL teams and the average number of points they allow per game.

14.39 Calculate the standard error of the estimate s_e.

$$s_e = \sqrt{\frac{\text{SSE}}{n-2}} = \sqrt{\frac{61.54}{7-2}} = \sqrt{12.308} = 3.51$$

Note: Problems 14.33–14.43 refer to the data set in Problem 14.33, the number of wins for seven NFL teams and the average number of points they allow per game.

14.40 Construct a 95% confidence interval for the average number of wins per season for a team that allows an average of 27.5 points per game.

Calculate the expected number of wins for a team that allows an average of 27.5 points per game.

$$\hat{y} = 20.63 - 0.549(27.5) = 20.63 - 15.0975 = 5.53$$

Compute the average points allowed per game for all seven teams.

$$\bar{x} = \frac{\sum x}{n} = \frac{172}{7} = 24.57$$

Given df = $n - 2 = 5$ degrees of freedom, the critical t-score for the 95% confidence interval is $t_c = 2.571$. Calculate the boundaries of the interval.

$$CI = \hat{y} \pm t_c s_e \sqrt{\frac{1}{n} + \frac{(x - \bar{x})^2}{(\sum x^2) - \frac{(\sum x)^2}{n}}}$$

$$= 5.53 \pm (2.571)(3.51) \sqrt{\frac{1}{7} + \frac{(27.5 - 24.57)^2}{4,344 - \frac{(172)^2}{7}}}$$

$$= 5.53 \pm 9.02 \sqrt{0.1429 + \frac{8.5849}{4,344 - 4,226.286}}$$

$$= 5.53 \pm 9.02 \sqrt{0.1429 + 0.0729}$$

$$= 5.53 \pm 9.02(0.4645)$$

$$= 5.53 \pm 4.19$$

$$= 1.34 \text{ and } 9.72$$

This confidence interval is very wide because, according to Problem 14.38, there isn't a significant relationship.

You are 95% confident that the average number of wins per season for a team that allows an average of 27.5 points per game is between 1.34 and 9.72.

Note: Problems 14.33–14.43 refer to the data set in Problem 14.33, the number of wins for seven NFL teams and the average number of points they allow per game.

14.41 Calculate the standard error of the slope s_b.

$$s_b = \frac{s_e}{\sqrt{\sum x^2 - n(\bar{x})^2}} = \frac{3.51}{\sqrt{4,344 - (7)(24.57)^2}} = \frac{3.51}{\sqrt{4,344 - 4,225.79}} = \frac{3.51}{10.87} = 0.323$$

Note: Problems 14.33–14.43 refer to the data set in Problem 14.33, the number of wins for seven NFL teams and the average number of points they allow per game.

14.42 Test the significance of the slope of the regression using $\alpha = 0.05$.

State the hypotheses.

$$H_0 : \beta = 0$$
$$H_1 : \beta \neq 0$$

Calculate the *t*-score for the test.

$$t = \frac{b - \beta}{s_b} = \frac{-0.549 - 0}{0.323} = -1.70$$

Given df = $n - 2 = 5$ degrees of freedom and $\alpha = 0.05$, the critical *t*-scores are $t_c = \pm 2.571$. Because $t = -1.70$ is neither less than -2.571 nor greater than 2.571, you fail to reject H_0. The slope of the regression equation is not different from zero, so the data does not support a linear relationship between wins per season and average points per game allowed.

Note: Problems 14.33–14.43 refer to the data set in Problem 14.33, the number of wins for seven NFL teams and the average number of points they allow per game.

14.43 Construct a 95% confidence interval for the slope of the regression equation.

The critical *t*-score t_c for this 95% confidence interval has $n - 2 = 5$ degrees of freedom, so $t_c = 2.571$. Calculate the boundaries of the interval.

$$\text{CI} = b \pm t_c s_b = -0.549 \pm (2.571)(0.323) = -0.549 \pm 0.830$$

You are 95% confident that the true population slope for the wins per season model is between $-0.549 - 0.830 = -1.379$ and $-0.549 + 0.830 = 0.281$.

Chapter 15
NONPARAMETRIC TESTS

Tests that do not require assumptions about the populations

Many of the statistical algorithms outlined in preceding chapters are classified as parametric tests because specific assumptions must be made, such as a normally distributed population or equal variances for two populations. If these assumptions are invalid, the results garnered from the statistical tests are invalid as well.

Nonparametric tests are not restricted by such assumptions, and they are relatively easy to perform. However, they tend to be less precise than parametric tests and require more compelling evidence to reject the null hypothesis.

This chapter will investigate a series of nonparametric tests that can be performed on one or more populations, including tests that compare both independent and dependent samples and multiple populations. The chapter also includes a hypothesis test for the median and a test for a correlation coefficient.

The Sign Test with a Small Sample Size

Test the median of a sample

15.1 Explain the purpose of nonparametric tests.

Nonparametric tests are used when a population distribution is unknown. They can also be used to test parameters other than the mean, proportion, and variance. For instance, nonparametric tests can examine the median of a population.

Basically all of the tests up to this point have been parametric tests. They use probability assumptions based on known distributions (like z-scores for normal distributions) to draw conclusions.

15.2 Identify the primary disadvantage of nonparametric tests.

Compared to parametric tests, the results of nonparametric tests are less precise. Given the same sample size and significance level, the confidence interval for a nonparametric test is wider that the corresponding confidence interval for a parametric test.

15.3 The following table shows the electrical usage in kilowatt-hours (kWh) of 12 homes during the month of February. Apply the sign test to investigate the claim that the median electrical usage is more than 1,000 kWh using $\alpha = 0.05$.

Number of kWh per House					
974	815	1320	1000	916	1,066
1,497	1,152	1,131	1,305	1,008	1,222

State the null and alternative hypotheses.

$$H_0 : \text{median} \leq 1,000 \text{ kWh}$$
$$H_1 : \text{median} > 1,000 \text{ kWh}$$

The claim is that the median is more than 1,000 kWh; it's the alternative hypothesis.

The sign table below replaces the data in the table with a negative sign when the value is below the hypothesized median of 1,000, a positive sign when the value is above 1,000, and a zero when the value is equal to 1,000.

Observations Above and Below the Median					
−	−	+	0	−	+
+	+	+	+	+	+

Leave out the zeros. Just count up the + and − signs.

There are eight positive signs and three negative signs, which constitute a sample size of $n = 8 + 3 = 11$ nonzero signs. When applying the sign test, any sample size n that is less than or equal to 25 is considered small. (See Problems 15.8–15.12 to investigate the sign test with larger sample sizes.)

Because $n \leq 25$, the test statistic x is defined as the smaller of the total number of positive or negative signs. In this case, there are 8 positive signs and 3 negative signs, so $x = 3$.

The critical value x_c for this test is identified using Reference Table 6. Given $\alpha = 0.05$ for a one-tailed test and $n = 11$, $x_c = 2$. The null hypothesis of a one-tailed sign test is rejected when $x \leq x_c$. ←

In this problem, $x = 3$ is greater than $x_c = 2$, so you fail to reject H_0. There is not enough evidence to support the claim, even though the majority of data points are above the hypothesized median.

> This may be the opposite of what you're used to. Usually, a test statistic has to be greater than the critical value to lie in the rejection region.

15.4 The following table lists the selling prices (in thousands of dollars) for 18 two-bedroom condominiums at a beach resort. Use the sign test to investigate the claim that the median selling price is less than $180,000, when $\alpha = 0.05$.

Selling Price (in thousands)					
$186	$144	$165	$180	$174	$177
$170	$191	$159	$165	$180	$172
$149	$155	$187	$173	$168	$175

State the null and alternative hypotheses.

$$H_0 : \text{median} \geq \$180,000$$
$$H_1 : \text{median} < \$180,000$$

Construct a sign table that substitutes "+" for all values over the hypothesized median of $180, "−" for all values below $180, and "0" for all values equal to $180.

Observations Above and Below the Median					
+	−	−	0	−	−
−	+	−	−	0	−
−	−	+	−	−	−

There are 3 positive and 13 negative signs, for a sample size of $n = 13 + 3 = 16$. Because $n \leq 25$, the test statistic x is the smaller of the number of positive or negative signs: $x = 3$.

According to Reference Table 6, the critical value for this test is $x_c = 4$. Because $x = 3$ is less than or equal to $x_c = 4$, you reject H_0 and conclude that there is enough evidence to support the claim.

> Only 3 of the data points are above the hypothesized median, so you reject the null hypothesis.

15.5 The following table lists the ages of 30 randomly selected M.B.A. students at a particular university. Test the claim that the median age of M.B.A. students is 32 using a sign test at the $\alpha = 0.05$ significance level.

Age of M.B.A. Students									
27	32	32	28	23	25	33	27	29	24
25	26	28	35	26	32	29	26	32	23
28	32	34	29	25	33	28	32	25	25

State the null and alternative hypotheses.

$$H_0: \text{ median} = 32 \text{ years old}$$
$$H_1: \text{ median} \neq 32 \text{ years old}$$

Construct a sign table that compares each data value to the median.

Observations Above and Below the Median									
−	0	0	−	−	−	+	−	−	−
−	−	−	+	−	0	−	−	0	−
−	0	+	−	−	+	−	0	−	−

When H_1 contains $\neq$, you use a two-tailed test.

The table contains 4 positive signs and 20 negative signs, so $n = 20 + 4 = 24$. Because $n \leq 25$, the test statistic x is the smaller of 4 and 20: $x = 4$.

According to Reference Table 6, a two-tailed test with $\alpha = 0.05$ and $n = 24$ has a critical value of $x_c = 6$. Because $x \leq x_c$, you reject H_0 and conclude that the median age of an M.B.A. student is not 32.

In this case, H_1 was the opposite of the claim, because the alternative hypothesis can never contain "equals." When you reject the null hypothesis, you're actually rejecting the original claim that the median age is 32.

15.6 The following table lists the daily high temperatures of a particular city (in degrees Fahrenheit) for 20 different days in March. Test the claim that the median temperature is *not* 45°F using a sign test at the $\alpha = 0.10$ significance level.

Daily High Temperature (in degrees Fahrenheit)									
48	50	36	44	40	45	32	38	45	46
47	53	45	40	36	49	47	45	54	48

State the null and alternative hypotheses.

$$H_0: \text{ median} = 45 \text{ degrees}$$
$$H_1: \text{ median} \neq 45 \text{ degrees}$$

Construct a sign table comparing the data values to the hypothesized median.

Observations Above and Below the Median									
+	+	–	–	–	0	–	–	0	+
+	+	0	–	–	+	+	0	+	+

The table contains 9 positive signs and 7 negative signs, so $n = 9 + 7 = 16$. Because $n \leq 25$, the test statistic x is the smaller of 7 and 9: $x = 7$.

According to Reference Table 6, a two-tailed test with $\alpha = 0.10$ and $n = 16$ has a critical x-value of $x_c = 4$. Because $x = 7$ is not less than or equal to $x_c = 4$, you fail to reject H_0 and conclude that the median temperature is not 45°F.

15.7 The following table lists the starting salaries, in thousands of dollars, of 20 business majors who recently graduated from college. Use a sign test to investigate a claim that the median starting salary for business majors exceeds $35,000 at the $\alpha = 0.01$ significance level.

Starting Salaries (thousands of dollars)									
$33	$30	$34	$36	$28	$35	$37	$33	$31	$30
$35	$29	$37	$30	$33	$30	$35	$27	$34	$33

State the null and alternative hypotheses.

$$H_0: \text{median} \leq \$35,000$$
$$H_1: \text{median} > \$35,000$$

Construct a sign table.

Observations Above and Below the Median									
–	–	–	+	–	0	+	–	–	–
0	–	+	–	–	–	0	–	–	–

The table contains 3 positive signs and 14 negative signs. In order to reject the null hypothesis, at the very least the number of positive signs should exceed the number of negative signs. This is not the case, so no further analysis is necessary. You fail to reject the null hypothesis and support the claim.

You're trying to prove that most of the salaries in the list are above $35,000, but only 3 of the 20 salaries actually are. There's no point in even continuing the test.

The Sign Test with a Large Sample Size
Test medians using z-scores

15.8 The following table lists the tips a waiter at a particular restaurant received from a sample of 40 customers. Apply the sign test to investigate management's claim that the median tip exceeds $15 using $\alpha = 0.05$.

Tips									
$18	$20	$19	$22	$14	$16	$20	$19	$17	$23
$13	$15	$8	$24	$12	$10	$11	$23	$17	$10
$18	$16	$16	$18	$25	$24	$9	$14	$19	$20
$12	$13	$19	$21	$14	$11	$20	$16	$18	$15

This problem works just like Problems 15.3–15.7 until it's time to calculate values for x and x_c.

State the null and alternative hypotheses.

$$H_0 : \text{median} \le \$15$$
$$H_1 : \text{median} > \$15$$

Construct a sign table comparing each data value to the hypothesized median.

If H_1 contained "less than," you'd be doing a one-tailed test on the left side of the distribution and you'd pick the smaller of the two numbers: x = 13.

Observations Above and Below the Median									
+	+	+	+	−	+	+	+	+	+
−	0	−	+	−	−	−	+	+	−
+	+	+	+	+	+	−	−	+	+
−	−	+	+	−	−	+	+	+	0

The table includes 25 positive signs and 13 negative signs, so $n = 25 + 13 = 38$. When $n > 25$, the test statistic for a one-tailed test on the right side of the distribution is the larger of 13 and 25: $x = 25$.

Left-tailed sign tests contain (x + 0.5) instead of (x − 0.5).

Calculate the z-score for the right-tailed sign test using the formula below.

$$z = \frac{(x-0.5)-(n/2)}{\sqrt{n}/2} = \frac{(25-0.5)-(38/2)}{\sqrt{38}/2} = \frac{24.5-19}{3.0822} = \frac{5.5}{3.0822} = 1.78$$

See Problem 10.8.

According to Reference Table 1, the critical z-score for a one-tailed test given $\alpha = 0.05$ is $z_c = 1.64$. Because $z = 1.78$ is greater than $z_c = 1.64$, you reject H_0 and conclude that the median tip is greater than $15.

Back to the familiar rejection rules: z has to be greater than a positive z_c or less than a negative z_c to reject H_0.

15.9 A commuter would like to test a hypothesis that her median commute time is less than 30 minutes. Of her last 45 commutes, 27 were shorter than 30 minutes and 16 were longer than 30 minutes. Apply the sign test to test her claim using $\alpha = 0.01$.

State the null and alternative hypotheses.

$$H_0 : \text{median} \geq 30 \text{ minutes}$$

$$H_1 : \text{median} < 30 \text{ minutes}$$

Because there are 27 observations above the hypothesized median (+) and 16 observations below it (–), the sample size is $n = 27 + 16 = 43$. ←

She traveled 45 times, so two of those trips must have lasted exactly 30 minutes: 45 – 43 = 2.

You are applying a one-tailed test on the left side of the distribution, so x is the smaller of 27 and 16: $x = 16$. As $n \geq 25$, the normal distribution is assumed. Calculate the z-score for the sign test.

$$z = \frac{(x + 0.5) - (n/2)}{\sqrt{n}/2} = \frac{(16 + 0.5) - (43/2)}{\sqrt{43}/2} = \frac{16.5 - 21.5}{3.2787} = \frac{-5}{3.2787} = -1.52$$

In a left-tailed test, z_c, like z, is negative.

The critical z-score for a left-tailed test with $\alpha = 0.01$ is $z_c = -2.33$. Because $z = -1.52$ is not less than $z_c = -2.33$, you fail to reject H_0 and conclude that the data is not sufficient to support the commuter's claim.

15.10 A sample of 75 lightbulbs contains 49 bulbs that lasted longer than 900 hours and 26 bulbs that lasted less than 900 hours. Use the sign test at the $\alpha = 0.05$ significance level to test the manufacturer's claim that the median life of the bulbs is 900 hours.

State the null and alternative hypotheses.

$$H_0 : \text{median} = 900 \text{ hours}$$

$$H_1 : \text{median} \neq 900 \text{ hours}$$

There are 49 observations above the hypothesized median and 26 observations below it, so the sample size is $n = 49 + 26 = 75$. You are applying a two-tailed test, so set the test statistic x equal to the larger of 26 and 49: $x = 49$. ←

One- and two-tailed tests ($n \leq 25$): x is the smaller number. One-tailed tests ($n > 25$): x depends on whether you are applying the test on the right or left side of the distribution. Two-tailed test ($n > 25$): x is the larger number.

The majority of the observations are above the median (49 > 26), so apply the z-score formula for a right-tailed test.

$$z = \frac{(x - 0.5) - (n/2)}{\sqrt{n}/2} = \frac{(49 - 0.5) - (75/2)}{\sqrt{75}/2} = \frac{48.5 - 37.5}{4.3301} = \frac{11}{4.3301} = 2.54$$

The critical z-scores for a two-tailed test with $\alpha = 0.05$ are $z_c = \pm 1.96$. Because $z = 2.54$ is greater than $z_c = 1.96$, you reject H_0 and conclude that the median life of a lightbulb is not 900 hours.

15.11 The owner of a car would like to test a hypothesis that the median number of miles per tankful of gas is 350. In a sample of 30 tanks of gas, 12 lasted more than 350 miles and 18 lasted less than 350 miles. Apply the sign test using $\alpha = 0.10$.

State the null and alternative hypotheses.

$$H_0 : \text{median} = 350 \text{ miles}$$
$$H_1 : \text{median} \neq 350 \text{ miles}$$

Twelve observations were greater than the hypothesized median and 18 observations were less than the median, so $n = 12 + 18 = 30$. You are applying a two-tailed test with $n > 25$, so x is the larger of 12 and 18: $x = 18$. The majority of the observations are below the median, so apply the z-score formula for a left-tailed test.

$$z = \frac{(x + 0.5) - (n/2)}{\sqrt{n}/2} = \frac{(18 + 0.5) - (30/2)}{\sqrt{30}/2} = \frac{18.5 - 15}{2.7386} = \frac{3.5}{2.7386} = 1.28$$

The critical z-scores for a two-tailed test with $\alpha = 0.10$ are $z_c = \pm1.64$. Because $z = 1.28$ is not less than -1.64, you fail to reject H_0; the car does not travel a median of 350 miles on a tank of gas.

15.12 A sample of 55 households in a particular community includes 34 with incomes less than \$60,000 and 21 with incomes greater than \$60,000. Apply the sign test to a researcher's claim that the median household income exceeds \$60,000 using $\alpha = 0.05$.

State the null and alternative hypotheses.

$$H_0 : \text{median} \leq \$60,000$$
$$H_1 : \text{median} > \$60,000$$

Don't bother applying a test for a claim that says "the median income is higher than \$60,000"; the median of the sample isn't even higher than \$60,000.

The majority of households (34 out of 55) have incomes below \$60,000, so there is no support for the alternative hypothesis. You fail to reject the null hypothesis and no further analysis is necessary.

The Paired-Sample Sign Test ($n \leq 25$)

Apply the sign test to two dependent data sets

15.13 The following table lists the customer service ratings of 10 employees before and after a training program. Apply the paired-sample sign test at an $\alpha = 0.05$ level of significance to investigate management's claim that the program improves employees' customer service ratings.

Employee	A	B	C	D	E	F	G	H	I	J
Before	7.9	8.0	6.5	7.1	7.9	8.4	7.7	8.0	8.2	9.0
After	8.2	8.3	7.0	7.1	7.7	8.6	7.8	8.5	8.4	8.7

The paired-sample sign test is similar to the hypothesis tests for the means of dependent samples explored in Chapter 11. The observations in one sample are related to the observations in the other.

In this problem, the paired samples represent employees' customer service ratings before and after a training program. The sign test is applied to determine whether a significant number of employees' ratings have improved. State the null and alternative hypotheses.

H_0: Customer service ratings will not increase after training.

H_1: Customer service ratings will increase after training.

Append a row to the data table containing signs that reflect the change in employee ratings. If an employee scores higher after training, place a positive sign below the paired data. If an employee's rating decreases after the training, record a negative sign. Unchanged ratings are indicated with a zero.

Employee	A	B	C	D	E	F	G	H	I	J
Before	7.9	8.0	6.5	7.1	7.9	8.4	7.7	8.0	8.2	9.0
After	8.2	8.3	7.0	7.1	7.7	8.6	7.8	8.5	8.4	8.7
Change	+	+	+	0	–	+	+	+	+	–

The appended row contains 7 positive signs and 2 negative signs, so the sample size is $n = 7 + 2 = 9$. You are applying a paired-sample sign test with $n \leq 25$, so the test statistic x is the smaller of 2 and 7: $x = 2$. ←

> When $n \leq 25$, you pick the smaller of the + and – numbers. This goes for one- and two-tailed tests.

According to Reference Table 6, a one-tailed test with $\alpha = 0.05$ and $n = 9$ has a critical x-value of $x_c = 1$. Because $x = 2$ is greater than $x_c = 1$, you fail to reject H_0; there is not enough evidence to support management's claim.

> When you use Reference Table 6, x has to be less than or equal to x_c to reject H_0.

15.14 The following table lists MCAT scores for 10 medical school applicants before and after an MCAT review course. Apply the paired-sample sign test using $\alpha = 0.05$ to investigate the program's advertised claim that it improves MCAT scores.

Student	A	B	C	D	E	F	G	H	I	J
Before	25	24	29	30	24	21	33	30	28	25
After	26	27	30	28	26	24	35	31	32	28

State the null and alternative hypotheses.

H_0 : MCAT scores will not increase after the course.

H_1 : MCAT scores will increase after the course.

Append a row to the table using signs to indicate the students' MCAT score changes upon completing the program.

Student	A	B	C	D	E	F	G	H	I	J
Before	25	24	29	30	24	21	33	30	28	25
After	26	27	30	28	26	24	35	31	32	28
Change	+	+	+	−	+	+	+	+	+	+

This table contains 9 positive signs and 1 negative sign, so $n = 9 + 1 = 10$. Because $n \leq 25$, x is equal to the smaller of 1 and 9: $x = 1$. According to Reference Table 6, a one-tailed test with $\alpha = 0.05$ and $n = 10$ has a critical x-value of $x_c = 1$.

Because $x \leq x_c$, you reject H_0 and conclude that the data supports the claim of increased scores.

> Note that 1 is greater than or equal to 1 because 1 equals 1.

15.15 An insurance company is comparing repair estimates from two body shops to determine whether they are different. Eighteen cars were sent to two different shops for estimates. Shop A's estimates exceeded Shop B's for 12 of the cars, and Shop B's estimates were larger for the remaining 6 cars.

Apply a paired-sample sign test at the $\alpha = 0.10$ significance level to determine whether the shops' repair estimates are different.

State the null and alternative hypotheses.

H_0 : Estimates from shop A and shop B are the same.

H_1 : Estimates from shop A and shop B are different.

Assign a positive sign to observations in which Shop A has a larger estimate and a negative sign when Shop B's estimates are larger. The sample size is $n = 12 + 6 = 18$. Because $n \leq 25$, x is the smaller of 6 and 12: $x = 6$.

According to Reference Table 6, a two-tailed test with $\alpha = 0.10$ and $n = 18$ has a critical x-value of $x_c = 5$. Because $x > x_c$, you fail to reject H_0. The data does not suggest that the repair shops give significantly different estimates.

15.16 An mortgage company is comparing 24 home appraisals conducted on the same homes by two different firms. Firm A's estimates exceeded Firm B's estimates on 17 homes, and Firm B's estimates exceeded Firm A's estimates on 5 homes.

Apply a paired-sample sign test using $\alpha = 0.05$ to investigate the mortgage company's claim that the firms are appraising the homes differently.

State the null and alternative hypotheses.

H_0 : Appraisals from Firm A and Firm B are the same.

H_1 : Appraisals from Firm A and Firm B are different.

Assign a positive sign to the homes for which Firm A's estimate is higher and a negative sign to the homes for which Firm B's was higher. This results in 17 positive signs and 5 negative signs, so $n = 17 + 5 = 22$. Because $n \le 25$, x is the smaller of 5 and 17: $x = 5$.

Because $x = 5$ is less than or equal to the critical x-value $x_c = 5$, you reject H_0. The data supports the mortgage company's claim.

The Paired-Sample Sign Test ($n > 25$)
Combine the sign test and z-scores to test paired data

Note: In Problems 15.17–15.18, a golf equipment company is testing a new club, comparing the distances golfers can hit with it versus the existing model of the same club. Sixty golfers were asked to use the old and new models of the club. Of that group, 38 hit the ball farther with the new model and 22 hit the ball farther with the old model.

15.17 Test the manufacturer's claim that the new club increases the distance the golf ball is hit using a right-tailed paired-sample sign test with $\alpha = 0.05$.

State the null and alternative hypotheses.

H_0 : Both models hit the ball the same distance.

H_1 : The new model hits the ball farther than the existing model.

Assign a positive sign to the golfers who hit the ball farther with the new model and a negative sign to the golfers who hit the ball farther with the existing model. This results in 38 positives and 22 negatives, so the sample size is $n = 38 + 22 = 60$.

In a right-tailed test with $n > 25$, the test statistic x is the larger of the sign totals (22 and 38): $x = 38$. Calculate the z-score.

$$z = \frac{(x - 0.5) - (n/2)}{\sqrt{n}/2} = \frac{(38 - 0.5) - (60/2)}{\sqrt{60}/2} = \frac{37.5 - 30}{3.87298} = \frac{7.5}{3.87298} = 1.94$$

The critical z-score for a right-tailed test using $\alpha = 0.05$ is 1.64. Because $z = 1.94$ is greater than $z_c = 1.64$, you reject H_0 and conclude that the new club model hits the ball farther than the existing model.

Paired-sample sign tests can be constructed to use a rejection region on the right or left tails of the normal distribution when n > 25.

Note: In Problems 15.17–15.18, a golf equipment company is testing a new club, comparing the distances golfers can hit with it versus the existing model of the same club. Sixty golfers were asked to use the old and new models of the club. Of that group, 38 hit the ball farther with the new model and 22 hit the ball farther with the old model.

15.18 Verify your answer to Problem 15.17 using a left-tailed paired-sample sign test with $\alpha = 0.05$.

The hypotheses and signs are defined in the same manner as in Problem 15.17. In a left-tailed paired-sample sign test with $n > 25$, however, you choose the smaller of the sign totals (22 and 38) to represent the test statistic: $x = 22$.

Notice that the z-score is the opposite of the z-score calculated in Problem 15.17.

Remember, the right-tailed test uses x − 0.5 and the left-tailed test uses x + 0.5 in the z-score formula.

$$z = \frac{(x+0.5)-(n/2)}{\sqrt{n}/2} = \frac{(22+0.5)-(60/2)}{\sqrt{60}/2} = \frac{22.5-30}{3.87298} = \frac{-7.5}{3.87298} = -1.94$$

The critical z-score for the left-tailed test is the opposite of the critical z-score calculated in Problem 15.17: $z_c = -1.64$. Because $z < z_c$, you reject H_0.

15.19 A group of 80 people joined a weight-loss program. Once they completed the program, 40 of them found they had lost weight, but 35 had gained weight. The remaining 5 participants maintained the same weight. Test the weight-loss company's claim that the program reduces weight using a paired-sample sign test given $\alpha = 0.10$.

State the null and alternative hypotheses.

H_0 : The new program does not effeectively reduce weight.

H_1 : The new program effectively reduces weight.

This book uses a left-tailed test because H₁ implies there are more − signs than + signs. Of course, you could use a right-tailed test if you wanted. Just look back at Problems 15.17–15.18 to see how that works.

If positive signs represent individuals who gained weight and negative signs represent people who lost weight, there are 35 positive signs and 40 negative signs. The sample size is $35 + 40 = 75$. Apply a left-tailed test and set the test statistic equal to the smaller of the two totals: $x = 35$.

Calculate the z-score.

$$z = \frac{(x+0.5)-(n/2)}{\sqrt{n}/2} = \frac{(35+0.5)-(75/2)}{\sqrt{75}/2} = \frac{35.5-37.5}{4.3301} = \frac{-2}{4.3301} = -0.46$$

The critical z-score for a left-tailed test using $\alpha = 0.10$ 1 is $z_c = -1.28$. Because $z = -0.46$ is not less than $z_c = -1.28$, you fail to reject H_0. The data does not support the company's claim.

15.20 An ice cream company has developed two new flavors and invites 90 customers to taste both and identify the flavor they prefer. At the conclusion of the taste test, 53 customers preferred Flavor A, 27 preferred Flavor B, and 10 could not decide.

Test the company's claim that there was a difference in customer preference using a paired-sample sign test with $\alpha = 0.01$.

> More people preferred Flavor A than Flavor B (52 versus 27 people). But is that enough to conclude that the general population will also prefer Flavor A?

State the null and alternative hypotheses.

H_0 : There is no difference in customer preference.

H_1 : There is a difference in customer preference.

There are 53 positive signs representing customers who preferred Flavor A and 27 negative signs representing customers who preferred Flavor B. Thus, the sample size is $n = 53 + 27 = 80$.

A two-tailed sign test with $n > 25$ uses the larger of the two totals as the test statistic: $x = 53$. Calculate the z-score.

> If you were trying to prove that more people preferred Flavor A, you'd use a one-tailed test. Instead, you're just trying to prove there's no difference in preference.

$$z = \frac{(x-0.5)-(n/2)}{\sqrt{n}/2} = \frac{(53-0.5)-(80/2)}{\sqrt{80}/2} = \frac{52.5-40}{4.4721} = \frac{12.5}{4.4721} = 2.80$$

The critical z-scores are $z_c = \pm 2.57$. Because $z = 2.80$ is greater than $z_c = 2.57$, you reject H_0 and conclude that there is a difference in customer preference.

The Wilcoxon Rank Sum Test for Small Samples

The magnitude of differences between two samples

15.21 The following table lists gasoline prices from a sample of gas stations in Delaware and New York. Use the Wilcoxon rank sum test to investigate the claim that gasoline prices differ between the two states at the $\alpha = 0.10$ significance level.

Gasoline Prices					
DE	$2.19	$2.15	$2.36	$2.25	$2.10	$2.29
NY	$2.27	$2.36	$2.45	$2.39	$2.28	

The Wilcoxon rank sum test determines whether two independent populations have the same distribution by accounting for the magnitude of the differences between the two samples. State the null and alternative hypotheses.

H_0 : Gasoline prices in New York and Delaware are equal.

H_1 : Gasoline prices in New York and Delaware are not equal.

Combine the data sets and rank the observations in order from least to greatest. For example, $2.10 is the lowest price, so it is assigned a rank of 1; $2.15 is the next lowest price, so it is assigned a rank of 2.

When two observations are the same, average the ranks. There are two observations of $2.36, so instead of assigning them ranks 8 and 9 in the table, assign them both a rank of $\frac{8+9}{2} = \frac{17}{2} = 8.5$.

Even though you combine the data sets, remember where each value came from. In the table, the gas prices are ranked and each price is accompanied by the state in which the price was recorded.

Price	State	Rank	Price	State	Rank
$2.10	DE	1	$2.29	DE	7
$2.15	DE	2	$2.36	NY	8.5
$2.19	DE	3	$2.36	DE	8.5
$2.25	DE	4	$2.39	NY	10
$2.27	NY	5	$2.45	NY	11
$2.28	NY	6			

There are two samples, $n_1 = 5$ prices from New York and $n_2 = 6$ prices from Delaware. If the samples are unbalanced, make sure that n_1 represents the smaller sample size. Add the ranks of the observations in the smaller sample. In this problem, R is the sum of the ranks of the New York gas prices.

If the sample sizes are equal you can use either sample to calculate R.

$$R = 5 + 6 + 8.5 + 10 + 11 = 40.5$$

When both sample sizes are less than or equal to 10, use Reference Table 7 to calculate the lower and upper critical values for one- and two-tailed Wilcoxon rank sum tests. Let a represent the lower critical value and b represent the upper critical value. If $a \leq R \leq b$, then do not reject H_0. Otherwise, reject H_0.

Some books use different values to define a "small" population for the Wilcoxon rank sum test. This book uses n = 10.

The critical values for a two-tailed test with sample sizes $n_1 = 5$ and $n_2 = 6$ and $\alpha = 0.10$ are 20 and 40. Because $R = 40.5$ is not between 20 and 40, you reject H_0 and conclude that the gasoline prices differ.

One-tailed tests prove that one thing is bigger than another. Two-tailed tests prove that things are not equal.

15.22 The following table lists the ages of a sample of men and women at a retirement community. Apply the Wilcoxon rank sum test to investigate the claim that women are older than men at the community at the $\alpha = 0.05$ significance level.

Community Member Ages							
Men	76	79	85	80	82	89	71
Women	84	76	88	70	85	90	

State the null and alternative hypotheses.

H_0: Women are not older than men in the community.

H_1: Women are older than men in the community.

Rank the observations in order from least to greatest, noting the population to which each observation belongs.

Age	Gender	Rank	Age	Gender	Rank
70	W	1	84	W	8
71	M	2	85	M	9.5
76	M	3.5	85	W	9.5
76	W	3.5	88	W	11
79	M	5	89	M	12
80	M	6	90	W	13
82	M	7			

The sample sizes are $n_1 = 6$ women and $n_2 = 7$ men. Add the ranks of the observations in the smaller sample. ⟵

$$R = 1 + 3.5 + 8 + 9.5 + 11 + 13 = 46$$

Add the ranks of the women's ages.

According to Reference Table 7, the critical values for a one-tailed test using $\alpha = 0.05$ with $n_1 = 6$ and $n_2 = 7$ are 30 and 54. Because $30 \le 46 \le 54$, you fail to reject H_0 and conclude that women are not older than men at the retirement community.

It's a one-tailed test because H_1 states that the women are older, not that the ages are unequal.

15.23 The following table lists the sizes of a sample of chemistry and physics classes at a university. Use the Wilcoxon rank sum test to investigate the claim that class sizes in the departments are different at the $\alpha = 0.05$ significance level.

Class Sizes					
Chemistry	23	26	41	15	28
Math	46	35	46	31	48

State the null and alternative hypotheses.

H_0 : Chemistry and physics class sizes are equal.

H_1 : Chemistry and physics class sizes are not equal.

Combine the data sets and rank the observations.

Size	Class	Rank	Size	Class	Rank
15	C	1	35	P	6
23	C	2	41	C	7
26	C	3	46	P	8.5
28	C	4	46	P	8.5
31	P	5	48	P	10

When sample sizes are equal, the assignment of n_1 and n_2 is arbitrary. In this problem, $n_1 = 5$ represents the sample size of chemistry classes and $n_2 = 5$ is the sample size of physics classes. Calculate the sum of the ranks of the chemistry classes. ⟵

R is always the sum of the group with sample size n_1, in this case, the chemistry classes.

$$R = 1 + 2 + 3 + 4 + 7 = 17$$

According to Reference Table 7, the critical values for a two-tailed test using $\alpha = 0.05$ with $n_1 = 5$ and $n_2 = 5$ in Reference Table 7 are 18 and 37. Because $R = 17$ does not lie between the critical values, you reject H_0 and conclude that there is a difference in class size.

The Wilcoxon Rank Sum Test for Large Samples
Use z-scores to measure rank differences

15.24 The following table lists salaries of several high school teachers in California and Florida. Apply the Wilcoxon rank sum test to determine whether California high school teachers earn more than Florida high school teachers at the $\alpha = 0.05$ significance level.

California	Florida	California	Florida
$47,700	$48,300	$59,900	$47,100
$60,500	$57,600	$49,600	$37,500
$40,900	$43,300	$48,400	$38,600
$40,700	$30,900	$53,600	$36,200
$57,100	$43,600	$47,700	$41,500
$35,500	$41,500	$46,000	$49,400

State the null and alternative hypotheses.

H_0: California teacher salaries are less than or equal to Florida salaries.

H_1: California teacher salaries are greater than Florida salaries.

Combine the observations and rank them in order, from least to greatest.

Salary	State	Rank	Salary	State	Rank
$30,900	FL	1	$47,100	FL	13
$35,500	CA	2	$47,700	CA	14.5
$36,200	FL	3	$47,700	CA	14.5
$37,500	FL	4	$48,300	FL	16
$38,600	FL	5	$48,400	CA	17
$40,700	CA	6	$49,400	FL	18
$40,900	CA	7	$49,600	CA	19
$41,500	FL	8.5	$53,600	CA	20
$41,500	FL	8.5	$57,100	CA	21
$43,300	FL	10	$57,600	FL	22
$43,600	FL	11	$59,900	CA	23
$46,000	CA	12	$60,500	CA	24

The sample sizes are equal, so $n_1 = n_2 = 12$. Recall that R is the sum of the ranks of the smaller sample. When the samples have the same size, either can be used to calculate R. The ranks of the California teachers are summed below.

$$R = 2 + 6 + 7 + 12 + 14.5 + 14.5 + 17 + 19 + 20 + 21 + 23 + 24 = 180$$

When both sample sizes are greater than or equal to 10, the normal distribution can be used to approximate the distribution of R. Before you can calculate the z-score, however, you must first calculate μ_R and σ_R.

$$\mu_R = \frac{n_1(n_1 + n_2 + 1)}{2} = \frac{(12)(12+12+1)}{2} = \frac{(12)(25)}{2} = \frac{300}{2} = 150$$

$$\sigma_R = \sqrt{\frac{n_1 n_2 (n_1 + n_2 + 1)}{12}} = \sqrt{\frac{(12)(12)(12+12+1)}{12}} = \sqrt{\frac{3,600}{12}} = \sqrt{300} = 17.321$$

Calculate the z-score by substituting the values of μ_R and σ_R into the formula below.

$$z = \frac{R - \mu_R}{\sigma_R} = \frac{180 - 150}{17.321} = \frac{30}{17.321} = 1.73$$

According to Reference Table 1, a one-tailed test using $\alpha = 0.05$ has a critical z-score of $z_c = 1.64$. Because $z = 1.73$ is greater than $z_c = 1.64$, you reject H_0 and conclude that California high school teachers earn more than Florida high school teachers.

15.25 The following table lists a sample of golf scores recorded by Bill and Steve. Apply the Wilcoxon rank sum test to determine whether their scores are different at the $\alpha = 0.05$ level of significance.

Bill	Steve	Bill	Steve
91	83	75	102
94	98	81	92
98	88	85	82
93	79	79	86
92	90	95	80
81	89	77	

State the null and alternative hypotheses.

H_0: Bill and Steve's golf scores are equal.

H_1: Bill and Steve's golf scores are not equal.

Rank all of the observations in order from least to greatest.

Score	Golfer	Rank	Score	Golfer	Rank
75	B	1	89	S	13
77	B	2	90	S	14
79	S	3.5	91	B	15
79	B	3.5	92	B	16.5
80	S	5	92	S	16.5
81	B	6.5	93	B	18
81	B	6.5	94	B	19
82	S	8	95	B	20
83	S	9	98	B	21.5
85	B	10	98	S	21.5
86	S	11	102	S	23
88	S	12			

Steve's scores

Steve lists $n_1 = 11$ scores and Bill lists $n_2 = 12$. Calculate R, the sum of the ranks of the smaller sample.

$$R = 3.5 + 5 + 8 + 9 + 11 + 12 + 13 + 14 + 16.5 + 21.5 + 23 = 136.5$$

Calculate μ_R, σ_R, and z.

$$\mu_R = \frac{n_1(n_1 + n_2 + 1)}{2} = \frac{(11)(11 + 12 + 1)}{2} = \frac{(11)(24)}{2} = \frac{264}{2} = 132$$

$$\sigma_R = \sqrt{\frac{n_1 n_2(n_1 + n_2 + 1)}{12}} = \sqrt{\frac{(11)(12)(11 + 12 + 1)}{12}} = \sqrt{\frac{3,168}{12}} = \sqrt{264} = 16.248$$

$$z = \frac{R - \mu_R}{\sigma_R} = \frac{136.5 - 132}{16.248} = \frac{4.5}{16.248} = 0.28$$

According to Reference Table 1, a two-tailed test at the $\alpha = 0.05$ significance level has critical z-scores $z_c = \pm 1.96$. Because $z = 0.28$ is neither less than -1.96 nor greater than 1.96, you fail to reject H_0 and conclude that Bill and Steve's golf scores are not different.

The Wilcoxon Signed-Rank Test

Difference in magnitude between dependent samples

15.26 The following table lists the cholesterol levels of patients before and after they tested a new cholesterol drug. Apply the Wilcoxon signed-rank test to determine whether the new drug effectively lowered cholesterol levels using $\alpha = 0.05$.

Patient	A	B	C	D	E	F	G
Before	190	175	189	160	184	178	184
After	176	176	189	171	173	163	170

The Wilcoxon signed-rank test is used to determine whether a difference exists between two dependent samples. Unlike the paired-sample sign test, this procedure considers the magnitudes of the differences between the samples. State the null and alternative hypotheses.

The samples in this problem are dependent because the data sets contain cholesterol levels for the same patients.

H_0 : The medication does not lower cholesterol levels.

H_1 : The medication lowers cholesterol levels.

Construct a table that includes columns for the data sets and the values described below:

- D, the difference between the pairs of data; in this problem D = before – after

You could reverse it if you want to: D = after – before.

- $|D|$, the absolute value of D, as defined above

- R, the rank of the nonzero values for $|D|$

- SR, the ranks defined in the R column accompanied by the corresponding sign in the D column

The smallest nonzero |D| value is 1, so Patient B gets rank R = 1. Because Patient B has a negative D value (D = –1), multiply the R value by –1 to get SR: SR = –1.

If two patients have the same value of $|D|$, average the ranks when you construct the table.

Patient	Before	After	D	$|D|$	R	SR
A	190	176	14	14	4.5	4.5
B	175	176	–1	1	1	–1
C	189	189	0	0	—	—
D	160	171	–11	11	2.5	–2.5
E	184	173	11	11	2.5	2.5
F	178	163	15	15	6	6
G	184	170	14	14	4.5	4.5

Calculate the sums of the positive signed ranks and negative signed ranks separately.

Sum of positive SR: $4.5 + 2.5 + 6 + 4.5 = 17.5$

Sum of negative SR: $(-1) + (-2.5) = -3.5$

The test statistic W is the smaller of the absolute values of those two sums. Because $3.5 < 17.5$, $W = 3.5$. Use Reference Table 8 to identify the critical value W_c for the Wilcoxon signed-rank test such that n is the number of nonzero ranks: $n = 6$.

A one-tailed test using $\alpha = 0.05$ with $n = 6$ has a critical value of $W_c = 2$. Because $W = 3.5$ is greater than $W_c = 2$, you fail to reject H_0 and conclude that the medication did not effectively reduce cholesterol levels.

> You reject H_0 in a one-tailed test when $W \leq W_c$.

15.27 The following table lists the times recorded by eight 50-yard freestyle swimmers before and after they participated in a new training program. Apply the Wilcoxon signed-rank test to determine whether the new program effectively reduced swimming times, using a significance level of $\alpha = 0.01$.

Swimmer	A	B	C	D	E	F	G	H
Before	31	29	34	31	32	35	36	32
After	28	30	31	29	27	29	31	30

State the null and alternative hypotheses.

H_0 : The training program does not lower swimming times.

H_1 : The training program lowers swimming times.

Construct a table that includes the difference between the data pairs, the absolute value of the difference, the ranks of the absolute values, and the signed ranks.

Swimmer	Before	After	D	|D|	R	SR
A	31	28	3	3	4.5	4.5
B	29	30	−1	1	1	−1
C	34	31	3	3	4.5	4.5
D	31	29	2	2	2.5	2.5
E	32	27	5	5	6.5	6.5
F	35	29	6	6	8	8
G	36	31	5	5	6.5	6.5
H	32	30	2	2	2.5	2.5

Calculate the sums of the positive and negative signed-rank values separately.

$$SR > 0: 4.5 + 4.5 + 2.5 + 6.5 + 8 + 6.5 + 2.5 = 35$$

$$SR < 0: -1$$

The test statistic W is the smaller of the absolute values of the above sums. Because $1 < 35$, $W = 1$. According to Reference Table 8, the critical value for a one-tailed test using $\alpha = 0.01$ with $n = 8$ is $W_c = 2$.

Because $W = 1$ is less than or equal to $W_c = 2$, you reject the null hypothesis and conclude that the new training program effectively reduces swimming times.

15.28 The following table lists the taste test results of eight individuals asked to rate two different sodas on a scale of 1 to 10. Apply the Wilcoxon signed-rank test to determine whether there is a difference between the ratings at the $\alpha = 0.10$ significance level.

Person	A	B	C	D	E	F	G	H
Soda A	4	3	3	7	6	5	9	8
Soda B	10	6	6	8	10	3	7	6

State the null and alternative hypotheses.

H_0 : There is no difference between the ratings.

H_1 : There is a difference between the ratings.

Construct the table below, defining D as the difference between the ratings of Soda A and Soda B.

Person	Soda A	Soda B	D	$|D|$	R	SR
A	4	10	−6	6	8	−8
B	3	6	−3	3	5.5	−5.5
C	3	6	−3	3	5.5	−5.5
D	7	8	−1	1	1	−1
E	6	10	−4	4	7	−7
F	5	3	2	2	3	3
G	9	7	2	2	3	3
H	8	6	2	2	3	3

Persons F, G, and H represent the second, third, and fourth largest absolute differences, so the ranks are the average of 2, 3, and 4:

$$\frac{2+3+4}{3} = \frac{9}{3} = 3$$

Add the positive and negative signed-rank values separately.

$$SR > 0: 3 + 3 + 3 = 9$$

$$SR < 0: = -8 + (-5.5) + (-5.5) + (-1) + (-7) = -27$$

Set the test statistic W equal to the smaller of the absolute values of the sums calculated above; $|9| < |-27|$, so $W = 9$. Because H_1 claims that the ratings are different (rather than claiming that one rating is greater than the other), you apply a two-tailed test with $\alpha = 0.10$ and $n = 8$. According to Reference Table 8, $W_c = 6$.

Because $W = 9$ is greater than $W_c = 6$, you fail to reject the null hypothesis and conclude that there is no difference in ratings.

15.29 An independent researcher is comparing the tread wear of two different brands of tires. She replaces the front tires of nine cars with a Brand X tire on the left and a Brand Y tire on the right.

The following table lists the percentage of remaining tread after each car was driven 30,000 miles. Apply the Wilcoxon signed-rank test to determine whether there is a difference between the tread wear of the brands using $\alpha = 0.05$.

Car	A	B	C	D	E	F	G	H	I
Brand X	45	60	40	55	45	50	60	40	45
Brand Y	40	40	45	30	30	50	30	40	35

State the null and alternative hypotheses.

H_0 : The brands exhibit the same remaining tread wear.

H_1 : The brands do not exhibit the same remaining tread wear.

Construct a table to identify the signed ranks of the differences between the data sets.

Car	Brand X	Brand Y	D	|D|	R	SR
A	45	40	5	5	1.5	1.5
B	60	40	20	20	5	5
C	40	45	−5	5	1.5	−1.5
D	55	30	25	25	6	6
E	45	30	15	15	4	4
F	50	50	0	0	—	—
G	60	30	30	30	7	7
H	40	40	0	0	—	—
I	45	35	10	10	3	3

Don't rank differences that equal zero.

There are 7 differences not equal to zero.

Add the positive and negative signed-rank values separately.

$$SR > 0: 1.5 + 5 + 6 + 4 + 7 + 3 = 26.5$$

$$SR < 0: = -1.5$$

The test statistic W is the smaller of 1.5 and 26.5: $W = 1.5$. A two-tailed Wilcoxon signed-rank test with $n = 7$ and $\alpha = 0.05$ has a critical score of $W_c = 2$. Because $W = 1.5$ is less than or equal to $W_c = 2$, you reject the null hypothesis and conclude that there is a difference in remaining tread.

15.30 A chemical company conducted a safety awareness program at ten different facilities. The table below shows the number of man-hours lost because of accidents over a one-month period at each facility, before and after the program. Apply the Wilcoxon signed-rank test to determine whether there is a reduction in lost man-hours after the program at the $\alpha = 0.05$ significance level.

Plant	A	B	C	D	E	F	G	H	I	J
Before	12	7	14	12	6	7	5	2	11	6
After	8	6	6	7	4	4	8	9	7	8

State the null and alternative hypotheses.

H_0 : There was no reduction in lost man-hours after the program.

H_1 : There was a reduction in lost man-hours after the program.

Construct a table to identify the signed ranks of the differences between the data sets.

Plant	Before	After	D	$|D|$	R	SR
A	12	8	4	4	6.5	6.5
B	7	6	1	1	1	1
C	14	6	8	8	10	10
D	12	7	5	5	8	8
E	6	4	2	2	2.5	2.5
F	7	4	3	3	4.5	4.5
G	5	8	–3	3	4.5	–4.5
H	2	9	–7	7	9	–9
I	11	7	4	4	6.5	6.5
J	6	8	–2	2	2.5	–2.5

Add the positive and negative signed-rank values separately.

$$SR > 0: 6.5 + 1 + 10 + 8 + 2.5 + 4.5 + 6.5 = 39$$

$$SR < 0: (-4.5) + (-9) + (-2.5) = -16$$

Let W be the smaller of the two absolute values $|37|$ and $|-16|$: $W = 16$. According to Reference Table 8, a one-tailed test with $n = 10$ and $\alpha = 0.05$ has a critical value of $W_c = 11$. Because $W = 16$ is greater than $W_c = 11$, you fail to reject the null hypothesis. There is no reduction in lost man-hours due to accidents after the safety program.

The Kruskal-Wallis Test
Comparing more than two populations

15.31 The following table lists math SAT scores for 15 high school seniors from three different high schools. Apply the Kruskal-Wallis test to determine whether there is a difference in SAT scores between the three high schools using $\alpha = 0.05$.

School 1	School 2	School 3
498	435	608
582	360	515
527	372	661
480	413	637
549	512	554

The Kruskal-Wallis test is used to determine whether three or more populations are identical. Unlike the ANOVA test discussed in Chapter 13, the Kruskal-Wallis test neither requires the populations to be normally distributed nor requires them to have equal variances. However, in order to apply the test, each sample size must be greater than or equal to 5.

State the null and alternative hypotheses.

H_0 : The schools have equal SAT scores.

H_1 : The schools do not have equal SAT scores.

Rank all 15 data values from least to greatest, as demonstrated in the table below.

The lowest of the 15 scores is 360 from High School 1. The highest is 661 from High School 3.

HS 1	Rank	HS 2	Rank	HS 3	Rank
498	6	435	4	608	13
582	12	360	1	515	8
527	9	372	2	661	15
480	5	413	3	637	14
549	10	512	7	554	11

Add the ranks of the data in each sample separately.

$$R_1 = 6+12+9+5+10 = 42$$
$$R_2 = 4+1+2+3+7 = 17$$
$$R_3 = 13+8+15+14+11 = 61$$

Calculate the test statistic H for the Kruskal-Wallis test using the following formula, in which $N = 15$ is the total number of observations, n_i is the size of the ith sample, and R_i is the rank of the ith sample.

$$H = \frac{12}{N(N+1)}\left(\frac{R_1^2}{n_1} + \frac{R_2^2}{n_2} + \frac{R_3^2}{n_3} + \ldots + \frac{R_k^2}{n_k}\right) - 3(N+1)$$

$$= \frac{12}{(15)(15+1)}\left(\frac{42^2}{5} + \frac{17^2}{5} + \frac{61^2}{5}\right) - 3(15+1)$$

$$= \frac{12}{240}(352.8 + 57.8 + 744.2) - 3(16)$$

$$= 0.05(1{,}154.8) - 48$$

$$= 9.74$$

The critical value H_c for the Kruskal-Wallis test uses the chi-square distribution—found in Reference Table 3—with $k-1$ degrees of freedom, where k is the number of populations tested. Thus, df = $3 - 1 = 2$. Given $\alpha = 0.05$ and df = 2, $H_c = 5.991$.

Because $H = 9.74$ is greater than $H_c = 5.991$, you reject H_0 and conclude that the schools' SAT scores differ.

> The rejection region is right of $H_c = 5.991$. Because $H > H_c$, you reject H_0.

15.32 The following table lists a sample of monthly car insurance premiums for adult males in Pennsylvania, Virginia, and Maryland. Use the Kruskal-Wallis test to determine whether there is a difference between the state premiums when $\chi = 0.10$.

PA	VA	MD
115	99	120
110	113	84
91	122	98
113	107	105
109	93	81
90	118	

> You can apply the Kruskal-Wallis test when the samples aren't the same size.

State the null and alternative hypotheses.

H_0 : The states have equal car insurance premiums.

H_1 : The states do not have equal car insurance premiums.

Rank all $N = 17$ data values from least to greatest, assigning averaged ranks when two or more ranks are the same.

PA	Rank	VA	Rank	MD	Rank
115	14	99	7	120	16
110	11	113	12.5	84	2
91	4	122	17	98	6
113	12.5	107	9	105	8
109	10	93	5	81	1
90	3	118	15		

The premium $113 appears twice (in the 12th and 13th positions), so both $113 premiums have a rank equal to the average of 12 and 13: 12.5.

Add the ranks assigned to each sample. Below, Pennsylvania is sample 1, Virginia is sample 2, and Maryland is sample 3.

$$R_1 = 14 + 11 + 4 + 12.5 + 10 + 3 = 54.5$$
$$R_2 = 7 + 12.5 + 17 + 9 + 5 + 15 = 65.5$$
$$R_3 = 16 + 2 + 6 + 8 + 1 = 33$$

Calculate the test statistic H.

If you round this number to 0.039, you get $H = 1.69$. Use six decimal places to get a more accurate answer.

$$H = \frac{12}{N(N+1)}\left(\frac{R_1^2}{n_1} + \frac{R_2^2}{n_2} + \frac{R_3^2}{n_3} + ... + \frac{R_k^2}{n_k}\right) - 3(N+1)$$

$$= \frac{12}{(17)(17+1)}\left(\frac{54.5^2}{6} + \frac{65.5^2}{6} + \frac{33^2}{5}\right) - 3(17+1)$$

$$= \frac{12}{306}(495.04 + 715.04 + 217.8) - 3(18)$$

$$= 0.039216(1,427.88) - 54$$

$$= 2.00$$

According to Reference Table 3, given $\alpha = 0.10$ and df $= k - 1 = 2$ degrees of freedom, $H_c = 4.605$. Because $H = 2.00$ is less than $H_c = 4.605$, you fail to reject H_0 and conclude that the car insurance premiums are not different.

15.33 A consumer research firm has recorded the prices of plasma, LCD, and DLP TVs of the same size in the table below. Apply the Kruskal-Wallis test to determine whether there is a price difference between the three samples when $\alpha = 0.01$.

Plasma	LCD	DLP
$1,399	$1,179	$1,019
$1,199	$999	$997
$1,075	$999	$947
$1,599	$1,145	$980
$1,399	$1,180	$939
$1,249	$1,150	$1,053

State the null and alternative hypotheses.

H_0: All three samples are the same price.

H_1: Not all of the samples are the same price.

Rank the $N = 18$ prices from least to greatest.

Plasma	Rank	LCD	Rank	DLP	Rank
$1,399	17	$1,179	12	$1,019	7
$1,199	14	$999	5.5	$997	4
$1,075	9	$999	5.5	$947	2
$1,599	18	$1,145	10	$980	3
$1,399	16	$1,180	13	$939	1
$1,249	15	$1,150	11	$1,053	8

Let sample 1 represent plasma, sample 2 represent LCD, and sample 3 represent DLP. Add the ranks of the samples separately.

$$R_1 = 17 + 14 + 9 + 18 + 16 + 15 = 89$$
$$R_2 = 12 + 5.5 + 5.5 + 10 + 13 + 11 = 57$$
$$R_3 = 7 + 4 + 2 + 3 + 1 + 8 = 25$$

Calculate the test statistic H.

$$H = \frac{12}{N(N+1)}\left(\frac{R_1^2}{n_1} + \frac{R_2^2}{n_2} + \frac{R_3^2}{n_3} + \ldots + \frac{R_k^2}{n_k}\right) - 3(N+1)$$

$$= \frac{12}{(18)(18+1)}\left(\frac{89^2}{6} + \frac{57^2}{6} + \frac{25^2}{6}\right) - 3(18+1)$$

$$= \frac{12}{342}(1{,}320.17 + 541.5 + 104.17) - 3(19)$$

$$= 0.035088(1{,}965.84) - 57$$

$$= 11.98$$

According to Reference Table 3, given $\alpha = 0.01$ and df $= k - 1 = 2$ degrees of freedom, $H_c = 9.210$. Because $H > H_c$, you reject H_0 and conclude that the prices are different.

15.34 The table below lists the prices of a sample of textbooks from three different publishing houses. Apply the Kruskal-Wallis test to determine whether the publishers charge different prices if $\alpha = 0.05$.

Publisher 1	Publisher 2	Publisher 3
$140	$154	$157
$117	$135	$163
$97	$169	$111
$143	$205	$172
$141	$203	$130
$160	$142	$155

State the null and alternative hypotheses.

H_0 : The publishers charge equal prices for the textbooks.

H_1 : The publishers do not charge equal prices for the textbooks.

Rank the $N = 18$ textbook prices from least to greatest.

Pub 1	Rank	Pub 2	Rank	Pub 3	Rank
$140	6	$154	10	$157	12
$117	3	$135	5	$163	14
$97	1	$169	15	$111	2
$143	9	$205	18	$172	16
$141	7	$203	17	$130	4
$160	13	$142	8		
		$155	11		

Add the ranks in each sample separately.

$$R_1 = 6 + 3 + 1 + 9 + 7 + 13 = 39$$
$$R_2 = 10 + 5 + 15 + 18 + 17 + 8 + 11 = 84$$
$$R_3 = 12 + 14 + 2 + 16 + 4 = 48$$

Calculate the test statistic H.

$$H = \frac{12}{N(N+1)}\left(\frac{R_1^2}{n_1} + \frac{R_2^2}{n_2} + \frac{R_3^2}{n_3} + ... + \frac{R_k^2}{n_k}\right) - 3(N+1)$$

$$= \frac{12}{(18)(18+1)}\left(\frac{39^2}{6} + \frac{84^2}{7} + \frac{48^2}{5}\right) - 3(18+1)$$

$$= \frac{12}{342}(253.5 + 1{,}008 + 460.8) - 3(19)$$

$$= 0.035088(1{,}722.3) - 57$$

$$= 3.43$$

According to Reference Table 3, given $\alpha = 0.05$ and df $= k - 1 = 2$, $H_c = 5.99$. Because $H < H_c$, you fail to reject H_0 and conclude that the data does not suggest that the publishers charge different prices.

15.35 A company has recorded customer satisfaction ratings (on a scale of 1 to 100) for four of its stores in the table below. Apply the Kruskal-Wallis test to determine whether the branches have different satisfaction ratings when $\alpha = 0.10$.

Store 1	Store 2	Store 3	Store 4
74	88	45	61
86	66	94	74
87	70	78	67
95	94	55	74
90	61	72	78
92	89	47	82

State the null and alternative hypotheses.

H_0 : The stores have equal ratings.

H_1 : The stores do not have equal ratings.

Rank all $N = 24$ data values from least to greatest.

Store 1	Rank	Store 2	Rank	Store 3	Rank	Store 4	Rank
74	11	88	18	45	1	61	4.5
86	16	66	6	94	22.5	74	11
87	17	70	8	78	13.5	67	7
95	24	94	22.5	55	3	74	11
90	20	61	4.5	72	9	78	13.5
92	21	89	19	47	2	82	15

> There's no rank 10 or 12 because there are three rankings of 74, which all get an average rank of 11.

Add the ranks in each sample separately.

$$R_1 = 11 + 16 + 17 + 24 + 20 + 21 = 109$$
$$R_2 = 18 + 6 + 8 + 22.5 + 4.5 + 19 = 78$$
$$R_3 = 1 + 22.5 + 13.5 + 3 + 9 + 2 = 51$$
$$R_4 = 4.5 + 11 + 7 + 11 + 13.5 + 15 = 62$$

Calculate the test statistic H.

$$H = \frac{12}{N(N+1)} \left(\frac{R_1^2}{n_1} + \frac{R_2^2}{n_2} + \frac{R_3^2}{n_3} + \ldots + \frac{R_k^2}{n_k} \right) - 3(N+1)$$

$$= \frac{12}{(24)(24+1)} \left(\frac{109^2}{6} + \frac{78^2}{6} + \frac{51^2}{6} + \frac{62^2}{6} \right) - 3(24+1)$$

$$= \frac{12}{600} (1,980.17 + 1,014 + 433.5 + 640.67) - 3(25)$$

$$= 0.02(4,068.34) - 75$$

$$= 6.37$$

According to Reference Table 3, given $\alpha = 0.10$ and $df = 4 - 1 = 3$ degrees of freedom, $H_c = 6.251$. Because $H = 6.37$ is greater than $H_c = 6.251$, you reject H_0 and conclude that the ratings for all four locations are not the same.

The Spearman Rank Correlation Coefficient Test

Correlating data sets according to rank differences

Note: Problems 15.36–15.37 refer to the data set below, the selling prices, in thousands of dollars, and square footage of seven randomly selected houses.

Selling Price (thousands)	Square Footage
$258	2,730
$191	1,860
$253	2,140
$168	2,180
$249	2,310
$245	2,450
$282	2,920

15.36 Calculate the Spearman rank correlation coefficient for the selling price and square footage.

The Spearman rank correlation coefficient r_s is similar to the correlation coefficient r described in Chapter 14; it measures the strength and the direction of a relationship between two variables. Unlike the correlation coefficient, the Spearman rank correlation coefficient does not require normally distributed variables.

Given a data set with n paired data points and a difference d between the ranks of the pairs, apply the formula below to calculate the Spearman rank correlation coefficient.

$$r_s = 1 - \frac{6 \sum d^2}{n(n^2 - 1)}$$

Values of r_s range between −1.0 (which represents a strong negative correlation) and 1.0 (which represents a strong positive correlation.

As the formula indicates, you will use ranks of data rather than the actual data values to calculate r_s. The variable d is defined as the rank difference; in this problem, d = rank of selling price − rank of square footage.

Calculate d by taking the difference between ranks for each data pair. For this problem, d = the rank of selling price minus the rank of square footage. The difference d is calculated for each pair of data values and is then squared in the table below.

> You could also set d = the rank of square footage minus the rank of selling price. The order doesn't matter.

Price	Rank	Square ft.	Rank	d	d^2
$258	6	2,730	6	6 − 6 = 0	$0^2 = 0$
$191	2	1,860	1	2 − 1 = 1	$1^2 = 1$
$253	5	2,140	2	5 − 2 = 3	$3^2 = 9$
$168	1	2,180	3	1 − 3 = −2	$(−2)^2 = 4$
$249	4	2,310	4	4 − 4 = 0	$0^2 = 0$
$245	3	2,450	5	3 − 5 = −2	$(−2)^2 = 4$
$282	7	2,920	7	7 − 7 = 0	$0^2 = 0$

Calculate the sum of the rightmost column, the squared differences d^2 for the paired data.

$$\sum d^2 = 0+1+9+4+0+4+0 = 18$$

With this value, you can now calculate the Spearman rank correlation coefficient for the $n = 7$ data pairs.

$$r_s = 1 - \frac{6\sum d^2}{n(n^2-1)} = 1 - \frac{(6)(18)}{(7)(7^2-1)} = 1 - \frac{108}{336} = 1 - 0.3214 = 0.679$$

Note: Problems 15.36–15.37 refer to the data set in Problem 15.36, the selling prices, in thousands of dollars, and square footage of seven randomly selected houses.

15.37 Test the significance of the Spearman rank correlation coefficient calculated in Problem 15.36 using $\alpha = 0.05$.

State the null and alternative hypotheses.

H_0 : There is no correlation between the variables.

H_1 : The correlation between the variables is significant.

According to Problem 15.36, $r_s = 0.679$. Use Reference Table 9 to identify r_{sc}, the critical value for the Spearman rank correlation coefficient. Given $n = 7$ data pairs and $\alpha = 0.05$, $r_{sc} = 0.786$.

You reject the null hypothesis H_0 when $|r_s| \geq r_{sc}$; otherwise you fail to reject H_0. In this problem, $|r_s| = 0.679$ is *not* greater than or equal to $r_{sc} = 0.786$. Thus, you fail to reject H_0 and conclude that there is no correlation between selling price and square footage.

Note: Problems 15.38–15.39 refer to the data set below, the rankings two judges assigned eight different figure skaters in a competition.

Skater	Judge 1	Judge 2
A	4	5
B	5	3
C	1	2
D	7	6
E	2	1
F	3	4
G	6	8
H	8	7

15.38 Calculate the Spearman rank correlation coefficient for the judges' rankings.

> The judges actually ranked the skaters 1 to 8, so you don't need to figure out the ranks yourself.

Let d = the rank of Judge 1 minus the rank of Judge 2. Calculate the difference of each data pair and square it, as shown in the table below.

Skater	Judge 1	Judge 2	d	d^2
A	4	5	$4 - 5 = -1$	$(-1)^2 = 1$
B	5	3	$5 - 3 = 2$	$2^2 = 4$
C	1	2	$1 - 2 = -1$	$(-1)^2 = 1$
D	7	6	$7 - 6 = 1$	$1^2 = 1$
E	2	1	$2 - 1 = 1$	$1^2 = 1$
F	3	4	$3 - 4 = -1$	$(-1)^2 = 1$
G	6	8	$6 - 8 = -2$	$(-2)^2 = 4$
H	8	7	$8 - 7 = 1$	$1^2 = 1$

Add the squared differences.

$$\sum d^2 = 1 + 4 + 1 + 1 + 1 + 1 + 4 + 1 = 14$$

Calculate the Spearman rank correlation coefficient of the $n = 8$ pairs of rankings.

$$r_s = 1 - \frac{6 \sum d^2}{n(n^2 - 1)} = 1 - \frac{(6)(14)}{(8)(8^2 - 1)} = 1 - \frac{84}{504} = 1 - 0.1667 = 0.833$$

Note: Problems 15.38–15.39 refer to the data set from Problem 15.38, the rankings two judges assigned eight different figure skaters in a competition.

15.39 Test the significance of the Spearman rank correlation coefficient calculated in Problem 15.38 using $\alpha = 0.05$.

State the null and alternative hypotheses.

H_0 : There is no correlation between the variables.

H_1 : The correlation between the variables is significant.

According to Problem 15.38, $r_s = 0.833$. Reference Table 9 indicates that the critical value r_{sc} of the Spearman rank correlation coefficient for the $n = 8$ data pairs is $r_{sc} = 0.738$. Because $r_s \geq r_{sc}$, you reject H_0 and conclude that there is a correlation between the judges' rankings.

> $0.833 > 0.738$

Note: Problems 15.40–15.41 refer to the data set below, the number of wins and runs scored by eight Major League Baseball teams during the 2008 season.

Team	Wins	Runs
Cubs	97	855
Mets	89	799
Phillies	92	799
Cardinals	86	779
Marlins	84	770
Braves	72	753
Brewers	90	750
Rockies	74	747

15.40 Calculate the Spearman rank correlation coefficient for wins and runs scored.

> Rank the win totals from 1 to 8 and rank the runs scored from 1 to 8.

Rank each set of data separately, calculate the difference d = rank of wins – rank of runs scored, and then square the differences.

Wins	Rank	Runs	Rank	d	d^2
97	8	855	8	0	0
89	5	799	6.5	–1.5	2.25
92	7	799	6.5	0.5	0.25
86	4	779	5	–1	1
84	3	770	4	–1	1
72	1	753	3	–2	4
90	6	750	2	4	16
74	2	747	1	1	1

Add the $n = 8$ values of d^2 in the rightmost column of the above table.

$$\sum d^2 = 0 + 2.25 + 0.25 + 1 + 1 + 4 + 16 + 1 = 25.5$$

Calculate the Spearman rank correlation coefficient.

$$r_s = 1 - \frac{6 \sum d^2}{n(n^2 - 1)} = 1 - \frac{(6)(25.5)}{(8)(8^2 - 1)} = 1 - \frac{153}{504} = 1 - 0.3036 = 0.696$$

Note: Problems 15.40–15.41 refer to the data set in Problem 15.40, the number of wins and runs scored by eight Major League Baseball teams during the 2008 season.

15.41 Test the significance of the Spearman rank correlation coefficient calculated in Problem 15.40 using $\alpha = 0.01$.

State the null and alternative hypotheses.

H_0 : There is no correlation between the variables.

H_1 : The correlation between the variables is significant.

According to Problem 15.40, $r_s = 0.696$. Reference Table 9 states that $r_{sc} = 0.881$ given $n = 8$ and $\alpha = 0.01$. Because $0.696 < 0.833$, you fail to reject H_0; there is no correlation between runs scored and wins.

Note: Problems 15.42–15.43 refer to the data set below, the weekly demand for a digital camera at various prices.

Demand	Price
16	$300
19	$310
14	$320
13	$330
11	$340
12	$350
8	$360

15.42 Calculate the Spearman rank correlation coefficient for price and demand.

Rank both variables, calculate the differences d in rank, and square the differences.

In this problem, d = demand rank – price rank. You're allowed to switch the order because you'll eventually square the difference: $(7 - 1)^2 = (1 - 7)^2 = 36$.

Demand	Rank	Price	Rank	d	d^2
16	6	$300	1	5	25
19	7	$310	2	5	25
14	5	$320	3	2	4
13	4	$330	4	0	0
11	2	$340	5	-3	9
12	3	$350	6	-3	9
8	1	$360	7	-6	36

The sum of the squared differences is $\sum d^2 = 108$. Calculate the Spearman rank correlation coefficient of the $n = 7$ data pairs.

$$r_s = 1 - \frac{6 \sum d^2}{n(n^2 - 1)} = 1 - \frac{(6)(108)}{(7)(7^2 - 1)} = 1 - \frac{648}{336} = 1 - 1.9286 = -0.929$$

> As the price of the camera increases, demand decreases, so the variables are negatively correlated.

Note: Problems 15.42–15.43 refer to the data set in Problem 15.42, showing the weekly demand for a digital camera at various prices.

15.43 Test the significance of the Spearman rank correlation coefficient calculated in Problem 15.42 using $\alpha = 0.05$.

State the null and alternative hypotheses.

H_0 : There is no correlation between the variables.

H_1 : The correlation between the variables is significant.

According to Problem 15.42, $r_s = -0.929$. Given $n = 7$ pairs of data and $\alpha = 0.05$, the critical value of the correlation coefficient is $r_{sc} = 0.786$. Because the absolute value of r_s is greater than or equal to r_{sc} ($0.929 \geq 0.786$), you reject H_0 and conclude that there is a correlation between demand and price of the digital camera.

Note: Problems 15.44–15.45 refer to the data set below, the mileage and selling price of eight used cars of the same model.

Mileage	Price	Mileage	Price
21,000	$16,000	65,000	$10,000
34,000	$11,000	72,000	$12,000
41,000	$13,000	76,000	$7,000
43,000	$14,000	84,000	$8,000

15.44 Calculate the Spearman rank correlation coefficient for price and mileage.

Rank the data sets individually, calculate the difference of each pair, and square the differences.

Mileage	Rank	Price	Rank	d	d²
21	1	$16	8	−7	49
34	2	$11	4	−2	4
41	3	$13	6	−3	9
43	4	$14	7	−3	9
65	5	$10	3	2	4
72	6	$12	5	1	1
76	7	$7	1	6	36
84	8	$8	2	6	36

The sum of the rank differences is $\sum d^2 = 148$. Calculate the Spearman rank correlation coefficient.

$$r_s = 1 - \frac{6\sum d^2}{n(n^2-1)} = 1 - \frac{(6)(148)}{(8)(8^2-1)} = 1 - \frac{888}{504} = 1 - 1.7619 = -0.762$$

Note: Problems 15.44–15.45 refer to the data set below, the mileage and selling price of eight used cars of the same model.

15.45 Test the significance of the Spearman rank correlation coefficient calculated in Problem 15.44 using $\alpha = 0.05$.

State the null and alternative hypotheses.

H_0 : There is no correlation between the variables.

H_1 : The correlation between the variables is significant.

According to Problem 15.44, $r_s = -0.762$. Given $\alpha = 0.05$ and $n = 8$, $r_{sc} = 0.738$. Because $|-0.762| \geq 0.738$, you reject H_0 and conclude that a correlation exists between mileage and price.

Chapter 16
FORECASTING

Predicting future values of random variables

This chapter investigates forecasting, applying a variety of mathematical techniques to predict future values of a random variable. Forecasts provide an integral foundation for organizational planning, as business decisions often rely on the dependable information garnered from these methods. In this chapter, you will explore some of the most common forecasting methods and the means by which to determine the accuracy of their predictions.

This chapter explores several different forecasting methods, including moving average, exponential smoothing, and the regression model that was introduced in Chapter 14. Forecasting accuracy is also measured, using mean absolute deviation and mean squared error to decide which forecasting parameters are most effective for different data sets.

Simple Moving Average

The most basic forecasting technique

Note: Problems 16.1–16.7 refer to the data set below, the net income (in millions of dollars) of a company over the last seven years.

Year	Net Income (millions)
1	$2.5
2	$3.0
3	$2.8
4	$2.6
5	$2.4
6	$3.1
7	$3.3

16.1 Calculate a two-period simple moving average forecast for Year 8.

A simple moving average forecast averages the n prior data points according to the following formula.

$$\text{forecast} = \frac{\sum \text{previous } n \text{ periods}}{n}$$

In this problem, F_8 (the forecast for year 8) is the average of the two prior years.

$$F_8 = \frac{\text{year 7} + \text{year 6}}{2} = \frac{3.1 + 3.3}{2} = \frac{6.4}{2} = 3.2$$

Note: Problems 16.1–16.7 refer to the data set in Problem 16.1, the net income (in millions of dollars) of a company over the last seven years.

16.2 Calculate the mean absolute deviation (MAD) for the two-period moving average forecast calculated in Problem 16.1.

The mean absolute deviation measures the accuracy of the forecasting method by applying it to the historical data. The third period is the first you can forecast using a two-period moving average.

Compute two-period simple moving average forecasts for Years 3 through 7 by averaging the data of the two years immediately preceding each. In the formulas below, A_n represents the actual data observed in the nth year.

Because the third period is the first period chronologically that has two prior periods of data. You can't forecast Year 2 using two periods of prior data because you only have information about Year 1.

$$F_3 = \frac{A_2 + A_1}{2} = \frac{3.0 + 2.5}{2} = 2.75 \qquad F_4 = \frac{A_3 + A_2}{2} = \frac{2.8 + 3.0}{2} = 2.9$$

$$F_5 = \frac{A_4 + A_3}{2} = \frac{2.6 + 2.8}{2} = 2.7 \qquad F_6 = \frac{A_5 + A_4}{2} = \frac{2.4 + 2.6}{2} = 2.5$$

$$F_7 = \frac{A_6 + A_5}{2} = \frac{3.1 + 2.4}{2} = 2.75$$

Calculate the error of the forecasts by subtracting the forecasted value for Years 3 through 7 from the actual values. Note that the error is reported as an absolute value.

> Whether the estimate was high or low doesn't matter in this instance. All that matters is that the forecast was inaccurate, so ignore the sign of the difference.

Year	Actual	Forecast	Error	Absolute Error
1	2.5			
2	3.0			
3	2.8	2.75	2.8 – 2.75 = 0.05	0.05
4	2.6	2.9	2.6 – 2.9 = –0.3	0.3
5	2.4	2.7	2.4 – 2.7 = –0.3	0.3
6	3.1	2.5	3.1 – 2.5 = 0.6	0.6
7	3.3	2.75	3.3 – 2.75 = 0.55	0.55

The mean absolute deviation (MAD) is the average absolute error. Add the values in the rightmost column of the above table and divide by $n = 5$, the total number of absolute errors calculated.

$$\text{MAD} = \frac{\sum |\text{actual} - \text{forecast}|}{n} = \frac{0.05 + 0.3 + 0.3 + 0.6 + 0.55}{5} = \frac{1.8}{5} = 0.36$$

The larger the mean absolute deviation, the less accurate the forecast technique is.

Note: Problems 16.1–16.7 refer to the data set in Problem 16.1, the net income (in millions of dollars) of a company over the last seven years.

16.3 Calculate a three-period simple moving average forecast for Year 8.

Average the observed values for the $n = 3$ years immediately preceding Year 8.

$$F_8 = \frac{A_7 + A_6 + A_5}{n} = \frac{3.3 + 3.1 + 2.4}{3} = \frac{8.8}{3} = 2.93$$

Note: Problems 16.1–16.7 refer to the data set in Problem 16.1, the net income (in millions of dollars) of a company over the last seven years.

16.4 Calculate the mean absolute deviation for the three-period moving average forecast calculated in Problem 16.3.

Year 4 is the first three-period moving average forecast you can calculate, as three prior years of data are required. Calculate F_4, F_5, F_6, and F_7.

$$F_4 = \frac{A_3 + A_2 + A_1}{3} = \frac{2.8 + 3.0 + 2.5}{3} = \frac{8.3}{3} = 2.77$$

$$F_5 = \frac{A_4 + A_3 + A_2}{3} = \frac{2.6 + 2.8 + 3.0}{3} = \frac{8.4}{3} = 2.8$$

$$F_6 = \frac{A_5 + A_4 + A_3}{3} = \frac{2.4 + 2.6 + 2.8}{3} = \frac{7.8}{3} = 2.6$$

$$F_7 = \frac{A_6 + A_5 + A_4}{3} = \frac{3.1 + 2.4 + 2.6}{3} = \frac{8.1}{3} = 2.7$$

Identify the absolute error of each forecast using the table below.

Year	Actual	Forecast	Error	Absolute Error
1	2.5			
2	3.0			
3	2.8			
4	2.6	2.77	−0.17	0.17
5	2.4	2.8	−0.4	0.4
6	3.1	2.6	0.5	0.5
7	3.3	2.7	0.6	0.6

Mean absolute deviation is the average of the values in the absolute error column above.

Increasing the moving average forecast from two to three periods results in one error term fewer in the MAD calculation—a denominator of 4 instead of 5.

$$MAD = \frac{0.17 + 0.4 + 0.5 + 0.6}{4} = \frac{1.67}{4} = 0.418$$

Note: Problems 16.1–16.7 refer to the data set in Problem 16.1, the net income (in millions of dollars) of a company over the last seven years.

16.5 Calculate a four-period simple moving average forecast for Year 8.

$$F_8 = \frac{A_7 + A_6 + A_5 + A_4}{n} = \frac{3.3 + 3.1 + 2.4 + 2.6}{4} = \frac{11.4}{4} = 2.85$$

16.6 Calculate the mean absolute deviation for the four-period moving average forecast calculated in Problem 16.5.

The first four-period moving average forecast you can calculate is F_5, as four prior years of data are required.

$$F_5 = \frac{A_4 + A_3 + A_2 + A_1}{4} = \frac{2.6 + 2.8 + 3.0 + 2.5}{4} = \frac{10.9}{4} = 2.73$$

$$F_6 = \frac{A_5 + A_4 + A_3 + A_2}{4} = \frac{2.4 + 2.6 + 2.8 + 3.0}{4} = \frac{10.8}{4} = 2.7$$

$$F_7 = \frac{A_6 + A_5 + A_4 + A_3}{4} = \frac{3.1 + 2.4 + 2.6 + 2.8}{4} = \frac{10.9}{4} = 2.73$$

Use the table below to calculate the absolute error of each forecast.

Year	Actual	Forecast	Error	Absolute Error
1	2.5			
2	3.0			
3	2.8			
4	2.6			
5	2.4	2.73	−0.33	0.33
6	3.1	2.7	0.4	0.4
7	3.3	2.73	0.57	0.57

Calculate the mean absolute deviation.

$$\text{MAD} = \frac{0.33 + 0.4 + 0.57}{3} = \frac{1.3}{3} = 0.433$$

16.7 Which of the three forecasts generated in Problems 16.1–16.6 most accurately predicts Year 8? Explain your answer.

The most accurate forecasting method has the lowest mean absolute deviation. Consider the table below, which summarizes the forecasts and MADs calculated in Problems 16.1–16.6.

Forecast Method	Forecast	MAD
Two-period moving average	3.2	0.360
Three-period moving average	2.93	0.418
Four-period moving average	2.85	0.433

The two-period moving average forecast 3.2 is the most accurate because 0.360 is the lowest MAD.

This doesn't mean that a two-period moving average forecast will always be better than a three-period or a four-period forecast. It's just true for this data set.

Weighted Moving Average

Recent data is weighted more heavily

Note: Problems 16.8–16.14 refer to the data set below, the weekly demand for a particular cell phone at a retail store.

Week	Demand
1	12
2	23
3	13
4	8
5	20
6	22
7	15

16.8 Calculate a two-period weighted moving average forecast for the demand for the cell phone in Week 8 using weights of 3 and 1.

A weighted moving average forecast applies a weighted average over the past n data points according to the formula below, in which A_n is the actual observed value of period n.

$$\text{forecast} = \frac{\sum \left(\text{weight of period } n \right) \left(A_n \right)}{\sum \text{weights}}$$

You're assuming that Week 7 provides a better forecast of sales than Week 6 because less time has passed. In fact, the weights say that the data from Week 7 is three times more valuable.

Conventionally, the highest weight is assigned to the most recent data. The problem instructs you to use weights of three and one, so the data immediately preceding Week 8 is multiplied by 3 and the week before that is multiplied by 1.

$$F_8 = \frac{3A_7 + 1A_6}{3+1} = \frac{3(15) + 1(22)}{4} = \frac{67}{4} = 16.75$$

Note: Problems 16.8–16.14 refer to the data set in Problem 16.8, the weekly demand for a particular cell phone at a retail store.

16.9 Calculate the mean squared error (MSE) for the two-period weighted moving average forecast calculated in Problem 16.8.

Just as the MAD applied the moving average forecast to past data and then figured out how accurate each forecast was.

The mean squared error is a measure of forecasting accuracy that applies the forecasting method to historical data. Calculate two-period weighted moving averages for as many of the previous weeks as possible.

Because two previous weeks of data are required, the first forecast you can calculate is F_3 for Week 3. Weight the data as described in Problem 16.8. For instance, to calculate F_5, add $3A_4$ and $1A_3$ and divide by the sum of the weights, $3 + 1 = 4$.

$$F_3 = \frac{3A_2 + 1A_1}{3+1} = \frac{3(23)+1(12)}{4} = \frac{81}{4} = 20.25$$

$$F_4 = \frac{3A_3 + 1A_2}{3+1} = \frac{3(13)+1(23)}{4} = \frac{62}{4} = 15.5$$

$$F_5 = \frac{3A_4 + 1A_3}{3+1} = \frac{3(8)+1(13)}{4} = \frac{37}{4} = 9.25$$

$$F_6 = \frac{3A_5 + 1A_4}{3+1} = \frac{3(20)+1(8)}{4} = \frac{68}{4} = 17$$

$$F_7 = \frac{3A_6 + 1A_5}{3+1} = \frac{3(22)+1(20)}{4} = \frac{86}{4} = 21.5$$

The error in each forecast is the difference of the forecasted and actual values. Square each of the errors, as illustrated in the table below.

Week	Actual	Forecast	Error	Squared Error
1	12			
2	23			
3	13	20.25	−7.25	52.56
4	8	15.5	−7.5	56.25
5	20	9.25	10.75	115.56
6	22	17	5	25
7	15	21.5	−6.5	42.25

The mean squared error is the average of the values in the right column of the table above.

$$\text{MSE} = \frac{52.56 + 56.25 + 115.56 + 25 + 42.25}{5} = \frac{291.62}{5} = 58.32$$

The MSE is usually much higher than the MAD because it squares the error term, heavily penalizing large forecasting errors.

Note: Problems 16.8–16.14 refer to the data set in Problem 16.8, the weekly demand for a particular cell phone at a retail store.

16.10 Calculate a three-period weighted moving average forecast for the demand for the cell phone in Week 8 using weights of 4, 2, and 1.

Assign the highest weight (4) to A_7 (the most recent data), assign A_6 the second highest weight (2), and assign A_5 the lowest weight (1). Divide by the sum of the weights: $4 + 2 + 1 = 7$.

$$F_8 = \frac{4A_7 + 2A_6 + 1A_5}{4+2+1} = \frac{4(15)+2(22)+1(20)}{7} = \frac{124}{7} = 17.71$$

Note: Problems 16.8–16.14 refer to the data set in Problem 16.8, the weekly demand for a particular cell phone at a retail store.

16.11 Calculate the mean squared error for the three-period weighted moving average forecast calculated in Problem 16.10.

The first period for which you can generate a three-period weighted moving average is Week 4, as it requires three previous weeks of data. Calculate F_4, F_5, F_6, and F_7.

$$F_4 = \frac{4A_3 + 2A_2 + 1A_1}{4+2+1} = \frac{4(13)+2(23)+1(12)}{7} = \frac{110}{7} = 15.71$$

$$F_5 = \frac{4A_4 + 2A_3 + 1A_2}{4+2+1} = \frac{4(8)+2(13)+1(23)}{7} = \frac{81}{7} = 11.57$$

$$F_6 = \frac{4A_5 + 2A_4 + 1A_3}{4+2+1} = \frac{4(20)+2(8)+1(13)}{7} = \frac{109}{7} = 15.57$$

$$F_7 = \frac{4A_6 + 2A_5 + 1A_4}{4+2+1} = \frac{4(22)+2(20)+1(8)}{7} = \frac{136}{7} = 19.43$$

Use the table below to compute the error for each forecast and the square of the error.

Week	Actual	Forecast	Error	Squared Error
1	12			
2	23			
3	13			
4	8	15.71	–7.71	59.44
5	20	11.57	8.43	71.06
6	22	15.57	6.43	41.34
7	15	19.43	–4.43	19.62

Complete the mean squared error.

$$\text{MSE} = \frac{59.44+71.06+41.34+19.62}{4} = \frac{191.46}{4} = 47.87$$

Note: Problems 16.8–16.14 refer to the data set in Problem 16.8, the weekly demand for a particular cell phone at a retail store.

16.12 Calculate a four-period weighted moving average forecast for the demand for the cell phone in Week 8 using weights of 0.4, 0.3, 0.2, and 0.1.

Assign the weights in descending order from the most recent demand data to the oldest demand data.

$$F_8 = \frac{0.4\left(A_7\right)+0.3\left(A_6\right)+0.2\left(A_5\right)+0.1\left(A_6\right)}{0.4+0.3+0.2+0.1} = \frac{0.4(15)+0.3(22)+0.2(20)+0.1(8)}{1} = 17.4$$

Note: Problems 16.8–16.14 refer to the data set in Problem 16.8, the weekly demand for a particular cell phone at a retail store.

16.13 Calculate the mean squared error for the four-period weighted moving average forecast calculated in Problem 16.12.

Calculate F_5, F_6, and F_7 using the weights identified in Problem 16.12.

$$F_5 = \frac{0.4(8)+0.3(13)+0.2(23)+0.1(12)}{0.4+0.3+0.2+0.1} = \frac{3.2+3.9+4.6+1.2}{1} = 12.9$$

$$F_6 = \frac{0.4(20)+0.3(8)+0.2(13)+0.1(23)}{0.4+0.3+0.2+0.1} = \frac{8+2.4+2.6+2.3}{1} = 15.3$$

$$F_7 = \frac{0.4(22)+0.3(20)+0.2(8)+0.1(13)}{0.4+0.3+0.2+0.1} = \frac{8.8+6+1.6+1.3}{1} = 17.7$$

Calculate the error for each forecast and its square.

Week	Actual	Forecast	Error	Squared Error
1	12			
2	23			
3	13			
4	8			
5	20	12.9	7.1	50.41
6	22	15.3	6.7	44.89
7	15	17.7	–2.7	7.29

The mean squared error is the mean of the values in the rightmost column above.

$$\text{MSE} = \frac{50.41+44.89+7.29}{3} = \frac{102.59}{3} = 34.20$$

Note: Problems 16.8–16.14 refer to the data set in Problem 16.8, showing the weekly demand for a particular cell phone at a retail store.

16.14 Which of the three forecasts generated in Problems 16.8–16.13 should most accurately predict Week 8? Explain your answer.

The more accurate the forecasting method, the lower the mean squared error should be. Consider the following table, which summarizes the results calculated in Problems 16.8–16.13.

Weighted Moving Average	Forecast	MSE
Two-period	16.75	58.32
Three-period	17.71	47.87
Four-period	17.4	34.20

The four-period weighted moving average forecast of 17.4 should most accurately predict Week 8 because its MSE of 34.20 is the least.

Exponential Smoothing

A self-correcting forecasting technique

Note: Problems 16.15–16.21 refer to the data set below, the number of customers per day who purchased items at a retail store.

Day	Number of Customers
1	74
2	87
3	62
4	72

16.15 Predict the number of paying customers on Day 5 using exponential smoothing with $\alpha = 0.6$.

Exponential smoothing predicts the value of period t by adjusting the forecast from the previous period (F_{t-1}) with a portion of the forecasting error from the previous period $\chi(A_{t-1} - F_{t-1})$. The term α is known as the smoothing constant—a value between 0 and 1 that determines how much of the forecasting error from the previous period is used to adjust the old forecast when calculating the current forecast.

Exponential smoothing is a self-correcting technique. The larger the error in the previous period, the larger the correction will be in the next forecast. Calculate F_t, the forecast for period t, using the formula below.

$$F_t = F_{t-1} + \alpha(A_{t-1} - F_{t-1})$$

The forecast for Day 5 requires a forecast from Day 4; the forecast for Day 4 requires a forecast from Day 3; and so on. Thus, you must begin with a seed value for F_1. You are given the actual data for Day 1, so set $F_1 = A_1 = 74$.

Apply the exponential smoothing formula to calculate F_2, the predicted number of paying customers on Day 2.

$$F_t = F_{t-1} + \alpha\left(A_{t-1} - F_{t-1}\right)$$
$$F_2 = F_1 + \alpha\left(A_1 - F_1\right)$$
$$F_2 = 74 + 0.6\left(74 - 74\right)$$
$$F_2 = 74 + 0.6\left(0\right)$$
$$F_2 = 74$$

There is no error in the Day 1 forecast, so the model predicts 74 paying customers on Day 2. However, there were $A_2 = 87$ customers on Day 2. Use this data to calculate F_3.

If you are forecasting period t, then the previous period is $t - 1$. The error of the previous period is the actual value of that period (A_{t-1}) minus the forecasted value (F_{t-1}).

There's no better predicted value for a period than the actual value itself.

Because the actual and forecasted values are exactly the same: $F_1 = A_1 = 74$.

$$F_3 = F_2 + \alpha \left(A_2 - F_2 \right)$$
$$= 74 + 0.6(87 - 74)$$
$$= 74 + (0.6)(13)$$
$$= 81.8$$

Each time you calculate a forecast, it enables you to calculate a forecast for the following period. Compute F_4.

$$F_4 = F_3 + \alpha \left(A_3 - F_3 \right)$$
$$= 81.8 + 0.6(62 - 81.8)$$
$$= 81.8 + 0.6(-19.8)$$
$$= 69.9$$

When the forecasting error is negative, the current forecast is less than the previous forecast. In this case, $F_4 < F_3$.

Finally, compute F_5.

$$F_5 = F_4 + \alpha \left(A_4 - F_4 \right)$$
$$= 69.9 + 0.6(72 - 69.9)$$
$$= 69.9 + 0.6(2.1)$$
$$= 71.2$$

Note: Problems 16.15–16.21 refer to the data set in Problem 16.15, the number of customers per day who purchased items at a retail store.

16.16 Compute the mean absolute deviation (MAD) for the exponential smoothing forecast calculated in Problem 16.15.

In the following table, the forecasted values from the first four days are compared to the actual values. The error is the actual value for each day minus the forecasted value, and the absolute error is the absolute value of each error.

Day	Actual	Forecast	Error	Absolute Error
1	74	74	—	—
2	87	74	13	13
3	62	81.8	–19.8	19.8
4	72	69.9	2.1	2.1

The forecasting and absolute errors for Day 1 are not included in the table because F_1 was arbitrarily set equal to the actual value on Day 1 ($F_1 = A_1 = 74$). Calculate the mean absolute deviation by averaging the three absolute errors.

You're measuring the error in the exponential smoothing technique, and you didn't use that technique to get F_1.

$$\text{MAD} = \frac{13 + 19.8 + 2.1}{3} = \frac{34.9}{3} = 11.63$$

Note: Problems 16.15–16.21 refer to the data set in Problem 16.15, the number of customers per day who purchased items at a retail store.

16.17 Predict the number of paying customers on Day 5 using exponential smoothing with $\alpha = 0.3$.

Like Problem 16.15, begin by setting the forecast for Day 1 equal to the actual observed value for Day 1: $F_1 = A_1 = 74$. Calculate F_2, F_3, F_4, and F_5 by substituting the forecasted and actual values for the previous period into the exponential smoothing formula.

$$F_2 = F_1 + \alpha\left(A_1 - F_1\right)$$
$$= 74 + 0.3(74 - 74)$$
$$= 74 + 0.3(0)$$
$$= 74$$

$$F_3 = F_2 + \alpha\left(A_2 - F_2\right)$$
$$= 74 + 0.3(87 - 74)$$
$$= 74 + 0.3(13)$$
$$= 77.9$$

$$F_4 = F_3 + \alpha\left(A_3 - F_3\right)$$
$$= 77.9 + 0.3(62 - 77.9)$$
$$= 77.9 + 0.3(-15.9)$$
$$= 73.1$$

$$F_5 = F_4 + \alpha\left(A_4 - F_4\right)$$
$$= 73.1 + 0.3(72 - 73.1)$$
$$= 73.1 + 0.3(-1.1)$$
$$= 72.8$$

According to the exponential smoothing technique with $\alpha = 0.03$, Day 5 should feature $F_5 = 72.8$ paying customers.

Note: Problems 16.15–16.21 refer to the data set in Problem 16.15, the number of customers per day who purchased items at a retail store.

16.18 Compute the mean absolute deviation for the exponential smoothing forecast calculated in Problem 16.17.

In the table below, the forecasted values for each day are subtracted from the actual values, resulting in the error and its absolute value.

Day	Actual	Forecast	Error	Absolute Error
1	74	74	—	—
2	87	74	13	13
3	62	77.9	−15.9	15.9
4	72	73.1	−1.1	1.1

Average the values in the right column to calculate the mean absolute deviation.

$$\text{MAD} = \frac{13 + 15.9 + 1.1}{3} = \frac{30}{3} = 10$$

Note: Problems 16.15–16.21 refer to the data set in Problem 16.15, the number of customers per day who purchased items at a retail store.

16.19 Predict the number of paying customers on Day 5 using exponential smoothing with $\alpha = 0.1$.

Once again set $F_1 = A_1 = 74$ and calculate F_2, F_3, F_4, and F_5 using the new value of α.

$$
\begin{aligned}
F_2 &= F_1 + \alpha\left(A_1 - F_1\right) \\
&= 74 + 0.1(74 - 74) \\
&= 74 + 0.1(0) \\
&= 74
\end{aligned}
\qquad
\begin{aligned}
F_3 &= F_2 + \alpha\left(A_2 - F_2\right) \\
&= 74 + 0.1(87 - 74) \\
&= 74 + 0.1(13) \\
&= 75.3
\end{aligned}
$$

$$
\begin{aligned}
F_4 &= F_3 + \alpha\left(A_3 - F_3\right) \\
&= 75.3 + 0.1(62 - 75.3) \\
&= 75.3 + 0.1(-13.3) \\
&= 74.0
\end{aligned}
\qquad
\begin{aligned}
F_5 &= F_4 + \alpha\left(A_4 - F_4\right) \\
&= 74.0 + 0.1(72 - 74.0) \\
&= 74.0 + 0.1(-2) \\
&= 73.8
\end{aligned}
$$

The store should host $F_5 = 73.8$ paying customers on Day 5.

Note: Problems 16.15–16.21 refer to the data set in Problem 16.15, the number of customers per day who purchased items at a retail store.

16.20 Compute the mean absolute deviation for the exponential smoothing forecast calculated in Problem 16.19.

Subtract the forecasted values from the actual values and list the absolute values of the errors.

Day	Actual	Forecast	Error	Absolute Error
1	74	74	—	—
2	87	74	13	13
3	62	75.3	–13.3	13.3
4	72	74.0	–2	2

Calculate the MAD.

$$
\text{MAD} = \frac{13 + 13.3 + 2}{3} = \frac{28.3}{3} = 9.43
$$

Note: Problems 16.15–16.21 refer to the data set in Problem 16.15, the number of customers per day who purchased items at a retail store.

16.21 Which of the three forecasts generated in Problems 16.15–16.20 should most accurately predicts Day 5? Explain your answer.

Consider the table below, which summarizes the results of Problems 16.15–16.20.

α	Forecast	MAD
0.6	71.2	11.63
0.3	72.8	10.00
0.1	73.8	9.43

The most accurate exponential smoothing forecast should be 73.8 customers on Day 5 because $\alpha = 0.1$ has the lowest mean absolute deviation, 9.43.

Exponential Smoothing with Trend Adjustment
Add trends to the self-correcting method

Note: Problems 16.22–16.30 refer to the data set below, daily high temperatures in a city (measured in degrees Fahrenheit) for four consecutive days.

Day	Temperature
1	48
2	56
3	53
4	64

16.22 Forecast the high temperature for Day 5 via exponential smoothing with trend adjustment using $\alpha = 0.2$ and $\beta = 0.4$.

> The increase or decrease does not have to be constant—a general upward or downward movement in the data constitutes a trend.

A trend is a general upward or downward movement in the actual data over time. The moving average and exponential smoothing techniques tend to perform poorly and lag behind the movement in actual data values when a trend is present. The exponential smoothing with trend adjustment method better models a trend in data.

A forecast that includes trend for a period t (FIT_t) is equal to the sum of an exponentially smoothed forecast for that period (F_t) and an exponentially smoothed trend (T_t): $\text{FIT} = F_t + T_t$. Calculate F_t and T_t using the formulas below, in which A_t is the actual data for period t, α is the exponential smoothing constant, and β is the trend smoothing constant.

$$F_t = \alpha(A_{t-1}) + (1-\alpha)(\text{FIT}_{t-1})$$
$$T_t = \beta(F_t - F_{t-1}) + (1-\beta)(T_{t-1})$$

Like the exponential smoothing method demonstrated in Problems 16.15–16.21, begin by setting the exponentially smoothed forecast for Day 1 to the actual data from Day 1: $F_1 = A_1 = 48$.

You also require an initial value for the exponentially smoothed trend. When no specific trend information is available, set $T_1 = 0$. Therefore, in this problem, $FIT_1 = F_1 + T_1 = 48 + 0 = 48$.

> From Day 1 to Day 4, the temperature mostly goes up. It dips a little on Day 3, but in general there is a warming trend. However, no specific trend information is given, so set $T_1 = 0$.

Calculate F_2, the exponentially smoothed forecast for Day 2.

$$F_t = \alpha(A_{t-1}) + (1-\alpha)(FIT_{t-1})$$
$$F_2 = \alpha(A_1) + (1-\alpha)(FIT_1)$$
$$F_2 = 0.2(48) + (1-0.2)(48)$$
$$F_2 = 9.6 + (0.8)(48)$$
$$F_2 = 48$$

Now calculate T_2, the exponentially smoothed trend for Day 2.

$$T_t = \beta(F_t - F_{t-1}) + (1-\beta)(T_{t-1})$$
$$T_2 = \beta(F_2 - F_1) + (1-\beta)T_1$$
$$T_2 = 0.4(48 - 48) + (1-0.4)(0)$$
$$T_2 = 0.4(0) + (0.6)(0)$$
$$T_2 = 0$$

FIT_2, the forecast including trend for Day 2, is the sum of F_2 and T_2: $FIT_2 = 48 + 0 = 48$. Use $F_2 = 48$, $T_2 = 0$, and $FIT_2 = 48$ to compute F_3 and T_3.

$$F_3 = \alpha(A_2) + (1-\alpha)(FIT_2) \qquad T_3 = \beta(F_3 - F_2) + (1-\beta)(T_2)$$
$$= 0.2(56) + (1-0.2)(48) \qquad\quad = 0.4(49.6 - 48) + (1-0.4)(0)$$
$$= 11.2 + (0.8)(48) \qquad\qquad\quad = 0.4(1.6) + (0.6)(0)$$
$$= 49.6 \qquad\qquad\qquad\qquad\quad = 0.64$$

The forecast including trend for Day 3 is the sum of F_3 and T_3: $FIT_3 = 49.6 + 0.64 = 50.24$. Calculate FIT_4.

> T_3 adjusts the exponentially smoothed forecast to account for the trend in the data. It boosts the original forecast from 49.60 to 50.24.

$$F_4 = \alpha(A_3) + (1-\alpha)(FIT_3) \qquad T_4 = \beta(F_4 - F_3) + (1-\beta)(T_3)$$
$$= 0.2(53) + (1-0.2)(50.24) \qquad = 0.4(50.79 - 49.6) + (1-0.4)(0.64)$$
$$= 10.6 + (0.8)(50.24) \qquad\qquad = (0.4)(1.19) + (0.6)(0.64)$$
$$= 50.79 \qquad\qquad\qquad\qquad\quad = 0.86$$

Substitute $FIT_4 = F_4 + T_4 = 50.79 + 0.86 = 51.65$ into the formula for F_5 (and the accompanying formula for T_5) to calculate FIT_5.

$$F_5 = \alpha(A_4) + (1-\alpha)(FIT_4) \qquad T_5 = \beta(F_5 - F_4) + (1-\beta)(T_4)$$
$$= 0.2(64) + (1-0.2)(51.65) \qquad = 0.4(54.12 - 50.79) + (1-0.4)(0.86)$$
$$= 12.8 + (0.8)(51.65) \qquad\qquad = 0.4(3.33) + (0.6)(0.86)$$
$$= 54.12 \qquad\qquad\qquad\qquad\quad = 1.85$$

The high temperature for Day 5 should be approximately
$FIT_5 = 54.12 + 1.85 = 55.97°$.

Note: Problems 16.22–16.30 refer to the data set in Problem 16.22, daily high temperatures in a city for four consecutive days.

16.23 Compute the mean squared error (MSE) for the Day 5 forecast FIT_5 calculated in Problem 16.22.

The table below summarizes the forecast including trend results calculated in Problem 16.22. Subtract the FIT value from the actual value to calculate the error in each forecast, and square the errors.

Like the exponential smoothing method, the first error term is not included in the MSE calculation.

Day	Actual	FIT	Error	Squared Error
1	48	48	—	—
2	56	48	8	64
3	53	50.24	2.76	7.62
4	64	51.65	12.35	152.52

The mean squared error is the average of the values in the rightmost column of the table above.

$$MSE = \frac{64 + 7.62 + 152.52}{3} = \frac{224.14}{3} = 74.71$$

Note: Problems 16.22–16.30 refer to the data set in Problem 16.22, daily high temperatures in a city for four consecutive days.

16.24 Forecast the high temperature for Day 5 via exponential smoothing with trend adjustment using $\alpha = 0.4$ and $\beta = 0.3$.

If there's no error, the model won't make any self-corrections. It will continue to predict values of 48 until that is no longer an accurate guess.

Set the exponentially smoothed forecast for Day 1 equal to the actual data from Day 1: $F_1 = A_1 = 48$. Therefore, Day 1 has no error and $F_2 = 48$ as well. No specific information is given about the data trend, so set $T_1 = 0$. Because $F_1 = F_2 = 48$ and $T_1 = 0$, T_2 will also equal 0.

Calculate the exponentially smoothed forecast F_3 and the exponentially smoothed trend T_3 for Day 3.

See the calculations for F_2 and T_2 in Problem 16.22. What does this mean? Basically, FIT_1 and FIT_2 are both equal to A_1, the first observed data value. It saves time not having to calculate F_2 and T_2 using the formulas.

$$F_3 = \alpha(A_2) + (1-\alpha)(FIT_2) \qquad T_3 = \beta(F_3 - F_2) + (1-\beta)(T_2)$$
$$= 0.4(56) + (1-0.4)(48) \qquad = 0.3(51.2 - 48) + (1-0.3)(0)$$
$$= 22.4 + (0.6)(48) \qquad\qquad = 0.3(3.2) + (0.7)(0)$$
$$= 51.2 \qquad\qquad\qquad\qquad = 0.96$$

Add F_3 and T_3 to get $FIT_3 = 52.16$. Use this value to calculate F_4 and T_4.

$$F_4 = \alpha(A_3) + (1-\alpha)(FIT_3) \qquad T_4 = \beta(F_4 - F_3) + (1-\beta)(T_3)$$
$$= 0.4(53) + (1-0.4)(52.16) \qquad = 0.3(52.50 - 51.2) + (1-0.3)(0.96)$$
$$= 21.2 + (.6)(52.16) \qquad = 0.3(1.3) + (0.7)(0.96)$$
$$= 52.50 \qquad = 1.06$$

Add F_4 and T_4 to get $FIT_4 = 52.50 + 1.06 = 53.56$. Finally, calculate F_5 and T_5.

$$F_5 = \alpha(A_4) + (1-\alpha)(FIT_4) \qquad T_5 = \beta(F_5 - F_4) + (1-\beta)(T_4)$$
$$= 0.4(64) + (1-0.4)(53.56) \qquad = 0.3(57.74 - 52.50) + (1-0.3)(1.06)$$
$$= 25.6 + (0.6)(53.56) \qquad = 0.3(5.24) + (0.7)(1.06)$$
$$= 57.74 \qquad = 2.31$$

The high temperature on Day 5 should be $FIT_5 = 57.74 + 2.31 = 60.05°$.

Note: Problems 16.22–16.30 refer to the data set in Problem 16.22, daily high temperatures in a city for four consecutive days.

16.25 Compute the mean squared error for the Day 5 forecast FIT_5 calculated in Problem 16.24.

Use the table below to compute the squared error for each forecasted temperature.

Day	Actual	FIT	Error	Squared Error
1	48	48	—	—
2	56	48	8	64
3	53	52.16	0.84	0.71
4	64	53.56	10.44	108.99

Calculate the average squared error.

$$MSE = \frac{64 + 0.71 + 108.99}{3} = \frac{173.7}{3} = 57.9$$

Note: Problems 16.22–16.30 refer to the data set in Problem 16.22, daily high temperatures in a city for four consecutive days.

16.26 Forecast the high temperature for Day 5 via exponential smoothing with trend adjustment using $\alpha = 0.6$ and $\beta = 0.8$.

The forecasts for Day 1 and Day 2 are both equal to the actual temperature for Day 1: $F_1 = F_2 = 48$. The corresponding trend values are also equal: $T_1 = T_2 = 0$. Thus, $FIT_1 = FIT_2 = 48 + 0 = 48$. Use these values to calculate F_3 and T_3.

Look back at Problem 16.24 if you're not sure how the book calculated FIT_1 and FIT_2 without lifting a finger.

$$F_3 = \alpha(A_2) + (1-\alpha)(FIT_2) \qquad T_3 = \beta(F_3 - F_2) + (1-\beta)(T_2)$$
$$= 0.6(56) + (1-0.6)(48) \qquad = 0.8(52.8 - 48) + (1-0.8)(0)$$
$$= 33.6 + (0.4)(48) \qquad = 0.8(4.8) + (0.2)(0)$$
$$= 52.8 \qquad = 3.84$$

Note that $FIT_3 = F_3 + T_3 = 52.8 + 3.84 = 56.64$. Substitute these values into the formulas for F_4 and T_4.

$$F_4 = \alpha(A_3) + (1-\alpha)(FIT_3) \qquad T_4 = \beta(F_4 - F_3) + (1-\beta)(T_3)$$
$$= 0.6(53) + (1-0.6)(56.64) \qquad = 0.8(54.46 - 52.8) + (1-0.8)(3.84)$$
$$= 31.8 + (0.4)(56.64) \qquad = 0.8(1.66) + (0.2)(3.84)$$
$$= 54.46 \qquad = 2.10$$

Add F_4 and T_4: $FIT_4 = 54.46 + 2.10 = 56.56$. Finally, calculate F_5 and T_5.

$$F_5 = \alpha(A_4) + (1-\alpha)(FIT_4) \qquad T_5 = \beta(F_5 - F_4) + (1-\beta)(T_4)$$
$$= 0.6(64) + (1-0.6)(56.56) \qquad = 0.8(61.02 - 54.46) + (1-0.8)(2.10)$$
$$= 38.4 + (0.4)(56.56) \qquad = 0.8(6.56) + (0.2)(2.10)$$
$$= 61.02 \qquad = 5.67$$

According to this model, the high temperature on Day 5 will be $FIT_5 = 61.02 + 5.67 = 66.69°$.

Note: Problems 16.22–16.30 refer to the data set in Problem 16.22, daily high temperatures in a city for four consecutive days.

16.27 Compute the mean squared error for the Day 5 forecast FIT_5 calculated in Problem 16.26.

Use the table below to calculate the error in each forecast and the square of each error.

Day	Actual	FIT	Error	Squared Error
1	48	48	—	—
2	56	48	8	64
3	53	56.64	−3.64	13.25
4	64	56.56	7.44	55.35

Calculate the MSE.

$$MSE = \frac{64 + 13.25 + 55.35}{3} = \frac{132.6}{3} = 44.2$$

Note: Problems 16.22–16.30 refer to the data set in Problem 16.22, daily high temperatures in a city for four consecutive days.

16.28 Forecast the high temperature for Day 5 via exponential smoothing with trend adjustment using $\alpha = 0.5$ and $\beta = 0.5$.

Note that $F_1 = F_2 = A_1 = 48$, $T_1 = T_2 = 0$, and $\text{FIT}_1 = \text{FIT}_2 = 0 + 48 = 48$. Calculate F_3 and T_3.

$$
\begin{aligned}
F_3 &= \alpha(A_2) + (1-\alpha)(\text{FIT}_2) \\
&= 0.5(56) + (1-0.5)(48) \\
&= 28 + (0.5)(48) \\
&= 52
\end{aligned}
\qquad
\begin{aligned}
T_3 &= \beta(F_3 - F_2) + (1-\beta)(T_2) \\
&= 0.5(52 - 48) + (1-0.5)(0) \\
&= 0.5(4) + (0.5)(0) \\
&= 2
\end{aligned}
$$

The forecast including trend for Day 3 is $\text{FIT}_3 = 54$. Now calculate F_4 and T_4.

$$
\text{FIT}_3 = F_3 + T_3 = 52 + 2
$$

$$
\begin{aligned}
F_4 &= \alpha(A_3) + (1-\alpha)(\text{FIT}_3) \\
&= 0.5(53) + (1-0.5)(54) \\
&= 26.5 + (0.5)(54) \\
&= 53.5
\end{aligned}
\qquad
\begin{aligned}
T_4 &= \beta(F_4 - F_3) + (1-\beta)(T_3) \\
&= 0.5(53.5 - 52) + (1-0.5)(2) \\
&= 0.5(1.5) + (0.5)(2) \\
&= 1.75
\end{aligned}
$$

This model predicts a high temperature of $\text{FIT}_4 = 53.5 + 1.75 = 55.25°$ for Day 4. Finally, calculate F_5 and T_5.

$$
\begin{aligned}
F_5 &= \alpha(A_4) + (1-\alpha)(\text{FIT}_4) \\
&= 0.5(64) + (1-0.5)(55.25) \\
&= 32 + (0.5)(55.25) \\
&= 59.63
\end{aligned}
\qquad
\begin{aligned}
T_5 &= \beta(F_5 - F_4) + (1-\beta)(T_4) \\
&= 0.5(59.63 - 53.5) + (1-0.5)(1.75) \\
&= 0.5(6.13) + (0.5)(1.75) \\
&= 3.94
\end{aligned}
$$

The high temperature on Day 5 should be $\text{FIT}_5 = 59.63 + 3.94 = 63.57°$.

Note: Problems 16.22–16.30 refer to the data set in Problem 16.22, daily high temperatures in a city for four consecutive days.

16.29 Compute the mean squared error for the Day 5 forecast FIT_5 calculated in Problem 16.28.

The following table calculates the error for each forecast and the square of each error.

Day	Actual	FIT	Error	Squared Error
1	48	48	—	—
2	56	48	8	64
3	53	54.00	−1	1
4	64	55.25	8.75	76.56

Calculate the mean squared error.

$$\text{MSE} = \frac{64 + 1 + 76.56}{3} = \frac{141.56}{3} = 47.19$$

Note: Problems 16.22–16.30 refer to the data set in Problem 16.22, daily high temperatures in a city for four consecutive days.

16.30 Which of the four forecasts generated in Problems 16.22–16.29 should most accurately predict the high temperature on Day 5? Explain your answer.

The table below summarizes the results calculated in Problems 16.22–16.29.

Exponential Smoothing	Forecast	MSE with Trend Adjustment
$\alpha = 0.2, \beta = 0.4$	55.97	74.71
$\alpha = 0.4, \beta = 0.3$	60.05	57.9
$\alpha = 0.6, \beta = 0.8$	66.69	44.2
$\alpha = 0.5, \beta = 0.5$	63.57	47.19

The most accurate exponential smoothing with trend adjustment forecast will have the lowest mean squared error. Thus, the best prediction for the temperature is 66.69 because 44.2 is the minimum MSE value in the table.

Trend Projection and Seasonality
Account for trends and seasonal influences

Note: Problems 16.31–16.37 refer to the data set below, the grade point average (GPA) of a college student over the last nine semesters.

Period	Year	Semester	GPA
1	Freshman	Fall	2.2
2	Freshman	Winter	2.7
3	Freshman	Spring	2.4
4	Sophomore	Fall	2.5
5	Sophomore	Winter	3.1
6	Sophomore	Spring	2.7
7	Junior	Fall	2.8
8	Junior	Winter	3.6
9	Junior	Spring	3.2

16.31 Construct a trend projection equation that describes the change in the student's GPA over time.

The trend projection equation for forecasting is the line of best fit $\hat{y} = a + bx$ for ordered pairs (x, y), where x is the independent variable and y is the dependent variable.

This trend projection method establishes a relationship between the variable to be forecasted and time. Thus, time is always the independent variable. In this problem, time is measured in semesters. The dependent variable is always the variable to be forecasted, in this case GPA.

The table below calculates the sum of the independent variables (and their squares), the sum of the dependent variables, and the sum of the products of the paired data.

> See Problems 14.12–14.13 to review linear regression, the line of best fit.

> See Problem 14.1 for a description of independent and dependent variables.

Semester x	GPA y	xy	x^2
1	2.2	2.2	1
2	2.7	5.4	4
3	2.4	7.2	9
4	2.5	10	16
5	3.1	15.5	25
6	2.7	16.2	36
7	2.8	19.6	49
8	3.6	28.8	64
9	3.2	28.8	81
Total 45	**25.2**	**133.7**	**285**

> The data is paired—each semester has a corresponding GPA. Multiply those xy pairs and add up the products.

Thus, $\sum x = 45$, $\sum y = 25.2$, $\sum xy = 133.7$, and $\sum x^2 = 285$ for the $n = 9$ data pairs. Calculate the slope b of the linear regression equation.

$$b = \frac{n\sum xy - \left(\sum x\right)\left(\sum y\right)}{n\sum x^2 - \left(\sum x\right)^2} = \frac{(9)(133.7) - (45)(25.2)}{(9)(285) - (45)^2} = \frac{69.3}{540} = 0.128$$

Note that $b > 0$, so the data trends upward over time; the student's grades improve over her time in school. Calculate a, the y-intercept of the linear regression.

$$a = \frac{\sum y}{n} - b\left(\frac{\sum x}{n}\right) = \frac{25.2}{9} - (0.128)\left(\frac{45}{9}\right) = 2.8 - (0.128)(5) = 2.16$$

The regression equation is $\hat{y} = 2.16 + 0.128x$; the trend in the data is a constant GPA increase of 0.128 each semester.

Note: Problems 16.31–16.37 refer to the data set in Problem 16.31, the grade point average (GPA) of a college student over the last nine semesters.

16.32 Predict the student's senior year fall semester GPA using the trend projection equation generated in Problem 16.31.

Freshman year = semesters 1, 2, and 3 (the order is fall, winter, then spring); sophomore year = semesters 4, 5, and 6; and junior year = semesters 7, 8, and 9. That means fall senior year is semester 10, winter is 11, and spring is 12.

The fall semester of senior year is the tenth semester, so substitute $x = 10$ into the trend projection.

$$\hat{y} = 2.16 + 0.128x$$
$$F_{10} = 2.16 + 0.128(10)$$
$$F_{10} = 2.16 + 1.28$$
$$F_{10} = 3.44$$

The forecast for this student's senior year fall semester GPA is $F_{10} = 3.44$.

Note: Problems 16.31–16.37 refer to the data set in Problem 16.31, the GPA of a college student over the last nine semesters.

16.33 Calculate the mean absolute deviation (MAD) for the trend projection equation constructed in Problem 16.31.

The independent variables are the integers from 1 to 9.

Substitute each of the $n = 9$ independent variables into the trend projection equation $\hat{y} = 2.16 + 0.128x$.

$$F_1 = 2.16 + 0.128(1) = 2.29 \qquad F_2 = 2.16 + 0.128(2) = 2.42$$
$$F_3 = 2.16 + 0.128(3) = 2.54 \qquad F_4 = 2.16 + 0.128(4) = 2.67$$
$$F_5 = 2.16 + 0.128(5) = 2.8 \qquad F_6 = 2.16 + 0.128(6) = 2.93$$
$$F_7 = 2.16 + 0.128(7) = 3.06 \qquad F_8 = 2.16 + 0.128(8) = 3.18$$
$$F_9 = 2.16 + 0.128(9) = 3.31$$

The following table calculates the error in each forecast by subtracting the forecast from the GPA for each semester. The absolute error is the absolute value of that difference.

Semester x	GPA y	Forecast	Error	Absolute Error
1	2.2	2.29	−0.09	0.09
2	2.7	2.42	0.28	0.28
3	2.4	2.54	−0.14	0.14
4	2.5	2.67	−0.17	0.17
5	3.1	2.8	0.3	0.3
6	2.7	2.93	−0.23	0.23
7	2.8	3.06	−0.26	0.26
8	3.6	3.18	0.42	0.42
9	3.2	3.31	−0.11	0.11

The mean absolute deviation is the average of the values in the absolute error column.

$$\text{MAD} = \frac{0.09 + 0.28 + 0.14 + 0.17 + 0.3 + 0.23 + 0.26 + 0.42 + 0.11}{9} = \frac{2}{9} = 0.222$$

Note: Problems 16.31–16.37 refer to the data set in Problem 16.31, the GPA of a college student over the last nine semesters.

16.34 Calculate the seasonal index for each semester in the GPA data.

Seasonality represents the influence that certain seasons have on the data; accounting for its influence often improves forecasts. Three literal seasons are present this problem: the fall, winter, and spring semesters. Calculate the average GPA earned by the student during each season.

$$\text{fall average} = \frac{2.2 + 2.5 + 2.8}{3} = \frac{7.5}{3} = 2.5$$

$$\text{winter average} = \frac{2.7 + 3.1 + 3.6}{3} = \frac{9.4}{3} = 3.13$$

$$\text{spring average} = \frac{2.4 + 2.7 + 3.2}{3} = \frac{8.3}{3} = 2.77$$

Now calculate the average of all $n = 9$ GPAs earned by the student, independent of season. Note that $\sum y = 25.2$, according to Problem 16.31.

$$\bar{y} = \frac{\text{sum of all GPAs}}{n} = \frac{\sum y}{n} = \frac{25.2}{9} = 2.8$$

A seasonal index SI is equal to the average of a season divided by the overall average $\bar{y}$.

$$\text{SI}_{\text{fall}} = \frac{\text{fall average}}{\bar{y}} = \frac{2.5}{2.8} = 0.893$$

$$\text{SI}_{\text{winter}} = \frac{\text{winter average}}{\bar{y}} = \frac{3.13}{2.8} = 1.118$$

$$\text{SI}_{\text{spring}} = \frac{\text{spring average}}{\bar{y}} = \frac{2.77}{2.8} = 0.989$$

The sum of the seasonal indices should be very close to the number of seasons. In this case, $0.893 + 1.118 + 0.989 = 3$.

Note: Problems 16.31–16.37 refer to the data set in Problem 16.31, the GPA of a college student over the last nine semesters.

16.35 Generate a seasonal forecast for the student's senior year fall semester GPA.

A seasonal forecast SF is equal to the product of F_t (the forecast as calculated by a trend projection equation) and the seasonal index: $\text{SF} = F_t(\text{SI})$. Recall that $x = 10$ represents the fall semester of the student's senior year. According to Problem 16.32, $F_{10} = 3.44$. Problem 16.34 determined that $\text{SI}_{\text{fall}} = 0.893$.

$$\text{SF}_{10} = F_{10}(\text{SI}_{\text{fall}}) = (3.44)(0.893) = 3.07$$

The student does not perform as well during fall semesters as she does during the other semesters. The seasonal forecast accounts for this, reducing the original forecast of 3.44 to the seasonal forecast of 3.07.

> Look at the seasonal indices calculated in Problem 16.34:
> $SI_{fall} < SI_{spring} < SI_{winter}$, so the student performs better during the winter semester and worse during the fall semester.

Note: Problems 16.31–16.37 refer to the data set in Problem 16.31, the GPA of a college student over the last nine semesters.

16.36 Calculate the mean absolute deviation for the seasonal forecast.

Generate seasonal forecasts for semesters 1 through 9 using the forecasts computed in Problem 16.33 and the seasonal indices computed in Problem 16.34.

$$SF_1 = F_1\left(SI_{fall}\right) = 2.29(0.893) = 2.04 \qquad SF_2 = F_2\left(SI_{winter}\right) = 2.42(1.118) = 2.71$$

$$SF_3 = F_3\left(SI_{spring}\right) = 2.54(0.989) = 2.51 \qquad SF_4 = F_4\left(SI_{fall}\right) = 2.67(0.893) = 2.38$$

$$SF_5 = F_5\left(SI_{winter}\right) = 2.8(1.118) = 3.13 \qquad SF_6 = F_6\left(SI_{spring}\right) = 2.93(0.989) = 2.90$$

$$SF_7 = F_7\left(SI_{fall}\right) = 3.06(0.893) = 2.73 \qquad SF_8 = F_8\left(SI_{winter}\right) = 3.18(1.118) = 3.56$$

$$SF_9 = F_9\left(SI_{spring}\right) = 3.31(0.989) = 3.27$$

Use the table below to calculate the error in each seasonal forecast and its absolute value.

Semester x	GPA y	Seasonal Forecast	Error	Absolute Error
1	2.2	2.04	0.16	0.16
2	2.7	2.71	−0.01	0.01
3	2.4	2.51	−0.11	0.11
4	2.5	2.38	0.12	0.12
5	3.1	3.13	−0.03	0.03
6	2.7	2.90	−0.2	0.2
7	2.8	2.73	0.07	0.07
8	3.6	3.56	0.04	0.04
9	3.2	3.27	−0.07	0.07

The mean absolute deviation is the mean of the values in the absolute error column.

$$\text{MAD} = \frac{0.16 + 0.01 + 0.11 + 0.12 + 0.03 + 0.2 + 0.07 + 0.04 + 0.07}{9} = \frac{0.81}{9} = 0.09$$

Note: Problems 16.31–16.37 refer to the data set in Problem 16.31, the GPA of a college student over the last nine semesters.

16.37 Which of the two forecasts generated in Problems 16.31–16.36 should most accurately predict the student's GPA in the fall semester of her senior year? Explain your answer.

Consider the table below, which summarizes the conclusions reached in Problems 16.31–16.36.

Forecast Method	Forecast	MAD
Trend projection	3.44	0.222
Trend projection with seasonality	3.07	0.09

The most accurate forecast should be 3.07, generated using trend projection with seasonality, because its mean absolute deviation value of 0.09 is lower than the MAD of the trend projection technique alone. Thus, the seasonal indices improve the trend projection forecast of this data, suggesting a consistent seasonal pattern.

Note: Problems 16.38–16.44 refer to the data set below, the number of new residential houses built (housing starts) in a particular region, in thousands of units, each quarter during 2007 and 2008.

Quarter	2007	2008
1	3.6	2.9
2	4.1	3.0
3	3.5	2.6
4	3.2	2.4

16.38 Construct a trend projection equation that describes the change in housing starts over time.

The independent variable is time, in this case expressed as $n = 8$ quarters, and the dependent variable is housing starts. Use the table below to calculate the sums necessary to compute the slope and y-intercept of a least-square regression model.

Quarter	Period x	Starts y	xy	x^2
2007 Q1	1	3.6	3.6	1
2007 Q2	2	4.1	8.2	4
2007 Q3	3	3.5	10.5	9
2007 Q4	4	3.2	12.8	16

(table continues)

(table continued)

Quarter	Period	Starts		
	x	*y*	*xy*	*x²*
2008 Q1	5	2.9	14.5	25
2008 Q2	6	3.0	18.0	36
2008 Q3	7	2.6	18.2	49
2008 Q4	8	2.4	19.2	64
Total	**36**	**25.3**	**105**	**204**

Substitute $\sum x = 36$, $\sum y = 25.3$, $\sum xy = 105$, and $\sum x^2 = 204$ into the formula for b, the slope of the linear regression model.

Housing starts trend downward over time, because b is negative.

$$b = \frac{n\sum xy - \left(\sum x\right)\left(\sum y\right)}{n\sum x^2 - \left(\sum x\right)^2} = \frac{(8)(105) - (36)(25.3)}{(8)(204) - (36)^2} = \frac{-70.8}{336} = -0.211$$

Substitute $b = -0.211$ into the formula for a, the y-intercept of the regression equation.

$$a = \frac{\sum y}{n} - b\left(\frac{\sum x}{n}\right) = \frac{25.3}{8} - (-0.211)\left(\frac{36}{8}\right) = 3.1625 + 0.9495 = 4.11$$

The original data is listed in thousands of units: 1,000(0.211) = 211.

The trend projection equation for the housing starts forecast is $\hat{y} = 4.11 - 0.211x$. Therefore, the number of housing starts decreases at a constant rate of 211 each quarter.

Note: Problems 16.38–16.44 refer to the data set in Problem 16.38, the number of new residential houses built (housing starts) in a particular region, in thousands of units, each quarter during 2007 and 2008.

16.39 Forecast the number of housing starts for the first quarter of 2009 using the trend projection equation.

The first quarter of 2009 immediately follows the $x = 8$ period (the fourth quarter of 2008). Substitute $x = 9$ into the trend projection equation to calculate F_9.

$$F_9 = 4.11 - 0.211(9) = 4.11 - 1.899 = 2.21$$

Approximately $1,000(2.21) = 2,210$ new residential houses will be built during the first quarter of 2009.

Note: Problems 16.38–16.44 refer to the data set in Problem 16.38, the number of new residential houses built (housing starts) in a particular region, in thousands of units, each quarter during 2007 and 2008.

16.40 Calculate the mean absolute deviation for the trend projection equation constructed in Problem 16.38.

Calculate housing starts forecasts for periods $x = 1$ through $x = 8$ using the trend projection equation $\hat{y} = 4.11 - 0.211x$.

$$F_1 = 4.11 - 0.211(1) = 3.90 \qquad F_2 = 4.11 - 0.211(2) = 3.69$$
$$F_3 = 4.11 - 0.211(3) = 3.48 \qquad F_4 = 4.11 - 0.211(4) = 3.27$$
$$F_5 = 4.11 - 0.211(5) = 3.06 \qquad F_6 = 4.11 - 0.211(6) = 2.84$$
$$F_7 = 4.11 - 0.211(7) = 2.63 \qquad F_8 = 4.11 - 0.211(8) = 2.42$$

The table below subtracts each of the forecasts from the actual housing starts to calculate the error and its absolute value.

Period x	Starts y	Forecast	Error	Absolute Error
1	3.6	3.90	−0.3	0.3
2	4.1	3.69	0.41	0.41
3	3.5	3.48	0.02	0.02
4	3.2	3.27	−0.07	0.07
5	2.9	3.06	−0.16	0.16
6	3.0	2.84	0.16	0.16
7	2.6	2.63	−0.03	0.03
8	2.4	2.42	−0.02	0.02

Average the absolute errors to calculate the mean absolute deviation.

$$\text{MAD} = \frac{0.3 + 0.41 + 0.02 + 0.07 + 0.16 + 0.16 + 0.03 + 0.02}{8} = \frac{1.17}{8} = 0.146$$

Note: Problems 16.38–16.44 refer to the data set in Problem 16.38, the number of new residential houses built (housing starts) in a particular region, in thousands of units, each quarter during 2007 and 2008.

16.41 Calculate the seasonal index for each quarter in the housing data.

This problem contains four seasons, as time is divided into four quarters. Calculate the average housing starts for each season separately.

$$\text{first quarter} = \frac{3.6 + 2.9}{2} = \frac{6.5}{2} = 3.25$$

$$\text{second quarter} = \frac{4.1 + 3.0}{2} = \frac{7.1}{2} = 3.55$$

$$\text{third quarter} = \frac{3.5 + 2.6}{2} = \frac{6.1}{2} = 3.05$$

$$\text{fourth quarter} = \frac{3.2 + 2.4}{2} = \frac{5.6}{2} = 2.8$$

The overall average for the $n = 8$ quarters is $\bar{y} = \dfrac{\sum y}{n} = \dfrac{25.3}{8} = 3.16$. Calculate the seasonal index for each quarter by dividing its average by the overall average.

$$SI_1 = \frac{\text{first quarter average}}{\bar{y}} = \frac{3.25}{3.16} = 1.028$$

$$SI_2 = \frac{\text{second quarter average}}{\bar{y}} = \frac{3.55}{3.16} = 1.123$$

$$SI_3 = \frac{\text{third quarter average}}{\bar{y}} = \frac{3.05}{3.16} = 0.965$$

$$SI_4 = \frac{\text{fourth quarter average}}{\bar{y}} = \frac{2.8}{3.16} = 0.886$$

Note: Problems 16.38–16.44 refer to the data set in Problem 16.38, the number of new residential houses built (housing starts) in a particular region, in thousands of units, each quarter during 2007 and 2008.

16.42 Generate a seasonal forecast for housing starts during the first quarter of 2009.

According to Problem 16.39, $F_9 = 2.21$. Problem 16.41 states that the seasonal index for the first quarter is $SI_1 = 1.028$. The seasonal forecast for the first quarter of 2009 is the product of those values.

$$SF_9 = F_9(SI_1) = 2.21(1.028) = 2.27$$

Approximately $1,000(2.27) = 2,270$ residential houses will be built during the first quarter of 2009.

Note: Problems 16.38–16.44 refer to the data set in Problem 16.38, the number of new residential houses built (housing starts) in a particular region, in thousands of units, each quarter during 2007 and 2008.

16.43 Calculate the mean absolute deviation for the seasonal forecast.

Generate seasonal forecasts for each of the eight quarters using the forecasts from Problem 16.40 and the seasonal indices from Problem 16.41.

$$SF_1 = F_1(SI_1) = 3.90(1.028) = 4.01 \qquad SF_2 = F_2(SI_2) = 3.69(1.123) = 4.14$$

$$SF_3 = F_3(SI_3) = 3.48(0.965) = 3.36 \qquad SF_4 = F_4(SI_4) = 3.27(0.886) = 2.90$$

$$SF_5 = F_5(SI_1) = 3.06(1.028) = 3.15 \qquad SF_6 = F_6(SI_2) = 2.84(1.123) = 3.19$$

$$SF_7 = F_7(SI_3) = 2.63(0.965) = 2.54 \qquad SF_8 = F_8(SI_4) = 2.42(0.886) = 2.14$$

Use the table below to calculate the absolute error in each of the $n = 8$ forecasts.

Period x	Starts y	Forecast	Error	Absolute Error
1	3.6	4.01	−0.41	0.41
2	4.1	4.14	−0.04	0.04
3	3.5	3.36	0.14	0.14
4	3.2	2.90	0.3	0.3

Period x	Starts y	Forecast	Error	Absolute Error
5	2.9	3.15	−0.25	0.25
6	3.0	3.19	−0.19	0.19
7	2.6	2.54	0.06	0.06
8	2.4	2.14	0.26	0.26

Calculate the MAD.

$$MAD = \frac{0.41 + 0.04 + 0.14 + 0.3 + 0.25 + 0.19 + 0.06 + 0.26}{8} = \frac{1.65}{8} = 0.206$$

Note: Problems 16.38–16.44 refer to the data set in Problem 16.38, showing the number of new residential houses built (housing starts) in a particular region, in thousands of units, each quarter during 2007 and 2008.

16.44 Which of the two forecasts generated in Problems 16.38–16.43 more accurately predicts the housing starts for the first quarter of 2009? Explain your answer.

Consider the following table, which summarizes the conclusions reached in Problems 16.38–16.43.

Forecast Method	Forecast	MAD
Trend projection	2.21	0.146
Trend projection with seasonality	2.27	0.206

The more accurate forecast should be 2,210 new homes, because the trend projection with seasonality has a higher MAD than the trend projection alone. This data, therefore, has no consistent seasonal pattern.

> Introducing a seasonal influence on the data actually makes the prediction less accurate in this example.

Causal Forecasting

The independent variable doesn't have to be time

Note: Problems 16.45–16.47 refer to the data set below, the monthly demand for a computer printer at various prices.

Demand	Price	Demand	Price
36	$70	14	$110
23	$80	10	$120
12	$90	5	$130
16	$100	2	$140

16.45 Construct a simple regression model to forecast demand based on price.

Simple regression can be used to forecast a dependent variable value based solely on an independent variable value. Unlike seasonal and trend forecasts, the independent variable is not restricted to time. In this problem, price is the independent variable and demand is the dependent variable.

Use the table below to compute the sums necessary to construct the regression model.

	Price x	Demand y	xy	x^2
	70	36	2,520	4,900
	80	23	1,840	6,400
	90	12	1,080	8,100
	100	16	1,600	10,000
	110	14	1,540	12,100
	120	10	1,200	14,400
	130	5	650	16,900
	140	2	280	19,600
Total	**840**	**118**	**10,710**	**92,400**

Substitute $\sum x = 840$, $\sum y = 118$, $\sum xy = 10,710$, and $\sum x^2 = 92,400$ into the formula for b, the slope of the linear regression model.

$$b = \frac{n\sum xy - \left(\sum x\right)\left(\sum y\right)}{n\sum x^2 - \left(\sum x\right)^2} = \frac{(8)(10,710) - (840)(118)}{(8)(92,400) - (840)^2} = \frac{-13,440}{33,600} = -0.4$$

Substitute $b = -0.4$ into the formula for a, the y-intercept of the regression model.

$$a = \frac{\sum y}{n} - b\left(\frac{\sum x}{n}\right) = \frac{118}{8} - (-0.40)\left(\frac{840}{8}\right) = 14.75 + 42 = 56.75$$

The regression equation is $\hat{y} = 56.75 - 0.4x$; each time the price is increased by $1, 0.4 fewer printers are sold.

Note: Problems 16.45–16.47 refer to the data set in Problem 16.45, the monthly demand for a computer printer at various prices.

16.46 Predict printer demand given a $95 price.

Substitute $x = 95$ into the regression model $\hat{y} = 56.75 - 0.4x$, constructed in Problem 16.45.

$$F_{95} = 56.75 - 0.4(95) = 56.75 - 38 = 18.75$$

The demand should be $18.75 \approx 19$ printers when the price is $95.

Note: Problems 16.45–16.47 refer to the data set in Problem 16.45, the monthly demand for a computer printer at various prices.

16.47 Calculate the mean squared error for the regression equation constructed in Problem 16.45.

Generate $n = 8$ demand forecasts, one for each price in the data.

$$F_{70} = 56.75 - 0.4(70) = 28.75 \qquad F_{80} = 56.75 - 0.4(80) = 24.75$$
$$F_{90} = 56.75 - 0.4(90) = 20.75 \qquad F_{100} = 56.75 - 0.4(100) = 16.75$$
$$F_{110} = 56.75 - 0.4(110) = 12.75 \qquad F_{120} = 56.75 - 0.4(120) = 8.75$$
$$F_{130} = 56.75 - 0.4(130) = 4.75 \qquad F_{140} = 56.75 - 0.4(140) = 0.75$$

Use the table below to compute the error in each forecast and its square.

Price x	Demand y	Forecast	Error	Squared Error
70	36	28.75	7.25	52.56
80	23	24.75	−1.75	3.06
90	12	20.75	−8.75	76.56
100	16	16.75	−0.75	0.56
110	14	12.75	1.25	1.56
120	10	8.75	1.25	1.56
130	5	4.75	0.25	0.06
140	2	0.75	1.25	1.56

> Price is the independent variable (you can affect it directly), so it goes in the x column.

Calculate the average of the squared errors.

$$\text{MSE} = \frac{52.56 + 3.06 + 76.56 + 0.56 + 1.56 + 1.56 + 0.06 + 1.56}{8} = \frac{137.48}{8} = 17.19$$

Note: Problems 16.48–16.50 refer to the data set below, the average attendance per game, in thousands of fans, for ten Major League Baseball teams and the number of games won by those teams during the 2008 season.

Team	Attendance	Wins	Team	Attendance	Wins
NY (AL)	53.1	89	ATL	31.2	72
STL	42.3	86	WAS	29.0	59
CHI (NL)	40.7	97	CLE	25.4	81
BOS	37.6	95	TEX	24.3	79
COL	33.1	74	PIT	20.1	67

"Based on" is a key phrase that identifies the independent variable. You want to forecast the dependent variable (attendance) based on the independent variable (wins).

16.48 Construct a simple regression model to forecast average attendance based on the number of wins.

Use the table below to compute the sums necessary to construct the linear regression model describing the relationship between wins (the independent variable) and attendance (the dependent variable).

	Wins	Attendance		
	x	y	xy	x^2
	89	53.1	4,725.9	7,921
	86	42.3	3,637.8	7,396
	97	40.7	3,947.9	9,409
	95	37.6	3,572	9,025
	74	33.1	2,449.4	5,476
	72	31.2	2,246.4	5,184
	59	29.0	1,711	3,481
	81	25.4	2,057.4	6,561
	79	24.3	1,919.7	6,241
	67	20.1	1,346.7	4,489
Total	**799**	**336.8**	**27,614.2**	**65,183**

Substitute $\sum x = 799$, $\sum y = 336.8$, $\sum xy = 27,614.2$, and $\sum x^2 = 65,183$ into the formula for b, the slope of the linear regression model. Note that there are $n = 10$ pairs of data.

$$b = \frac{n \sum xy - \left(\sum x \right)\left(\sum y \right)}{n \sum x^2 - \left(\sum x \right)^2} = \frac{(10)(27,614.2) - (799)(336.8)}{(10)(65,183) - (799)^2} = \frac{7,038.8}{13,429} = 0.524$$

Calculate the y-intercept of the regression model.

$$a = \frac{\sum y}{n} - b\left(\frac{\sum x}{n}\right) = \frac{336.8}{10} - (0.524)\left(\frac{799}{10}\right) = 33.68 - 41.8676 = -8.19$$

The simple regression model is $\hat{y} = -8.19 + 0.524x$; each win increases attendance by approximately $1{,}000(0.524) = 524$ fans.

Note: Problems 16.48–16.50 refer to the data set in Problem 16.48, the average attendance per game, in thousands of fans, for ten Major League Baseball teams and the number of games won by those teams during the 2008 season.

16.49 Predict average attendance for a team that wins 85 games during the season.

Substitute $x = 85$ into the regression model $\hat{y} = -8.19 + 0.524x$ generated in Problem 16.48.

$$F_{85} = -8.19 + 0.524(85) = -8.19 + 44.54 = 36.35$$

According to the regression model, an average of $1{,}000(36.35) = 36{,}350$ fans will attend the games of a team that wins 85 games.

Note: Problems 16.48–16.50 refer to the data set in Problem 16.48, the average attendance per game, in thousands of fans, for ten Major League Baseball teams and the number of games won by those teams during the 2008 season.

16.50 Calculate the mean squared error for the regression equation constructed in Problem 16.48.

Generate attendance forecasts for each of the win totals in the data set.

$$F_{89} = -8.19 + 0.524(89) = 38.45 \qquad F_{86} = -8.19 + 0.524(86) = 36.87$$
$$F_{97} = -8.19 + 0.524(97) = 42.64 \qquad F_{95} = -8.19 + 0.524(95) = 41.59$$
$$F_{74} = -8.19 + 0.524(74) = 30.59 \qquad F_{72} = -8.19 + 0.524(72) = 29.54$$
$$F_{59} = -8.19 + 0.524(59) = 22.73 \qquad F_{81} = -8.19 + 0.524(81) = 34.25$$
$$F_{79} = -8.19 + 0.524(79) = 33.21 \qquad F_{67} = -8.19 + 0.524(67) = 26.92$$

Use the table below to calculate the squared error for each forecast.

Wins x	Attendance y	Forecast	Error	Squared Error
89	53.1	38.45	14.65	214.62
86	42.3	36.87	5.43	29.48
97	40.7	42.64	−1.94	3.76
95	37.6	41.59	−3.99	15.92
74	33.1	30.59	2.51	6.30
72	31.2	29.54	1.66	2.76

(table continues)

(table continued)

Wins x	Attendance y	Forecast	Error	Squared Error
59	29.0	22.73	6.27	39.31
81	25.4	34.25	−8.85	78.32
79	24.3	33.21	−8.91	79.38
67	20.1	26.92	−6.82	46.51

Calculate the MSE.

$$\text{MSE} = \frac{214.62 + 29.48 + 3.76 + 15.92 + 6.3 + 2.76 + 39.31 + 78.32 + 79.38 + 46.51}{10} = \frac{516.36}{10} = 51.64$$

Chapter 17
STATISTICAL PROCESS CONTROL

Using statistics to measure quality

Statistical process control objectively evaluates the performance of a process by measuring the current state of the process and drawing conclusions based on statistical analysis. You can both ensure that a process falls within particular specifications and determine whether a process is capable of meeting its design specifications.

Control charts establish lower and upper limits that are used to decide whether a process is operating satisfactorily. Process capability compares the actual performance of a process to the expected performance to determine whether the process is capable of performing as expected.

Introduction to Statistical Process Control
Exploring the different types of quality measurement

> "Out of statistical control" means something is probably wrong with the process.

17.1 Explain how statistical process control determines whether a process is in or out of statistical control.

Statistical process control is used to measure the performance of a process by sampling the process at predetermined intervals. The results of the sample are reported in control charts that plot time on the x-axis and the measurement of interest on the y-axis. The chart contains lower and upper control limits used to establish the condition of the process. If the result of the sample falls between these limits, the process is considered to be in statistical control. If the sample result falls above the upper limit or below the lower limit, the process is said to be out of statistical control.

> Buttons produced at a manufacturing plant may look the same to the naked eye, but given precise instruments, you could find variation in things like the diameter or the thickness of the buttons.

17.2 Describe the two types of variation a process can exhibit.

Variation is part of every process, whether a manufacturing or a service process. Differences, however small, are always present between items produced or services rendered. *Natural variation* describes the differences that are normal for the process and are to be expected. As long as the process variation falls into this category, no corrective steps are necessary.

Assignable variation is due to a breakdown in the process, and is therefore more significant than natural variation. Assignable variation may be the result of poorly trained workers, machine wear, or low-quality raw materials. Once assignable variation becomes a factor in a process, its cause should be identified and addressed.

> An example of a variation in a service process would be the number of minutes a person waits on hold when calling for customer service.

17.3 Describe the two types of measurements used in statistical process control.

The first type of statistical process control is *variable measurement*, which measures quality on a continuous scale. Processes using this type of statistical control include weight, height, speed, and thickness measurements. Depending on the precision of the measuring devices used, there may be infinitely many unique measurements.

The second type of statistical process control is *attribute measurement*, which measures quality on a discrete scale. Counting the number of defective items in a production batch is an example of attribute measurement.

> Measuring things = variable measurement
> Counting things = attribute measurement

Statistical Process Control for Variable Measurement
Mean and range control charts

17.4 Explain how mean and range control charts are constructed and describe the role they play in statistical process control.

At predetermined time intervals, n items from the process are sampled. The mean of the sample is calculated and plotted on a mean control chart.

The range of the sample is also measured; it is plotted on the range control chart. If the sample mean is within the lower and upper limits of the mean chart and the sample range is within the lower and upper limits of the range chart, the process is considered to be in control.

> The range of the sample is the highest value of the sample minus the lowest value.

The mean and range charts must both lie within the control limits to consider the process in control. If one chart indicates the process is within limits while the other chart concludes the process is outside limits, you conclude that the process is out of control.

Note: Problems 17.5–17.7 refer to the data set below, the weights in ounces of cereal boxes sampled from a filling process over a three-hour period.

Sample	Time	Box 1	Box 2	Box 3
1	6 A.M.	16.4	16.4	15.8
2	7 A.M.	16.0	16.2	16.4
3	8 A.M.	15.6	16.1	16.3
4	9 A.M.	16.0	15.9	16.1

17.5 Calculate the lower and upper control limits for a 3-sigma mean chart.

Calculate the mean and range for each of the four samples.

$$\bar{x}_1 = \frac{16.4 + 16.4 + 15.8}{3} = \frac{48.6}{3} = 16.2 \qquad R_1 = 16.4 - 15.8 = 0.6$$

$$\bar{x}_2 = \frac{16.0 + 16.2 + 16.4}{3} = \frac{48.6}{3} = 16.2 \qquad R_2 = 16.4 - 16.0 = 0.4$$

$$\bar{x}_3 = \frac{15.6 + 16.1 + 16.3}{3} = \frac{48}{3} = 16 \qquad R_3 = 16.3 - 15.6 = 0.7$$

$$\bar{x}_4 = \frac{16.0 + 15.9 + 16.1}{3} = \frac{48}{3} = 16 \qquad R_4 = 16.1 - 15.9 = 0.2$$

The grand average $\bar{\bar{x}}$ of the data is the average of the sample means; the average range $\bar{R}$ is the average of the sample ranges.

$$\bar{\bar{x}} = \frac{16.2 + 16.2 + 16 + 16}{4} = \frac{64.4}{4} = 16.1$$

$$\bar{R} = \frac{0.6 + 0.4 + 0.7 + 0.2}{4} = \frac{1.9}{4} = 0.475$$

A 3-sigma mean control chart sets limits three standard deviations above and below the process mean. Calculate the lower control limit $LCL_{\bar{x}}$ and upper control limit $UCL_{\bar{x}}$ for the mean chart using the following formulas, where A_2 is the mean factor constant from Reference Table 10. In this problem, each sample consists of $n = 3$ values, so $A_2 = 1.023$.

> The variable n represents the size of the samples (3 in this problem) not the number of samples (4).

$$LCL_{\bar{x}} = \bar{\bar{x}} - A_2 \bar{R} \qquad\qquad UCL_{\bar{x}} = \bar{\bar{x}} + A_2 \bar{R}$$
$$= 16.1 - 1.023(0.475) \qquad = 16.1 + 1.023(0.475)$$
$$= 15.61 \qquad\qquad = 16.59$$

Note: Problems 17.5–17.7 refer to the data set in Problem 17.5, the weights in ounces of cereal boxes sampled from a filling process over a three-hour period.

17.6 Calculate the lower and upper control limits for a 3-sigma range chart.

The data consists of four samples of size $n = 3$. According to Reference Table 10, this sample size corresponds to constant values $D_3 = 0$ and $D_4 = 2.574$. Substitute D_3 and D_4 into the formulas for the lower control limit LCL_R and the upper control limit UCL_R.

According to Problem 17.5, $\bar{R} = 0.475$.

$$LCL_R = D_3 \bar{R} \qquad\qquad UCL_R = D_4 \bar{R}$$
$$= 0(0.475) \qquad\qquad = 2.574(0.475)$$
$$= 0 \qquad\qquad\qquad = 1.22$$

Note: Problems 17.5–17.7 refer to the data set in Problem 17.5, the weights in ounces of cereal boxes sampled from a filling process over a three-hour period.

17.7 Three more boxes are sampled at 10 A.M., and the following weights (in ounces) are recorded: 15.2, 15.9, and 16.6. Determine whether the filling process is in control based on the control charts generated in Problems 17.5 and 17.6.

In order for the process to be considered in control, the 10 A.M. sample mean must be within the limits of the mean chart and the 10 A.M. sample range must be within the limits of the range chart. Calculate the mean and range of the 10 A.M. sample.

$$\bar{x} = \frac{15.2 + 15.9 + 16.6}{3} = \frac{47.7}{3} = 15.9$$
$$R = 16.6 - 15.2 = 1.4$$

The range chart measures the variability of the process. Too much variability is an indication of poor process quality.

The 10 A.M. sample mean $\bar{x} = 15.9$ lies within the mean control limits of $LCL_{\bar{x}} = 15.61$ and $UCL_{\bar{x}} = 16.59$. However, the sample range $R = 1.4$ is greater than the range upper control limit $UCL_R = 1.22$. Because the sample range lies outside the allowable limits, the process is considered out of control.

Note: Problems 17.8–17.10 refer to the data set below, the weights (in pounds) of fertilizer bags sampled from a filling process over a three-hour period.

Sample	Time	Bag 1	Bag 2	Bag 3	Bag 4
1	1 P.M.	49.0	48.7	50.0	50.7
2	2 P.M.	51.2	49.1	50.1	50.4
3	3 P.M.	52.1	51.4	49.6	49.7
4	4 P.M.	50.1	49.5	51.8	51.8

17.8 Calculate the lower and upper control limits for a 3-sigma mean chart.

Calculate the mean and range of each sample.

$$\bar{x}_1 = \frac{49.0 + 48.7 + 50.0 + 50.7}{4} = \frac{198.4}{4} = 49.6 \qquad R_1 = 50.7 - 48.7 = 2$$

$$\bar{x}_2 = \frac{51.2 + 49.1 + 50.1 + 50.4}{4} = \frac{200.8}{4} = 50.2 \qquad R_2 = 51.2 - 49.1 = 2.1$$

$$\bar{x}_3 = \frac{52.1 + 51.4 + 49.6 + 49.7}{4} = \frac{202.8}{4} = 50.7 \qquad R_3 = 52.1 - 49.6 = 2.5$$

$$\bar{x}_4 = \frac{50.1 + 49.5 + 51.8 + 51.8}{4} = \frac{203.2}{4} = 50.8 \qquad R_4 = 51.8 - 49.5 = 2.3$$

Compute the grand average $\bar{\bar{x}}$ and the average range $\bar{R}$.

$$\bar{\bar{x}} = \frac{49.6 + 50.2 + 50.7 + 50.8}{4} = \frac{201.3}{4} = 50.33$$

$$\bar{R} = \frac{2 + 2.1 + 2.5 + 2.3}{4} = \frac{8.9}{4} = 2.23$$

Four bags are sampled each hour, so $n = 4$. According to Reference Table 10, $A_2 = 0.729$. Apply the control limit formulas for the mean chart.

$$LCL_{\bar{x}} = \bar{\bar{x}} - A_2 \bar{R} \qquad\qquad UCL_{\bar{x}} = \bar{\bar{x}} + A_2 \bar{R}$$

$$= 50.33 - 0.729(2.23) \qquad\qquad = 50.33 + 0.729(2.23)$$

$$= 48.70 \qquad\qquad\qquad = 51.96$$

Note: Problems 17.8–17.10 refer to the data set in Problem 17.8, the weights (in pounds) of fertilizer bags sampled from a filling process over a three-hour period.

17.9 Calculate the lower and upper control limits for a 3-sigma range chart.

According to Reference Table 10, $D_3 = 0$ and $D_4 = 2.282$ when $n = 4$. Calculate the control limits for the range chart.

$$LCL_R = D_3 \bar{R} \qquad\qquad UCL_R = D_4 \bar{R}$$

$$= 0(2.23) \qquad\qquad = 2.282(2.23)$$

$$= 0 \qquad\qquad\qquad = 5.09$$

Note: Problems 17.8–17.10 refer to the data set in Problem 17.8, the weights (in pounds) of fertilizer bags sampled from a filling process over a three-hour period.

17.10 Determine whether the filling process is in control based on a sample taken at 5 P.M. that consists of the following weights (in pounds): 50.9, 50.4, 48.4, and 51.7.

Calculate the mean and range of the 5 P.M. sample.

$$\bar{x} = \frac{50.9 + 50.4 + 48.4 + 51.7}{4} = \frac{201.4}{4} = 50.35$$

$$R = 51.7 - 48.4 = 3.3$$

The 5 P.M. sample mean lies within the mean chart limits calculated in Problem 17.8 (48.70 < 50.35 < 51.96); the range of the new sample lies within the range limits as well (0 < 3.3 < 5.09).

Because the sample mean and sample range both lie within the control limits, the process is considered in control.

One of the 5 P.M. sample weights (48.4) is below the lower limit of the mean chart (48.70). Even if individual measurements lie outside the limits, the process isn't necessarily out of control.

Note: Problems 17.11–17.13 refer to the data set below, the time (in minutes) that customers waited to be greeted by their server in a restaurant over a three-night time period.

Evening	Table 1	Table 2	Table 3	Table 4	Table 5
1	3.7	5.2	2.9	4.0	3.5
2	4.6	5.7	3.4	4.5	5.0
3	3.0	2.3	3.6	5.1	5.6

17.11 Calculate the lower and upper control limits for a 3-sigma mean chart.

Calculate the mean and range of each sample.

$$\bar{x}_1 = \frac{3.7 + 5.2 + 2.9 + 4.0 + 3.5}{5} = \frac{19.3}{5} = 3.86 \qquad R_1 = 5.2 - 2.9 = 2.3$$

$$\bar{x}_2 = \frac{4.6 + 5.7 + 3.4 + 4.5 + 5.0}{5} = \frac{23.2}{5} = 4.64 \qquad R_2 = 5.7 - 3.4 = 2.3$$

$$\bar{x}_3 = \frac{3.0 + 2.3 + 3.6 + 5.1 + 5.6}{5} = \frac{19.6}{5} = 3.92 \qquad R_3 = 5.6 - 2.3 = 3.3$$

The grand average $\bar{\bar{x}}$ and the average range $\bar{R}$ are the averages of the sample means and sample ranges, respectively.

$$\bar{\bar{x}} = \frac{3.86 + 4.64 + 3.92}{3} = \frac{12.42}{3} = 4.14$$

$$\bar{R} = \frac{2.3 + 2.3 + 3.3}{3} = \frac{7.9}{3} = 2.63$$

The data contains three samples, each of size $n = 5$. According to Reference Table 10, $A_2 = 0.577$. Calculate the control limits for the 3-sigma mean chart.

$$\text{LCL}_{\bar{x}} = \bar{\bar{x}} - A_2 \bar{R} \qquad \text{UCL}_{\bar{x}} = \bar{\bar{x}} + A_2 \bar{R}$$
$$= 4.14 - 0.577(2.63) \qquad = 4.14 + 0.577(2.63)$$
$$= 2.62 \qquad = 5.66$$

You're not actually drawing a chart; imagine two horizontal lines representing the lower and upper control limits. Any time a sample mean spikes above or below those lines, the process is out of control.

Note: Problems 17.11–17.13 refer to the data set in Problem 17.11, the time (in minutes) that customers waited to be greeted by their server in a restaurant over a three-night time period.

17.12 Calculate the lower and upper control limits for a 3-sigma range control chart.

According to Reference Table 10, given $n = 5$, $D_3 = 0$ and $D_4 = 2.115$. Compute the control limits for the range.

$$\text{LCL}_R = D_3 \bar{R} \qquad \text{UCL}_R = D_4 \bar{R}$$
$$= 0(2.63) \qquad = 2.115(2.63)$$
$$= 0 \qquad = 5.56$$

Note: Problems 17.11–17.13 refer to the data set in Problem 17.11, the time (in minutes) that customers waited to be greeted by their server in a restaurant over a three-night time period.

17.13 A sample taken on the fourth evening had the following times: 5.2, 5.8, 6.3, 6.4, and 5.5. Determine whether the service process in the restaurant is in control.

Calculate the mean and range of the new sample.

$$\bar{x} = \frac{5.2 + 5.8 + 6.3 + 6.4 + 5.5}{5} = \frac{29.2}{5} = 5.84$$
$$R = 6.4 - 5.2 = 1.2$$

Whereas the sample range lies within its control limits of 0 and 5.56, the sample mean does not—5.84 is greater than the upper limit of $\text{UCL}_{\bar{x}} = 5.66$. Because both sample statistics do not lie within their control limits, the service process is considered out of control.

Servers were not greeting customers fast enough during the fourth evening.

Note: Problems 17.14–17.16 refer to the table below, the data recorded by a manufacturing plant that randomly selected a sample of nine pistons per day for five days and measured the diameter of the pistons (in millimeters).

Day	Sample Mean	Sample Range
1	80.6	0.9
2	82.1	1.4
3	81.5	1.0
4	82.9	1.9
5	81.0	1.5

17.14 Calculate the lower and upper control limits for a 3-sigma mean chart.

The sample means and ranges are provided, so begin by calculating the grand mean and average range.

$$\bar{\bar{x}} = \frac{80.6 + 82.1 + 81.5 + 82.9 + 81.0}{5} = \frac{408.1}{5} = 81.62$$

$$\bar{R} = \frac{0.9 + 1.4 + 1.0 + 1.9 + 1.5}{5} = \frac{6.7}{5} = 1.34$$

Each sample consisted of $n = 9$ pistons; according to Reference Table 10, $A_2 = 0.337$. Calculate the control limits.

$$\text{LCL}_{\bar{x}} = \bar{\bar{x}} - A_2 \bar{R} \qquad\qquad \text{UCL}_{\bar{x}} = \bar{\bar{x}} + A_2 \bar{R}$$
$$= 81.62 - 0.337(1.34) \qquad\qquad = 81.62 + 0.337(1.34)$$
$$= 81.17 \qquad\qquad\qquad\quad = 82.07$$

Note: Problems 17.14–17.16 refer to the table in Problem 17.14, the data recorded by a manufacturing plant that randomly selected a sample of nine pistons per day for five days and measured the diameter of the pistons (in millimeters).

17.15 Calculate the lower and upper control limits for a 3-sigma range chart.

Given $n = 9$, Reference Table 10 states that $D_3 = 0.184$ and $D_4 = 1.816$. Calculate the control limits.

$$\text{LCL}_R = D_3 \bar{R} \qquad\qquad \text{UCL}_R = D_4 \bar{R}$$
$$= 0.184(1.34) \qquad\qquad = 1.816(1.34)$$
$$= 0.25 \qquad\qquad\qquad = 2.43$$

Note: Problems 17.14–17.16 refer to the table in Problem 17.14, the data recorded by a manufacturing plant that randomly selected a sample of nine pistons per day for five days and measured the diameter of the pistons (in millimeters).

17.16 A sample of nine pistons randomly selected on the sixth day consisted of the following measurements: 80.7, 82.7, 82.4, 82.1, 82.2, 80.4, 81.9, 81.9, and 82.1. Determine whether the manufacturing process is in control.

Calculate the mean and range of the new sample.

$$\bar{x} = \frac{80.7 + 82.7 + 82.4 + 82.1 + 82.2 + 80.4 + 81.9 + 81.9 + 82.1}{9} = \frac{736.4}{9} = 81.82$$

$$R = 82.7 - 80.4 = 2.3$$

This sample mean $\bar{x} = 81.82$ lies between the mean charts limits 81.17 and 82.07 that are computed in Problem 17.4. The sample range $R = 2.3$ lies within the range limits of 0.25 and 2.43, also computed in Problem 17.15. Because the sample mean and sample range are both within control limits, the process is considered in control.

Statistical Process Control for Attribute Measurement Using p-charts

Calculate the proportion of defective items

17.17 Explain how to develop and apply *p*-charts as an attribute measurement statistical process control technique.

Attribute measurement is applied when classifying items as defective or nondefective. One such technique uses a *p*-chart to measure the percent defective in a sample. Attribute measurement requires only one chart to determine whether a process is in control. ←

Unlike variable measurement, which has control limits for the mean and the range (as described in Problems 17.4– 17.16).

The standard deviation for the control chart limits σ_p is calculated using the following formula, in which $\bar{p}$ is the average percent defective in the sample and n is the sample size.

$$\sigma_p = \sqrt{\frac{\bar{p}(1-\bar{p})}{n}}$$

A 3-sigma *p*-chart assigns control limits that are three standard deviations above and below the average percent defective $\bar{p}$. Thus, it is common to set $z = 3$ in the control limit formulas below.

$$\text{LCL}_p = \bar{p} - z\sigma_p \qquad \text{UCL}_p = \bar{p} + z\sigma_p$$

Note: Problems 17.18–17.19 refer to the data set below, the number of defective lightbulbs from 10 samples of size n = 100.

Sample	Number of Defects	Sample	Number of Defects
1	7	6	5
2	5	7	5
3	6	8	2
4	3	9	1
5	3	10	4

17.18 Compute the lower and upper control limits for a 3-sigma *p*-chart.

Calculate the average percent defective by dividing the total number of defective lightbulbs by the total number of lightbulbs in the samples.

> *Each of the 10 samples contained 100 lightbulbs, for a total of 10(100) = 1,000 lightbulbs tested.*

$$\bar{p} = \frac{7+5+6+3+3+5+5+2+1+4}{10(100)} = \frac{41}{1,000} = 0.041$$

Substitute $\bar{p} = 0.041$ into the formula for the standard deviation of the control limits.

$$\sigma_p = \sqrt{\frac{\bar{p}(1-\bar{p})}{n}} = \sqrt{\frac{0.041(1-0.041)}{100}} = \sqrt{\frac{0.0393}{100}} = \sqrt{0.000393} = 0.0198$$

Substitute $z = 3$ into the formulas below to calculate the lower control limit LCL_p and the upper control limit UCL_p for a 3-sigma *p*-chart.

> *A p-chart can't have a negative limit. If LCL_p is less than zero, round it up to zero.*

$$\begin{aligned}
\text{LCL}_p &= \bar{p} - z\sigma_p \\
&= 0.041 - 3(0.0198) \\
&= -0.0184 \\
&\approx 0
\end{aligned}
\qquad
\begin{aligned}
\text{UCL}_p &= \bar{p} + z\sigma_p \\
&= 0.041 + 3(0.0198) \\
&= 0.041 + 0.0594 \\
&= 0.1004
\end{aligned}$$

Note: Problems 17.18–17.19 refer to the data set in Problem 17.18, the number of defective lightbulbs from 10 samples of size n = 100.

17.19 A new sample of 100 lightbulbs includes 6 that are defective. Determine whether the process is in control.

Calculate the percent defective of the new sample.

$$\bar{p} = \frac{\text{number of defective items}}{\text{number of items tested}} = \frac{6}{100} = 0.06$$

Because $\bar{p} = 0.06$ lies between the control limits $\text{LCL}_p = 0$ and $\text{UCL}_p = 0.1004$, the process is in control.

Note: Problems 17.20–17.21 refer to the table below, the number of free throws missed by a basketball player attempting 150 free throws every day for 12 days.

Day	Missed Free Throws	Day	Missed Free Throws
1	40	7	38
2	36	8	40
3	35	9	37
4	41	10	43
5	40	11	40
6	46	12	41

17.20 Calculate the lower and upper control limits for a 3-sigma p-chart.

Calculate the average percent defective $\bar{p}$ for all 12 days. In this instance, "percent defective" refers to the percentage of free throws missed. The player shot $n = 150$ free throws a day for 12 days, a total of $12(150) = 1,800$ free throws.

$$\bar{p} = \frac{40+36+35+41+40+46+38+40+37+43+40+41}{12(150)} = \frac{477}{1,800} = 0.265$$

Use $\bar{p} = 0.265$ to compute σ_p, the standard deviation for the control chart limits.

$$\sigma_p = \sqrt{\frac{\bar{p}(1-\bar{p})}{n}} = \sqrt{\frac{(0.265)(1-0.265)}{150}} = \sqrt{\frac{0.19478}{150}} = \sqrt{0.00130} = 0.0361$$

When you're constructing a 3-sigma p-chart, z = 3.

Apply the 3-sigma p-chart upper and lower control limit formulas.

$$\text{LCL}_p = \bar{p} - z\sigma_p \qquad\qquad \text{UCL}_p = \bar{p} + z\sigma_p$$
$$= 0.265 - 3(0.0361) \qquad\qquad = 0.265 + 3(0.0361)$$
$$= 0.157 \qquad\qquad\qquad = 0.373$$

Note: Problems 17.20–17.21 refer to the table in Problem 17.20, the number of free throws missed by a basketball player attempting 150 free throws every day for 12 days.

17.21 Today, the player attempts 150 free throws. He misses 58. Determine whether the player's free throw process is in control.

Calculate today's percentage of missed free throws.

$$\bar{p} = \frac{58}{150} = 0.387$$

The player's free throw process is considered out of control because the sample proportion for today (0.387) is greater than the upper control limit $\text{UCL}_p = 0.373$ calculated in Problem 17.20.

Note: Problems 17.22–17.23 refer to the data set below, the total number of defective shirts produced per day based on a daily random sample of 125 shirts.

Sample	Number of Defects	Sample	Number of Defects
1	2	8	2
2	5	9	7
3	8	10	2
4	1	11	4
5	3	12	3
6	5	13	8
7	3	14	3

17.22 Calculate the lower and upper control limits for a 3-sigma p-chart.

Calculate the average percent defective and the standard deviation for the control chart limits.

$$\bar{p} = \frac{2+5+8+1+3+5+3+2+7+2+4+3+8+3}{14(125)} = \frac{56}{1,750} = 0.032$$

$$\sigma_p = \sqrt{\frac{\bar{p}(1-\bar{p})}{n}} = \sqrt{\frac{(0.032)(1-0.032)}{125}} = \sqrt{\frac{0.03098}{125}} = \sqrt{0.000248} = 0.0157$$

Compute the lower and upper control limits for a 3-sigma p-chart.

Remember, the lower limit can't be negative. When it is, set it equal to zero instead.

$$
\begin{aligned}
LCL_p &= \bar{p} - z\sigma_p \\
&= 0.032 - 3(0.0157) \\
&= -0.0151 \\
&\approx 0
\end{aligned}
\qquad
\begin{aligned}
UCL_p &= \bar{p} + z\sigma_p \\
&= 0.032 + 3(0.0157) \\
&= 0.032 + 0.0471 \\
&= 0.079
\end{aligned}
$$

Note: Problems 17.22–17.23 refer to the data set in Problem 17.22, the total number of defective shirts produced per day based on a daily random sample of 125 shirts.

17.23 The next sample of 125 shirts includes 11 that are defective. Determine whether the process is in control.

Calculate the percent defective of the new sample.

$$\bar{p} = \frac{11}{125} = 0.088$$

The percent defective is greater than the upper control limit determined in Problem 17.22 (0.088 > 0.079), so the process is considered out of control.

Statistical Process Control for Attribute Measurement Using c-charts

Counting the number of defective items

17.24 Explain how to develop and apply *c*-charts as an attribute measurement statistical process control technique.

> So set $z = 3$ in the control limit formulas.

Whereas a *p*-chart (described in Problems 17.17–17.23) is applied when an item is classified as either defective or nondefective, a *c*-chart is applied when an item may have more than one defect. Like *p*-charts, *c*-chart control limits are set three standard deviations above and below $\bar{c}$, the average number of defects per unit.

Apply the following formulas to calculate the upper and lower control limits for a *c*-chart.

$$\text{LCL}_c = \bar{c} - z\sqrt{\bar{c}} \qquad \text{UCL}_c = \bar{c} + z\sqrt{\bar{c}}$$

> A c-chart counts the number of defective items per unit of output, like the number of dead pixels on a new computer monitor or the number of typos in a newspaper article.

Note: Problems 17.25–17.26 refer to the data set below, the number of customer complaints a company receives per week over a 16-week period.

Week	Number of Complaints	Week	Number of Complaints
1	3	9	2
2	3	10	9
3	4	11	9
4	4	12	1
5	6	13	8
6	9	14	5
7	3	15	1
8	3	16	8

17.25 Calculate the lower and upper control limits for a 3-sigma *c*-chart.

Compute the average number of complaints received per week.

$$\begin{aligned}
\bar{c} &= \frac{\text{total number of complaints}}{\text{number of weeks}} \\
&= \frac{3+3+4+4+6+9+3+3+2+9+9+1+8+5+1+8}{16} \\
&= \frac{78}{16} \\
&= 4.9
\end{aligned}$$

Substitute $\bar{c} = 4.9$ into the lower and upper control limit c-chart formulas.

$$\begin{aligned} \text{LCL}_c &= \bar{c} - z\sqrt{\bar{c}} & \text{UCL}_c &= \bar{c} + z\sqrt{\bar{c}} \\ &= 4.9 - 3\sqrt{4.9} & &= 4.9 + 3\sqrt{4.9} \\ &= -1.7 & &= 4.9 + 6.64 \\ &\approx 0 & &= 11.5 \end{aligned}$$

The lower control limits of c-charts, like the lower control limits of p-charts, cannot be negative numbers. If the calculated value of LCL_c is less than zero, set $\text{LCL}_c = 0$.

Note: Problems 17.25–17.26 refer to the data set in Problem 17.25, the number of customer complaints a company receives per week over a 16-week period.

17.26 The company receives 9 complaints this week. Determine whether the process is in control.

According to Problem 17.25, the company receives between $\text{LCL}_c = 0$ and $\text{UCL}_c = 11.5$ complaints per week. If 12 or more complaints are recorded, assignable variation is present and the company should investigate the cause for the increase. This week elicited 9 complaints, which lies between the control boundaries and indicates the presence of natural variation for the company this week.

That doesn't mean nine complaints should be acceptable for the company. (Most companies shoot for zero complaints.) It's better to say that receiving nine complaints in a week is not unusual for the company.

Note: Problems 17.27–17.28 refer to the data set below, the number of typos per chapter in a draft of a new book.

Chapter	Number of Typos	Chapter	Number of Typos
1	5	7	6
2	14	8	4
3	5	9	9
4	13	10	12
5	7	11	6
6	6	12	15

17.27 Calculate the lower and upper control limits for a 3-sigma c-chart.

Calculate the average number of typos per chapter.

$$\bar{c} = \frac{5 + 14 + 5 + 13 + 7 + 6 + 6 + 4 + 9 + 12 + 6 + 15}{12} = \frac{102}{12} = 8.5$$

Calculate the lower and upper control limits for a 3-sigma c-chart.

$$\text{LCL}_c = \bar{c} - z\sqrt{c} \qquad \text{UCL}_c = \bar{c} + z\sqrt{c}$$
$$= 8.5 - 3\sqrt{8.5} \qquad\qquad = 8.5 + 3\sqrt{8.5}$$
$$= -0.25 \qquad\qquad\qquad = 8.5 + 8.75$$
$$\approx 0 \qquad\qquad\qquad\quad = 17.3$$

Note: Problems 17.27–17.28 refer to the data set in Problem 17.27, the number of typos per chapter in a draft of a new book.

17.28 Determine whether the process is in control if the next chapter contains 19 typos.

According to the data recorded for the first 12 chapters, natural variation accounts for between $\text{LCL}_c = 0$ and $\text{UCL}_c = 17.3$ typos per chapter. The newest chapter contains 19 typos, which exceeds the control limits and indicates the presence of assignable variation. ◄

> Something has changed in the manuscript preparation process. Based on the pattern of the first 12 chapters, 19 typos is unusually high.

Note: Problems 17.29–17.31 refer to the data set below, the number of errors committed by Major League Baseball player Derek Jeter every season for 10 seasons.

Season	Errors	Season	Errors
1998	9	2003	14
1999	14	2004	13
2000	24	2005	15
2001	15	2006	15
2002	14	2007	18

17.29 Calculate the lower and upper control limits for a 3-sigma c-chart.

Calculate the average number of errors per season.

$$\bar{c} = \frac{9 + 14 + 24 + 15 + 14 + 14 + 13 + 15 + 15 + 18}{10} = \frac{151}{10} = 15.1$$

Compute the lower and upper control limits for a 3-sigma c-chart.

$$\text{LCL}_c = \bar{c} - z\sqrt{c} \qquad \text{UCL}_c = \bar{c} + z\sqrt{c}$$
$$= 15.1 - 3\sqrt{15.1} \qquad\qquad = 15.1 + 3\sqrt{15.1}$$
$$= 3.4 \qquad\qquad\qquad = 26.8$$

Note: Problems 17.29–17.31 refer to the data set in Problem 17.29, the number of errors committed by Major League Baseball player Derek Jeter every season for 10 seasons.

17.30 During the 2008 season, Derek Jeter committed 12 errors. Determine whether his fielding process is in control.

> UCL_c is 26.8, so he doesn't quite have 27 errors as his upper control boundary; 27 would be out of control

Under normal circumstances, Jeter commits between $LCL_c \approx 4$ and $UCL_c \approx 26$ errors per season. Twelve errors lies between these limits and indicates the presence of only natural variation. His fielding process appears to be in control for the 2008 season.

Note: Problems 17.29–17.31 refer to the data set in Problem 17.29, the number of errors committed by Major League Baseball player Derek Jeter every season for 10 seasons.

17.31 If Derek Jeter committed only 2 errors during the 2008 season, would his fielding process be considered in control?

> The process change could be due to something more obvious if you examine the data. For example, he might have played fewer games due to an injury.

Two errors is below the lower limit of $LCL_c = 3.4$, which indicates the presence of assignable variation. Although this variation isn't an indication of a problem—given the choice, players would certainly prefer committing an unusually low number of errors to an unusually high number of errors—the results still indicate that something has happened to his fielding process to cause such a drastic change. You would conclude that his fielding process was out of control because this season had far fewer errors than previous seasons.

Process Capability Ratio

Is a process capable of performing according to design?

17.32 Describe the purpose of the process capability ratio C_p.

The process capability ratio measures the ability of a process to meet its design specifications. It is equal to the ratio of the process design range and the actual process range. In the formula for C_p below, σ represents the observed process standard deviation.

$$C_p = \frac{\text{process design range}}{\text{actual range of process}} = \frac{\text{upper specification limit} - \text{lower specification limit}}{6\sigma}$$

If $C_p \geq 1.0$, the process has the capability to meet its design specifications; if $C_p < 1.0$, the process cannot meet the design specifications.

17.33 A process that packages pretzels in bags is designed to fill each bag with 12.0 ounces of pretzels with a design range of ±0.3 ounces. The process exhibits a standard deviation of 0.08 ounces. Determine whether the process is capable of meeting the design specifications.

The process is designed to fill the bags with 12.0 ounces of pretzels with a design range of ±3 ounces. Use this information to calculate the lower and upper specification limits.

$$\text{lower specification limit} = 12.0 - 0.3 = 11.7$$

$$\text{upper specification limit} = 12.0 + 0.3 = 12.3$$

The problem states that $\sigma = 0.08$ ounces. Calculate the process capability ratio C_p.

$$C_p = \frac{\text{upper specification limit} - \text{lower specification limit}}{6\sigma} = \frac{12.3 - 11.7}{6(0.08)} = \frac{0.6}{0.48} = 1.25$$

The process is capable of meeting the design specifications because $C_p \geq 1.0$.

17.34 A lightbulb manufacturer produces bulbs designed to average 1,100 hours ±125 hours of life. The life of the bulbs actually produced has a standard deviation of 56 hours. Determine whether the process is capable of meeting the design specifications.

Calculate the lower and upper specification limits.

$$\text{lower specification limit} = 1,100 - 125 = 975 \text{ hours}$$

$$\text{upper specification limit} = 1,100 + 125 = 1,225 \text{ hours}$$

Given $\sigma = 56$ hours, compute the process capability ratio.

$$C_p = \frac{\text{upper specification limit} - \text{lower specification limit}}{6\sigma} = \frac{1,225 - 975}{6(56)} = \frac{250}{336} = 0.74$$

The process is not capable of meeting the design specifications because $C_p = 0.74$ is less than 1.0.

17.35 A manufacturer of laptop batteries is producing a new battery that is designed to last between 5.7 and 6.4 hours. The life of the batteries produced by the manufacturing process has a standard deviation of 0.09 hours. Determine whether the process is capable of meeting the design specifications.

The problem states that the lower and upper specification limits are 5.7 and 6.4 hours, respectively. Given $\sigma = 0.09$ hours, compute the process capability ratio.

$$C_p = \frac{6.4 - 5.7}{6(0.09)} = \frac{0.7}{0.54} = 1.30$$

Because $C_p = 1.30$ is greater than or equal to 1.0, the process is capable of meeting the design specifications.

Process Capability Index

Measuring capability for a process that has shifted

17.36 Describe the purpose of the process capability index C_{pk}.

The process capability index C_{pk} measures the ability of a process to meet its design specifications when the process mean is not the same as the design mean. In the formula for C_{pk} below, σ is the observed process standard deviation and $\bar{\bar{x}}$ is the overall process mean.

C_{pk} is equal to the lesser of the two fractions in the brackets.

$$C_{pk} = \text{minimum}\left[\frac{\text{upper specification limit} - \bar{\bar{x}}}{3\sigma}, \frac{\bar{\bar{x}} - \text{lower specification limit}}{3\sigma}\right]$$

If $C_{pk} \geq 1.0$, the process has the capability to meet its design specifications; if $C_{pk} < 1.0$, it does not.

If the process mean and design mean are equal, the C_{pk} formula will produce the same value as the C_p formula in Problems 17.32–17.35.

17.37 A process that packages toothpaste is designed to fill tubes with 5.0 ounces of toothpaste with a design range of ±0.15 ounces. The process has been averaging 5.08 ounces per tube with a standard deviation of 0.06 ounces.

Determine whether the process is capable of meeting the design specifications.

The process is designed with a lower specification limit of $5.0 - 0.15 = 4.85$ ounces and an upper specification limit of $5.0 + 0.15 = 5.15$ ounces. However, the process has an actual mean of $\bar{\bar{x}} = 5.08$ ounces and a standard deviation of $\sigma = 0.06$ ounces. Calculate the process capability index C_{pk}.

$$C_{pk} = \text{minimum}\left[\frac{\text{upper specification limit} - \bar{\bar{x}}}{3\sigma}, \frac{\bar{\bar{x}} - \text{lower specification limit}}{3\sigma}\right]$$

$$= \text{minimum}\left[\frac{5.15 - 5.08}{3(0.06)}, \frac{5.08 - 4.85}{3(0.06)}\right]$$

$$= \text{minimum}\left[\frac{0.07}{0.18}, \frac{0.23}{0.18}\right]$$

$$= \text{minimum}[0.39, 1.28]$$

$$= 0.39$$

The process is not capable of meeting the design specifications because $C_{pk} = 0.39$ is less than 1.0.

17.38 A commercial refrigeration system is designed to maintain an average temperature of 40°F with a design range of ±3.0°F. Several temperature readings from these systems average 40.8°F with a standard deviation of 0.65°F.

Determine whether these refrigeration systems are capable of meeting the design specifications.

The upper and lower specification limits are 43.0°F and 37.0°F, respectively. The actual mean is $\bar{\bar{x}} = 40.8°$F and the actual standard deviation is $\sigma = 0.65°$F. Calculate the process capability index.

40 + 3 = 43 and 40 – 3 = 37

$$C_{pk} = \text{minimum} \left[\frac{\text{upper specification limit} - \bar{\bar{x}}}{3\sigma}, \frac{\bar{\bar{x}} - \text{lower specification limit}}{3\sigma} \right]$$

$$= \text{minimum} \left[\frac{43.0 - 40.8}{3(0.65)}, \frac{40.8 - 37.0}{3(0.65)} \right]$$

$$= \text{minimum} \left[\frac{2.2}{1.95}, \frac{3.8}{1.95} \right]$$

$$= \text{minimum} [1.13, 1.95]$$

$$= 1.13$$

Because $C_{pk} = 1.13$ is greater than or equal to 1.0, the refrigeration systems are capable of meeting the design specifications.

17.39 A machine that manufactures buttons for overcoats is designed to maintain an average button diameter of 25 mm with a design range of ±0.7 mm. Buttons produced by this machine have an average diameter of 24.6 mm with a standard deviation of 0.090 mm.

Determine whether this machine is capable of meeting the design specifications.

The lower and upper specification limits for diameter are 25 – 0.7 = 24.3 mm and 25 + 0.7 = 25.7 mm, respectively. Calculate the process capability index.

$$C_{pk} = \text{minimum} \left[\frac{\text{upper specification limit} - \bar{\bar{x}}}{3\sigma}, \frac{\bar{\bar{x}} - \text{lower specification limit}}{3\sigma} \right]$$

$$= \text{minimum} \left[\frac{25.7 - 24.6}{3(0.090)}, \frac{24.6 - 24.3}{3(0.090)} \right]$$

$$= \text{minimum} \left[\frac{1.1}{0.27}, \frac{0.3}{0.27} \right]$$

$$= \text{minimum} [4.07, 1.11]$$

$$= 1.11$$

The machine is capable of meeting the design specifications because $1.11 \geq 1.0$.

In Problem 17.36, I told you that the C_p and C_{pk} formulas gave you the same number when the process and design means were equal. It's time to see if I was lying.

17.40 A machine that manufactures glass panes for windows is designed to maintain an average thickness of 9.0 mm with a design range of ±0.25 mm. Glass panes produced by the machine have an average thickness of 9.0 mm with a standard deviation of 0.070 mm. (Note that the process and design means are equal.)

Calculate the process capability index C_{pk} to determine whether the machine is capable of meeting the design specifications, and verify that the process capability ratio C_p results in the same conclusion.

The lower and upper specification limits of the production process are $9.0 - 0.25 = 8.75$ mm and $9.0 + 0.25 = 9.25$ mm, respectively. Calculate C_{pk}.

$$C_{pk} = \text{minimum}\left[\frac{\text{upper specification limit} - \bar{\bar{x}}}{3\sigma}, \frac{\bar{\bar{x}} - \text{lower specification limit}}{3\sigma}\right]$$

$$= \text{minimum}\left[\frac{9.25 - 9.0}{3(0.070)}, \frac{9.0 - 8.75}{3(0.070)}\right]$$

$$= \text{minimum}\left[\frac{0.25}{0.21}, \frac{0.25}{0.21}\right]$$

$$= \text{minimum}[1.19, 1.19]$$

$$= 1.19$$

The machine is capable of meeting the design specifications because $C_{pk} = 1.19$ is greater than or equal to 1.0. Verify that the process capability ratio is also greater than or equal to 1.0.

$$C_p = \frac{\text{upper specification limit} - \text{lower specification limit}}{6\sigma} = \frac{9.25 - 8.75}{6(0.070)} = \frac{0.50}{0.42} = 1.19$$

I was telling the truth!

The actual mean is centered at the design mean of 9.0 mm, so $C_{pk} = C_p = 1.19$, indicating that the machine is capable of meeting the design specifications.

Chapter 18
CONTEXTUALIZING STATISTICAL CONCEPTS

Figuring out when to use what formula

Determining which algorithm to apply to a given problem is one of the most daunting challenges to a student of statistics. In this chapter, you are provided with a collection of review examples based on the concepts investigated in this book, with a particular focus on Chapters 6–14. The problems are randomized, which requires that you not only understand the underlying statistical procedures but that you be able to identify which procedure should be applied in each circumstance.

Most of the practice problems in this book so far have focused on helping you understand *how* to use formulas, but it's also important that you understand *when* to use them. Clusters of problems in this book all use the same procedures, so you rarely end up asking yourself, "What test should I use here?" The answer is either the title of the chapter you're working on or the section the problem is in.

There's nothing new in this chapter. All of the problems are very similar to problems already covered in the book. However, they're arranged in a random order, so it's up to you to figure out which concepts, formulas, tables, and hypotheses to use. You're not totally on your own, though. At the end of every problem, I've included a margin note that will direct you to a similar problem in case you need help.

Try to do these problems without using my notes (and flipping back to past examples) right away. This chapter is a good way to figure out how ready you are for a stats final exam—study harder on the concepts you had to go back and look up.

Note: In Problems 18.1–18.3, a university professor claims that business students average more than 12 hours of studying per week. A sample of 50 students studied an average of 13.4 hours. Assume the population standard deviation is 4.6 hours.

18.1 Using $\alpha = 0.10$, test the professor's claim by comparing the calculated z-score to the critical z-score.

$$H_0 : \mu \leq 12 \text{ hours}$$
$$H_1 : \mu > 12 \text{ hours}$$

According to Reference Table 1, $z_c = 1.28$. Calculate the standard error of the mean.

$$\sigma_{\bar{x}} = \frac{\sigma}{\sqrt{n}} = \frac{4.6}{\sqrt{50}} = 0.6505$$

Calculate the z-score for $\bar{x} = 13.4$.

$$z_{13.4} = \frac{\bar{x} - \mu}{\sigma_{\bar{x}}} = \frac{13.4 - 12}{0.6505} = 2.15$$

See Problem 10.11.

Because $z_{13.4} = 2.15$ is greater than $z_c = 1.28$, you reject H_0 and conclude that business students average more than 12 hours of studying per week.

Note: In Problems 18.1–18.3, a university professor claims that business students average more than 12 hours of studying per week. A sample of 50 students studied an average of 13.4 hours. Assume the population standard deviation is 4.6 hours.

18.2 Using $\alpha = 0.10$, test the professor's claim by comparing the calculated sample mean to the critical sample mean.

The following information is available from Problem 18.1.

$$H_0 : \mu \leq 12 \text{ hours}$$
$$H_1 : \mu > 12 \text{ hours}$$
$$\bar{x} = 13.4 \text{ hours}$$
$$z_c = 1.28$$
$$\sigma_{\bar{x}} = 0.6505 \text{ hours}$$

Calculate the critical sample mean.

$$\bar{x}_c = \mu + z_c \sigma_{\bar{x}} = 12 + 1.28(0.6505) = 12.83$$

See Problem 10.13.

Because $\bar{x} = 13.4$ is greater than $\bar{x}_c = 12.83$, you reject H_0 and conclude that business students average more than 12 hours of studying per week.

Note: In Problems 18.1–18.3, a university professor claims that business students average more than 12 hours of studying per week. A sample of 50 students studied an average of 13.4 hours. Assume the population standard deviation is 4.6 hours.

18.3 Using $\alpha = 0.10$, test the professor's claim by comparing the p-value to the level of significance.

According to Problem 18.1, $z_{13.4} = 2.15$.

$$p\text{-value} = P\left(z_{\bar{x}} > 2.15\right) = 0.50 - 0.4842 = 0.0158$$

Because the p-value is less than $\alpha = 0.10$, you reject the null hypothesis and support the professor's claim.

See Problem 10.14.

Note: Problems 18.4–18.5 refer to the data set below, the ages and systolic blood pressure of eight women.

Age	Blood Pressure	Age	Blood Pressure
34	125	29	132
51	145	38	118
25	115	55	136
46	122	48	140

18.4 Calculate the correlation coefficient between age and systolic blood pressure.

The following table summarizes the correlation calculations.

Age x	Blood Pressure y	xy	x^2	y^2
34	125	4,250	1,156	15,625
51	145	7,395	2,601	21,025
25	115	2,875	625	13,225
46	122	5,612	2,116	14,884
29	132	3,828	841	17,424
38	118	4,484	1,444	13,924
55	136	7,480	3,025	18,496
48	140	6,720	2,304	19,600
Total $\sum x = 326$	$\sum y = 1{,}033$	$\sum xy = 42{,}644$	$\sum x^2 = 14{,}112$	$\sum y^2 = 134{,}203$

Calculate the correlation coefficient for the $n = 8$ pairs of data.

$$r = \frac{n\sum xy - (\sum x)(\sum y)}{\sqrt{\left[n\sum x^2 - (\sum x)^2\right]\left[n\sum y^2 - (\sum y)^2\right]}}$$

$$= \frac{(8)(42,644) - (326)(1,033)}{\sqrt{\left[(8)(14,112) - (326)^2\right]\left[(8)(134,203) - (1,033)^2\right]}}$$

$$= \frac{4,394}{\sqrt{(6,620)(6,535)}}$$

$$= \frac{4,394}{\sqrt{43,261,700}}$$

$$= 0.668$$

See Problem 14.6.

Note: Problems 18.4–18.5 refer to the data set in Problem 18.4, the ages and systolic blood pressure of eight women.

18.5 Test the significance of the correlation coefficient between age and systolic blood pressure using $\alpha = 0.10$.

$$H_0 : \rho = 0$$
$$H_1 : \rho \neq 0$$
$$r = 0.668$$
$$n = 8$$

Use Reference Table 2 to identify the critical t-score.

$$\text{df} = n - 2 = 8 - 2 = 6$$
$$t_c = 2.447$$

Calculate the t-score.

$$t = \frac{r}{\sqrt{\dfrac{1 - r^2}{n - 2}}} = \frac{0.668}{\sqrt{\dfrac{1 - (0.668)^2}{8 - 2}}} = \frac{0.668}{\sqrt{\dfrac{0.55378}{6}}} = \frac{0.668}{\sqrt{0.09230}} = 2.20$$

See Problem 14.7.

Because $t = 2.20$ is less than $t_c = 2.447$, you fail to reject the null hypothesis; the data does not support a relationship between age and systolic blood pressure.

Note: In Problems 18.6–18.7, assume that 15% of the customers who visit a particular website make a purchase. Assume the number of customers who make a purchase is binomially distributed.

18.6 Calculate the probability that none of the next nine visitors to the website will make a purchase.

$$n = 9$$
$$r = 0$$
$$p = 0.15$$
$$q = 1 - p = 1 - 0.15 = 0.85$$

Apply the binomial formula.

$$P(r) = \binom{n}{r} p^r q^{n-r}$$

$$P(0) = \binom{9}{0} (0.15)^0 (0.85)^{9-0}$$

$$P(0) = (1)(1)(0.85)^9$$

$$P(0) = 0.2316$$

See Problem 6.14.

Note: In Problems 18.6–18.7, assume that 15% of the customers who visit a particular website make a purchase. Assume the number of customers who make a purchase is binomially distributed.

18.7 Calculate the probability that fewer than two of the next nine website visitors will make a purchase.

The following information can be garnered from Problem 18.6.

$$n = 9$$
$$p = 0.15$$
$$q = 0.85$$
$$P(0) = 0.2316$$

Note that $P(r < 2) = P(0) + P(1)$. Calculate the probability of one customer making a purchase.

$$P(1) = \binom{9}{1} (0.15)^1 (0.85)^{9-1} = (9)(0.15)(0.27249) = 0.3679$$

Calculate the probability that fewer than two visitors will make a purchase.

$$P(r < 2) = 0.2316 + 0.3679 = 0.5995$$

See Problem 6.15.

Note: Problems 18.8–18.9 refer to a claim that the average lifespan of a fruit fly is different than 32 days. A random sample of 15 fruit flies had an average lifespan of 30.6 days with a sample standard deviation of 7.0 days. Assume the life spans of fruit flies are normally distributed.

18.8 Using $\alpha = 0.02$, test the claim by comparing the calculated t-score to the critical t-score.

$$H_0 : \mu = 32 \text{ days}$$
$$H_1 : \mu \neq 32 \text{ days}$$

Use Reference Table 2 to identify t_c.

$$\text{df} = n - 1 = \overset{15}{\cancel{18}} - 1 = \cancel{17}\ 14$$
$$t_c = \pm 2.624$$

correct for df = 14

Calculate the standard error of the mean and the t-score for $\bar{x} = 30.6$.

$$\hat{\sigma}_{\bar{x}} = \frac{s}{\sqrt{n}} = \frac{7.0}{\sqrt{15}} = 1.807$$

$$t_{30.6} = \frac{\bar{x} - \mu}{\hat{\sigma}_{\bar{x}}} = \frac{30.6 - 32}{1.807} = -0.77$$

See Problem 10.40.

Because $t_{30.6} = -0.77$ is not less than $t_c = -2.624$, you fail to reject H_0 and conclude that you cannot support the claim.

Note: Problems 18.8–18.9 refer to a claim that the average lifespan of a fruit fly is different than 32 days. A random sample of 15 fruit flies had an average lifespan of 30.6 days with a sample standard deviation of 7.0 days. Assume the life spans of fruit flies are normally distributed.

18.9 Using $\alpha = 0.02$, test the claim by comparing the calculated sample mean to the critical sample mean.

$$H_0 : \mu = 32 \text{ days}$$
$$H_1 : \mu \neq 32 \text{ days}$$

Use Reference Table 2 to identify the critical t-scores.

$$\text{df} = n - 1 = \overset{15}{\cancel{18}} - 1 = \cancel{17}\ 14$$
$$t_c = \pm 2.624$$

correct for df = 14

Calculate the lower and upper bounds of the rejection regions. Recall that $\hat{\sigma}_{\bar{x}} = 1.807$, according to Problem 18.8.

$$\bar{x}_c = \mu - t_c \hat{\sigma}_{\bar{x}} \qquad\qquad \bar{x}_c = \mu + t_c \hat{\sigma}_{\bar{x}}$$
$$= 32 + (-2.624)(1.807) \qquad = 32 + (2.624)(1.807)$$
$$= 27.26 \qquad\qquad\qquad = 36.74$$

Because the sample mean $\bar{x} = 30.6$ is neither less than 27.26 nor greater than 36.74, you fail to reject the null hypothesis and cannot support the claim.

See Problem 10.41.

18.10 The following table lists the number of family members in 10 randomly selected households. Assume that the population is normally distributed. Construct a 95% confidence interval for the sample.

Number of Residents per Household									
3	7	4	6	4	5	5	4	2	4

Calculate the sample mean.

$$\bar{x} = \frac{\sum x}{n} = \frac{3+7+4+6+4+5+5+4+2+4}{10} = \frac{44}{10} = 4.4$$

Calculate the sum of the squares of the data values.

x	x^2
3	9
7	49
4	16
6	36
4	16
5	25
5	25
4	16
2	4
4	16
Total 44	212

Compute the sample standard deviation and the standard error of the mean.

$$s = \sqrt{\frac{\sum x^2 - \frac{\left(\sum x\right)^2}{n}}{n-1}} = \sqrt{\frac{212 - \frac{(44)^2}{10}}{10-1}} = \sqrt{\frac{18.4}{9}} = 1.430$$

$$\hat{\sigma}_{\bar{x}} = \frac{s}{\sqrt{n}} = \frac{1.430}{\sqrt{10}} = 0.452$$

Use Reference Table 2 to identify the critical t-score.

$$\text{df} = n - 1 = 10 - 1 = 9$$
$$t_c = 2.262$$

Determine the boundaries of the 95% confidence interval.

$$\text{lower limit} = \bar{x} - t_c \hat{\sigma}_{\bar{x}} \qquad\qquad \text{upper limit} = \bar{x} + t_c \hat{\sigma}_{\bar{x}}$$
$$= 4.4 - (2.262)(0.452) \qquad\qquad = 4.4 + (2.262)(0.452)$$
$$= 3.38 \qquad\qquad\qquad\qquad = 5.42$$

See Problem 9.28.

Based on this sample, you are 95% confident that the average number of residents per household is between 3.38 and 5.42.

Note: Problems 18.11–18.12 refer to a company that administers a screening test to applicants seeking employment. The scores are normally distributed with a mean of 74.5 and a standard deviation of 4.4.

18.11 Calculate the probability that a randomly chosen applicant will score less than 80.

Calculate the z-score of $x = 80$ and then compute the probability that $x < 80$, using Reference Table 1.

See Problem 7.12.

$$z_{80} = \frac{x - \mu}{\sigma} = \frac{80 - 74.5}{4.4} = \frac{5.5}{4.4} = 1.25$$
$$P(x < 80) = P(z < 1.25) = 0.3944 + 0.5 = 0.8944$$

Note: Problems 18.11–18.12 refer to a company that administers a screening test to applicants seeking employment. The scores are normally distributed with a mean of 74.5 and a standard deviation of 4.4.

18.12 Calculate the probability that a randomly chosen applicant will score between 75 and 79.

Calculate the z-scores of the boundaries.

$$z_{75} = \frac{75 - 74.5}{4.4} \qquad z_{79} = \frac{79 - 74.5}{4.4}$$
$$= \frac{0.5}{4.4} \qquad\qquad = \frac{4.5}{4.4}$$
$$= 0.11 \qquad\qquad = 1.02$$

See Problem 7.13.

Calculate $P(75 < x < 79)$.

$$P(75 < x < 79) = P(0.11 < z < 1.02)$$
$$= 0.3461 - 0.0438$$
$$= 0.3023$$

Note: Problems 18.13–18.14 refer to the data set below, the scores shot by members of a golf course on a particular day and the percentage of its members that the golf course claims shoot in each score category.

Golf Scores	Percentage	Observed
70–79	15	7
80–89	25	12
90–99	35	32
100–109	15	7
110–119	10	12
Total		**70**

18.13 Calculate the expected number of golfers in each category using the distribution identified by the golf course.

See Problem 12.16.

Golf Scores	Percentage	Number	Expected
70–79	15	70	10.5
80–89	25	70	17.5
90–99	35	70	24.5
100–109	15	70	10.5
110–119	10	70	7

Note: Problems 18.13–18.14 refer to the data set in Problem 18.13, the scores shot by members of a golf course on a particular day and the percentage of its members that the golf course claims shoot in each score category.

18.14 Test the hypothesis that the golf scores follow the stated distribution using $\alpha = 0.05$.

H_0 : The golf scores follow the stated distribution.

H_1 : The golf scores do not follow the stated distribution.

Using data from Problem 18.13, the following table summarizes the calculations for the chi-squared statistic.

Scores	O	E	$O - E$	$(O - E)^2$	$\dfrac{(O - E)^2}{E}$
70–79	7	10.5	−3.5	12.25	1.17
80–89	12	17.5	−5.5	30.25	1.73
90–99	32	24.5	7.5	56.25	2.30
100–109	7	10.5	−3.5	12.25	1.17
110–119	12	7	5	25	3.57
Total					**9.94**

Use Reference Table 3 to identify the critical chi-squared score.

$$\chi^2 = \sum \frac{(O-E)^2}{E} = 9.94$$
$$df = k - 1 = 5 - 1 = 4$$
$$\chi_c^2 = 9.488$$

See Problem 12.17.

Because $\chi^2 = 9.94$ is greater than $\chi_c^2 = 9.488$, you reject H_0 and conclude that the golf scores do not follow the stated distribution.

18.15 In a random sample of 400 people, 122 had blue eyes. Construct a 90% confidence interval to estimate the true proportion of people with blue eyes.

Calculate the proportion p_s of the sample, which has size $n = 400$.

$$p_s = \frac{\text{number of successes in the sample}}{n} = \frac{122}{400} = 0.305$$

The critical z-score (according to Reference Table 1) is $z_c = 1.64$. Calculate the standard error of the proportion.

$$\hat{\sigma}_p = \sqrt{\frac{p_s(1-p_s)}{n}} = \sqrt{\frac{(0.305)(1-0.305)}{400}} = \sqrt{0.000530} = 0.0230$$

Calculate the boundaries of the 90% confidence interval.

$$\text{lower limit} = p_s - z_c \hat{\sigma}_p \qquad \text{upper limit} = p_s + z_c \hat{\sigma}_p$$
$$= 0.305 - (1.64)(0.023) \qquad\qquad = 0.305 + (1.64)(0.023)$$
$$= 0.267 \qquad\qquad\qquad\qquad = 0.343$$

See Problem 9.51.

Based on this sample, you are 90% confident that the true proportion of people with blue eyes is between 26.7% and 34.3%.

18.16 A pilot sample of 50 people included 18 with brown eyes. How many additional people must be sampled to construct a 90% confidence interval with a margin of error equal to 0.04?

Calculate the sample proportion.

$$p_s = \frac{\text{number of successes in the sample}}{n} = \frac{18}{50} = 0.36$$

According to Reference Table 1, $z_c = 1.64$. Calculate the required sample size given $E = 0.04$.

$$n = \frac{p_s(1-p_s)z_c^2}{E^2} = \frac{(0.36)(1-0.36)(1.64)^2}{(0.04)^2} = \frac{0.619684}{0.0016} = 387.3 \approx 388$$

An additional $388 - 50 = 338$ people should be sampled to construct a 90% confidence interval with a margin of error equal to 0.04. ⟵

See Problem 9.52.

Note: Problems 18.17–18.19 refer to the table below, the average bill per customer for two competing grocery stores. Assume the population of customer bills is normally distributed.

	Store A	Store B
Sample mean	$182	$192
Sample size	20	23
Population standard deviation	$24	$21

18.17 Test the hypothesis that the average customer bill at Store A is lower than the average bill at Store B by comparing the calculated z-score to the critical z-score using $\alpha = 0.05$.

Assign Store A to population 1 and Store B to population 2.

$$H_0 : \mu_1 - \mu_2 \geq 0$$
$$H_1 : \mu_1 - \mu_2 < 0$$

According to Reference Table 1, the critical z-score is $z_c = 1.64$. Calculate the standard error of the proportion.

$$\sigma_{\bar{x}_1 - \bar{x}_2} = \sqrt{\frac{\sigma_1^2}{n_1} + \frac{\sigma_2^2}{n_2}} = \sqrt{\frac{24^2}{20} + \frac{21^2}{23}} = \sqrt{\frac{576}{20} + \frac{441}{23}} = \sqrt{47.97391} = 6.93$$

Compute the difference of the sample means and the corresponding z-score.

$$\bar{x}_1 - \bar{x}_2 = 182 - 192 = -10$$

$$z_{\bar{x}_1 - \bar{x}_2} = \frac{(\bar{x}_1 - \bar{x}_2) - (\mu_1 - \mu_2)}{\sigma_{\bar{x}_1 - \bar{x}_2}} = \frac{-10 - 0}{6.93} = -1.44$$

Because $z_{\bar{x}_1 - \bar{x}_2} = -1.44$ is not less than $z_c = -1.64$, you fail to reject H_0 and conclude that the average customer bill in Store A is not lower than the average bill in Store B. ⟵

See Problem 11.14.

Note: Problems 18.17–18.19 refer to the table in Problem 18.17, the average bill per customer for two competing grocery stores. Assume the population of customer bills is normally distributed.

18.18 Verify your answer to Problem 18.17 by comparing the p-value to the level of significance $\alpha = 0.05$.

According to Problem 18.17, $z_{\bar{x}_1 - \bar{x}_2} = -1.44$. Calculate the p-value.

$$p\text{-value} = P\left(z_{\bar{x}_1 - \bar{x}_2} < -1.44\right) = 0.50 - 0.4251 = 0.0749$$

See Problem 11.15.

Because the *p*-value 0.0749 is greater than $\alpha = 0.05$, you fail to reject the null hypothesis.

Note: Problems 18.17–18.19 refer to the table in Problem 18.17, the average bill per customer for two competing grocery stores. Assume the population of customer bills is normally distributed.

18.19 Construct a 95% confidence interval for the difference between customer bills at the stores.

A 95% confidence interval has a corresponding critical *z*-score of $z_c = 1.96$. According to Problem 18.17, $\bar{x}_1 - \bar{x}_2 = -10$ and $\sigma_{\bar{x}_1 - \bar{x}_2} = 6.93$. Calculate the 95% confidence interval.

$$\text{lower limit} = \left(\bar{x}_1 - \bar{x}_2\right) - z_c\sigma_{\bar{x}_1 - \bar{x}_2} \qquad \text{upper limit} = \left(\bar{x}_1 - \bar{x}_2\right) + z_c\sigma_{\bar{x}_1 - \bar{x}_2}$$
$$= -10 - (1.96)(6.93) \qquad\qquad\qquad = -10 + (1.96)(6.93)$$
$$= -23.58 \qquad\qquad\qquad\qquad\qquad = 3.58$$

Based on these samples, you are 95% confident that the average difference in customer bills at Stores A and B is between –$23.58 and $3.58. Because this confidence interval includes zero, it supports the conclusion drawn in Problems 18.17–18.18: the average bill at Store A is not lower than the average bill at Store B.

See Problem 11.16.

Note: Problems 18.20–18.30 refer to the data set below, the pitching staff earned run averages (ERAs) for eight Major League Baseball teams and the number of games the teams won during the 2008 season.

Team	ERA	Wins
Cleveland	4.43	81
Tampa Bay	3.82	97
Philadelphia	3.88	92
Pittsburgh	5.08	67
Texas	5.37	79
Cincinnati	4.55	74
St. Louis	4.19	86
Washington	4.66	59

18.20 Construct the linear equation that best fits the data and interpret the results.

Calculate the sums of *x*, *y*, *xy*, x^2, and y^2 using the following table.

	ERA	Wins			
	x	y	xy	x^2	y^2
	4.43	81	358.83	19.62	6,561
	3.82	97	370.54	14.59	9,409
	3.88	92	356.96	15.05	8,464
	5.08	67	340.36	25.81	4,489
	5.37	79	424.23	28.84	6,241
	4.55	74	336.70	20.70	5,476
	4.19	86	360.34	17.56	7,396
	4.66	59	274.94	21.72	3,481
Total	**35.98**	**635**	**2,822.9**	**163.89**	**51,517**

Calculate the slope and y-intercept of the regression equation.

$$b = \frac{n\sum xy - \left(\sum x\right)\left(\sum y\right)}{n\sum x^2 - \left(\sum x\right)^2} = \frac{(8)(2,822.9) - (35.98)(635)}{(8)(163.89) - (35.98)^2} = \frac{-264.1}{16.56} = -15.948$$

$$a = \frac{\sum y}{n} - b\left(\frac{\sum x}{n}\right) = \frac{635}{8} - (-15.948)\left(\frac{35.98}{8}\right) = 79.375 - (-71.726) = 151.101$$

The regression equation is $\hat{y} = 151.101 - 15.948x$. An ERA increase of 1 corresponds to a loss of 16 additional games during the season. ← See Problem 14.22.

Note: Problems 18.20–18.30 refer to the data set in Problem 18.20, the pitching staff earned run averages (ERAs) for eight Major League Baseball teams and the number of games the teams won during the 2008 season.

18.21 Predict the number of games a team will win if its pitching staff has an ERA of 4.0.

Substitute $x = 4$ into the regression equation constructed in Problem 18.20.

$$\hat{y} = 151.101 - 15.948x$$
$$= 151.101 - 15.948(4)$$
$$= 87.309$$
$$\approx 87 \text{ games}$$

← See Problem 14.23.

Note: Problems 18.20–18.30 refer to the data set in Problem 18.20, the pitching staff earned run averages (ERAs) for eight Major League Baseball teams and the number of games the teams won during the 2008 season.

18.22 Calculate the total sum of squares for the linear regression.

Recall that $\sum y = 635$, $\sum y^2 = 51{,}517$, and $n = 8$.

See Problem 14.24.

$$\text{SST} = \sum y^2 - \frac{\left(\sum y\right)^2}{n} = 51{,}517 - \frac{(635)^2}{8} = 51{,}517 - 50{,}403.125 = 1{,}113.88$$

Note: Problems 18.20–18.30 refer to the data set in Problem 18.20, the pitching staff earned run averages (ERAs) for eight Major League Baseball teams and the number of games the teams won during the 2008 season.

18.23 Partition the total sum of squares computed in Problem 18.22 into the sum of squares regression and the sum of squares error.

The following values were computed in Problems 18.20–18.22.

$$\sum y = 635 \qquad \sum y^2 = 51{,}517 \qquad \sum xy = 2{,}822.9$$
$$a = 151.101 \qquad b = -15.948 \qquad \text{SST} = 1{,}113.88$$

Calculate the sum of squares error.

$$\begin{aligned}
\text{SSE} &= \sum y^2 - a\sum y - b\sum xy \\
&= 51{,}517 - (151.101)(635) - (-15.948)(2{,}822.9) \\
&= 51{,}517 - 95{,}949.135 - (-45{,}019.609) \\
&= 587.47
\end{aligned}$$

See Problem 14.25.

Calculate the sum of squares regression.

$$\text{SSR} = \text{SST} - \text{SSE} = 1{,}113.88 - 587.47 = 526.41$$

Note: Problems 18.20–18.30 refer to the data set in Problem 18.20, the pitching staff earned run averages (ERAs) for eight Major League Baseball teams and the number of games the teams won during the 2008 season.

18.24 Calculate the coefficient of determination for the model.

Recall that SSR = 526.41 and SST = 1,113.88. Calculate R^2.

See Problem 14.26.

$$R^2 = \frac{\text{SSR}}{\text{SST}} = \frac{526.41}{1{,}113.88} = 0.473$$

Note: Problems 18.20–18.30 refer to the data set in Problem 18.20, the pitching staff earned run averages (ERAs) for eight Major League Baseball teams and the number of games the teams won during the 2008 season.

18.25 Test the significance of the coefficient of determination using $\alpha = 0.10$.

$$H_0 : \rho^2 = 0$$
$$H_1 : \rho^2 \neq 0$$

Recall that SSR = 526.41, SSE = 587.47, and $n = 8$. Calculate the F-score.

$$F = \frac{SSR}{\left(\dfrac{SSE}{n-2}\right)} = \frac{526.41}{\left(\dfrac{587.47}{8-2}\right)} = \frac{526.41}{97.912} = 5.38$$

$D_1 = 1$ and $D_2 = n - 2 = 8 - 2 = 6$. Identify the critical F-score using Reference Table 4: $F_c = 3.776$. Because $F = 5.38$ is greater than $F_c = 3.776$, you reject H_0 and conclude that the coefficient of determination is not different from zero. A relationship exists between ERA and wins during the 2008 season.

See Problem 14.27.

Note: Problems 18.20–18.30 refer to the data set in Problem 18.20, the pitching staff earned run averages (ERAs) for eight Major League Baseball teams and the number of games the teams won during the 2008 season.

18.26 Calculate the standard error of the estimate s_e for the regression model.

$$s_e = \sqrt{\frac{SSE}{n-2}} = \sqrt{\frac{587.47}{8-2}} = \sqrt{97.91167} = 9.90$$

See Problem 14.28.

Note: Problems 18.20–18.30 refer to the data set in Problem 18.20, the pitching staff earned run averages (ERAs) for eight Major League Baseball teams and the number of games the teams won during the 2008 season.

18.27 Construct a 90% confidence interval for the average number of wins for a team with an ERA of 4.7.

Calculate the expected number of wins using the regression model.

$$\hat{y} = 151.101 - 15.948x = 151.101 - (15.948)(4.7) = 151.101 - 74.956 = 76.145$$

Calculate the average ERA for all eight teams.

$$\bar{x} = \frac{\sum x}{n} = \frac{35.98}{8} = 4.4975$$

According to Reference Table 2, given df = $n - 2 = 8 - 2 = 6$, the critical t-score is $t_c = 1.943$. Calculate the 90% confidence interval.

$$CI = \hat{y} \pm t_c s_e \sqrt{\frac{1}{n} + \frac{\left(x - \bar{x}\right)^2}{\left(\sum x^2\right) - \frac{\left(\sum x\right)^2}{n}}}$$

$$= 76.145 \pm 1.943(9.90)\sqrt{\frac{1}{8} + \frac{(4.7 - 4.4975)^2}{163.89 - \frac{(35.98)^2}{8}}}$$

$$= 76.145 \pm 19.236\sqrt{0.125 + \frac{0.0410}{2.0700}}$$

$$= 76.145 \pm 19.236\sqrt{0.14481}$$

$$= 76.145 \pm 7.320$$

$$= 68.8 \text{ and } 83.5$$

See Problem 14.29.

You are 90% confident that the average number of wins for a team with an ERA of 4.7 is between 68.8 and 83.5.

Note: Problems 18.20–18.30 refer to the data set in Problem 18.20, the pitching staff earned run averages (ERAs) for eight Major League Baseball teams and the number of games the teams won during the 2008 season.

18.28 Calculate the standard error of the slope s_b for the regression model.

See Problem 14.30.

$$s_b = \frac{s_e}{\sqrt{\sum x^2 - n\left(\bar{x}\right)^2}} = \frac{9.90}{\sqrt{163.89 - (8)(4.4975)^2}} = \frac{9.90}{\sqrt{163.89 - 161.82}} = \frac{9.90}{\sqrt{2.07}} = 6.88$$

Note: Problems 18.20–18.30 refer to the data set in Problem 18.20, the pitching staff earned run averages (ERAs) for eight Major League Baseball teams and the number of games the teams won during the 2008 season.

18.29 Test for the significance of the slope b of the regression equation using $\alpha = 0.10$.

$$H_0 : \beta = 0$$
$$H_1 : \beta \neq 0$$

Calculate the t-score.

$$t = \frac{b - \beta}{s_b} = \frac{-15.948 - 0}{6.88} = -2.32$$

See Problem 14.31.

According to Reference Table 2, given df = 6, the critical t-scores are $t_c = \pm 1.943$. Because $t = -2.32$ is less than $t_c = -1.943$, you reject H_0 and conclude that the slope of the regression equation is different from zero.

Note: Problems 18.20–18.30 refer to the data set in Problem 18.20, the pitching staff earned run averages (ERAs) for eight Major League Baseball teams and the number of games the teams won during the 2008 season.

18.30 Construct a 90% confidence interval for the slope of the regression equation.

Given df $= n - 2 = 6$, the critical t-score is $t_c = \pm 1.943$. Calculate the 90% confidence interval.

$$CI = b \pm t_c s_b$$
$$= -15.948 \pm (1.943)(6.88)$$
$$= -15.948 \pm 13.368$$
$$= -29.316 \text{ and } -2.58$$

You are 90% confident that the true population slope for the baseball model is between –2.58 and –29.316. Because this confidence interval does not include zero, you can conclude that there is a significant relationship between ERA and team wins during the 2008 season. ←

See Problem 14.32.

Note: Problems 18.31–18.32 refer to a process designed to fill bottles with 16 ounces of soda.

18.31 A sample of 40 bottles contained an average of 16.2 ounces of soda with a sample standard deviation of 1.7 ounces. Construct a 96% confidence interval to estimate the average volume of soda per bottle from this process.

According to the problem, $\bar{x} = 16.2$, $s = 1.7$, and $n = 40$. The critical z-score is $z_c = 2.05$. Calculate the standard error of the mean.

$$\hat{\sigma}_{\bar{x}} = \frac{s}{\sqrt{n}} = \frac{1.7}{\sqrt{40}} = 0.269$$

Calculate the 96% confidence interval.

$$\text{lower limit} = \bar{x} - z_c \sigma_{\bar{x}} \qquad \text{upper limit} = \bar{x} + z_c \sigma_{\bar{x}}$$
$$= 16.2 - (2.05)(0.269) \qquad\qquad = 16.2 + (2.05)(0.269)$$
$$= 15.65 \qquad\qquad\qquad\qquad = 16.75$$

You are 96% confident that the average volume of soda in a bottle filled by this process is between 15.65 and 16.75 ounces. ←

See Problem 9.43.

Note: Problems 18.31–18.32 refer to a process designed to fill bottles with 16 ounces of soda.

18.32 Determine the minimum sample size needed to compute a 98% confidence interval for the average volume of soda per bottle with a margin of error of 0.06 ounces, given the minimum volume in this sample was 15.4 ounces and the maximum volume was 16.6 ounces.

Calculate the estimated population standard deviation.

$$\hat{\sigma} = \frac{R}{6} = \frac{16.6 - 15.4}{6} = \frac{1.2}{6} = 0.2$$

The critical z-score is $z_c = 2.33$. Calculate the sample size given $E = 0.06$.

See
Problem
9.44.

$$n = \left(\frac{z_c \hat{\sigma}}{E}\right)^2 = \left[\frac{(2.33)(0.2)}{0.06}\right]^2 = \left(\frac{0.466}{0.06}\right)^2 = (7.767)^2 = 60.3 \approx 61$$

Note: Problems 18.33–18.35 refer to a sample of 60 tax returns from New York that included 12 with errors. A sample of 75 tax returns from Pennsylvania included 18 that had errors.

18.33 Test the claim that there is no difference in the proportion of tax returns with errors between New York and Pennsylvania residents by comparing the calculated z-score to the critical z-score using $\alpha = 0.02$.

Let New York be population 1 and Pennsylvania be population 2.

$$H_0 : p_1 - p_2 = 0$$
$$H_1 : p_1 - p_2 \neq 0$$

The critical z-scores are $z_c = \pm 2.33$. Calculate the sample proportions, the difference of the sample proportions, and the estimated overall proportion.

$$\bar{p}_1 = \frac{x_1}{n_1} = \frac{12}{60} = 0.2$$

$$\bar{p}_2 = \frac{x_2}{n_2} = \frac{18}{75} = 0.24$$

$$\bar{p}_1 - \bar{p}_2 = 0.2 - 0.24 = -0.04$$

$$\hat{p} = \frac{x_1 + x_2}{n_1 + n_2} = \frac{12 + 18}{60 + 75} = \frac{30}{135} = 0.222$$

Compute the estimated standard error of the difference between the proportions.

$$\hat{\sigma}_{\bar{p}_1 - \bar{p}_2} = \sqrt{(\hat{p})(1 - \hat{p})\left(\frac{1}{n_1} + \frac{1}{n_2}\right)} = \sqrt{(0.222)(1 - 0.222)\left(\frac{1}{60} + \frac{1}{75}\right)} = \sqrt{0.005181} = 0.0720$$

Calculate $z_{\bar{p}_1-\bar{p}_2}$:

$$z_{\bar{p}_1-\bar{p}_2} = \frac{(\bar{p}_1-\bar{p}_2)-(p_1-p_2)}{\hat{\sigma}_{\bar{p}_1-\bar{p}_2}} = \frac{-0.04-0}{0.0720} = -0.56$$

Because $z_{\bar{p}_1-\bar{p}_2} = -0.56$ is neither less than $z_c = -2.33$ nor greater than $z_c = 2.33$, you fail to reject H_0. There is no difference in the proportion of tax returns with errors.

See Problem 11.57.

Note: Problems 18.33–18.35 refer to a sample of 60 tax returns from New York that included 12 with errors. A sample of 75 tax returns from Pennsylvania included 18 that had errors.

18.34 Test the claim that there is no difference in the proportion of tax returns with errors between New York and Pennsylvania residents by comparing the *p*-value to the level of significance $\alpha = 0.02$.

$$\begin{aligned} p\text{-value} &= 2 \cdot P\left(z_{\bar{p}_1-\bar{p}_2} < -0.56\right) \\ &= 2\left[P\left(z_{\bar{p}_1-\bar{p}_2} < 0\right) - P\left(-0.56 < z_{\bar{p}_1-\bar{p}_2} < 0\right)\right] \\ &= 2(0.50 - 0.2123) \\ &= 0.5754 \end{aligned}$$

Because the *p*-value 0.5754 is greater than $\alpha = 0.02$, you fail to reject H_0 and conclude that there is no difference in the proportions.

See Problem 11.58.

Note: Problems 18.33–18.35 refer to a sample of 60 tax returns from New York that included 12 with errors. A sample of 75 tax returns from Pennsylvania included 18 that had errors.

18.35 Construct a 95% confidence interval for the difference in proportions of tax returns with errors.

$$\text{lower limit} = (\bar{p}_1-\bar{p}_2) - z_c\hat{\sigma}_{\bar{p}_1-\bar{p}_2} \qquad \text{upper limit} = (\bar{p}_1-\bar{p}_2) + z_c\hat{\sigma}_{\bar{p}_1-\bar{p}_2}$$
$$= -0.04 - (1.96)(0.0720) \qquad\qquad = -0.04 + (1.96)(0.0720)$$
$$= -0.1811 \qquad\qquad\qquad\qquad = 0.1011$$

Based on these samples, you are 95% confident that the difference in the proportions is between −0.1811 and 0.1011.

See Problem 11.59

Note: Problems 18.36–18.37 refer to the table below, the percentage of tread remaining on two brands of tires after both are installed on nine different cars that are then driven 30,000 miles.

Car	1	2	3	4	5	6	7	8	9
Brand A	45	60	40	55	45	50	60	40	45
Brand B	40	40	45	30	30	50	30	40	35

18.36 Test the claim that the remaining tread for Brand A is greater than the remaining tread for Brand B using $\alpha = 0.05$.

Let Brand A be population 1 and Brand B be population 2.

$$d = \text{brand A} - \text{brand B}$$
$$H_0 : \mu_d \leq 0$$
$$H_1 : \mu_d > 0$$

Car	Brand A	Brand B	d	d²
A	45	40	5	25
B	60	40	20	400
C	40	45	–5	25
D	55	30	25	625
E	45	30	15	225
F	50	50	0	0
G	60	30	30	900
H	40	40	0	0
I	45	35	10	100
Total			**100**	**2,300**

Calculate the standard deviation of the difference.

$$s_d = \sqrt{\frac{\sum d^2 - \frac{\left(\sum d\right)^2}{n}}{n-1}} = \sqrt{\frac{2,300 - \frac{(100)^2}{9}}{9-1}} = \sqrt{\frac{2,300 - 1,111.11}{8}} = \sqrt{148.61125} = 12.19$$

Compute the average difference $\bar{d}$ and the corresponding t-score $t_{\bar{d}}$.

$$\bar{d} = \frac{\sum d}{n} = \frac{100}{9} = 11.11$$

$$t_d = \frac{\bar{d} - \mu_d}{s_d / \sqrt{n}} = \frac{11.11 - 0}{12.19 / \sqrt{9}} = \frac{11.11}{4.06} = 2.74$$

According to Reference Table 2, given df = $n - 1 = 9 - 1 = 8$ degrees of freedom, $t_c = 1.860$. Because $t_{\bar{d}} = 2.74$ is greater than $t_c = 1.860$, you reject H_0 and support the claim that Brand A has a higher percentage of tread remaining.

See Problem 11.39.

Note: Problems 18.36–18.37 refer to the table in Problem 18.36, the percentage of tread remaining on two brands of tires after both are installed on nine different cars that are then driven 30,000 miles.

18.37 Construct a 95% confidence interval for the population mean paired difference between the brands.

According to Reference Table 2, given $n = 9$ and df = $n - 1 = 8$, $t_c = 2.306$. Calculate the 95% confidence interval.

$$\text{lower limit} = \bar{d} - t_c \frac{s_d}{\sqrt{n}} \qquad\qquad \text{upper limit} = \bar{d} - t_c \frac{s_d}{\sqrt{n}}$$

$$= 11.11 - (2.306)\frac{12.19}{\sqrt{9}} \qquad\qquad = 11.11 + (2.306)\frac{12.19}{\sqrt{9}}$$

$$= 1.74 \qquad\qquad\qquad\qquad = 20.48$$

You are 95% confident that the difference in remaining tread for Brands A and B is between 1.74% and 20.48%.

See Problem 11.40.

Note: Problems 18.38–18.42 refer to the data set below, customers' satisfaction ratings (on a scale of 1 to 10) for three different energy drinks.

Drink 1	Drink 2	Drink 3
6	5	9
3	5	8
2	7	6
6	5	9
4	7	8

18.38 Calculate the total sum of squares of the data values.

x_i	x_i^2	x_i	x_i^2	x_i	x_i^2
6	36	5	25	9	81
3	9	5	25	8	64
2	4	7	49	6	36
6	36	5	25	9	81
4	16	7	49	8	64

The sums of the $n_T = 15$ data values are $\sum x_i = 90$ and $\sum x_i^2 = 600$. Calculate SST.

See Problem 13.3.

$$SST = \sum x_i^2 - \frac{\left(\sum x_i\right)^2}{n_T} = 600 - \frac{(90)^2}{15} = 600 - 540 = 60$$

Note: Problems 18.38–18.42 refer to the data set in Problem 18.38, customers' satisfaction ratings (on a scale of 1 to 10) for three different energy drinks.

18.39 Partition the total sum of squares into the sum of squares within and the sum of squares between.

Calculate the mean of each sample and the grand mean $\overline{\overline{x}}$.

$$\overline{x}_1 = \frac{6+3+2+6+4}{5} = \frac{21}{5} = 4.2$$

$$\overline{x}_2 = \frac{5+5+7+5+7}{5} = \frac{29}{5} = 5.8$$

$$\overline{x}_3 = \frac{9+8+6+9+8}{5} = \frac{40}{5} = 8$$

$$\overline{\overline{x}} = \frac{\overline{x}_1 + \overline{x}_2 + \overline{x}_3}{3} = \frac{4.2+5.8+8}{3} = \frac{18}{3} = 6$$

Calculate the sum of squares between.

$$SSB = \sum_{i=1}^{k} n_i \left(\overline{x}_i - \overline{\overline{x}}\right)^2$$
$$= 5(4.2-6)^2 + 5(5.8-6)^2 + 5(8-6)^2$$
$$= 5(-1.8)^2 + 5(-0.2)^2 + 5(2)^2$$
$$= 5(3.24) + 5(0.04) + 5(4)$$
$$= 36.4$$

The sum of squares within is the difference of the total sum of squares and the sum of squares between.

See Problem 13.4.

$$SSW = SST - SSB$$
$$= 60 - 36.4$$
$$= 23.6$$

Note: Problems 18.38–18.42 refer to the data set in Problem 18.38, customers' satisfaction ratings (on a scale of 1 to 10) for three different energy drinks.

18.40 Perform a hypothesis test to determine whether there is a difference in customer satisfaction between the three energy drinks using $\alpha = 0.05$.

$$H_0 : \mu_1 = \mu_2 = \mu_3$$
$$H_1 : \text{At least two population means are different.}$$

Calculate the mean square between, the mean square within, and the corresponding F-score. Note that the data contains $k = 3$ populations.

$$MSB = \frac{SSB}{k-1} = \frac{36.4}{3-1} = 18.2$$

$$MSW = \frac{SSW}{n_T - k} = \frac{23.6}{15-3} = 1.97$$

$$F = \frac{MSB}{MSW} = \frac{18.2}{1.97} = 9.24$$

According to Reference Table 4, given $D_1 = k - 1 = 2$, $D_2 = n_T - k = 12$, and $\alpha = 0.05$, the critical F-score is $F_c = 3.885$. Because $F = 9.24$ is greater than $F_c = 3.885$, you reject H_0 and conclude that at least two of the sample means are different. ←

See Problem 13.5.

Note: Problems 18.38–18.42 refer to the data set in Problem 18.38, customers' satisfaction ratings (on a scale of 1 to 10) for three different energy drinks.

18.41 Construct a one-way ANOVA table for completely randomized design summarizing the findings in Problems 18.38–18.40.

Source of Variation	SS	df	MS	F
Between Samples	36.4	2	18.2	9.24
Within Samples	23.6	12	1.97	
Total	60	14		

See Problem 13.6.

Note: Problems 18.38–18.42 refer to the data set in Problem 18.38, customers' satisfaction ratings (on a scale of 1 to 10) for three different energy drinks.

18.42 Perform Scheffé's pairwise comparison test at the $\alpha = 0.05$ significance level to identify the unequal population means.

The calculations below compare the sample means of Drink 1 and Drink 2, Drink 1 and Drink 3, and Drink 2 and Drink 3, respectively.

$$F_S = \frac{\left(\bar{x}_1 - \bar{x}_2\right)^2}{MSW\left(\frac{1}{n_1} + \frac{1}{n_2}\right)} = \frac{(4.2 - 5.8)^2}{1.97\left(\frac{1}{5} + \frac{1}{5}\right)} = \frac{2.56}{0.788} = 3.25$$

$$F_S = \frac{\left(\bar{x}_1 - \bar{x}_3\right)^2}{MSW\left(\frac{1}{n_1} + \frac{1}{n_3}\right)} = \frac{(4.2 - 8)^2}{1.97\left(\frac{1}{5} + \frac{1}{5}\right)} = \frac{14.44}{0.788} = 18.32$$

$$F_S = \frac{\left(\bar{x}_2 - \bar{x}_3\right)^2}{MSW\left(\frac{1}{n_2} + \frac{1}{n_3}\right)} = \frac{(5.8 - 8)^2}{1.97\left(\frac{1}{5} + \frac{1}{5}\right)} = \frac{(-2.2)^2}{0.788} = 6.14$$

Note that $F_{sc} = (k-1)(F_c) = (3-1)(3.885) = 7.770$.

Sample Pair	F_s	F_{sc}	Conclusion
1 and 2	3.25	7.770	No difference
1 and 3	18.32	7.770	Difference
2 and 3	6.14	7.770	No difference

See Problem 13.7.

According to Scheffé's pairwise comparison test, only Drinks 1 and 3 have significantly different customer satisfaction ratings.

18.43 A process that fills bags with 40 pounds of mulch uses two different filling machines. The table below summarizes the results of random samples taken from each machine. Perform a hypothesis test to determine whether the machines have different variations at the $\alpha = 0.10$ significance level.

	Machine A	Machine B
Sample standard deviation	0.7 pounds	1.2 pounds
Sample size	18	19

Let Machine B represent population 1 and Machine A represent population 2.

$$H_0 : \sigma_1^2 = \sigma_2^2$$
$$H_1 : \sigma_1^2 \neq \sigma_2^2$$

Calculate the F-score.

$$F = \frac{s_1^2}{s_2^2} = \frac{(1.2)^2}{(0.7)^2} = \frac{1.44}{0.49} = 2.939$$

See Problem 12.44.

According to Reference Table 4, given $D_1 = n_1 - 1 = 18$ and $D_2 = n_2 - 1 = 17$, the critical F-score at the $\alpha = 0.10$ significance level is $F_c = 2.257$. Because $F = 2.939$ is greater than $F_c = 2.257$, you reject H_0 and conclude that the variability in the machines is different.

Note: Problems 18.44–18.45 refer to the table below, the results of a taste test in which randomly selected respondents were asked to rate two brands of cookies on a scale of 1 to 10. Assume the customer ratings are normally distributed and the population variances are equal.

	Cookie A	Cookie B
Sample mean	8.7	6.9
Sample size	12	10
Sample standard deviation	2.3	2.1

18.44 Test the hypothesis that Cookie A is preferred over Cookie B by comparing the calculated z-score to the critical z-score using $\alpha = 0.05$.

Let population 1 be Cookie A and population 2 be Cookie B.

$$H_0 : \mu_1 - \mu_2 \leq 0$$
$$H_1 : \mu_1 - \mu_2 > 0$$

According to Reference Table 2, given df $= n_1 + n_2 - 2 = 20$, the critical t-score is $t_c = 1.725$. Calculate the pooled standard deviation.

$$s_p = \sqrt{\frac{(n_1 - 1)s_1^2 + (n_2 - 1)s_2^2}{n_1 + n_2 - 2}} = \sqrt{\frac{(12-1)(2.3)^2 + (10-1)(2.1)^2}{12 + 10 - 2}} = \sqrt{\frac{58.19 + 39.69}{20}} = 2.212$$

Calculate the standard error for the difference between the means.

$$\hat{\sigma}_{\bar{x}_1 - \bar{x}_2} = s_p \sqrt{\frac{1}{n_1} + \frac{1}{n_2}} = 2.212 \sqrt{\frac{1}{12} + \frac{1}{10}} = 2.212\sqrt{0.18333} = 0.9471$$

Compute the difference between the sample means and the corresponding t-score.

$$\bar{x}_1 - \bar{x}_2 = 8.7 - 6.9 = 1.8$$

$$t_{\bar{x}_1 - \bar{x}_2} = \frac{(\bar{x}_1 - \bar{x}_2) - (\mu_1 - \mu_2)}{\hat{\sigma}_{\bar{x}_1 - \bar{x}_2}} = \frac{1.8 - 0}{0.9471} = 1.90$$

Because $t_{\bar{x}_1 - \bar{x}_2} = 1.90$ is greater than $t_c = 1.725$, you reject H_0 and conclude that Cookie A is preferred over Cookie B. ←

See Problem 11.25.

Note: Problems 18.44–18.45 refer to the table in Problem 18.44, the results of a taste test in which randomly selected respondents were asked to rate two brands of cookies on a scale of 1 to 10. Assume the customer ratings are normally distributed and the population variances are equal.

18.45 Construct a 90% confidence interval for the difference in the average customer ratings.

$$\text{lower limit} = \left(\bar{x}_1 - \bar{x}_2\right) - t_c\,\hat{\sigma}_{\bar{x}_1 - \bar{x}_2} \qquad \text{upper limit} = \left(\bar{x}_1 - \bar{x}_2\right) + t_c\,\hat{\sigma}_{\bar{x}_1 - \bar{x}_2}$$
$$= 1.8 - 1.725(0.9471) \qquad\qquad = 1.8 + 1.725(0.9471)$$
$$= 0.17 \qquad\qquad\qquad\qquad = 3.43$$

See Problem 11.26.

You are 90% confident that the difference in the average customer ratings of Cookie A and Cookie B is between 0.17 and 3.43.

Note: Problems 18.46–18.47 refer to a precinct in New York City that averages 4.2 car accidents per week. Assume the number of weekly accidents follows the Poisson distribution.

18.46 Calculate the probability that no more than two accidents will occur in this precinct next week.

$$P(x \le 2) = P(x = 0 \text{ or } 1 \text{ or } 2)$$
$$= P(0) + P(1) + P(2)$$

Compute $P(0)$, $P(1)$, and $P(2)$ separately.

$$P(0) = \frac{(4.2)^0 e^{-4.2}}{0!} \qquad P(1) = \frac{(4.2)^1 e^{-4.2}}{1!} \qquad P(2) = \frac{(4.2)^2 e^{-4.2}}{2!}$$
$$= \frac{1(0.014996)}{1} \qquad = \frac{4.2(0.014996)}{1} \qquad = \frac{17.64(0.014996)}{2}$$
$$= 0.0150 \qquad\qquad = 0.0630 \qquad\qquad = 0.1323$$

See Problem 6.23.

Thus, $P(x \le 2) = 0.0150 + 0.0630 + 0.1323 = 0.2103$.

Note: Problems 18.46–18.47 refer to a precinct in New York City that averages 4.2 car accidents per week. Assume the number of weekly accidents follows the Poisson distribution.

18.47 Calculate the probability that more than three accidents will occur in this precinct next week.

$$P(x > 3) = 1 - P(0) - P(1) - P(2) - P(3)$$

According to Problem 18.46, $P(0) = 0.0150$, $P(1) = 0.0630$, and $P(2) = 0.1323$. Calculate $P(3)$.

$$P(3) = \frac{(4.2)^3 e^{-4.2}}{3!} = \frac{74.088(0.014996)}{6} = \frac{1.11102}{6} = 0.1852$$

Compute $P(x > 3)$.

$$P(x > 3) = 1 - 0.0150 - 0.0630 - 0.1323 - 0.1852 = 0.6045$$

See Problem 6.19.

Note: Problems 18.48–18.50 refer to a researcher's claim that less than 40% of households in the United States watched the first game of the most recent World Series. A random sample of 160 households included 54 that watched the game.

18.48 Using $\alpha = 0.10$, test the claim by comparing the calculated z-score to the critical z-score.

$$H_0 : p \geq 0.40$$
$$H_1 : p < 0.40$$

According to Reference Table 1, the critical z-score is $z_c = -1.28$. Calculate the sample proportion.

$$p_s = \frac{54}{160} = 0.3375$$

Calculate the standard error of the proportion.

$$\sigma_p = \sqrt{\frac{p(1-p)}{n}} = \sqrt{\frac{0.40(1-0.40)}{160}} = \sqrt{0.0015} = 0.0387$$

Calculate z_p.

$$z_{0.3375} = \frac{0.3375 - 0.40}{0.0387} = -1.61$$

Because $z_{0.3375} = -1.61$ is less than $z_c = -1.28$, you reject H_0 and support the researcher's claim.

See Problem 10.65

Note: Problems 18.48–18.50 refer to a researcher's claim that less than 40% of households in the United States watched the first game of the most recent World Series. A random sample of 160 households included 54 that watched the game.

18.49 Verify your answer to Problem 18.48 by comparing the calculated sample proportion to the critical sample proportion.

$$H_0 : p \geq 0.40$$
$$H_1 : p < 0.40$$

Calculate the critical sample proportion.

$$p_c = p + z_c \sigma_p$$
$$= 0.40 + (-1.28)(0.0387)$$
$$= 0.350$$

See
Problem
10.66.

Because the sample proportion $p_s = 0.3375$ is less than $p_c = 0.350$, you reject the null hypothesis.

Note: *Problems 18.48–18.50 refer to a researcher's claim that less than 40% of households in the United States watched the first game of the most recent World Series. A random sample of 160 households included 54 that watched the game.*

18.50 Verify your answer to Problem 18.48 by comparing the *p*-value to the level of significance $\alpha = 0.10$.

According to Problem 18.48, $z_{0.3375} = -1.61$. Calculate the *p*-value.

$$p\text{-value} = P(z_p < -1.61) = 0.50 - 0.4463 = 0.0537$$

See
Problem
10.67.

The *p*-value 0.0537 is less than $\alpha = 0.10$, so you reject the null hypothesis.

18.51 Assume that men have an average height of 69.3 inches with a standard deviation of 5.7 inches. Calculate the probability that the average height of a sample of 36 men will be between 70 and 71 inches.

Calculate the standard error of the mean.

$$\sigma_{\bar{x}} = \frac{\sigma}{\sqrt{n}} = \frac{5.7}{\sqrt{36}} = \frac{5.7}{6} = 0.95$$

Compute z_{70} and z_{71}.

$$z_{70} = \frac{70 - 69.3}{0.95} = \frac{0.7}{0.95} = 0.74 \qquad z_{71} = \frac{71 - 69.3}{0.95} = \frac{1.7}{0.95} = 1.79$$

Calculate the probability that the average height of a sample of 36 men will be between 70 and 71 inches.

See
Problem
8.10.

$$
\begin{aligned}
P(70 < \bar{x} < 71) &= P(0.74 < z_{\bar{x}} < 1.79) \\
&= P(0 < z_{\bar{x}} < 1.79) - P(0 < z_{\bar{x}} < 0.74) \\
&= 0.4633 - 0.2704 \\
&= 0.1929
\end{aligned}
$$

					Second digit of z					
Z	**0.00**	**0.01**	**0.02**	**0.03**	**0.04**	**0.05**	**0.06**	**0.07**	**0.08**	**0.09**
0.0	0.0000	0.0040	0.0080	0.0120	0.0160	0.0199	0.0239	0.0279	0.0319	0.0359
0.1	0.0398	0.0438	0.0478	0.0517	0.0557	0.0596	0.0636	0.0675	0.0714	0.0753
0.2	0.0793	0.0832	0.0871	0.0910	0.0948	0.0987	0.1026	0.1064	0.1103	0.1141
0.3	0.1179	0.1217	0.1255	0.1293	0.1331	0.1368	0.1406	0.1443	0.1480	0.1517
0.4	0.1554	0.1591	0.1628	0.1664	0.1700	0.1736	0.1772	0.1808	0.1844	0.1879
0.5	0.1915	0.1950	0.1985	0.2019	0.2054	0.2088	0.2123	0.2157	0.2190	0.2224
0.6	0.2257	0.2291	0.2324	0.2357	0.2389	0.2422	0.2454	0.2486	0.2517	0.2549
0.7	0.2580	0.2611	0.2642	0.2673	0.2704	0.2734	0.2764	0.2794	0.2823	0.2852
0.8	0.2881	0.2910	0.2939	0.2967	0.2995	0.3023	0.3051	0.3078	0.3106	0.3133
0.9	0.3159	0.3186	0.3212	0.3238	0.3264	0.3289	0.3315	0.3340	0.3365	0.3389
1.0	0.3413	0.3438	0.3461	0.3485	0.3508	0.3531	0.3554	0.3577	0.3599	0.3621
1.1	0.3643	0.3665	0.3686	0.3708	0.3729	0.3749	0.3770	0.3790	0.3810	0.3830
1.2	0.3849	0.3869	0.3888	0.3907	0.3925	0.3944	0.3962	0.3980	0.3997	0.4015
1.3	0.4032	0.4049	0.4066	0.4082	0.4099	0.4115	0.4131	0.4147	0.4162	0.4177
1.4	0.4192	0.4207	0.4222	0.4236	0.4251	0.4265	0.4279	0.4292	0.4306	0.4319
1.5	0.4332	0.4345	0.4357	0.4370	0.4382	0.4394	0.4406	0.4418	0.4429	0.4441
1.6	0.4452	0.4463	0.4474	0.4484	0.4495	0.4505	0.4515	0.4525	0.4535	0.4545
1.7	0.4554	0.4564	0.4573	0.4582	0.4591	0.4599	0.4608	0.4616	0.4625	0.4633
1.8	0.4641	0.4649	0.4656	0.4664	0.4671	0.4678	0.4686	0.4693	0.4699	0.4706
1.9	0.4713	0.4719	0.4726	0.4732	0.4738	0.4744	0.4750	0.4756	0.4761	0.4767
2.0	0.4772	0.4778	0.4783	0.4788	0.4793	0.4798	0.4803	0.4808	0.4812	0.4817
2.1	0.4821	0.4826	0.4830	0.4834	0.4838	0.4842	0.4846	0.4850	0.4854	0.4857
2.2	0.4861	0.4864	0.4868	0.4871	0.4875	0.4878	0.4881	0.4884	0.4887	0.4890
2.3	0.4893	0.4896	0.4898	0.4901	0.4904	0.4906	0.4909	0.4911	0.4913	0.4916
2.4	0.4918	0.4920	0.4922	0.4925	0.4927	0.4929	0.4931	0.4932	0.4934	0.4936
2.5	0.4938	0.4940	0.4941	0.4943	0.4945	0.4946	0.4948	0.4949	0.4951	0.4952
2.6	0.4953	0.4955	0.4956	0.4957	0.4959	0.4960	0.4961	0.4962	0.4963	0.4964
2.7	0.4965	0.4966	0.4967	0.4968	0.4969	0.4970	0.4971	0.4972	0.4973	0.4974
2.8	0.4974	0.4975	0.4976	0.4977	0.4977	0.4978	0.4979	0.4979	0.4980	0.4981
2.9	0.4981	0.4982	0.4982	0.4983	0.4984	0.4984	0.4985	0.4985	0.4986	0.4986
3.0	0.4987	0.4987	0.4987	0.4988	0.4988	0.4989	0.4989	0.4989	0.4990	0.4990
3.1	0.4990	0.4991	0.4991	0.4991	0.4992	0.4992	0.4992	0.4992	0.4993	0.4993
3.2	0.4993	0.4993	0.4994	0.4994	0.4994	0.4994	0.4994	0.4995	0.4995	0.4995
3.3	0.4995	0.4995	0.4995	0.4996	0.4996	0.4996	0.4996	0.4996	0.4996	0.4997
3.4	0.4997	0.4997	0.4997	0.4997	0.4997	0.4997	0.4997	0.4997	0.4997	0.4998
3.5	0.4998	0.4998	0.4998	0.4998	0.4998	0.4998	0.4998	0.4998	0.4998	0.4998

Reference Table 2   Student's t-Distribution

	Probabilities Under the t–Distribution Curve								
1-Tail	0.2000	0.1500	0.1000	0.0500	0.0250	0.0100	0.0050	0.0010	0.0005
2-Tail	0.4000	0.3000	0.2000	0.1000	0.0500	0.0200	0.0100	0.00200	0.0010
Conf Lev.	0.6000	0.7000	0.8000	0.9000	0.9500	0.9800	0.9900	0.9980	0.9990
df									
1	1.376	1.963	3.078	6.314	12.706	31.821	63.657	318.31	636.62
2	1.061	1.386	1.886	2.920	4.303	6.965	9.925	22.327	31.599
3	0.978	1.250	1.638	2.353	3.182	4.541	5.841	10.215	12.924
4	0.941	1.190	1.533	2.132	2.776	3.747	4.604	7.173	8.610
5	0.920	1.156	1.476	2.015	2.571	3.365	4.032	5.893	6.869
6	0.906	1.134	1.440	1.943	2.447	3.143	3.707	5.208	5.959
7	0.896	1.119	1.415	1.895	2.365	2.998	3.499	4.785	5.408
8	0.889	1.108	1.397	1.860	2.306	2.896	3.355	4.501	5.041
9	0.883	1.100	1.383	1.833	2.262	2.821	3.250	4.297	4.781
10	0.879	1.093	1.372	1.812	2.228	2.764	3.169	4.144	4.587
11	0.876	1.088	1.363	1.796	2.201	2.718	3.106	4.025	4.437
12	0.873	1.083	1.356	1.782	2.179	2.681	3.055	3.930	4.318
13	0.870	1.079	1.350	1.771	2.160	2.650	3.012	3.852	4.221
14	0.868	1.076	1.345	1.761	2.145	2.624	2.977	3.787	4.140
15	0.866	1.074	1.341	1.753	2.131	2.602	2.947	3.733	4.073
16	0.865	1.071	1.337	1.746	2.120	2.583	2.921	3.686	4.015
17	0.863	1.069	1.333	1.740	2.110	2.567	2.898	3.646	3.965
18	0.862	1.067	1.330	1.734	2.101	2.552	2.878	3.610	3.922
19	0.861	1.066	1.328	1.729	2.093	2.539	2.861	3.579	3.883
20	0.860	1.064	1.325	1.725	2.086	2.528	2.845	3.552	3.850
21	0.859	1.063	1.323	1.721	2.080	2.518	2.831	3.527	3.819
22	0.858	1.061	1.321	1.717	2.074	2.508	2.819	3.505	3.792
23	0.858	1.060	1.319	1.714	2.069	2.500	2.807	3.485	3.768
24	0.857	1.059	1.318	1.711	2.064	2.492	2.797	3.467	3.745
25	0.856	1.058	1.316	1.708	2.060	2.485	2.787	3.450	3.725
26	0.856	1.058	1.315	1.706	2.056	2.479	2.779	3.435	3.707
27	0.855	1.057	1.314	1.703	2.052	2.473	2.771	3.421	3.690
28	0.855	1.056	1.313	1.701	2.048	2.467	2.763	3.408	3.674
29	0.854	1.055	1.311	1.699	2.045	2.462	2.756	3.396	3.659
30	0.854	1.055	1.310	1.697	2.042	2.457	2.750	3.385	3.646
40	0.851	1.050	1.303	1.684	2.021	2.423	2.704	3.307	3.551
50	0.849	1.047	1.299	1.676	2.009	2.403	2.678	3.261	3.496

				Area in Right Tail of Distribution						
df	0.995	0.99	0.975	0.95	0.90	0.10	0.05	0.025	0.01	0.005
1	—	—	0.001	0.004	0.016	2.706	3.841	5.024	6.635	7.879
2	0.010	0.020	0.051	0.103	0.211	4.605	5.991	7.378	9.210	10.597
3	0.072	0.115	0.216	0.352	0.584	6.251	7.815	9.348	11.345	12.838
4	0.207	0.297	0.484	0.711	1.064	7.779	9.488	11.143	13.277	14.860
5	0.412	0.554	0.831	1.145	1.610	9.236	11.070	12.833	15.086	16.750
6	0.676	0.872	1.237	1.635	2.204	10.645	12.592	14.449	16.812	18.548
7	0.989	1.239	1.690	2.167	2.833	12.017	14.067	16.013	18.475	20.278
8	1.344	1.646	2.180	2.733	3.490	13.362	15.507	17.535	20.090	21.955
9	1.735	2.088	2.700	3.325	4.168	14.684	16.919	19.023	21.666	23.589
10	2.156	2.558	3.247	3.940	4.865	15.987	18.307	20.483	23.209	25.188
11	2.603	3.053	3.816	4.575	5.578	17.275	19.675	21.920	24.725	26.757
12	3.074	3.571	4.404	5.226	6.304	18.549	21.026	23.337	26.217	28.300
13	3.565	4.107	5.009	5.892	7.042	19.812	22.362	24.736	27.688	29.819
14	4.075	4.660	5.629	6.571	7.790	21.064	23.685	26.119	29.141	31.319
15	4.601	5.229	6.262	7.261	8.547	22.307	24.996	27.488	30.578	32.801
16	5.142	5.812	6.908	7.962	9.312	23.542	26.296	28.845	32.000	34.267
17	5.697	6.408	7.564	8.672	10.085	24.769	27.587	30.191	33.409	35.718
18	6.265	7.015	8.231	9.390	10.865	25.989	28.869	31.526	34.805	37.156
19	6.844	7.633	8.907	10.117	11.651	27.204	30.144	32.852	36.191	38.582
20	7.434	8.260	9.591	10.851	12.443	28.412	31.410	34.170	37.566	39.997
21	8.034	8.897	10.283	11.591	13.240	29.615	32.671	35.479	38.932	41.401
22	8.643	9.542	10.982	12.338	14.041	30.813	33.924	36.781	40.289	42.796
23	9.260	10.196	11.689	13.091	14.848	32.007	35.172	38.076	41.638	44.181
24	9.886	10.856	12.401	13.848	15.659	33.196	36.415	39.364	42.980	45.559
25	10.520	11.524	13.120	14.611	16.473	34.382	37.652	40.646	44.314	46.928
26	11.160	12.198	13.844	15.379	17.292	35.563	38.885	41.923	45.642	48.290
27	11.808	12.879	14.573	16.151	18.114	36.741	40.113	43.195	46.963	49.645
28	12.461	13.565	15.308	16.928	18.939	37.916	41.337	44.461	48.278	50.993
29	13.121	14.256	16.047	17.708	19.768	39.087	42.557	45.722	49.588	52.336
30	13.787	14.953	16.791	18.493	20.599	40.256	43.773	46.979	50.892	53.672

Reference Table 4   F–Distribution

Area in the Right Tail of Distribution = 0.10

D_1

D_2	1	2	3	4	5	6	7	8	9	10
1	39.863	49.500	53.593	55.833	57.240	58.204	58.906	59.439	59.858	60.195
2	8.526	9.000	9.162	9.243	9.293	9.326	9.349	9.367	9.381	9.392
3	5.538	5.462	5.391	5.343	5.309	5.285	5.266	5.252	5.240	5.230
4	4.545	4.325	4.191	4.107	4.051	4.010	3.979	3.955	3.936	3.920
5	4.060	3.780	3.619	3.520	3.453	3.405	3.368	3.339	3.316	3.297
6	3.776	3.463	3.289	3.181	3.108	3.055	3.014	2.983	2.958	2.937
7	3.589	3.257	3.074	2.961	2.883	2.827	2.785	2.752	2.725	2.703
8	3.458	3.113	2.924	2.806	2.726	2.668	2.624	2.589	2.561	2.538
9	3.360	3.006	2.813	2.693	2.611	2.551	2.505	2.469	2.440	2.416
10	3.285	2.924	2.728	2.605	2.522	2.461	2.414	2.377	2.347	2.323
11	3.225	2.860	2.660	2.536	2.451	2.389	2.342	2.304	2.274	2.248
12	3.177	2.807	2.606	2.480	2.394	2.331	2.283	2.245	2.214	2.188
13	3.136	2.763	2.560	2.434	2.347	2.283	2.234	2.195	2.164	2.138
14	3.102	2.726	2.522	2.395	2.307	2.243	2.193	2.154	2.122	2.095
15	3.073	2.695	2.490	2.361	2.273	2.208	2.158	2.119	2.086	2.059
16	3.048	2.668	2.462	2.333	2.244	2.178	2.128	2.088	2.055	2.028
17	3.026	2.645	2.437	2.308	2.218	2.152	2.102	2.061	2.028	2.001
18	3.007	2.624	2.416	2.286	2.196	2.130	2.079	2.038	2.005	1.977
19	2.990	2.606	2.397	2.266	2.176	2.109	2.058	2.017	1.984	1.956
20	2.975	2.589	2.380	2.249	2.158	2.091	2.040	1.999	1.965	1.937

Area in the Right Tail of Distribution = 0.10

D_1

D_2	11	12	13	14	15	16	17	18	19	20
1	60.473	60.705	60.903	61.073	61.220	61.350	61.464	61.566	61.658	61.740
2	9.401	9.408	9.415	9.420	9.425	9.429	9.433	9.436	9.439	9.441
3	5.222	5.216	5.210	5.205	5.200	5.196	5.193	5.190	5.187	5.184
4	3.907	3.896	3.886	3.878	3.870	3.864	3.858	3.853	3.849	3.844
5	3.282	3.268	3.257	3.247	3.238	3.230	3.223	3.217	3.212	3.207
6	2.920	2.905	2.892	2.881	2.871	2.863	2.855	2.848	2.842	2.836
7	2.684	2.668	2.654	2.643	2.632	2.623	2.615	2.607	2.601	2.595
8	2.519	2.502	2.488	2.475	2.464	2.455	2.446	2.438	2.431	2.425
9	2.396	2.379	2.364	2.351	2.340	2.329	2.320	2.312	2.305	2.298
10	2.302	2.284	2.269	2.255	2.244	2.233	2.224	2.215	2.208	2.201
11	2.227	2.209	2.193	2.179	2.167	2.156	2.147	2.138	2.130	2.123
12	2.166	2.147	2.131	2.117	2.105	2.094	2.084	2.075	2.067	2.060
13	2.116	2.097	2.080	2.066	2.053	2.042	2.032	2.023	2.014	2.007
14	2.073	2.054	2.037	2.022	2.010	1.998	1.988	1.978	1.970	1.962
15	2.037	2.017	2.000	1.985	1.972	1.961	1.950	1.941	1.932	1.924
16	2.005	1.985	1.968	1.953	1.940	1.928	1.917	1.908	1.899	1.891
17	1.978	1.958	1.940	1.925	1.912	1.900	1.889	1.879	1.870	1.862
18	1.954	1.933	1.916	1.900	1.887	1.875	1.864	1.854	1.845	1.837
19	1.932	1.912	1.894	1.878	1.865	1.852	1.841	1.831	1.822	1.814
20	1.913	1.892	1.875	1.859	1.845	1.833	1.821	1.811	1.802	1.794

Area in the Right Tail of Distribution = 0.05

D_2	1	2	3	4	5	6	7	8	9	10
					D_1					
1	161.448	199.500	215.707	224.583	230.162	233.986	236.768	238.883	240.543	241.882
2	18.513	19.000	19.164	19.247	19.296	19.330	19.353	19.371	19.385	19.396
3	10.128	9.552	9.277	9.117	9.013	8.941	8.887	8.845	8.812	8.786
4	7.709	6.944	6.591	6.388	6.256	6.163	6.094	6.041	5.999	5.964
5	6.608	5.786	5.409	5.192	5.050	4.950	4.876	4.818	4.772	4.735
6	5.987	5.143	4.757	4.534	4.387	4.284	4.207	4.147	4.099	4.060
7	5.591	4.737	4.347	4.120	3.972	3.866	3.787	3.726	3.677	3.637
8	5.318	4.459	4.066	3.838	3.687	3.581	3.500	3.438	3.388	3.347
9	5.117	4.256	3.863	3.633	3.482	3.374	3.293	3.230	3.179	3.137
10	4.965	4.103	3.708	3.478	3.326	3.217	3.135	3.072	3.020	2.978
11	4.844	3.982	3.587	3.357	3.204	3.095	3.012	2.948	2.896	2.854
12	4.747	3.885	3.490	3.259	3.106	2.996	2.913	2.849	2.796	2.753
13	4.667	3.806	3.411	3.179	3.025	2.915	2.832	2.767	2.714	2.671
14	4.600	3.739	3.344	3.112	2.958	2.848	2.764	2.699	2.646	2.602
15	4.543	3.682	3.287	3.056	2.901	2.790	2.707	2.641	2.588	2.544
16	4.494	3.634	3.239	3.007	2.852	2.741	2.657	2.591	2.538	2.494
17	4.451	3.592	3.197	2.965	2.810	2.699	2.614	2.548	2.494	2.450
18	4.414	3.555	3.160	2.928	2.773	2.661	2.577	2.510	2.456	2.412
19	4.381	3.522	3.127	2.895	2.740	2.628	2.544	2.477	2.423	2.378
20	4.351	3.493	3.098	2.866	2.711	2.599	2.514	2.447	2.393	2.348

Area in the Right Tail of Distribution = 0.05

D_2	11	12	13	14	15	16	17	18	19	20
					D_1					
1	242.983	243.906	244.690	245.364	245.950	246.464	246.918	247.323	247.686	248.013
2	19.405	19.413	19.419	19.424	19.429	19.433	19.437	19.440	19.443	19.446
3	8.763	8.745	8.729	8.715	8.703	8.692	8.683	8.675	8.667	8.660
4	5.936	5.912	5.891	5.873	5.858	5.844	5.832	5.821	5.811	5.803
5	4.704	4.678	4.655	4.636	4.619	4.604	4.590	4.579	4.568	4.558
6	4.027	4.000	3.976	3.956	3.938	3.922	3.908	3.896	3.884	3.874
7	3.603	3.575	3.550	3.529	3.511	3.494	3.480	3.467	3.455	3.445
8	3.313	3.284	3.259	3.237	3.218	3.202	3.187	3.173	3.161	3.150
9	3.102	3.073	3.048	3.025	3.006	2.989	2.974	2.960	2.948	2.936
10	2.943	2.913	2.887	2.865	2.845	2.828	2.812	2.798	2.785	2.774
11	2.818	2.788	2.761	2.739	2.719	2.701	2.685	2.671	2.658	2.646
12	2.717	2.687	2.660	2.637	2.617	2.599	2.583	2.568	2.555	2.544
13	2.635	2.604	2.577	2.554	2.533	2.515	2.499	2.484	2.471	2.459
14	2.565	2.534	2.507	2.484	2.463	2.445	2.428	2.413	2.400	2.388
15	2.507	2.475	2.448	2.424	2.403	2.385	2.368	2.353	2.340	2.328
16	2.456	2.425	2.397	2.373	2.352	2.333	2.317	2.302	2.288	2.276
17	2.413	2.381	2.353	2.329	2.308	2.289	2.272	2.257	2.243	2.230
18	2.374	2.342	2.314	2.290	2.269	2.250	2.233	2.217	2.203	2.191
19	2.340	2.308	2.280	2.256	2.234	2.215	2.198	2.182	2.168	2.155
20	2.310	2.278	2.250	2.225	2.203	2.184	2.167	2.151	2.137	2.124

Area in the Right Tail of Distribution = 0.25

D_1

D_2	1	2	3	4	5	6	7	8	9	10
1	647.789	799.500	864.163	899.583	921.848	937.111	948.217	956.656	963.285	968.627
2	38.506	39.000	39.165	39.248	39.298	39.331	39.355	39.373	39.387	39.398
3	17.443	16.044	15.439	15.101	14.885	14.735	14.624	14.540	14.473	14.419
4	12.218	10.649	9.979	9.605	9.364	9.197	9.074	8.980	8.905	8.844
5	10.007	8.434	7.764	7.388	7.146	6.978	6.853	6.757	6.681	6.619
6	8.813	7.260	6.599	6.227	5.988	5.820	5.695	5.600	5.523	5.461
7	8.073	6.542	5.890	5.523	5.285	5.119	4.995	4.899	4.823	4.761
8	7.571	6.059	5.416	5.053	4.817	4.652	4.529	4.433	4.357	4.295
9	7.209	5.715	5.078	4.718	4.484	4.320	4.197	4.102	4.026	3.964
10	6.937	5.456	4.826	4.468	4.236	4.072	3.950	3.855	3.779	3.717
11	6.724	5.256	4.630	4.275	4.044	3.881	3.759	3.664	3.588	3.526
12	6.554	5.096	4.474	4.121	3.891	3.728	3.607	3.512	3.436	3.374
13	6.414	4.965	4.347	3.996	3.767	3.604	3.483	3.388	3.312	3.250
14	6.298	4.857	4.242	3.892	3.663	3.501	3.380	3.285	3.209	3.147
15	6.200	4.765	4.153	3.804	3.576	3.415	3.293	3.199	3.123	3.060
16	6.115	4.687	4.077	3.729	3.502	3.341	3.219	3.125	3.049	2.986
17	6.042	4.619	4.011	3.665	3.438	3.277	3.156	3.061	2.985	2.922
18	5.978	4.560	3.954	3.608	3.382	3.221	3.100	3.005	2.929	2.866
19	5.922	4.508	3.903	3.559	3.333	3.172	3.051	2.956	2.880	2.817
20	5.871	4.461	3.859	3.515	3.289	3.128	3.007	2.913	2.837	2.774

Area in the Right Tail of Distribution = 0.25

D_1

D_2	11	12	13	14	15	16	17	18	19	20
1	973.025	976.708	979.837	982.528	984.867	986.919	988.733	990.349	991.797	993.103
2	39.407	39.415	39.421	39.427	39.431	39.435	39.439	39.442	39.445	39.448
3	14.374	14.337	14.304	14.277	14.253	14.232	14.213	14.196	14.181	14.167
4	8.794	8.751	8.715	8.684	8.657	8.633	8.611	8.592	8.575	8.560
5	6.568	6.525	6.488	6.456	6.428	6.403	6.381	6.362	6.344	6.329
6	5.410	5.366	5.329	5.297	5.269	5.244	5.222	5.202	5.184	5.168
7	4.709	4.666	4.628	4.596	4.568	4.543	4.521	4.501	4.483	4.467
8	4.243	4.200	4.162	4.130	4.101	4.076	4.054	4.034	4.016	3.999
9	3.912	3.868	3.831	3.798	3.769	3.744	3.722	3.701	3.683	3.667
10	3.665	3.621	3.583	3.550	3.522	3.496	3.474	3.453	3.435	3.419
11	3.474	3.430	3.392	3.359	3.330	3.304	3.282	3.261	3.243	3.226
12	3.321	3.277	3.239	3.206	3.177	3.152	3.129	3.108	3.090	3.073
13	3.197	3.153	3.115	3.082	3.053	3.027	3.004	2.983	2.965	2.948
14	3.095	3.050	3.012	2.979	2.949	2.923	2.900	2.879	2.861	2.844
15	3.008	2.963	2.925	2.891	2.862	2.836	2.813	2.792	2.773	2.756
16	2.934	2.889	2.851	2.817	2.788	2.761	2.738	2.717	2.698	2.681
17	2.870	2.825	2.786	2.753	2.723	2.697	2.673	2.652	2.633	2.616
18	2.814	2.769	2.730	2.696	2.667	2.640	2.617	2.596	2.576	2.559
19	2.765	2.720	2.681	2.647	2.617	2.591	2.567	2.546	2.526	2.509
20	2.721	2.676	2.637	2.603	2.573	2.547	2.523	2.501	2.482	2.464

Area in the Right Tail of Distribution = 0.01

D_1

D_2	1	2	3	4	5	6	7	8	9	10
1	4052.2	4999.5	5403.4	5624.6	5763.6	5859.0	5928.4	5981.1	6022.5	6055.8
2	98.503	99.000	99.166	99.249	99.299	99.333	99.356	99.374	99.388	99.399
3	34.116	30.817	29.457	28.710	28.237	27.911	27.672	27.489	27.345	27.229
4	21.198	18.000	16.694	15.977	15.522	15.207	14.976	14.799	14.659	14.546
5	16.258	13.274	12.060	11.392	10.967	10.672	10.456	10.289	10.158	10.051
6	13.745	10.925	9.780	9.148	8.746	8.466	8.260	8.102	7.976	7.874
7	12.246	9.547	8.451	7.847	7.460	7.191	6.993	6.840	6.719	6.620
8	11.259	8.649	7.591	7.006	6.632	6.371	6.178	6.029	5.911	5.814
9	10.561	8.022	6.992	6.422	6.057	5.802	5.613	5.467	5.351	5.257
10	10.044	7.559	6.552	5.994	5.636	5.386	5.200	5.057	4.942	4.849
11	9.646	7.206	6.217	5.668	5.316	5.069	4.886	4.744	4.632	4.539
12	9.330	6.927	5.953	5.412	5.064	4.821	4.640	4.499	4.388	4.296
13	9.074	6.701	5.739	5.205	4.862	4.620	4.441	4.302	4.191	4.100
14	8.862	6.515	5.564	5.035	4.695	4.456	4.278	4.140	4.030	3.939
15	8.683	6.359	5.417	4.893	4.556	4.318	4.142	4.004	3.895	3.805
16	8.531	6.226	5.292	4.773	4.437	4.202	4.026	3.890	3.780	3.691
17	8.400	6.112	5.185	4.669	4.336	4.102	3.927	3.791	3.682	3.593
18	8.285	6.013	5.092	4.579	4.248	4.015	3.841	3.705	3.597	3.508
19	8.185	5.926	5.010	4.500	4.171	3.939	3.765	3.631	3.523	3.434
20	8.096	5.849	4.938	4.431	4.103	3.871	3.699	3.564	3.457	3.368

Area in the Right Tail of Distribution = 0.01

D_1

D_2	11	12	13	14	15	16	17	18	19	20
1	6083.3	6106.3	6125.9	6142.7	6157.3	6170.1	6181.4	6191.5	6200.6	6208.7
2	99.408	99.416	99.422	99.428	99.433	99.437	99.440	99.444	99.447	99.449
3	27.133	27.052	26.983	26.924	26.872	26.827	26.787	26.751	26.719	26.690
4	14.452	14.374	14.307	14.249	14.198	14.154	14.115	14.080	14.048	14.020
5	9.963	9.888	9.825	9.770	9.722	9.680	9.643	9.610	9.580	9.553
6	7.790	7.718	7.657	7.605	7.559	7.519	7.483	7.451	7.422	7.396
7	6.538	6.469	6.410	6.359	6.314	6.275	6.240	6.209	6.181	6.155
8	5.734	5.667	5.609	5.559	5.515	5.477	5.442	5.412	5.384	5.359
9	5.178	5.111	5.055	5.005	4.962	4.924	4.890	4.860	4.833	4.808
10	4.772	4.706	4.650	4.601	4.558	4.520	4.487	4.457	4.430	4.405
11	4.462	4.397	4.342	4.293	4.251	4.213	4.180	4.150	4.123	4.099
12	4.220	4.155	4.100	4.052	4.010	3.972	3.939	3.909	3.883	3.858
13	4.025	3.960	3.905	3.857	3.815	3.778	3.745	3.716	3.689	3.665
14	3.864	3.800	3.745	3.698	3.656	3.619	3.586	3.556	3.529	3.505
15	3.730	3.666	3.612	3.564	3.522	3.485	3.452	3.423	3.396	3.372
16	3.616	3.553	3.498	3.451	3.409	3.372	3.339	3.310	3.283	3.259
17	3.519	3.455	3.401	3.353	3.312	3.275	3.242	3.212	3.186	3.162
18	3.434	3.371	3.316	3.269	3.227	3.190	3.158	3.128	3.101	3.077
19	3.360	3.297	3.242	3.195	3.153	3.116	3.084	3.054	3.027	3.003
20	3.294	3.231	3.177	3.130	3.088	3.051	3.018	2.989	2.962	2.938

Critical Values of the Studentized Range (0.05 level)

D_1

D_2	2	3	4	5	6	7	8	9	10
2	6.085	8.331	9.798	10.881	11.734	12.435	13.027	13.538	13.988
3	4.501	5.910	6.825	7.502	8.037	8.478	8.852	9.177	9.462
4	3.927	5.040	5.757	6.287	6.707	7.053	7.347	7.602	7.826
5	3.635	4.602	5.219	5.673	6.033	6.330	6.582	6.801	6.995
6	3.461	4.339	4.896	5.305	5.629	5.895	6.122	6.319	6.493
7	3.344	4.165	4.681	5.060	5.359	5.606	5.815	5.998	6.158
8	3.261	4.041	4.529	4.886	5.167	5.399	5.596	5.767	5.918
9	3.199	3.949	4.415	4.755	5.024	5.244	5.432	5.595	5.738
10	3.151	3.877	4.327	4.654	4.912	5.124	5.304	5.461	5.598
11	3.113	3.820	4.256	4.574	4.823	5.028	5.202	5.353	5.486
12	3.081	3.773	4.199	4.508	4.748	4.947	5.116	5.263	5.395
13	3.055	3.734	4.151	4.453	4.690	4.884	5.049	5.192	5.318
14	3.033	3.701	4.111	4.407	4.639	4.829	4.990	5.130	5.253
15	3.014	3.673	4.076	4.367	4.595	4.782	4.940	5.077	5.198
16	2.998	3.649	4.046	4.333	4.557	4.741	4.896	5.031	5.150
17	2.984	3.628	4.020	4.303	4.524	4.705	4.858	4.991	5.108
18	2.971	3.609	3.997	4.276	4.494	4.673	4.824	4.955	5.071
19	2.960	3.593	3.977	4.253	4.469	4.645	4.794	4.924	5.038
20	2.950	3.578	3.958	4.232	4.445	4.620	4.768	4.895	5.008
21	2.943	3.566	3.943	4.214	4.425	4.599	4.745	4.871	4.982
22	2.935	3.554	3.928	4.197	4.407	4.578	4.723	4.848	4.958
23	2.927	3.543	3.915	4.182	4.389	4.559	4.703	4.827	4.936
24	2.920	3.533	3.902	4.167	4.374	4.542	4.685	4.808	4.916

Critical Values of the Studentized Range (0.01 level)

D_1

D_2	2	3	4	5	6	7	8	9	10
2	14.035	19.019	22.294	24.717	26.628	28.199	29.528	30.677	31.687
3	8.263	10.616	12.170	13.324	14.240	14.997	15.640	16.198	16.689
4	6.511	8.118	9.173	9.958	10.582	11.099	11.539	11.925	12.264
5	5.702	6.976	7.806	8.422	8.913	9.321	9.669	9.971	10.239
6	5.243	6.331	7.033	7.556	7.974	8.318	8.611	8.869	9.097
7	4.948	5.919	6.543	7.006	7.373	7.678	7.940	8.167	8.368
8	4.745	5.635	6.204	6.625	6.960	7.238	7.475	7.681	7.864
9	4.596	5.428	5.957	6.347	6.658	6.915	7.134	7.326	7.495
10	4.482	5.270	5.769	6.136	6.428	6.669	6.875	7.055	7.214
11	4.392	5.146	5.621	5.970	6.247	6.476	6.671	6.842	6.992
12	4.320	5.046	5.502	5.836	6.101	6.321	6.507	6.670	6.814
13	4.261	4.964	5.404	5.727	5.981	6.192	6.372	6.528	6.666
14	4.210	4.895	5.322	5.634	5.881	6.085	6.258	6.410	6.543
15	4.167	4.836	5.252	5.556	5.796	5.994	6.162	6.309	6.438
16	4.131	4.786	5.192	5.489	5.722	5.915	6.079	6.222	6.348
17	4.099	4.742	5.140	5.430	5.659	5.847	6.007	6.147	6.270
18	4.071	4.703	5.094	5.379	5.603	5.787	5.944	6.081	6.201
19	4.046	4.669	5.054	5.333	5.553	5.735	5.889	6.022	6.141
20	4.024	4.639	5.018	5.293	5.509	5.688	5.839	5.970	6.086
21	4.014	4.619	4.992	5.264	5.476	5.651	5.800	5.929	6.043
22	3.995	4.594	4.963	5.231	5.440	5.613	5.759	5.887	5.999
23	3.979	4.572	4.936	5.202	5.408	5.579	5.723	5.848	5.959
24	3.964	4.552	4.912	5.175	5.379	5.547	5.690	5.814	5.923

Source: E.S. Pearson and H.O. Hartley, *Biometrika Tables for Statisticians*, New York: Cambridge University Press, 1954

Reference Table 6   *Critical Values for the Sign Test*

One–Tailed α	0.05	0.025	0.01	0.005
Two–Tailed α	0.10	0.05	0.02	0.01
n				
8	1	0	0	0
9	1	1	0	0
10	1	1	0	0
11	2	1	1	0
12	2	2	1	1
13	3	2	1	1
14	3	3	2	1
15	3	3	2	2
16	4	3	2	2
17	4	4	3	2
18	5	4	3	3
19	5	4	4	3
20	5	5	4	3
21	6	5	4	4
22	6	5	5	4
23	7	6	5	4
24	7	6	5	5
25	7	6	6	5

Source: From *Journal of American Statistical Association* Vol. 41 (1946) pp. 557–66. W.J. Dixon and A.M. Mood.

Reference Table 7   *Lower and Upper Critical Values for Wilcoxon Rank Sum Test*

$\alpha = 0.025$ (one–tail) or $\alpha = 0.05$ (two–tail)								
n_1	3	4	5	6	7	8	9	10
n_2								
3	5,16	6,18	6,21	7,23	7,26	8,28	8,31	9,33
4	6,18	11,25	12,28	12,32	13,35	14,38	15,41	16,44
5	6,21	12,28	18,37	19,41	20,45	21,49	22,53	24,56
6	7,23	12,32	19,41	26,52	28,56	29,61	31,65	32,70
7	7,26	13,35	20,45	28,56	37,68	39,73	41,78	43,83
8	8,28	14,38	21,49	29,61	39,73	49,87	51,93	54,98
9	8,31	15,41	22,53	31,65	41,78	51,93	63,108	66,114
10	9,33	16,44	24,56	32,70	43,83	54,98	66,114	79,131

(Note: n_1 is the smaller of the two samples – i.e., $n_1 \leq n_2$.)

$\alpha = 0.05$ (one–tail) or $\alpha = 0.10$ (two–tail)								
n_1	3	4	5	6	7	8	9	10
n_2								
3	6,15	7,17	7,20	8,22	9,24	9,27	10,29	11,31
4	7,17	12,24	13,27	14,30	15,33	16,36	17,39	18,42
5	7,20	13,27	19,36	20,40	22,43	24,46	25,50	26,54
6	8,22	14,30	20,40	28,50	30,54	32,58	33,63	35,67
7	9,24	15,33	22,43	30,54	39,66	41,71	43,76	46,80
8	9,27	16,36	24,46	32,58	41,71	52,84	54,90	57,95
9	10,29	17,39	25,50	33,63	43,76	54,90	66,105	69,111
10	11,31	18,42	26,54	35,67	46,80	57,95	69,111	83,127

(Note: $n_1 \leq n_2$.)

Source: F. Wilcoxon and R. A. Wilcox, *Some Approximate Statistical Procedures* (New York: American Cyanamid Company, 1964), pp. 20-23.

Reference Table 8   Critical Values W_c for the Wilcoxon Signed–Rank Test

One–Tailed α	0.05	0.025	0.01	0.005
Two–Tailed α	0.10	0.05	0.02	0.01
n				
5	1			
6	2	1		
7	4	2	0	
8	6	4	2	0
9	8	6	3	2
10	11	8	5	3
11	14	11	7	5
12	17	14	10	7

Source: *Some Rapid Approximate Statistical Procedures.* Copyright 1949, 1964 Lerderle Laboratories, American Cyanamid Co., Wayne, N.J.

Reference Table 9   Critical Values for the Spearman Rank Correlation

α	0.10	0.05	0.01
n			
5	0.900	—	—
6	0.829	0.886	—
7	0.714	0.786	0.929
8	0.643	0.738	0.881
9	0.600	0.700	0.833
10	0.564	0.648	0.794
11	0.536	0.618	0.818
12	0.497	0.591	0.780

Source: N.L. Johnson and F.C. Leone, *Statistical and Experimental Design*, Vol. 1 (1964), p. 412.

Reference Table 10   Factors for 3-Sigma Control Chart Limits

Sample Size n	Mean Factor A_2	Lower Range D_3	Upper Range D_4
2	1.880	0	3.268
3	1.023	0	2.574
4	0.729	0	2.282
5	0.577	0	2.115
6	0.483	0	2.004
7	0.419	0.076	1.924
8	0.373	0.136	1.864
9	0.337	0.184	1.816
10	0.308	0.223	1.777

Source: E.S. Pearson, *The Percentage Limits for the Distribution Range in Samples from a Normal Population*, Biometrika 24 (1932): 416.

Appendix A
Critical Values and Confidence Intervals

Confidence Intervals

Type	Sample	Population	σ	Confidence Interval
mean	$n \geq 30$	any	known	$\bar{x} \pm z_c \dfrac{\sigma}{\sqrt{n}}$
mean	$n \geq 30$	any	unknown	$\bar{x} \pm z_c \dfrac{s}{\sqrt{n}}$
mean	$n < 30$	normal	known	$\bar{x} \pm z_c \dfrac{\sigma}{\sqrt{n}}$
mean	$n < 30$	normal	unknown	$\bar{x} \pm t_c \dfrac{s}{\sqrt{n}}$
proportion	$np \geq 5$ $nq \geq 5$	any		$P_s \pm z_c \sqrt{\dfrac{P_s(1 - P_s)}{n}}$

Critical z-scores

Confidence Interval	Alpha	Tail	Critical z-score
0.99	0.01	two	± 2.57
	0.01	one	± 2.33
0.98	0.02	two	± 2.33
	0.02	one	± 2.05
0.95	0.05	two	± 1.96
	0.05	one	± 1.64
0.90	0.10	two	± 1.64
	0.10	one	± 1.28

Sample Size for Confidence Intervals

Type	Sample size
mean	$n = \left(\dfrac{z\sigma}{E}\right)^2$
proportion	$n = pq\left(\dfrac{z_c}{E}\right)^2$

Appendix B
Hypothesis Testing

One-Sample Hypothesis Test

Type	Sample	Population	σ	Test Statistic
mean	$n \geq 30$	any	known	$z_{\bar{x}} = \dfrac{\bar{x} - \mu}{\sigma / \sqrt{n}}$
mean	$n \geq 30$	any	unknown	$z_{\bar{x}} = \dfrac{\bar{x} - \mu}{s / \sqrt{n}}$
mean	$n < 30$	normal	known	$z_{\bar{x}} = \dfrac{\bar{x} - \mu}{\sigma / \sqrt{n}}$
mean	$n < 30$	normal	unknown	$t_{\bar{x}} = \dfrac{\bar{x} - \mu}{s / \sqrt{n}}$
proportion	$np \geq 5$ $nq \geq 5$	any		$z_p = \dfrac{\bar{p} - p}{\sqrt{\dfrac{p(1-p)}{n}}}$

Two-Sample Hypothesis Test

Type	Sample	Population	σ_1, σ_2	Test Statistic
mean	$n_1, n_2 \geq 30$ Independent samples	any	known	$z_{\bar{x}} = \dfrac{(\bar{x}_1 - \bar{x}_2) - (\mu_1 - \mu_2)}{\sqrt{\dfrac{\sigma_1^2}{n_1} + \dfrac{\sigma_2^2}{n_2}}}$
mean	$n_1, n_2 \geq 30$ Independent samples	any	unknown	$z_{\bar{x}} = \dfrac{(\bar{x}_1 - \bar{x}_2) - (\mu_1 - \mu_2)}{\sqrt{\dfrac{s_1^2}{n_1} + \dfrac{s_2^2}{n_2}}}$
mean	$n_1, n_2 < 30$ Independent samples	normal	known	$z_{\bar{x}} = \dfrac{(\bar{x}_1 - \bar{x}_2) - (\mu_1 - \mu_2)}{\sqrt{\dfrac{\sigma_1^2}{n_1} + \dfrac{\sigma_2^2}{n_2}}}$
mean	$n_1, n_2 < 30$ Independent samples	normal	unknown and equal	$t_{\bar{x}} = \dfrac{(\bar{x}_1 - \bar{x}_2) - (\mu_1 - \mu_2)}{\sqrt{\dfrac{(n_1 - 1)s_1^2 + (n_2 - 1)s_2^2}{n_1 + n_2 - 2}} \sqrt{\dfrac{1}{n_1} + \dfrac{1}{n_2}}} \qquad df = n_1 + n_2 - 2$
mean	$n_1, n_2 < 30$ Independent samples	normal	unknown and unequal	$t_{\bar{x}} = \dfrac{(\bar{x}_1 - \bar{x}_2) - (\mu_1 - \mu_2)}{\sqrt{\dfrac{s_1^2}{n_1} + \dfrac{s_2^2}{n_2}}} \qquad df = \dfrac{\left(\dfrac{s_1^2}{n_1} + \dfrac{s_2^2}{n_2}\right)^2}{\dfrac{\left(\dfrac{s_1^2}{n_1}\right)^2}{n_1 - 1} + \dfrac{\left(\dfrac{s_2^2}{n_2}\right)^2}{n_2 - 1}}$
proportion	$np \geq 5$ $nq \geq 5$ Independent samples	any		$z_p = \dfrac{(\bar{p}_1 - \bar{p}_2) - (p_1 - p_2)}{\sqrt{(\hat{p})(1 - \hat{p})\left(\dfrac{1}{n_1} + \dfrac{1}{n_2}\right)}} \qquad \hat{p} = \dfrac{x_1 + x_2}{n_1 + n_2}$

Appendix C
Regression and ANOVA Equations

Correlation Equations

$$r = \frac{n\sum xy - \left(\sum x\right)\left(\sum y\right)}{\sqrt{\left[n\sum x^2 - \left(\sum x\right)^2\right]\left[n\sum y^2 - \left(\sum y\right)^2\right]}}$$

$$t = \frac{r}{\sqrt{\frac{1-r^2}{n-2}}}$$

Regression slope and y-intercept

$$b = \frac{n\sum xy - \left(\sum x\right)\left(\sum y\right)}{n\sum x^2 - \left(\sum x\right)^2}$$

$$a = \frac{\sum y}{n} - b\left(\frac{\sum x}{n}\right)$$

Sum of Square Equations

$$SST = \sum y^2 - \frac{\left(\sum y\right)^2}{n}$$

$$SSE = \sum y^2 - a\sum y - b\sum xy$$

$$SSR = SST - SSE$$

Coefficient of Determination Equations

$$R^2 = \frac{SSR}{SST} \qquad F = \frac{SSR}{\left(\frac{SSE}{n-2}\right)}$$

Significance of slope equations

$$s_e = \frac{SSE}{n-2} \qquad t = \frac{b - \beta}{s_b}$$

$$s_b = \frac{s_e}{\sqrt{\sum x^2 - n(\bar{x})^2}} \qquad CI = b \pm t_c s_b$$

ANOVA Equations (completely randomized design)

$$SST = \sum_{i=1}^{n_T} x_i^2 - \frac{\left(\sum_{i=1}^{n_T} x_i\right)^2}{n_T} \qquad SSB = \sum_{i=1}^{k} n_i\left(\bar{x}_i - \bar{\bar{x}}\right)^2 \qquad SSW = SST - SSB$$

$$MSB = \frac{SSB}{k-1} \qquad MSW = \frac{SSW}{n_T - k} \qquad F = \frac{MSB}{MSW}$$

The Humongous Book of Statistics Problems

Index
ALPHABETICAL LIST OF CONCEPTS WITH PROBLEM NUMBERS

This comprehensive index organizes the concepts and skills discussed within the book alphabetically. Each entry is accompanied by one or more problem numbers, in which the topics are most prominently featured.

All these numbers refer to problems, not pages, in the book. For example, 8.2 is the second problem in Chapter 8.

D

N–O

P

T–U–V

W–X–Y–Z